Experimental Techniques in Mineral and Rock Physics

The Schreiber Volume

**Edited by
Robert C. Liebermann
Carl H. Sondergeld**

1994

Springer Basel AG

Reprint from Pure and Applied Geophysics
(PAGEOPH), Volume 141 (1993), No. 2–4

The Editors:

Dr. Robert C. Liebermann
Center for High Pressure Research
Department of Earth and Space Sciences
University at Stony Brook
Stony Brook, NY 1194
USA

Dr. Carl H. Sondergeld
Rock Properties
Theoretical Geophysical Research
Amoco Production Company Research Center
Tulsa, OK 74102
USA

Library of Congress Cataloging-in-Publication Data

Experimental techniques in mineral and rock physics: the Schreiber
 volume / edited by Robert C. Liebermann, Carl H. Sondergeld.
 Papers from a symposium held during the fall 1992 meeting of the
 American Geophysical Union in memory of Edward Schreiber.
 "Reprinted from Pure and applied geophysics (PAGEOPH), volume 141
 (1993), no. 2/4" – – T.p. verso.

 1. Petrology–Congresses.   2. Mineralogy–Congresses.   3. Rock
 mechanics–Congresses.   I. Schreiber, Edward, 1930–
 II. Liebermann, R. C.   III. Sondergeld, Carl H.
 QE431.5.E97   1994
 552'.06–dc20

Deutsche Bibliothek Cataloging-in-Publication Data

Experimental techniques in mineral and rock physics: the
Schreiber volume; [reprint from Pure and applied geophysics,
Vol. 141 (1993)] / ed. by Robert C. Liebermann; Carl H.
Sondergeld – Basel ; Boston ; Berlin : Birkhäuser, 1994.
ISBN 978-3-7643-5028-4   ISBN 978-3-0348-5108-4 (eBook)
DOI 10.1007/978-3-0348-5108-4

NE: Liebermann, Robert C. [Hrsg.]

© Springer Basel AG 1994
Originally published by Birkhäuser Verlag, Basel in 1994
Printed on acid-free paper produced from chlorine-free pulp

9 8 7 6 5 4 3 2 1

Contents

PAGEOPH, Vol. 141, No. 2/3/4 (1993)

0033–4553/93/040209–01$1.50 + 0.20/0

Introduction

On December 10–11, 1992, a symposium entitled "Experimental Techniques in Mineral and Rock Physics" was held during the Fall 1992 Meeting of the American Geophysical Union in memory of Edward Schreiber. This symposium dramatized the growth and importance of the field of experimental geophysics which was nurtured in its infancy by researchers like Edward Schreiber. The symposium was convened by two former students and colleagues of Schreiber, Robert C. Liebermann and Carl H. Sondergeld.

A total of fifty papers was contributed to this symposium, including a tribute to Schreiber's career by William A. Bassett. These papers were invited and solicited for their focus on frontier research in the development of experimental technqiues in mineral and rock physics, because this was the realm of geophysics on which Ed Schreiber left his most indelible mark. We were honored by the attendance at the symposium of Ed's wife, Charlotte Schreiber who, with her graduate student colleague, also contributed a paper to the proceedings.

On the basis of the success of the Schreiber Symposium, we decided to publish a topical issue of the journal *Pure and Applied Geophysics* (*PAGEOPH*) which would include as many of the symposium papers as possible. Authors were permitted to include results and discussion of applications to illustrate their experimental work, but they were explicitly encouraged to submit papers which consisted totally, or largely, of a detailed description of the new experimental techniques in their laboratory.

This volume contains 22 papers submitted and reviewed under our auspices as guest Editors of PAGEOPH. These include papers on rock formation and rock properties, acoustic studies of the elasticity and equation of state of minerals, diamond-anvil cell experiments, rheological investigations, and new advances in high-pressure calorimetry, diffusion, and sealing and calibration techniques. We would like to thank Ms. Ann Lattimore for her editorial assistance in helping us to process these papers.

The papers in this volume emphasize the merit and the need for careful and precise measurements of the physical properties of rocks and minerals in resolving fundamental problems of global geophysics. It is our final tribute to our dear colleague and friend, Edward Schreiber.

Robert C. Liebermann
Center for High Pressure Research
Mineral Physics Institute
University at Stony Brook
Stony Brook, NY 11794, U.S.A.

Carl H. Sondergeld
Rock Properties Theoretical Geophysical Research
Amoco Production Company Research Center
Tulsa, OK 74102, U.S.A.

PAGEOPH, Vol. 141, No. 2/3/4 (1993)

0033-4553/93/040211-07$1.50 + 0.20/0

Reflections on the Career of Edward Schreiber

ROBERT C. LIEBERMANN[1] and WILLIAM A. BASSETT[2]

Figure 1
Photo of Edward Schreiber.

[1] Center for High Pressure Research (An NSF Science and Technology Center), Mineral Physics Institute, University at Stony Brook, Stony Brook, NY 11794, U.S.A.
[2] Department of Geological Sciences, Cornell University, Ithaca, NY 14853-1503, U.S.A.

Ed Schreiber's life was taken suddenly and tragically on the morning of November 11, 1991 as he was driving to work on the Bronx River Parkway. He was on his way to Queens College, an institution with which he had had a long and valuable association. Ed was born in Brooklyn in 1930. He attended the New York State College of Ceramic Engineering at Alfred University where he received his B.S. degree Magna Cum Laude in 1956 and his Ph.D. degree in 1963. As a graduate student, Ed was inspired by both Taro Takahashi and Orson Anderson.

After receiving his Ph.D. degree, Ed joined the staff of Lamont Geological Observatory of Columbia University where he worked with Orson Anderson to establish a new laboratory devoted to the measurement of the physical properties of earth minerals, especially the elastic properties by ultrasonic interferometry techniques. Over the next 7 years, this Mineral Physics Laboratory (the name coined by Ed) published more than 70 research papers on ultrasonic measurements at high pressures and temperatures, development of new experimental techniques, equations of state, and applications to the Earth and moon. The name Mineral Physics was subsequently adopted by the AGU's Committee on Mineral Physics and has come to be a part of the vocabulary of researchers around the world.

Ed was the consummate "indoor geologist" who brought a remarkable combination of background in ceramics and experimental physics to the task of performing careful and precise measurements of the elastic and other thermophysical properties of minerals and publishing data whose significance is demonstrated by the fact that they are still quoted and used more than 25 years later. However, he also enjoyed geological field work, whether it was on the diatremes of the Four Corners area looking for fragments of the mantle or in Italy assisting his wife and colleague Charlotte on her sedimentological investigations.

Perhaps one of the most famous papers to appear in SCIENCE magazine was written by Ed Schreiber and Orson Anderson in 1970 and entitled: 'Properties and composition of lunar materials: Earth analogies." It is a classic and we would like to share parts of it with you; even if you are familiar with this paper, we hope you will enjoy this recollection. Seismographs were left behind on the lunar surface by the Apollo missions to record the travel times of sound waves through lunar rocks. Measurements by these seismographs yielded extraordinarily low velocities for lunar rocks. Puzzled by these data, Schreiber and Anderson made some laboratory studies of rocks and other terrestrial materials. We quote:

"To account for this very low velocity, we decided to consider materials other than those listed initially by Birch or the later more detailed compilations of Anderson and Liebermann. The search was aided by considerations of much earlier speculations concerning the nature of the moon (Erasmus, 1542), and a significant group of materials was found which have velocities that cluster around those actually observed for lunar rocks (Fig. 2)."

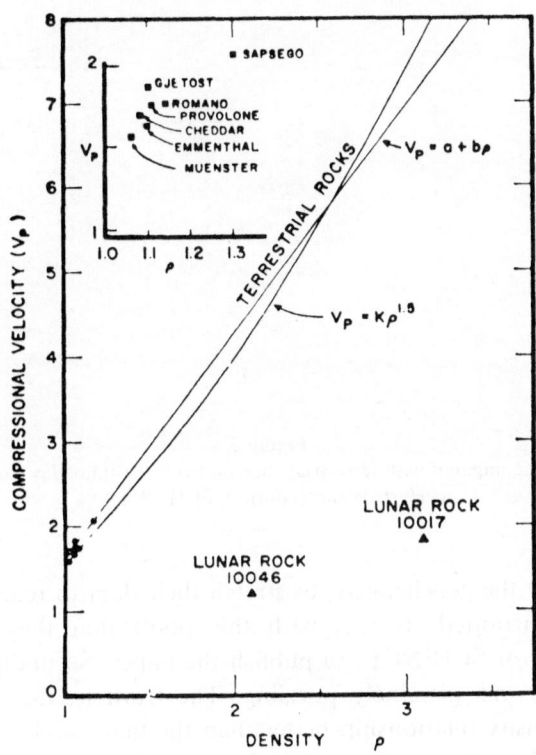

Figure 2

From Schreiber and Anderson (1970) showing the comparison between sound velocities for the lunar rocks and for various earth materials.

They go on to say:

"The materials were chosen so as to represent a broad geographic distribution in order to preclude any bias that might be introduced by regional sampling. It is seen that these materials exhibit compressional velocities that are in consonance with those measured for the lunar rocks—which leads us to suspect that perhaps old hypotheses are best after all and should not be lightly discarded."

Finally, they conclude that:

"This apparent inconsistency... may be readily accounted for when one considers how much better aged the lunar materials are."

This masterful spoof, of which Schreiber was the instigator, also served as the basis for the Season's Greetings card sent by Ed to Herb Wang in 1970 (Fig. 3).

What a delicious bit of satire! But wait, there are some serious messages here. Firstly, Ed and Orson were taking a dig at the rather unsettling tendency for lunar

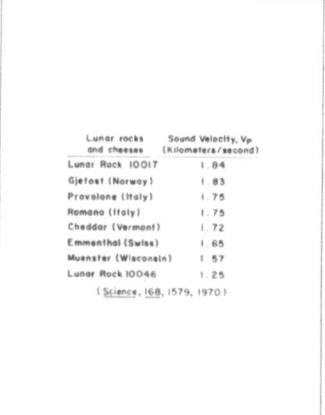

Figure 3
Velocities of lunar rocks compared with terrestrial cheeses. From 1970 holiday card sent by Ed Schreiber
to Herb Wang (courtesy of H. Wang).

scientists, especially the geochemists, to stretch their data to reach conclusions that were not quite warranted. It was with this point that they persuaded Philip Abelson, the editor of SCIENCE, to publish the paper. Secondly, the low velocity of the lunar rocks was genuinely puzzling. The truth is, the various cheeses fit Birch's velocity-density relationship better than the lunar rocks. It was years later, and with an entirely new technique, that Orson Anderson and his colleagues were able to find a logical explanation, namely, that sound waves in lunar rocks must travel greater distances as they are reflected by vacuum-filled cracks.

While conducting his research at Lamont-Doherty, Ed joined the faculty of Queens College of the City University of New York as an Associate Professor, and later Professor of Geology. He is remembered there not only for his research endeavors, but more importantly for the dedication and warmth he brought to the classroom, nurturing the careers of aspiring graduate students (including Carl Sondergeld and Jay Bass) and his work in the undergraduate admissions program and the Academic Senate. Ed also found time to serve on several review and advisory panels for NASA's Lunar and Planetary Science Program and served two terms (1981–83 and 1985–86) with the Geosciences Program of the DOE's Department of Basic Energy Sciences. He continued as an Adjunct Senior Research Associate at Lamont-Doherty, and in 1974 published an important book on "Elastic Constants and Their Measurement" with Orson Anderson and Naohiro Soga.

Ed wrote several papers with his wife Charlotte, who was also a Professor at Queens College. These were for the most part papers on sedimentary processes. Some of these displayed a rather whimsical sense of humor as well. For instance, the title of a paper they wrote with Denys Smith starts off "Spring peas from New

York State." Contrary to what one might expect, this paper does not deal with agricultural pursuits, but rather some fascinating observations on nucleation and growth of hollow ooliths and pisoliths. The authors suggest that they may have formed by precipitation of calcite on the surfaces of drops of water.

Ed was a silent leader, often operating out of the limelight, but was the epitome of collegiality. He did much to help others who wanted to adopt ultrasonic techniques for measuring the elastic wave velocities of materials. Individuals such as Hartmut Spetzler, Murli Manghnani, and Bob Liebermann learned from Ed and then went on to establish productive labs of their own. In this way, his influence in building this new and promising science of mineral physics reached even beyond what he accomplished with his numerous papers on the subject. This was just one way in which Ed helped others to pursue their research interests.

Ed was a warm and generous person, both professionally and personally, and a great experimental scientist in whom humility was indeed a virtue. Although Ed was not one to blow his own horn, he was quietly and generously a powerful moving force who had a profound influence on our science. His colleagues could always rely on his counsel. He is survived by his wife Charlotte and two daughters. His scientific influence lives on in the work of his students, protégés and colleagues whose lives and careers he touched.

Many new experimental techniques for studying minerals were furthered directly or indirectly by Ed. It is most appropriate to honor him by this Topical Issue and we are sure that Ed would be pleased to know that his memory is being honored by his friends and colleagues in this way.

REFERENCES

ERASMUS, "*With this merry toy, he...made his friends believe the moon to be made of green cheese*," in Adagia, 1542.

SCHREIBER, E., and ANDERSON, O. L. (1970), *Properties and Composition of Lunar Materials: Earth Analogies*, Science *168*, 1579–1580.

SCHREIBER, C., SMITH, D., and SCHREIBER, E. (1981), *Spring Peas from from New York State: Nucleation and Growth of Hollow Ooliths and Pisoliths*, J. Sedimentary Petrology *51*, 1341–1346.

PUBLICATIONS OF EDWARD SCHREIBER

Publications

Book

Elastic Constants and Their Measurement, with O. L. Anderson and Naohiro Soga, McGraw-Hill, New York, 1974.

Papers

Application of thermodynamics to pneumatolytic process-system H_2O-CO_2 in the supercritical region. With T. Takahashi, *Discussion in symposium on Problems of Post-Magmatic Ore Deposition, Vol. II*, 51–523, Prague 1965.

The relations between refractive index and density of minerals related to the earth's mantle. With O. L. Anderson, *J. Geophys. Res. 70*, 1463–1471, 1965.

The pressure derivatives of the sound velocities of polycrystalline alumina, with O. L. Anderson, *J. Amer. Ceram. Soc. 49*, 184–190, 1966.

Temperature dependence of the velocity derivatives of periclase, with O. L. Anderson, *J. Geophys. Res. 71*, 3007–3012, 1966.

Elastic constants of silicon carbide, with N. Soga, *J. Amer. Ceram. Soc. 49*, 341, 1966.

Variable air transformer for impedance matching, with P. Mattaboni, *Rev. Sci. Instr. 37*, 1616–1625, 1966.

Estimation of bulk modulus and sound velocities of oxides at very high temperatures, with N. Soga and O. L. Anderson, *J. Geophys. Res. 71*, 5315–5320, 1966.

Pressure derivatives of the sound velocities of polycrystalline forsterite, with 6% porosity, with O. L. Anderson, *J. Geophys. Res. 72*, 762–764, 1967.

The elastic moduli of single crystal spinel at 25°C and at 2 Kbar, *J. Appl. Phys. 38*, 2508–2511, 1967.

Method of pulse transmission measurements for determining sound velocities, with P. Mattaboni, *J. Geophys. Res. 72*, 5160–5162, 1967.

Use of ultrasonic interferometry technique for studying elastic properties of rocks, with M. Manghnani and N. Soga, *J. Geophys. Res. 73*, 824–826, 1968.

Porosity dependence of sound velocity and Poisson's ratio for polycrystalline MgO determined by resonant sphere method, with N. Soga, *J. Amer. Ceram. Soc. 51*, 465–466, 1968.

Revised data on polycrystalline MgO, with O. L. Anderson, *J. Geophys. Res. 73*, 2837–2833, 1968.

Comment on the elastic modulus porosity relationship, *J. Amer. Ceram. Soc. 51*, 541–542, 1968.

Elastic constants of polycrystalline hematite as a function of pressure to 3 Kbar, with R. C. Liebermann, *J. Geophys. Res. 73*, 6585–6590, 1968.

Some elastic constant data on mineral relevant to geophysics, with O. L. Anderson, R. C. Liebermann and N. Soga, *Rev. Geophys. 6*, 491–524, 1968.

Leak detection in high pressure gas systems, with H. Spetzler and D. Newbigging, *Rev. Sci. Instr. 40*, 179, 1969.

Coupling of ultrasonic energy through lapped surfaces, with H. Spetzler and L. Peselnick, *J. Acoust. Soc. Amer. 45*, 520, 1969.

Coupling of ultrasonic energy through lapped surfaces at high temperature and pressure, with H. Spetzler and D. Newbigging, *J. Acoust. Soc. Amer. 45*, 1057–1058, 1969.

Critical thermal gradients in the mantle, with R. C. Liebermann, *Earth Planet. Sci. Lett. 7*, 77–81, 1969.

The effect of solid solutions upon the bulk modulus and its pressure derivatives: Implications for equations of state, *Earth Planet. Sci. Lett. 7*, 137–140, 1969.

Sound velocity, compressibility for lunar rocks 17 and 46 and for glass spheres from the lunar soil, with O. L. Anderson, N. Soga, N. Warren and C. Scholz, *Science 167*, 732–734, 1970.

Elastic properties of a microbrecca, igneous rock, and lunar fines from Apollo 11 mission, with O. L. Anderson, C. Scholz, N. Soga and N. Warren, *Proceedings, Apollo 11 Lunar Science Conference, Vol. 3*, 1959–1973, Geochim. Cosmochim. Acta, Suppl., Pergamon Press, 1970.

Properties of lunar materials, earth analogies, with O. L. Anderson, *Science 168*, 1579–1580, 1970.

Elastic properties of minerals, with R. C. Liebermann, *EOS 52*, 142–146, 1971.

The geology of the Caribbean crust: Tertiary sediments, granitic and basic rocks from the Aves. Ridge, with P. J. Fox and B. C. Heezen, *Tectonophysics 12*, 80–109, 1971.

Elastic and thermal properties of Apollo 22 and Apollo 12 rocks, with N. Warren, C. Scholz, J. A. Morreson, P. R. Norton, M. Kumazawa and O. L. Anderson, *Proceedings, Second Lunar Conference, Vol. 3*, 2345–2360, M.I.T. Press, 1971.

Compressional sound velocities in semi-indurated sediments and basalts from DSDP Leg XI, with P. J. Fox and J. J. Peterson, in: *Initial Reports of the Deep Sea Drilling Project, Vol. XI*, 723–727, U.S. Govt. Printing Office, 1972.

Compressional wave velocities in basalt and altered basalt recovered during Leg XIV, with P. J. Fox and J. J. Peterson, in: *Initial Reports of the Deep Sea Drilling Project, Vol. XIV*, 773–775, U.S. Govt. Printing Office, 1972.

Effects of stress-induced anisotropy and porosity on elastic properties of polycrystals, with H. Spetzler and R. O'Connell, *J. Geophys. Res. 25*, 4930–4944, 1972.

Compressional wave velocities in selected samples of gabbro, schist, limestone, anhydrite, gypsum and halite, with P. J. Fox and J. J. Peterson, in: *Initial Reports of the Deep Sea Drilling Project, Vol. XIII, Part 2*, 595–597, U.S. Govt. Printing Office, 1973.

Compressional wave velocities in basalt and dolerite samples recovered during Leg XV, with P. J. Fox, in: *Initial Reports of the Deep Sea Drilling Project, Vol. XV*, 1013–1016, U.S. Govt. Printing Office, 1973.

An alumina standard reference material for resonance frequency and dynamic elastic moduli measurements, II: For use from 25C to 1000C, with R. W. Dickson, *J. Nat. Bur. Std. 77A*, 391–396, 1973.

The geology of the oceanic crust compressional wave velocities of oceanic rocks, with P. J. Fox and J. J. Peterson, *J. Geophys. Res. 78*, 5155–5172, 1973.

Compressional wave velocities of oceanic rocks and the geology of the oceanic crust: A brief summary, with P. J. Fox, *EOS 54* (11), 1033–1035, 1973.

Newfoundland ophiolites and the oceanic layer, with J. J. Peterson and P. J. Fox, *Nature 247*, 194–196, 1974.

Compressional wave velocities in basalts from DSDP, Leg 24, with M. Perfit and P. J. Cernock, in: *Initial Reports of the Deep Sea Drilling Project, Vol. XXIV*, 787–790, U.S. Govt. Printing Office, 1974.

The compressional wave velocities of some DSDP, Leg 34, basalts: A brief report, in: *Initial Reports of the Deep Sea Drilling Project, Vol. XXXIV*, 547–548, U.S. Govt. Printing Office, 1976.

The geology of the oceanographer fracture zone: A model for fracture zones, with P. J. Fox, K. McCamy and H. Rowlett, *J. Geophys. Res. 81*, 4117–4128, 1976.

Depositional environments of Upper Miocene (Messinian) evaporite deposits on the Sicilian Basin, with B. C. Schreiber, G. M. Friedman and A. Decima, *Sedimentol. 23*, 729–760, 1976.

Density and P-wave velocity of rocks from the Famous Region and their implication to the structure of the oceanic crust, with P. J. Fox, *Bull. Geol. Soc. Am. 88*, 600–608, 1977.

An evaporitic lithofacies continuum: Evaporitic facies observed in the Upper Miocene (Messinian) deposits of the Salemi Basin (Sicily) and a modern analog, with B. C. Schreiber and R. Catalano, in: *Reefs and Evaporites*, AAPG Special Publication # 5, 169–180, 1977.

The moon and Q, *Proc. Eighth Lunar Science Conference*, 1201–1208, 1977.

A thermodynamic study of manganese minerals and its application to the formation of some manganese mineral deposits in Japan, with T. Takahashi, in: *Geological Studies of the Mineral Deposits of Japan and East Asia*, (Hideki Amai, ed.) pp. 335–349, University of Tokyo Press, 1978.

The salt that was, with B. C. Schreiber, *Geology 5*, 527–528, 1978.

Acoustic wave velocity measurements of oceanic crustal samples, with P. D. Rabinowitz, *Initial Reports of the Deep Sea Drilling Project, Vol. 45*, 383–386, 1978.

Effects of annealing on the properties of hot pressed MgO polycrystals, with C. Sondergeld, *J. Amer. Ceram. Soc. 61*, 535–536, 1978.

An Examination of the elasticity of hot-pressed MgO, with C. Sondergeld, *Phys. Chem. Minerals 5*, 21–31, 1979.

Spring Peas from New York State: Nucleation and growth of fresh water hollow ooliths and pisoliths, with B. C. Schreiber and D. Smith, *J. Sedimentary Petrology 51*, 1341–1346, 1981.

Mass mortality and its environmental and evolutionary consequences, with 19 other authors, *Science 216*, 249–256, 1982.

DSDP Leg 73: Contributions to paleogene stratigraphy in nomenclature, chronology, and sedimentation rates, with 12 other authors, *Paleogeography, Paleoclimatology, Paleoecology 42*, 91–125, 1983.

Continental drilling, with R. Andrews, *Geotimes 32*, 14–16, 1987.

Big science versus little science: Continental scientific drilling programs in the United States, with D. Speidel, *Geotimes 35*, 27–29, 1990.

Rocks and Rock Properties

PAGEOPH, Vol. 141, No. 2/3/4 (1993)   0033 4553/93/040221 27$1.50 + 0.20/0
(c) 1993 Birkhäuser Verlag, Basel

Experimental Simulation of Plagioclase Diagenesis at *P-T* Conditions of 3.5 km Burial Depth

STEPHEN L. KARNER[1,3] and B. CHARLOTTE SCHREIBER[1,2]

Abstract — Dissolution of plagioclase under the physical conditions at shallow to intermediate burial depths is a prime candidate for secondary porosity generation in feldspathic siliciclastic sediments. The diagenetic behavior of granular aggregates of plagioclase feldspar and quartz has been investigated by experimentation performed in a Bridgeman-type pressure vessel. The experiments, each of two weeks duration, simulated pressure-temperature conditions approximating 3.5 km burial depth. By using a double-acting pore-fluid reservoir, solutions of various chemistries were cycled through samples composed of oligoclase or labradorite feldspar and quartz (90:10 wt% respectively).

Scanning electron microscope analysis of the post-experiment samples reveals dissolution features and precipitated products. Dissolution voids of ~10 microns occur typically in areas of maximum stress such as crack-tips and grain contacts. Dissolution on a larger scale is exemplified by topographical smoothing of grain surfaces. The dissolved species are subsequently reprecipitated as Ca-enriched overgrowths (possibly zeolites) and clays. These precipitates are found individually on the scale of 10 microns and collectively as surface coatings on both feldspar and quartz grains. Atomic absorption spectroscopic analyses of the pore fluid suggest that the fluid chemistry is consistent with the observed experimental precipitates.

These experiments show that clay coatings are unnecessary precursors to grain surface dissolution and that the diagenetic precipitation is not mineral selective. Also, the mass transfer of the dissolved species appears to be localized because grains displaying both dissolution and precipitation features are commonplace. Volume changes due to mineral transformation/alteration may increase secondary porosity if the dissolved species produced from dissolution are only partially involved in reprecipitation and the remaining dissolved material is flushed out by the pore fluids. However, if the mass transfer is primarily local then permeability would significantly decrease as precipitates may choke the pore throats.

Key words: Experimental, plagioclase, diagenesis, porosity, permeability, dissolution, precipitates.

Introduction

The hydrocarbon industry has yielded vast amounts of subsurface data regarding the geology of sedimentary basins. Despite this ever increasing data-bank, our understanding of porosity in these sedimentary rocks is often very speculative due

[1] Department of Geology, Queens College, Flushing, NY 11367, U.S.A.
[2] Lamont Doherty Earth Observatory, Palisades, NY 10964, U.S.A.
[3] Now at Department of Earth, Atmospheric, and Planetary Sciences, M.I.T., Cambridge, MA 02139, U.S.A.

to the relatively small diameter of recovered core samples and the large lateral distances between exploration wells. The calculated pressure-depth curves for sedimentary basins may yield first-order approximations of porosity within these rocks. However, studies of the subsurface environment demonstrate that abnormal physical and chemical conditions commonly exist at depth which may have considerable implications for the development of sandstone porosity. Alteration of relatively unstable minerals (such as plagioclase feldspar) may be responsible for the creation, or destruction, of sandstone porosity at depth.

Laboratory experiments using a Bridgeman-type pressure apparatus provide a mechanism by which the subsurface environment can be simulated. More importantly, the experiments can be run in a controlled chemical environment—a definite advantage when studying natural geological problems. The data from these experimental studies potentially add more realistic constraints that may be employed for basin modeling where subsurface information is lacking. Thus the purpose of this study is to experimentally investigate the chemo-mechanical behavior of plagioclase feldspar and determine the implications for the formation and/or destruction of porosity in feldspathic sandstones.

The Effect of Diagenesis on Porosity Development

Obtaining an understanding of the factors that result in the alteration of sandstone porosity and permeability is not straightforward. The physical and petrologic observations that have been attributed to the destruction or enhancement of sandstone porosity are displayed in Figure 1. These diagenetic features can be broken down into two major, mutually inclusive categories—mechanical and chemical. These processes, responsible for all the observed morphological and compositional alteration of sedimentary particles, are active during weathering, transportation, deposition, burial, and exhumation.

Mechanical diagenesis is the most simple to explain as it is principally a function of the directed stresses acting to mechanically push grains closer together. Vertical stress due to the weight of the overlying rock (overburden), or nonvertical stresses produced through active tectonism (such as in compressional or extensional basins) significantly affect the amount of porosity in sedimentary rocks. Tectonic stresses are generally less important than overburden stresses and may generate fracture networks that complicate the general diagenetic sequence. This study is concerned with the development of intergranular porosity and permeability, both of which are primarily affected by vertical overburden stresses.

The degree of physical compaction in sandstones is chiefly dependent on the physical parameters of the grains that are inherited from the initial depositional environments for the sandstone and immediately adjacent lithologies. The angularity of these clastic particles inhibits grain reorientation as the overburden pressures

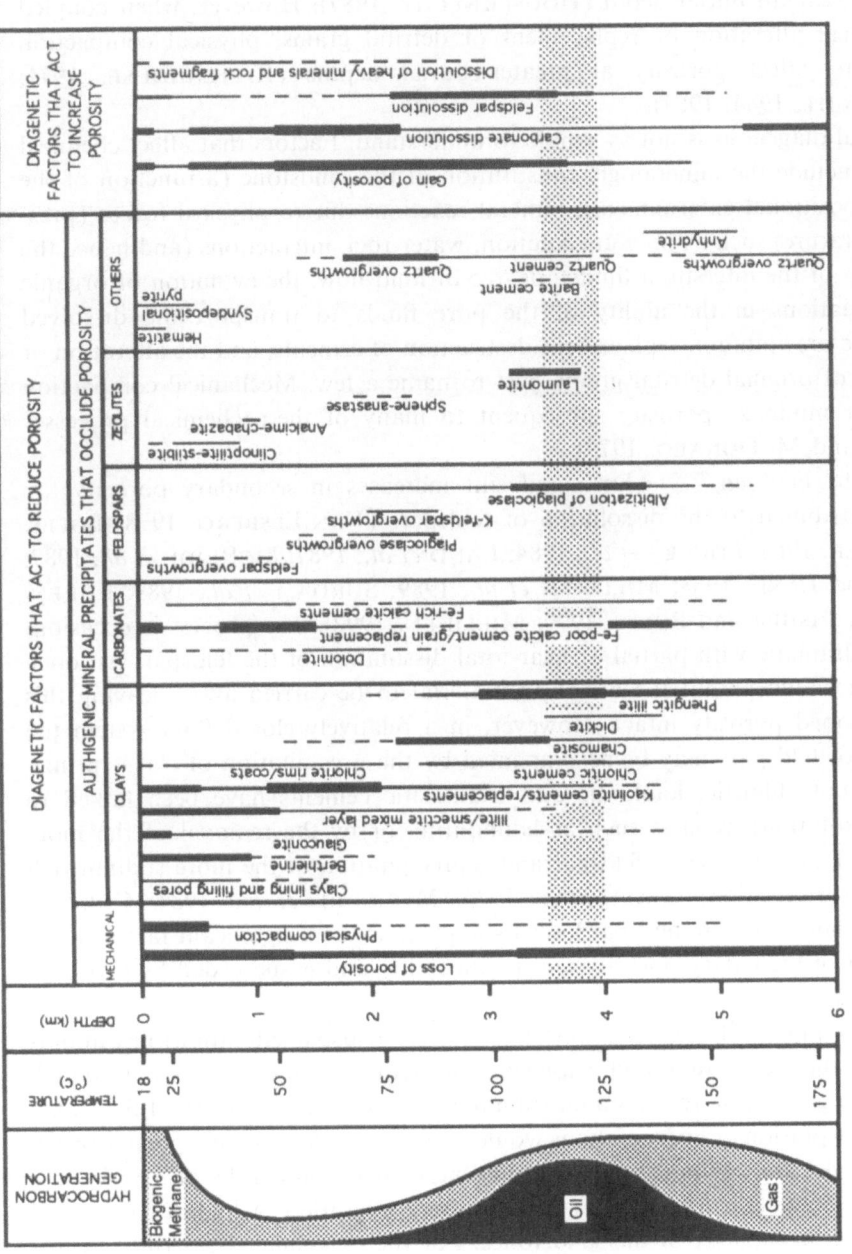

Figure 1

Generalized schematic diagram displaying a compendium of diagenetic data reported for sandstones. The thickness of the vertical lines is indicative of the significance of the associated diagenetic phase (thicker lines being more significant). The stippled pattern represents the optimal depth of plagioclase feldspar diagenesis related to hydrocarbon generation. Data from the Gulf Coast (LOUCKS et al., 1984; KAISER, 1984), Alberta Basin (LONGSTAFFE et al., 1992); Michigan Basin (BARNES et al., 1992); Santa Ynez Mtns, Calif. (HELMOLD and VAN DE KAMP, 1984); North Sea (BURLEY and MACQUAKER, 1992; SCHMIDT and McDONALD, 1979); Prudhoe Bay (SCHMIDT and McDONALD, 1979); Cayagan Basin, Philippines (MATHISEN, 1984); also SURDAM et al. (1989); SELLEY (1985).

increase (RICHARDSON and MCSWEEN, 1989). Hence, physical compaction alone is only effective to relatively shallow depths of burial and minimizes porosity at around 1.5–2.5 km burial depth (HOUSEKNECHT, 1987). However, when coupled with chemical alteration or replacement of detrital grains, physical compaction continues to affect porosity at greater burial depths (RITTENHOUSE, 1971; HOUSEKNECHT, 1984, 1987).

Chemical diagenesis is not as simple to understand. Factors that affect chemical diagenesis include the mineralogic constitution of the sandstone (a function of the original depositional environment), mineral reactions due to physical forces (pressure, temperature) such as pressure solution, water-rock interactions (and hence the composition of the interstitial fluids), degree of fluid flow, the evolution of organic matter, variations in the ability of the pore fluids to transport the dissolved material, the precipitation-replacement-destruction of cements, and the alteration or destruction of original detrital grains, just to name a few. Mechanical compaction may further minimize porosity subsequent to many of these chemical processes (SCHMIDT and MCDONALD, 1979).

At depths between 2.5–4 km significant increases in secondary porosity are generally attributed to the dissolution of feldspars (VAN ELSBERG, 1978; BOLES, 1984; KAISER, 1984; LOUCKS et al., 1984; LAND et al., 1987; MCBRIDE et al., 1987; DUTTON and LAND, 1988; MILLIKEN et al., 1989; SURDAM et al., 1989; EHRENBERG, 1990; FISHER and BOLES, 1990; MILLIKEN, 1992). The process is occasionally rather dramatic with partial or near-total dissolution of the feldspar. An open fluid system would permit the dissolved material to be carried away, leaving this newly developed porosity intact. However, in a relatively closed fluid system the dissolution of feldspars may be accompanied by the precipitation of clays (principally illite, but chloritic, kaolinitic and chamositic cements have been linked to feldspar dissolution), zeolites such as laumontite, or by the removal of the more calcic phase from plagioclase feldspar and reprecipitation of the more sodium rich plagioclase (albitization) (GALLOWAY, 1974; MONCURE et al., 1984; CURTISS, 1985). By about 4–4.5 km feldspar loses its importance as a significant factor in the development of secondary porosity and permeability and is succeeded by clay-zeolite alterations.

The development of sandstone porosity (due to feldspar alteration) has importance when the generation and expulsion of hydrocarbons is also considered. Hydrocarbon generation reaches a maximum at an approximate depth of 3.5–4 km whereas the expulsion of hydrocarbons would be maximized at slightly greater burial depths (~4–4.5 km). Hence, sandstone diagenesis at burial depths of 3.5–4 km is a significant factor that affects both the secondary migration of hydrocarbons and the accumulation capacity of the sandstones. For the experiments described in this paper, the physical and chemical conditions present at depths of 3.5–4 km were selected to reflect the optimal depth range in which hydrocarbon migration and accumulation are dependent on sandstone porosity and, hence, diagenesis.

Calculations of Experimental Parameters

The experimental operating conditions were determined assuming a lithostatic gradient of 24.3 MPa/km, hydrostatic gradient of 10.3 MPa/km, geothermal gradient of 26.1°C/km, and surface temperature of 18.2°C. By considering these general depth-pressure-temperature relationships, assuming a high Poisson ratio for consolidated sand ($v = 0.4$), and satisfying the no-failure condition for unconsolidated sand, the operating parameters for these experiments were calculated to be:

$$\text{Axial load (overburden pressure)} = 93 \text{ MPa}$$

$$\text{Confining pressure} = 70 \text{ MPa}$$

$$\text{Pore-fluid pressure} = 40 \text{ MPa}$$

$$\text{Temperature} = 129°C$$

Experimental Technique

The Apparatus

The fundamental operational aspects of this type of pressure vessel has been described by SCHOLZ and KOCZYNSKI (1979), MARONE et al. (1988), CHESTER and HIGGS (1992). The apparatus consists of four pressure vessels which are hydraulically operated in closed-loop servo-control (Fig. 2). The sample column is top-loaded into the main vessel where it rests on an external loadcell. Bottled argon gas is boosted to the desired confining pressure by an air-driven Haskel pump. During an experiment confining pressure is regulated by a secondary intensifier vessel. The sample column has two ports (top plug and piston) to which the two pore-pressure vessels may be connected. These pore-pressure vessels may be controlled independently to induce fluid flow through the sample.

Sample Stack Configuration

The sample column designed for these experiments generally followed that described by MARONE et al. (1988) and CHESTER and HIGGS (1992). However, these authors established sample columns for use in experiments that were run with solid rock samples in deionized water. Hence, the assembly had to be modified to accommodate the granular nature of our starting material, corrosiveness of the pore fluids, and maintenance of temperature during cycling.

The sample stack configuration shown in Figure 3 provides the optimal thermal conditions for the sample while under experiment. The stainless steel spacers act as a heat sink for the furnaces and expedite heat transfer to the sample. They also mininize the thermal variance around the sample. On the other hand, the ceramic

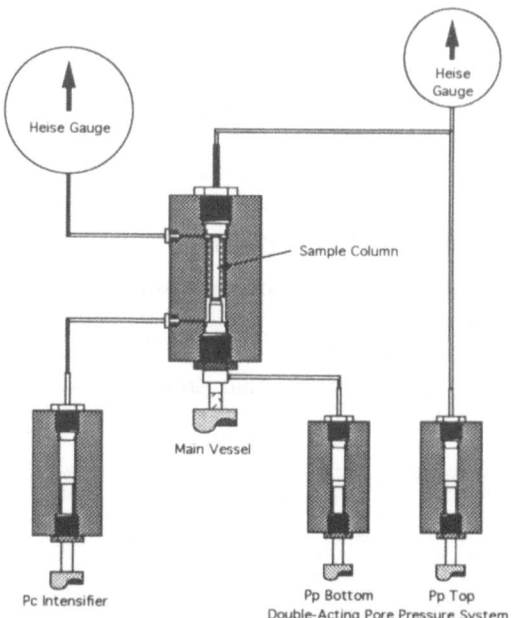

Figure 2
Schematic illustration of the interaction between the four pressure vessels of the triaxial pressure apparatus used for this study. The sample, located within the Main Vessel, is exposed to the three pressure regimes (axial load, confining pressure, pore pressure) all of which are independently controlled. The double-acting pore pressure system permits fluid flow through the sample.

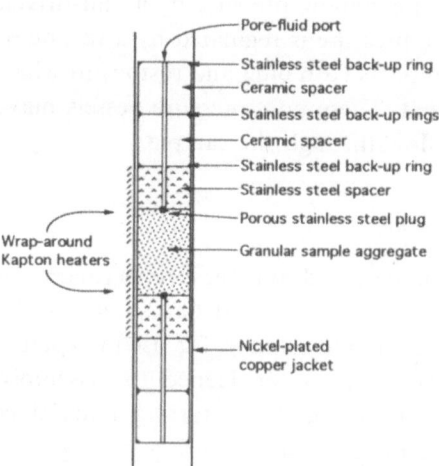

Figure 3
Sample stack configuration showing positions of the granular aggregate and spacers. Two foil heaters are wrapped around the jacket positioned over the steel spacers and granular aggregate. The entire jacketed sample stack is then wrapped in a fibrous material which provides further thermal insulation.

spacers provide thermal insulation minimizing the vertical dissipation of heat from the sample. The stainless-steel back-up rings take up the stress at the corners of the ceramic spacers thus preventing chipping. The sample jacketing consisted of ex-truded copper tubing (2.41 cm I.D.) which was electrochemically nickel plated.

Heating for each experiment was provided by two flexible Kapton foil heaters wrapped around the jacketed sample and adjacent steel spacers. The sample temperature was measured by a thermocouple located in the pore fluid system inside the upper steel spacer. An axial thermal profile of the sample could not be achieved because the granular nature of the sample prohibited the positioning of a thermocouple within the sample. However, the thermal variance of a cylinder of dry Westerly granite having the same dimensions as the granular aggregate was approximately 10°C from the center of the sample to either end.

Preparation of Experimental Mineral Sample

The plagioclase feldspar used in these experiments was obtained from Wards Natural Science Establishment whereas the quartz was obtained from Fisher Scientific. All mineral samples were crushed and sieved to 202–508 μm grain size. Ferromagnesian material was removed from the crushed sample by magnetic separation in a Franz Magnetic Separator. Finally, the granular sample was washed and ultrasonically cleaned with distilled water to remove any fine-grained material. For each experiment quartz and feldspar were weighed with the ratio 1:9, respectively (approximately 36 g total solids), and thoroughly mixed.

Preparation of Experimental Pore Fluids

All pore fluids for this study were artificially generated in the laboratory using salts dissolved in demineralized, deionized clean water. The composition of the brine used in these experiments (Table 1) was determined by averaging data for formation brines at approximately 3–4 km burial depth (DE SITTER, 1947; WHITE, 1965; DICKEY, 1969; OVERTON, 1973; COLLINS, 1975; DREVER, 1982; SELLEY, 1985; CONNOLLY et al., 1990; BÅRTH, 1991). In order to minimize the quantity of dissolved atmospheric gases in the pore fluid, a vacuum was drawn on the pore pressure system prior to pore fluid injection.

Also, various authors have suggested that formation waters bearing carbon dioxide, or organic acids, can significantly alter aluminosilicate minerals, such as plagioclase feldspar (BJØRLYKKE et al., 1979; AL-SHAIEB and SHELTON, 1981; YIN and SURDAM, 1985; STOESSELL and PITTMAN, 1990; BOWKER and SHULER, 1991). Thus, experiments were also run with carbon dioxide as a pore fluid component. To achieve this the two pore-pressure vessels were initially isolated from each other. A predetermined volume of carbon dioxide was injected into one of the pore-pressure vessels and servo-controlled to the operating pressure. The brine (without CO_2) was

Table 1

Ionic concentrations for the simulated brine used as a pore fluid in these experiments

ION	CONCENTRATION (ppm)	CONCENTRATION (wt % in solution)
Cl^-	33470	3.252
Na^+	14000	1.361
Ca^{2+}	5000	0.486
K^+	950	0.044
Mg^{2+}	600	0.087
Br^-	300	0.029
SO_4^{2-}	200	0.019
Total dissolved solid	54520	5.279

pumped into the second vessel and individually servo-controlled to pressure before the two fluids were thoroughly mixed.

The volumetric ratio between the total pore-pressure system and the sample void space was 100 cc to 7 cc, respectively. Fluid flow through the sample was achieved by establishing a servo-controlled pressure gradient between the two pore-pressure vessels. Typical flow rates through the sample were to the order of 100 cc per hour. The relatively high fluid flow rates for the experiments of this study may be considered to be unrealistic when compared to the natural subsurface environment. The natural time-span of geologic reactions makes it necessary to scale down the reactive time-frame so that reactions occur within the space of days, weeks, or months. This decrease in reaction time is commonly achieved by experimenting at elevated pressures or temperatures, or by manipulation of the initial chemical conditions for the experiments. For diagenetic studies, any increase in pressure and/or temperature may lead to reactions that typify a diagenetic realm different to that which is to be investigated (as can be seen by Fig. 1). Furthermore, altering the initial chemical conditions may result in chemistries that depart considerably from those observed in the natural subsurface environment. Hence, of all the experimental variables (pressure, temperature, sample chemistry, fluid chemistry, fluid flow, and time) the latter two would be the factors least likely to affect the type of diagenetic reactions. As time is the variable which ideally should be minimized, fluid flow rates provide the only viable means by which diagenetic reactions may be investigated within a sensible experimental time period.

Experimental Procedure

The confining pressure and axial load were applied prior to pore fluid pumping and the establishment of elevated temperature (dry sample at room temperature). This sequence of experimental start-up allows for easy detection of confining fluid leakage into the pore pressure system. Prior to heating the sample, the pore fluids were pumped into the pore-pressure system (reservoirs, tubing and sample) at start-up with no subsequent fluid replenishment through the course of an experiment. All experiments were run with the pressured pore fluids continuously cycling through the sample. The experiments were shut-down in the reverse order to start-up (i.e., furnaces turned off, pore fluids extracted, axial load and confining pressure decreased).

The physical and chemical conditions for each experiment are shown in Table 2. The experiments were designed to run for approximately two weeks under constant axial load, confining pressure, pore pressure, fluid flow, and temperature. During this period the axial displacement, axial load, confining pressure, pore pressure (ambient and differential), pore fluid cycling rate, and sample temperature were monitored on analog strip chart records.

The pore fluids and the granular sample aggregate were recovered for analysis following the completion of each experiment. Experiments 4 and 7 ended abruptly

Table 2

The operating parameters (sample type, fluid type, pressures, temperature, and duration) for all experiments of this study

Experiment Number	Sample (weight %)	Pore Fluid Composition	Axial Load	Confining Pressure	Fluid Pressure	Temperature (°C)	Experiment Duration (days)
1B	Oligoclase (90%) Quartz (10%)	0.15 molal NaCl	53 MPa (7700 psi)	40 MPa (5800 psi)	20 MPa (2900 psi)	135 (115-155)	11.74
2	Labradorite (90%) Quartz (10%)	Brine (54500 ppm TDS)	93 MPa (13500 psi)	70 MPa (10150 psi)	40 MPa (5800 psi)	129.5 (125-134)	14.44
3	Oligoclase (90%) Quartz (10%)	Brine (54500 ppm TDS)	93 MPa (13500 psi)	70 MPa (10150 psi)	40 MPa (5800 psi)	130 (126-134)	17.42
4	Oligoclase (90%) Quartz (10%)	Brine/CO_2 mix (54500 ppm TDS) (CO_2 ~2 mole%)	93 MPa (13500 psi)	70 MPa (10150 psi)	40 MPa (5800 psi)	129 (127-131)	7.16
5	Oligoclase (90%) Quartz (10%)	Brine/CO_2 mix (54500 ppm TDS) (CO_2 ~2 mole%)	93 MPa (13500 psi)	70 MPa (10150 psi)	40 MPa (5800 psi)	130 (129-131)	15.14
7	Labradorite (90%) Quartz (10%)	Brine/CO_2 mix (54500 ppm TDS) (CO_2 ~2 mole%)	93 MPa (13500 psi)	70 MPa (10150 psi)	40 MPa (5800 psi)	129.5 (126-133)	7.11
8	Labradorite (90%) Quartz (10%)	Brine/CO_2 mix (54500 ppm TDS) (CO_2 ~2 mole%)	93 MPa (13500 psi)	70 MPa (10150 psi)	40 MPa (5800 psi)	128.5 (125-132)	17.3

when the seal between the confining system and the pore-pressure system was breached. However, the pore fluids and samples from these experiments were recovered for analysis. All pore fluids were analyzed by atomic absorption spectroscopy (AAS) for the major cations (Na, Ca, Mg, and K). Post-experiment concentrations were compared to the initial brine concentrations and the differential cation concentrations for each experiment were calculated (Fig. 4).

The sample aggregates were studied on a scanning electron microscope (SEM). This analysis provided a visual account of the grain morphologies with elemental scans from the energy dispersive X-ray system, EDX(SEM), yielding a qualitative information regarding elemental concentrations. SEM photomicrographs and EDX(SEM) scans showing the observed features of samples recovered from the diagenetic simulations are presented in Figures 5–12.

X-ray diffraction techniques did not assist in the identification of the precipitates as the diffraction patterns were dominated by the feldspar and quartz. Qualitative data was readily obtained by use of an electron microprobe but quantitative elemental determinations were not achieved by the microprobe as the precipitates could not withstand the intensity of the microprobe beam (perhaps a function of the precipitate mineralogy, or size, or both). Hence, the ensuing discussion of the precipitates is based on the SEM morphological data.

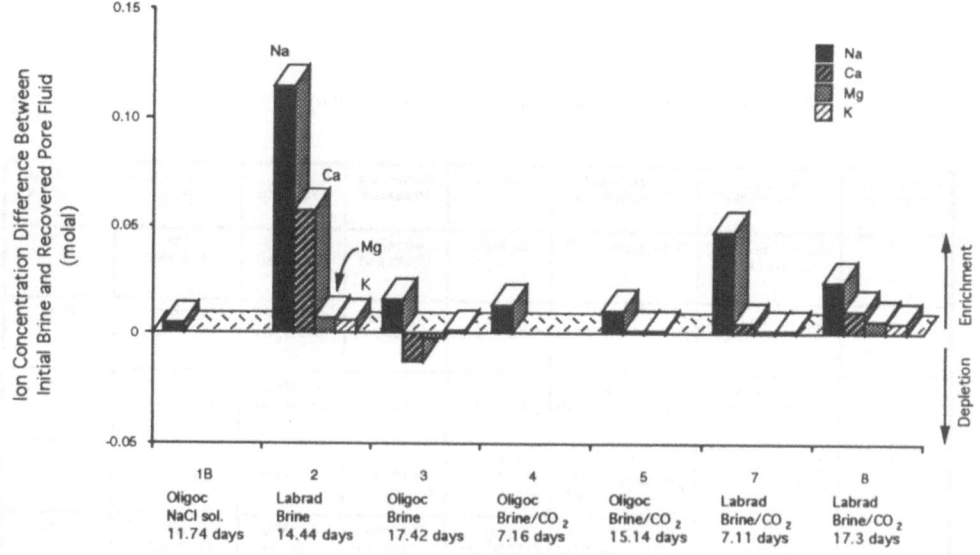

Figure 4

Histogram summarizing the data from the AAS analysis. Pre- and post-experimental fluid concentrations have been compared, and the concentration differences are plotted here. A positive differential concentration reflects a concentration increase of the post-experiment fluid relative to the initial fluid concentration. The sequence of vertical bars for each experiment represents concentration differences for (from left to right) sodium, calcium, magnesium and potassium (as is illustrated for experiment 2).

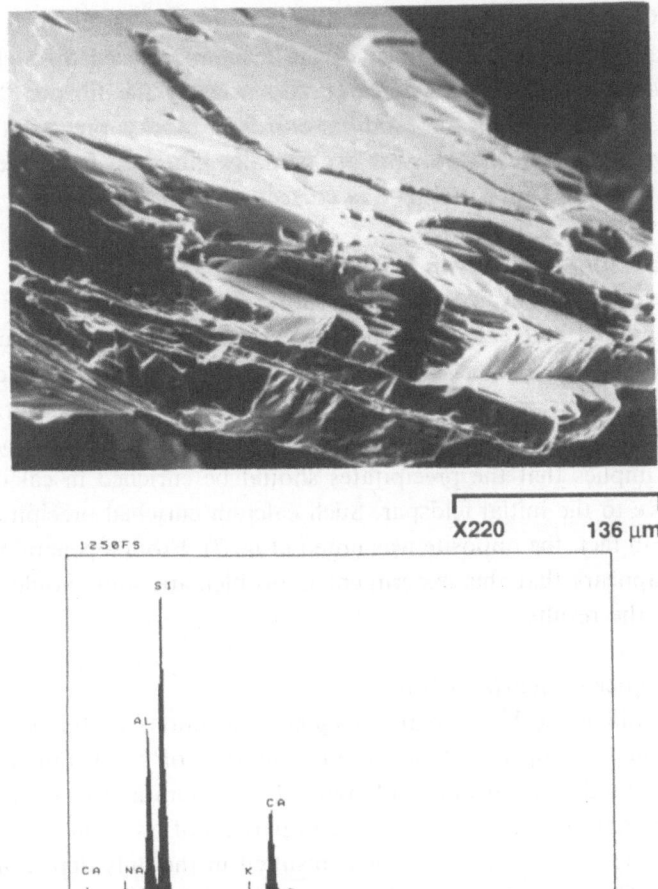

Figure 5

SEM photomicrograph showing the typical morphology (cleavage and irregular fracture) of labradorite grains. Although not presented here the morphology of pre-experiment oligoclase grains was similar. Scale bar at bottom right of photo is 136 μm. EDX(SEM) scan displays relative peak intensities for the labradorite grain in photo (1250 counts full logarithmic scale). Compare EXD(SEM) scan to the typical peak intensities for oligoclase (EDX(SEM) scan, Fig. 6) noting differences between sodium, aluminum, and calcium peaks.

Results

Comparison of Authigenic Precipitate Chemistry to AAS Analysis

The relative changes in the elemental concentrations of the AAS analyses are shown in Figure 4. Correlations between the fluid and precipitate elemental trends are apparent (Table 3) and are discussed below. For all experiments the dissolution of feldspar liberated considerable quantities of sodium and calcium, although the relative amounts differed depending on the feldspar that was used.

Exp. 1B: Oligoclase/quartz in NaCl solution

Virtually all the calcium and most of the sodium released through dissolution was incorporated into the precipitates (predominantly the fibrous "Fe-smectite" (Fig. 6) and to some extent the sodium-enriched blocky precipitate). The iron content of the fibrous precipitate was accidentally inherited from the steel of the pore pressure system. This problem was corrected for the subsequent experiments.

Exp. 2: Labradorite/quartz in brine

Dissolution of feldspar was volumetrically significant, liberating equal molal volumes of sodium and calcium. The fluid concentration of sodium increased dramatically while the calcium concentration increased only moderately (approximately 2:1 molal difference between Na and Ca). The Na:Ca concentration of the initial labradorite (almost 1:1) coupled with the Na:Ca in the post-experiment pore fluid (Fig. 4) implies that the precipitates should be enriched in calcium, and not sodium, relative to the initial feldspar. Such calcium enriched precipitates were not observed and, in fact, the opposite was noted (Fig. 7). From a general mass balance viewpoint, it appears that this experiment is problematic and should be rerun for verification of the results.

Exp. 3: Oligoclase/quartz in brine

The major difference between the diagenetic reactions of this experiment and that of experiment 2 appears to be the precipitation of a calcium-enriched phase (Fig. 8) with a fluid concentration difference for sodium and calcium that fits the general mass balance considerations. The comparison of pre- and post-experimental fluids (Fig. 4) shows that this experiment resulted in the only depletion of calcium from the fluid. This may be governed by the lower sodium to calcium ratio ($\sim 7{:}3$) of the oligoclase used in this experiment relative to the labradorite used in experiment 2, the high calcium content of the unidentified precipitates, or a lesser degree of feldspar dissolution. A similar depletion of calcium from the pore fluid was not observed for the other oligoclase experiments (1B, 4 and 5) which may be explained by the difference in physical experimental conditions (Exp. 1B) and/or the difference in the pore-fluid chemistries (Exps. 1B, 4, 5).

Table 3

General observations derived from the scanning electron microscope, and atomic absorption spectroscopy analyses. Arrows next to elements, in the Authigenic Precipitate chemistry column, and arrows in the AAS columns, represent increases (up arrow) or decreases (down arrow) in the concentration of the indicated element. SEM element concentration differences are relative to the initial feldspar, while the AAS concentration differences are relative to the initial pore fluid chemistry. Precipitates in the Authigenic Precipitate Chemistry column are approximately listed by observed abundance

EXPERIMENT NUMBER	SCANNING ELECTRON MICROSCOPE ANALYSES				A.A.S. ANALYSIS Trend Of Differential Concentration Between The Pre- And Post-Experiment Pore-fluids			
	GRAIN ALTERATION		PRECIPITATIVE PHASES					
	Mechanical	Chemical	Authigenic Precipitate Chemistry	Precipitate Morphology	Na	Ca	Mg	K
1B	Fracturing of all grains Extensive fracturing and cleaving of feldspar grains	Surficial etching of feldspar grains Dissolution voids on feldspar grains	Ferrian Sepiolite (or smectite) Na↑,Ca↑,Fe↑ Albitic feldspar overgrowths Unknown precipitate Na↑,Ca↑,Al↑,K↑,Fe↑	Fibrous (honeycomb) Blocky - tabular Rounded globular mass	↑ (slight)	Same	Same	Same
2	Fracturing of all grains Extensive fracturing and cleaving of feldspar grains	Surficial etching of feldspar grains Dissolution voids on feldspar grains	Unknown precipitate Na↓,Ca↓,Al↓,Mg↑ Albitic feldspar overgrowths	Platy - tabular Blocky - tabular	↑ (large)	↑(moderate)	↑ (slight)	↑ (slight)
3	Fracturing of all grains Extensive fracturing and cleaving of feldspar grains	Surficial etching of feldspar grains Dissolution voids on feldspar grains	Unknown precipitate Na↓,Ca↑,Al↑ Albitic feldspar overgrowths	Tabular Blocky - tabular	↑(moderate)	↓ (moderate)	↓ (slight)	↓ (slight)
4	Fracturing of all grains Extensive fracturing and cleaving of feldspar grains	Surficial etching of feldspar grains Dissolution voids on feldspar grains	Unknown precipitate Ca↑, Al↓	Tabular	↑(moderate)	Same	Same	Same
5	Fracturing of all grains Extensive fracturing and cleaving of feldspar grains	Surficial etching of feldspar grains Dissolution voids on feldspar grains	Unknown precipitate Ca↑, Al↑	Tabular (layered)	↑(moderate)	Same	Same	Same
7	Fracturing of all grains Extensive fracturing and cleaving of feldspar grains	Surficial etching of feldspar grains Dissolution voids on feldspar grains	Unknown precipitates Ca↑, Al↑, K↑ (Na same) Ca↓, Al↓ (Na depleted) Albitic feldspar overgrowths	Tabular (platy) Bulbous globular mass Blocky - tabular	↑ (large)	↑(moderate)	Same	↑(slight)
8	Fracturing of all grains Extensive fracturing and cleaving of feldspar grains	Surficial etching of feldspar grains Dissolution voids on feldspar grains	Unknown precipitates Ca↑, K↑, Al↓ (Na same) Ca↑, K↑, Al↑ (Na same) Ca↑, K↑ (Na, Al same) Na↓,Ca↑,Al↑,Fe↑	Tabular (platy) Fibrous globular mass Half Bow-tie Blocky - tabular	↑(moderate)	↑(moderate)	↑(slight)	↑(slight)

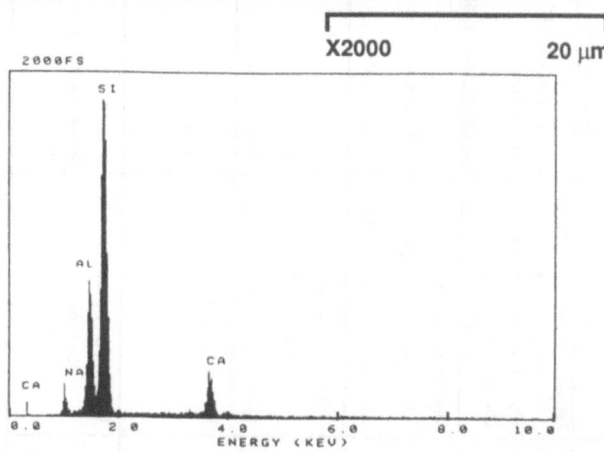

Figure 6

SEM photomicrograph from Exp. 1B of an oligoclase grain showing the development of a dissolution void at a crack-tip (arrowed). Also visible on the right side of the grain is an anastomosing fibrous precipitate crust (probably an iron-rich smectite) with blocky precipitates. Scale bar at bottom of photo is 20 μm. EDX(SEM) scan is of the oligoclase feldspar grain (2000 counts full logarithmic scale).

Exps. 4, 5: Oligoclase/quartz in brine/CO_2

It appears that virtually all of the calcium released by the dissolution of oligoclase was incorporated back into the precipitates. Dissolved sodium, for the most part, did enter into the reprecipitation process and only a small fraction remained in solution. The preferential precipitation of calcium-enriched phases (Fig. 9) for these experiments and for experiments 7, 8 may be a function of calcium stability in the more acidic brine/CO_2 fluids. It would appear that for the physical and chemical conditions of experiments 4, 5, 7, and 8 that calcium is more stable in

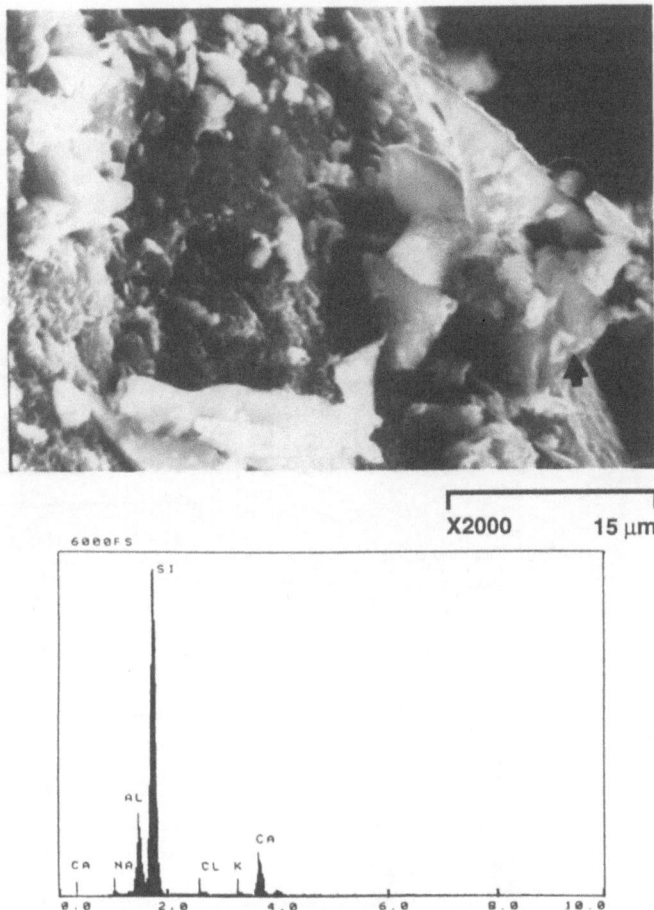

Figure 7

SEM photomicrograph from Exp. 2 of a calcium-diminished precipitate on a quartz grain. Scale bar at bottom right of photo is 15 μm. EDX(SEM) scan (6000 counts full logarithmic scale) shows the relative peak intensities of the calcium-reduced precipitate at the right-hand side of the photo (arrowed). Compare with EDX(SEM) scan, Figure 5, for labradorite.

the solid phases of the precipitates (suspected to be zeolites) than as plagioclase feldspar.

Exps. 7, 8: Labradorite/quartz in brine/CO_2

The labradorite dissolution (Fig. 10) released more calcium into the fluid than experiments 4 and 5 and is reflected by the increase in calcium content of the pore fluid and precipitates (Fig. 11). Not all of the liberated sodium was used in the reprecipitation and, hence, the fluid concentrations increased to a level greater than that for experiments 4 and 5.

Figure 8

SEM micrograph from Exp. 3 of an oligoclase grain showing surface etching and a coating of calcium-enriched precipitate. Scale bar at bottom right of photo is 100 μm. EDX(SEM) scan (2500 counts full logarithmic scale) shows the relative peak intensities of the calcium-enriched precipitate cluster at right side of grain (arrowed). Compare with EDX(SEM) scan, Figure 6, for oligoclase.

Discussion

Previous experimental studies related to sandstone diagenesis (PETROVIC et al., 1976; STOESSELL and PITTMAN, 1990; NESBITT et al., 1991; SCHUTJENS, 1991; SMALL et al., 1992a,b) differed from this study in that the fluids used were not characteristic of the natural subsurface environment, and that the pressure-temperature conditions did not reflect the optimal depths for hydrocarbon generation. HAJASH and BLOOM (1991) used seawater as the pore fluid but experimented at a temperature more characteristic of 6–6.5 km burial depth.

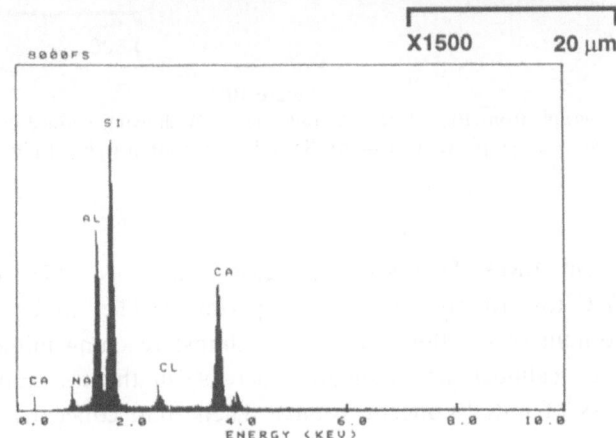

Figure 9

SEM photomicrograph from Exp. 5 of an oligoclase grain showing dissolution voids (D) and precipitate enriched in calcium and aluminum (P). Scale bar at bottom right of photo is 20 μm. EDX(SEM) scan (8000 counts full logarithmic scale) is of the blocky, calcium-enriched precipitate at the center of the photo (arrowed). Compare with EDX(SEM) scan, Figure 6, for oligoclase.

The major differences in the unidentified precipitate chemistries relate to the EDX(SEM) peak intensities for sodium, calcium, and aluminum. Magnesium and potassium, although present in the pore fluid, did not appear to have played a major role in the reprecipitation process as they were not detected by EDX(SEM). This observation for magnesium disagrees with the experiments of HAJASH and BLOOM (1991) who noted a dramatic decrease in their fluid concentration of Mg within the first two days. This disparity may be due to the difference in the initial Mg concentration, or result from the difference in the experiment temperature (Mg 1370 ppm at 200°C, HAJASH and BLOOM; Mg 600 ppm at 129°C, this study). In

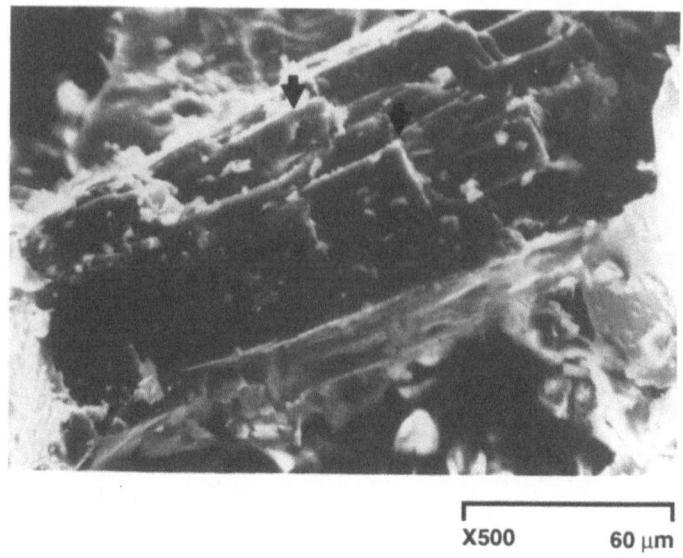

X500   **60 μm**

Figure 10

SEM photomicrograph from Exp. 7 of a labradorite grain showing surface etching that follows the feldspar cleavage planes (arrowed). Scale bar at bottom right of photo is 60 μm.

their experiment BM3, HAJASH and BLOOM dropped their temperature from 202°C to 127°C towards the end of the experiment. They noted a gradual increase in the Mg content of the fluid over time (almost reaching initial concentrations), and a dramatic (almost instantaneous) increase in the Ca content of their pore fluid (in excess of initial concentrations). Their fluid concentration differences at the lower temperature are, for the most part, in agreement with this study. The comparison of this study with that from HAJASH and BLOOM suggests that the classic trade-off between experimental time and temperature may result in diagenetic reactions that do not correctly reflect the burial depth of interest. Thus, other experimental parameters (differential stress, fluid volume, and fluid flow rates) will provide the only means by which diagenetic reactions may be hastened in the laboratory.

Grain Alteration

A review of the A.A.S. and S.E.M. data presented in the previous section illustrates the general diagenetic processes that occurred during each experiment (summarized in Table 3). These processes have been divided into the two categories of *in situ* grain alteration (mechanical and chemical), and authigenic mineral precipitation.

X1200   25 μm

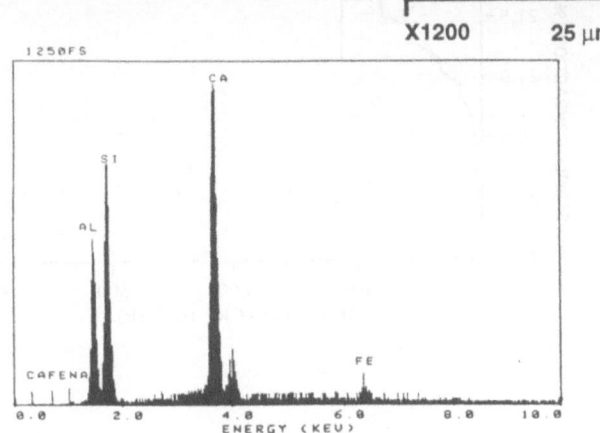

Figure 11

SEM photomicrograph from Exp. 8 showing a blocky precipitate (Ca, Al, Fe enriched; Na diminished) on a labradorite grain. Scale bar at bottom right of photo is 25 μm. EDX(SEM) scan (1250 counts full logarithmic scale) shows the relative peak intensities of the blocky precipitate at the center of the photo (arrowed). Note the dramatic increase in calcium content of the precipitate relative to the initial labradorite (EDX(SEM) scan, Fig. 5). Iron is a contaminant from the pore pressure plumbing system.

Mechanical Grain Alteration

In response to the pressure conditions, both feldspar and quartz grains have mechanically adjusted by brittle failure with no apparent dependence on mineralogy or fluid composition. Quartz and feldspar differ in their response to stress. Experimentally induced failure of quartz grains was not commonplace, but was observed in the form of conchoidal or irregular fracturing. On the other hand, the feldspar grains were extensively crushed with failure evident as cleavage or as irregular fracture (similar to the pre-experiment feldspar shown in Fig. 5).

Although any given sample compacted throughout the entire experiment, most of this compaction (up to 18–19% shortening) occurred within the initial establishment of confining pressure, axial load, pore pressure, and temperature. The data from experiment 8 is displayed for the effective experimental period in which the sample was not only loaded, but also had pore fluids continuously cycling through at temperature (Fig. 12). The noteworthy observation from Figure 12 is that the sample continues to compact an additional 2–3%, reflecting the loss of pore volume. The data also shows an approximately four-fold decrease in permeability

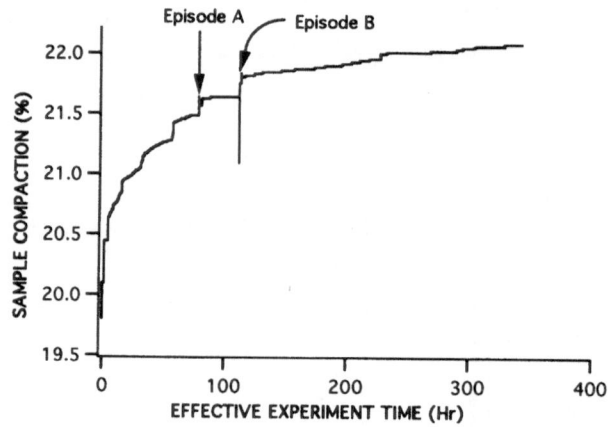

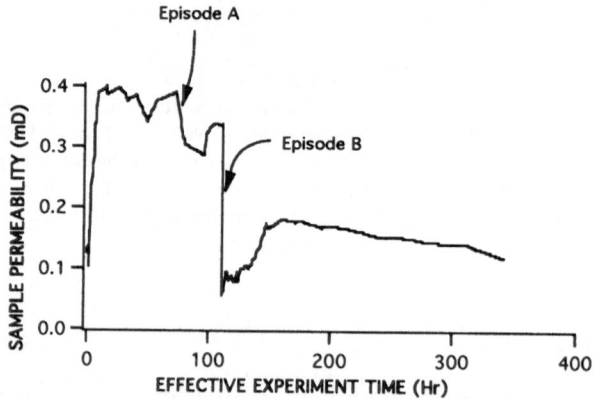

Figure 12

Plots showing compaction (Fig. 12a) and permeability (Fig. 12b) of the sample from experiment 8. Data represents the experimental time period in which the sample was pressured, at temperature, and had fluid pumping. The two transients marked "Episode A" and "Episode B" were induced by power outages which resulted in significant disturbances of the porosity-permeability network.

(relative to the initial values) as the experiment proceeded. The loss of pore volume (related to deformation) combined with chemical diagenesis may have resulted in the continued alteration of the porosity-permeability structure and "lithification" of the sample (SCHOLZ, pers. comm.).

The data presented for experiment 8 (Fig. 12) is of particular interest due to the two episodic power failures that affected the mechanical operation of the triaxial pressure vessel (labeled Episode A and Episode B). Episode B almost had drastic consequences because the axial load dropped to approximately 62 MPa (confining and pore pressures remained constant). The reestablishment of axial load to 93 MPa resulted in a small, but significant, increase in the sample compaction, accompanied by a sharp decrease in permeability. Approximately 50 hours after this event, the sample permeability had recovered to values almost equivalent to those of the overall decreasing trend. Episode A marked a less dramatic event where the axial load increased from 91 MPa to 100 MPa prior to being reset to 93 MPa. These events are considered to have mechanically altered the porosity structure of the sample and consequently affected the permeability network. However, the subsequent improvement in permeability is surprising and is considered to result from the development of a new porosity network through which fluid flow proceeded (SCHOLZ, pers. comm.). This reestablishment of the permeability network may have a more general bearing on studies involving episodic stress fluctuations (such as rapid switching of depositional lobes at the deltas of major river systems; or the sudden slip along a permeable fault network). To adequately study the mechanical aspects of porosity-permeability networks, experiments should be designed that measure the physical parameters of both experiment and sample with extreme accuracy (such as pressures, sample shortening, temperature, pore fluid volume, fluid flow rates).

Chemical Grain Alteration

The lack of significant dissolution features on quartz grains suggests that they did not enter into any *in situ* chemical reactions. Dissolution features, such as surficial etching and dissolution voids, were commonly observed on feldspar grains. The small-scale dissolution voids (only up to 10 microns in diameter) are commonly associated with structural features such as fractures and cleavage (Figs. 6, 10). It is suspected that dissolution voids result from the exploitation of inherent structural and/or chemical weaknesses (sites where the dissolution process is initiated).

The surficial etching of feldspars (Figs. 8, 10) is evident for partial areas of grain faces and commonly accentuates the feldspar cleavage planes (Fig. 10). Typically these regions are relatively barren of precipitative phases, suggesting that precipitation is physically restricted by the presence of contacting grains. However, other grains do not display this mutual exclusion of precipitation and dissolution (Fig. 8). It is suggested, in the case where precipitates and dissolution coincide, that the relatively broad-scale surface etching is initiated as dissolution voids, and becomes more extensive as a response to intragranular stresses.

Dissolution appears to be the only process for *in situ* grain alteration. EDX(SEM) scans from any given experiment do not show a marked difference in the feldspar chemistry. Also, preliminary microprobe analysis of the feldspars (from core to rim) has not defined any major chemical zonation, suggesting that *in situ* alteration to other mineralogic phases has not occurred.

Authigenic Precipitates

Altered mineralogic phases were precipitated on both quartz and feldspar grain surfaces for all experiments (Figs. 6–9, 11). The exact nature of these precipitates differs, depending on the experimental conditions. The precise mineralogy of these precipitates is, for the large part, unknown due to their relatively low abundance and small size. The only meaningful mineralogic constraints for these precipitates have been derived from the SEM analysis. Qualitative precipitate chemistry has been obtained from the EDX(SEM) scans (by comparing elemental peak intensities), and the precipitate morphology is visually apparent from SEM photomicrographs. The concentration differences between the precipitates and the feldspars are indicated in Table 3 by the arrows located next to each element (an up arrow represents a concentration increase relative to the initial feldspar; conversely, a down arrow represents a concentration decrease).

SMITH's (1983) discussion of the subsolidus phase reactions involving plagioclase feldspars was significant in that they provided a means by which plagioclase feldspar could be studied from a geochemical viewpoint. While the subsolidus phase relations outlined by SMITH are applicable to metamorphic reactions, they may not be truly representative of reactions that occur in the diagenetic realm because the relations are primarily concerned with higher temperatures and pressures than the *P-T* conditions of diagenesis. As the literature is lacking phase relation data for the *P-T* conditions of diagenesis, studies of phase relations at higher *P-T* conditions provide the only insights into the chemical diagenesis of plagioclase feldspar at the atomic scale.

SMITH (1983) suggests that low-albite and P-anorthite are the only truly ordered plagioclase phases. He also considers that all other plagioclase phases represent intergrowths of the two stable end-member phases (which exist as stable domains). The "stable-domain" concept implies that plagioclase diagenesis involves domains of albite and anorthite where only one end-member remains stable in the solid form. Hence, plagioclase dissolution should result in the removal of the unstable end-member, leaving the stable end-member intact (thereby enriching the original plagioclase grain in the stable phase). However, the microprobe analyses of the post-experiment feldspar grains did not indicate any noticeable chemical zonation.

This lack of end-member enrichment towards the rim of the feldspar grains may suggest that:

1. neither low-albite nor P-anorthite were stable forms during these experiments,
2. plagioclase feldspar cannot be regarded as simple domains of the two end-member phases,
3. the end-member enrichment towards the rim of feldspar grains is beyond the resolution of the microprobe.

Similarly, the "stable-domain" concept of SMITH (1983) implies that authigenic precipitation of plagioclase feldspar during diagenesis would result in the growth of crystals having the chemical composition of the stable end-member form. A strict consideration of the concepts outlined by SMITH suggests that none of the authigenic precipitates of this study is albite or anorthite as no one precipitate exhibited the chemical character of the end-members. However, the application of relatively high P-T phase relations to the low P-T conditions typical for diagenesis is questionable and the possibility that plagioclase precipitation may involve both end-members cannot be ruled out.

With the possible exception of the fibrous precipitate of Experiment 1B (Fig. 6), no other precipitative phases have been definitely identified. The general chemistries and morphologies of the unidentified phases are different to that of the initial feldspars, suggesting that they belong to a different mineral group. Although the SEM data for these precipitates are not indicative of the precise mineralogy, they are suspected to be either clays or zeolites.

Perhaps the most noteworthy correlation observed from these experiments is that for calcium. Experiments 4, 5, 7, and 8, which were run with the brine/CO_2 pore fluid mixture, are dominated by precipitates significantly enriched in calcium relative to the feldspar (Figs. 9, 11) with no net decrease in the calcium content of the pore fluids (experiments 7 and 8 recorded moderate increases). Although some calcium-enriched precipitates were observed from experiment 3 (Fig. 8), the majority of the unidentified precipitates of experiments 1B, 2, and 3 (run without CO_2 in the pore fluid) were not dominated by such a calcium-enriched phase and were, in fact, typically calcium-diminished (Figs. 6, 7). The precipitation of these calcium-enriched phases (that are not reactive to dilute HCl) appears to be favored where pore fluids may be more acidic (as determined by the brine/CO_2 experiments).

Although not common, one type of precipitate observed on grain surfaces as a result of many of these experiments is a sodium-enriched, calcium-diminished, tabular/blocky mineral (Fig. 7) suspected to be either a zeolite or clay (too fine-grained for positive identification). As with the other unidentified precipitate phases, further experimentation designed to increase the bulk concentration of these authigenic minerals may shed light on the phase relations of plagioclase feldspar in the diagenetic realm. Should the "stable-domain" concept of SMITH (1983) provide an insufficient boundary condition for plagioclase diagenesis, then the sodium-enriched precipitate described above may be representative of an intermediate stage in a process that may ultimately lead to complete alteration of the plagioclase to an end-member phase.

Increases in the potassium content of the precipitates were commonly associated with chlorine peaks. The appearance of the potassium and chlorine peaks together is considered to be an artifact produced from potassium-chloride salts precipitated from residual brines trapped in the sample during the post-experiment drying process.

Implications for Porosity-permeability Development

Closed systems: transfer of dissolved material would be local (perhaps within the same pore space). Authigenic minerals would be precipitated in the pore spaces apparently occluding porosity. However, the nature of the precipitative phases should be considered together with the extent of feldspar dissolution (feldspar dissolution serving to enhance porosity). Precipitative phases that are denser than the original feldspar would not occupy the same volume. Hence, the net change in volume would result in an increase in porosity. Conversely, less dense precipitates would reduce the overall porosity. Authigenic clays may not only reduce the overall porosity, but also choke the pore throats thereby reducing permeability.

Open system: the dissolution of plagioclase feldspars in a sandstone would effectively create secondary pore space. No significant reduction in this newly generated porosity would occur because the dissolved material would be carried away. Should this dissolved material be transported up-section and be precipitated so as to occlude porosity, then the process of feldspar diagenesis may partially provide its own seal.

Summary

1. Further experimentation is required to refine the study of plagioclase feldspar diagenesis. It may be advantageous to design experiments that concentrate on one aspect of the plagioclase diagenesis (mechanical or chemical) prior to studies that involve both aspects. Accurate measurements of pressures, temperature, sample and fluid volumes will assist research into the mechanical diagenesis with elevated P-T conditions potentially accelerating the process. The chemical component of plagioclase diagenesis may be investigated by experiments that are designed to increase the bulk concentration and size of precipitates (perhaps by constantly replenishing the pore fluids and by flowing fluid through the sample at a relatively high rate).

2. The comparison of the pre- and post-experiment SEM micrographs indicates that the plagioclase feldspar grains are significantly crushed and fractured due to the rapid change in pressure during experimental start-up. The feldspars also

exhibit partial dissolution. Quartz grains are not as extensively fractured and crushed and do not exhibit dissolution features.

3. Plagioclase dissolution occurs in the form of:

a) general grain surface etching (topographic dissolution) occasionally related to the plagioclase cleavage,

b) dissolution voids (or pits) that persist to some depth within the plagioclase grains, occasionally related to structural imperfections (e.g., cracks).

4. Preliminary microprobe analysis of feldspar grains, along a line from the core to the rim, does not suggest that grain alteration in the form of *in situ* chemical zonation has occurred.

5. The dissolved material derived from the plagioclase dissolution is incorporated into the pore fluid. Much of this solute material is reprecipitated onto both feldspar and quartz grain surfaces as a different mineralogic phase. This suggests that, for these experiments, the fluid system behaved more like a closed system even though the average experimental fluid flow rate was approximately 1.92 cc/min.

6. A correlation between pore-fluid type and calcium content of the precipitates is observed. Experiments that were run using a pore fluid containing carbon dioxide resulted in precipitates that were calcium-enriched relative to the initial plagioclase composition. Those experiments run without carbon dioxide as a part of the pore fluid resulted in precipitates that were calcium-diminished relative to the original feldspar composition.

Acknowledgments

The authors are grateful to C. H. Scholz for providing the laboratory in which this work was completed, and for the constructive criticism of this manuscript. We are also grateful to F. M. Chester and E. Schreiber for their experimental guidance. Finally, our appreciation is expressed to all who reviewed this manuscript.

REFERENCES

AL-SHAIEB, Z., and SHELTON, J. (1981), *Migration of Hydrocarbons and Secondary Porosity in Sandstones*, Am. Assoc. Petrol. Geol. Bull. *65*, 2433–2436.

BARNES, D. A., GIRARD, J.-P., and ARONSON, J. L., *K-Ar Dating of illite diagenesis in the Middle Ordovician St. Peters Sandstone, central Michigan Basin, USA: Implications for thermal history.* In *Origin, Diagenesis, and Petrophysics of Clay Minerals in Sandstones* (Houseknecht, D. W., and Pittman, E. D., eds.) (S.E.P.M. Spec. Pub. *47*, 1992) pp. 35–48.

BÅRTH, T. (1991), *Organic Acids and Inorganic Ions in Waters from Petroleum Reservoirs, Norwegian Continental Shelf: A Multivariate Statistical Analysis and Comparison with American Reservoir Formation Waters*, Appl. Geochem. *6*, 1–15.

BJØRLYKKE, K., BERGEN, A., ELVERHØI, O., and MALM, A. O. (1979), *Diagenesis in the Mesozoic Sandstones from Spitzbergen and the North Sea*, Geol. Rindschau *68*, 1151–1171.

BOLES, J. R., *Secondary porosity reactions in the Stevens Sandstone, San Joaquin Valley, California.* In *Clastic Diagenesis* (McDonald, D. A., and Surdam, R. C., eds.) (Am. Assoc. Petrol. Geol. Mem. *37*, 1984) pp. 217–224.

BOWKER, K. A., and SHULER, P. J. (1991), *Carbon Dioxide Injection and Resultant Alteration of the Weber Sandstone, Rangely Field, Colorado,* Am. Assoc. Petrol. Geol. Bull. *75*, 1489–1499.

BURLEY, S. D., and MACQUAKER, J. H. S., *Authigenic clays, diagenetic sequences and conceptual diagenetic models in contrasting basin-margin and basin-center North Sea Jurassic sandstones and mudstones.* In *Origin, Diagenesis, and Petrophysics of Clay Minerals in Sandstones* (Houseknecht, D. W., and Pittman, E. D., eds.) (S.E.P.M. Spec. Pub. *47*, 1992) pp. 81–110.

CHESTER, F. M., and HIGGS, N. G. (1992), *Multimechanism Friction Constitutive Model for Ultrafine Quartz Gouge at Hypocentral Conditions,* JGR (B) *97*, 1859–1870.

COLLINS, A. G., *Geochemistry of Oilfield Waters* (Elsevier, New York 1975) 495 pp.

CONNOLLY, C. A., WALTER, L. M., BAADSGAARD, H., and LONGSTAFFE, F. J. (1990), *Origin and Evolution of Formation Waters, Alberta Basin, Western Canada Sedimentary Basin. I. Chemistry,* Appl. Geochem. *5*, 375–395.

CURTISS, C. D. (1985), *Clay Mineral Precipitation and Transformation during Burial Diagenesis,* Phil. Trans. Roy. Soc. Lond. *A315*, 91–105.

DE SITTER, L. U. (1947), *Diagenesis of Oil-field Brines,* Am. Assoc. Petrol. Geol. Bull. *31*, 2030–2040.

DICKEY, P. A. (1969), *Increasing Concentration of Subsurface Brines with Depth,* Chem. Geol. *4*, 361–370.

DREVER, J. I., *Geochemistry of Natural Waters* (Prentice-Hall, Englewood Cliffs 1982) 388 pp.

DUTTON, S. P., and LAND, L. S. (1988), *Cementation and Burial History of a Low-permeability Quartarenite, Lower Cretaceous Travis Peak Formation, East Texas,* Geol. Soc. Am. Bull. *100*, 1271–1282.

EHRENBERG, S. N. (1990), *Relationship between Diagenesis and Reservoir Quality in Sandstones of the Garn Formation, Haltenbanken, Mid-Norwegian Continental Shelf,* Am. Assoc. Petrol. Geol. Bull. *74*, 1538–1558.

FISHER, J. B., and BOLES, J. R. (1990), *Water-rock Interactions in Tertiary Sandstones, San Joaquin Basin, California,* Chem. Geol. *82*, 83–101.

HELMOLD, K. P., and VAN DE KAMP, P. C., *Diagenetic mineralogy and controls on albitization and laumontite formation in Paleogene arkoses, Santa Ynez Mountains, California.* In *Clastic Diagenesis* (McDonald, D. A., and Surdam, R. C., eds.) (Amer. Assoc. Petrol. Geol. Mem. *37*, 1984) pp. 239–275.

GALLOWAY, W. E. (1974), *Deposition and Diagenetic Alteration of Sandstone in Northeast Pacific Arc-related Basins: Implications for Graywacke Genesis,* Geol. Soc. Am. Bull. *85*, 379–390.

HAJASH, A., and BLOOM, M. A. (1991), *Marine Diagenesis of Feldspathic Sand: A Flow-through Experimental Study at 200°C, 1 Kbar,* Chem. Geol. *89*, 359–377.

HOUSEKNECHT, D. W. (1984), *Influence of Grain Size and Temperature on Intergranular Pressure Solution, Quartz Cementation, and Porosity in a Quartzose Sandstone,* J. Sedim. Petrol. *54*, 348–361.

HOUSEKNECHT, D. W. (1987), *Assessing the Relative Importance of Compaction Processes and Cementation to Reduction of Porosity in Sandstones,* Am. Assoc. Petrol. Geol. Bull. *71*, 633–642.

KAISER, W. R., *Predicting reservoir quality and diagenetic history in the Frio Formation (Oligocene) of Texas.* In *Clastic Diagenesis* (McDonald, D. A., and Surdam, R. C., eds.) (Amer. Assoc. Petrol. Geol. Mem. *37*, 1984) pp. 195–215.

LAND, L. S., MILLIKEN, K. L., and MCBRIDE, E. F. (1987), *Diagenetic Evolution of Cenozoic Sandstones, Gulf of Mexico Sedimentary Basin,* Sedim. Geol. *50*, 195–225.

LONGSTAFFE, F. J., TILLEY, B. J., AYALON, A., and CONNELLY, C. A., *Controls on porewater evolution during sandstone diagenesis, western Canada sedimentary basin: An oxygen isotope perspective.* In *Origin, Diagenesis, and Petrophysics of Clay Minerals in Sandstones* (Houseknecht, D. W., and Pittman, E. D., eds.) (S.E.P.M. Spec. Pub. *47*, 1992) pp. 13–34.

LOUCKS, R. G., DODGE, M. M., and GALLOWAY, W. E., *Regional controls on diagenesis and reservoir quality in lower Tertiary sandstones along the Texas Gulf Coast.* In *Clastic Diagenesis* (McDonald, D. A., and Surdam, R. C., eds.) (Amer. Assoc. Petrol. Geol. Mem. *37*, 1984) pp. 15–45.

MARONE, C., RUBENSTONE, J., and ENGELDER, T. (1988), *An Experimental Study of Permeability and Fluid Chemistry in an Artificially Jointed Marble,* J. Geol. Res. *93* (B), 13763–13775.

MATHISEN, M. E., *Diagenesis of Plio-pleistocene nonmarine sandstones, Cayagan Basin, Philippines: Early development of secondary porosity in volcanic sandstones*. In *Clastic Diagenesis* (McDonald, D. A., and Surdam, R. C., eds.) (Am. Assoc. Petrol. Geol. Mem. *37*, 1984) pp. 177–193.

McBRIDE, E. F., LAND, L. S., and MACK, L. E. (1987), *Diagenesis of Eolian and Fluvial Feldspathic Sandstones, Norphlet Formation (Upper Jurassic), Rankin County, Mississippi, and Mobile County, Alabama*, Am. Assoc. Petrol. Geol. Bull. *71*, 1019–1034.

MILLIKEN, K. L., McBRIDE, E. F., and LAND, L. S. (1989), *Numerical Assessment of Dissolution versus Replacement in the Subsurface Destruction of Detrital Feldspars, Oligocene Frio Formation, South Texas*, J. Sedim. Petrol. *59*, 740–757.

MILLIKEN, K. L. (1992), *Chemical Behavior of Detrital Feldspars in Mudrocks versus Sandstones, Frio Formation (Oligocene), South Texas*, J. Sedim. Petrol. *62*, 790–801.

MONCURE, G. K., LAHANN, R. W., and SIEBERT, R. M., *Origin of secondary porosity and cement distribution in a sandstone/shale sequence from the Frio Formation (Oligocene)*. In *Clastic Diagenesis* (McDonald, D. A., and Surdam, R. C., eds.) (Amer. Assoc. Petrol. Geol. Mem. *37*, 1984) pp. 151–161.

NESBITT, H. W., MACRAE, N. D., and SHOTYK, W. (1991), *Congruent and Incongruent Dissolution of Labradorite in Dilute, Acidic, Salt Solutions*, J. Geol. *99*, 429–442.

OVERTON, H. L. (1973), *Water Chemistry Analysis in Sedimentary Basins*, Soc. Prof. Well Log Anal. Annual Logging Symp., Trans No. *14* (L), 22 pp.

PETROVIC, R., BERNER, R. A., and GOLDHABER, M. B. (1976), *Rate Control in Dissolution of Alkali Feldspars — I: Study of Residual Feldspar Grains by X-ray Photoelectron Spectroscopy*, Geochim. et Cosmochim. Acta *40*, 537–548.

RICHARDSON, S. M., and McSWEEN Jr., H. Y., *Geochemistry: Pathways and Processes* (Prentice-Hall, Englewood Cliffs 1989) 488 pp.

RITTENHOUSE, G. (1971), *Pore-space Reduction by Solution and Cementation*, Amer. Assoc. Petrol. Geol. Bull. *55*, 80–91.

SCHMIDT, V., and McDONALD, D. A., *The role of secondary porosity in the course of sandstone diagenesis*. In *Aspects of Diagenesis* (Scholle, P. A., and Schluger, P. R., eds.) (S.E.P.M. Spec. Pub. *26*, 1979) pp. 175–201.

SCHOLZ, C. H., and KOCZYNSKI, T. A. (1979), *Dilatancy Anisotropy and the Response of Rock to Large Cyclic Loads*, JGR (B) *84*, 5525–5534.

SCHUTJENS, P. M. T. M. (1991), *Experimental Compaction of Quartz Sand at Low Effective Stress and Temperature Conditions*, J. Geol. Soc. Lond. *148*, 527–539.

SELLEY, R. C., *Elements of Petroleum Geology* (W. H. Freeman and Co., New York 1985) 449 pp.

SMALL, J. S., HAMILTON, D. L., and HABESCH, S. (1992a), *Experimental Simulation of Clay Precipitation within Reservoir Sandstones I: Techniques and Examples*, J. Sedim. Petrol. *62*, 508–519.

SMALL, J. S., HAMILTON, D. L., and HABESCH, S. (1992b), *Experimental Simulation of Clay Precipitation within Reservoir Sandstones II: Mechanism of Illite Formation and Controls on Morphology*, J. Sedim. Petrol. *62*, 520–529.

SMITH, J. V., *Phase equilibria of plagioclase*. In *Feldspar Mineralogy* (Ribbe, P. H., ed.) (Mineral. Soc. Amer. Rev. in Mineral Vol. 2, 1983) pp. 223–239.

STOESSELL, R. K., and PITTMAN, E. D. (1990), *Secondary Porosity Revisited: The Chemistry of Feldspar Dissolution by Carboxylic Acids and Anions*, Am. Assoc. Petrol. Geol. Bull. *74*, 1795–1805.

SURDAM, R. C., CROSSEY, L. J., HAGEN, E. S., and HEASLER, H. P. (1989), *Organic-inorganic Interactions and Sandstone Diagenesis*, Am. Assoc. Petrol. Geol. Bull. *73*, 1–23.

VAN ELSBERG, J. M. (1978), *A New Approach to Sediment Diagenesis. Part I: An Observed Relationship between Sonic Transit-time and Depth in the Tertiary Sediments of the Mackenzie Delta; A Potential Exploration Tool. Part II: A Revised Concept of Sediment Diagenesis*, Can. Petrol. Geol. Bull. *26*, 57–86.

WHITE, D. E., *Saline waters of sedimentary rocks*. In *Fluids in Subsurface Environments* (Young, A., and Galley, J. E., eds.) (Am. Assoc. Petrol. Geol. Mem. *4*, 1965) pp. 342–366.

YIN, P., and SURDAM, R. C. (1985), *Naturally Enhanced Porosity and Permeability in the Hydrocarbon Reservoirs of the Gippsland Basin, Australia*, Proc. of the First Enhanced Oil Recovery Symposium, 79–109.

(Received April 20, 1993, revised/accepted November, 1993)

PAGEOPH, Vol. 141, No. 2/3/4 (1993)

0033–4553/93/040249–20$1.50 + 0.20/0

A New Exploration Tool: Quantitative Core Characterization

CARL H. SONDERGELD[1] and CHANDRA S. RAI[1]

Abstract — We will describe a new laboratory system which was designed to be highly automated and portable while maintaining quality. Driving this design was the recognition of the temporal dependence of physical properties. It becomes apparent that some sedimentary rocks, particularly shales, degrade and disaggregate so completely that mechanical or elastic properties cannot be measured. This temporal dependence displays a time scale much shorter than normal weathering but greater than the time for stress relief. A system was designed to permit field characterization of freshly recovered core material. A benefit of automation and portability is a marked increase in measurement efficiency. The attributes of this system permit rapid characterization of a large number of fresh cores in remote, frontier exploration areas. This feature can significantly reduce prospect evaluation time. Statistically significant rock property databases can be created in a short period of time.

Key words: Exploration tool, physical properties, core characterization.

Introduction

Geophysical emphasis in exploration and exploitation is focusing more and more on using powerful 3D seismic data acquisition and processing (ROBERTSON, 1989), analysis of Amplitude Variation with Offset (AVO) (OSTRANDER, 1984) tomography (JUSTICE *et al.*, 1989; LINES, 1991), and direct shear wave studies (MUELLER, 1991) to generate more accurate interpretations of the subsurface geology. Absent in all these approaches is the direct capability to quantify a specific geological quantity, such as lithology or pore fluid. It is only through knowledge of how seismically mappable parameters such as amplitudes and velocities relate to rock properties that the tie between seismics and geology can be made. Various approaches are utilized to achieve this end and perhaps the most common is to utilize data taken from nearby wells, typically in the form of wireline logs. While great improvements have been made in logging capabilities and data quality, there remains a number of questions in utilizing wireline derived data. Among the most pressing issues are: (1) the fact that different logging tools integrate in different manners over different volumes of the subsurface (ENDERLIN *et al.*, 1991); (2) no

---

[1] Amoco Production Research, 4502 E. 41st Street, Tulsa, OK 74135, U.S.A.

logging tool determines mineralogy directly; (3) the presence of a drill hole is a perturbation of *in situ* temperature and stresses; (4) influence of formation damage both known and unknown on measured properties, and finally (5) the expense of collecting a sufficient suite of logs in terms of both rig time and money to properly address the problems of interest. GEM, the Geophysical Evaluation Modules, can be viewed as a tool permitting us to bridge geology and geophysics. An extensive suite of measurements are performed on the same small controlled volume of a sample. This permits us to make reliable empirical correlations between various measured quantities. Included in the suite of properties are the following: grain (ρ_g), dry (ρ_d) and saturated (ρ_s) densities; Boyle's law and effective porosities; compressional and polarized shear wave velocities as a function of effective pressure; shear wave birefringence as a function of effective pressure; qualitative and quantitative mineralogy; magnetic susceptibility; compressive strength; static moduli (Young's modulus and Poisson's ratio); and total organic carbon content (TOC). The extrapolation of these measurements to field problems becomes a focus of this paper.

Capabilities and Characteristics

The system developed six years ago consists of a sample preparation module and a measurement module. Figure 1 shows the deployment of the system at a well site. Each module is roughly 8 ft wide by 12 ft long and 8 ft high and when outfitted with equipment weighs about 4500 lbs. Associated with the system but not housed within the modules are an air compressor, cut-off saw, and weatherproof boxes of supplies and replacement components.

Central to the efficiency of GEM is the automation of both sample tracking and measurements. The operational philosophy is one of parallel sample processing. Field deployment necessarily requires more consideration than normally encountered in a stationary truck mounted and/or shipboard facility. Some of these considerations are reflected in the hardware selection, mounting and configuration and others in the software. A higher degree of automation places rigid demands on sample quality and would thus limit the applicability of the system. Specifically, the measurement procedures and techniques we describe are not appropriate to unconsolidated material in general.

Typical Sample Measurement Flow

Core is plugged if need be to produce a suitable sample diameter for measurement: acceptable diameters are 1 in., $1\frac{7}{8}$ in., $2\frac{1}{2}$ in. or $3\frac{3}{8}$ in. The required length is normally $2\frac{1}{4}$ in. although recent modifications permit us to work with plugs as short as $\frac{1}{2}$ in. The normal length plug is cut into two specimens: (1) a disc $\frac{1}{2}$ in. long and

Figure 1
One of the two modules comprising GEM. Two modules are employed to provide separation of the
sample preparation and measurement processes. This helps to preserve a cleaner measurement environment. Modules can be transported easily by most means including helicopters.

(2) a core $1\frac{1}{2}$ in. long. The $1\frac{1}{2}$ in. long core is ground parallel and flat on both ends.
Subsequent processing is deployment dependent. At a well site, fresh core is
available therefore wet or saturated measurements are performed first. Saturation is
preserved at all processing steps. At a core facility, dry measurements are performed
first. Procedures are carried out to vacuum dry fresh core and saturate old core to
collect a complete data set for each saturation state. Core weights (dry and
saturated) are used along with volumes determined through Mercury immersion to
calculate both dry and saturated densities of both the core and disc samples. Discs
are then used to measure magnetic susceptibility. Discs are crushed, inserted in
helium pycnometers to determine grain densities. The samples are ball milled and
hand ground to be suitable for infrared mineralogy and total organic carbon (TOC)
determination. The infrared mineralogy is determined using transmission Fourier
Transform Infrared Spectroscopy (FTIR) (GRIFFITHS and DE HASETH, 1986). The
FTIR defines which samples require acid treatment prior to TOC determination.
TOC is determined with the aid of an elemental analyzer. A sample is vaporized at
high temperature to decompose both the CO_3 radicals and organic carbon. The
gases are scrubbed and analyzed for carbon, hydrogen and nitrogen. The associated

core sample (1.5 in. length) is used to determine seismic velocities, V_p and V_s as a function of effective pressure. We use a pulse transmission technique to determine velocities (SCHREIBER *et al.*, 1973). P waves are measured at approximately 500 KHz and shear at about 300 KHz. Two mutually orthogonally polarized shear wave transducers are used. This configuration allows us to examine shear wave birefringence (SONDERGELD and RAI, 1992). Typical waveforms for P and S waves are shown in Figures 2a and b as a function of effective pressure. Note the decrease in travel time and the improved signal quality at higher effective pressures. Cores are typically run at two saturations, dry and $\sim 100\%$ brine saturated. The measurements at two saturations can be used to estimate velocity dispersion (WINKLER, 1986; MAVKO and JIZBA, 1991).

If compressive strength is desired, another core plug is taken, measuring 1 in. in diameter and 3 in. in length. The sample ends are ground flat and parallel. After measuring dimensions and weight, the specimen is jacketed with heat-shrink tubing and outfitted with two displacement transducers (LVDTs), one axial and one radial. This specimen is then run through a standard test i.e., strain rate, saturation,

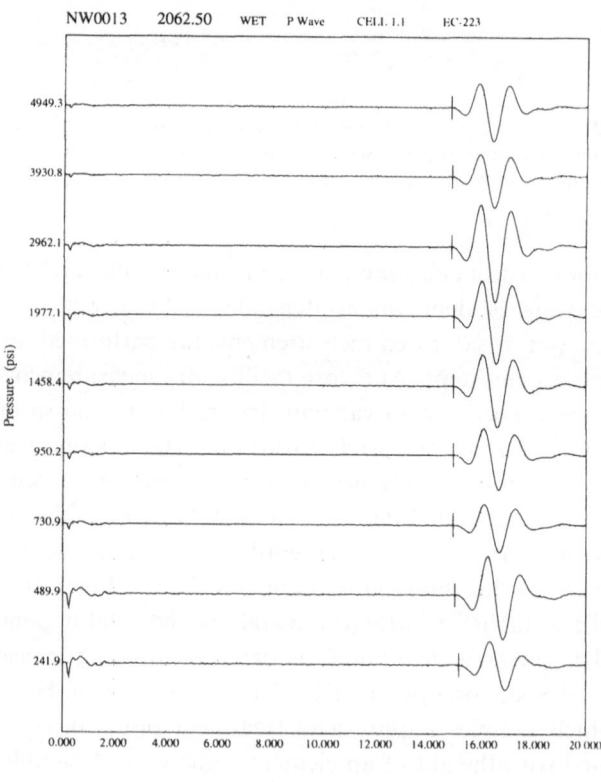

Figure 2(a)

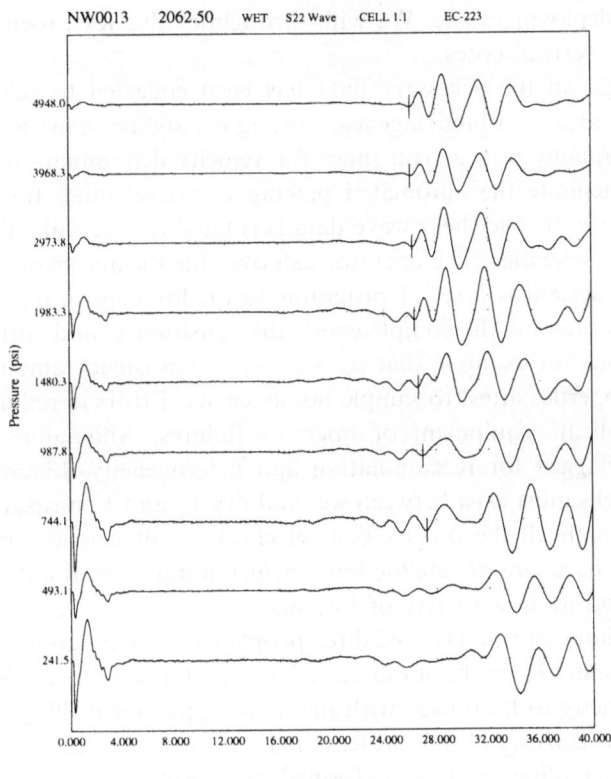

Figure 2(b)

Figure 2

Digitized velocity time series from a rock being subjected to increasing effective pressure. The horizontal axes are both time in microseconds (note scale differences). Vertical offsets are not scaled in pressure. The effective pressure associated with each waveform is labeled to the left. Figure (a) shows compressional waveforms and (b) shear waveforms. Note the strong birefringence in (b) is pressure dependent.

pore and confining pressure to determine the failure strength and static (large strain amplitude) values of Young's modulus and Poisson's ratio.

Another core plug is taken for air permeability measurements. The plug is 1 in. in diameter and up to 2 in. long. Permeability is measured using a standard flow through technique (API, 1960). Maintaining a constant flow rate, one uses the pressure differential to calculate permeability. Both the velocity measurements and the permeability require sample cleaning as a part of the preparation process. A Soxhlet extractor is used with a distillate of toluene and methanol for this purpose.

Geophysical exploration problems require data be obtained on vertical core samples while production and reservoir studies focus on horizontal core properties. Since we carry our own machining and plugging capabilities, both requirements can be met at any deployment site. We will, throughout this text, focus the discussion on results from vertical cores.

At this stage, all the necessary data has been collected to calculate densities, porosities, velocities, and birefringence. Editing capabilities exist to alter any input values or to manually pick arrival times for velocity determinations. Software was developed to facilitate the automated picking of arrivel times for P and S wave energy (see Figure 2). The shear wave data is rotated to determine the true fast and slow shear wave velocities. The operator can override the automatic picks. Included in the software system is a set of programs to quality control data collected. The quality controls examine the completeness, the consistency, and correctness of data sets. It is obvious for example that $\rho_g > \rho_s > \rho_d$; consistency among the core and disc-derived properties attest to sample homogeneity. Errors in redundant measures are used to indicate equipment or operator failures. Anomalous disc and core properties are flagged for reexamination and heterogeneity. Likewise, a series of consistency checks must exist between wet and dry V_p and V_s measurements. A data set passing through all the quality control checks is then ready for transfer to a database over a hardwire or satellite link, inclusion into a local database for on-site analysis and plotting in a variety of formats.

In a twelve-hour shift, a crew of three people can process about fifty samples of one saturation state. It has been our experience that at a well site deployment, this is sufficient capacity to keep pace with the most aggressive drilling and coring rigs. An optimal crew consists of one professional and two technicians. The judgements of quality and sampling are best performed by a professional.

Mineralogy is determined using a transmission infrared technique (GRIFFITHS and DE HASETH, 1986). This requires careful weighing of both salt (KBr) and specimen. The relative weights are kept constant for all specimens. Thus, the background loss i.e., bulk absorption due to the KBr is the same from specimen to specimen. However, when minerals are present, one observes peaks in absorption at characteristic frequencies or wave numbers (see Figure 3). Absorption peaks are due to energy losses associated with excitation of vibrational modes of chemical bonds in minerals. Since each bond can have more than one vibrational mode, even simple minerals display multiple peaks. The spectra displayed in Figure 4 depict how easy it is to determine predominant clay types. The portion of the spectra at wave numbers greater than 3200 contains very diagnostic clay peaks.

In the bandwidth of investigation, minerals have very diagnostic spectra (see Figures 3 and 4). So it is quite simple to compare an unknown combination of mineral spectra (Figure 3) to those in a library of standard minerals, e.g., Figure 4, and determine qualitatively the species present. This is important information in its own right, though qualitative. As practiced in GEM, this data is available as raw spectra and as color coded waterfall plot. We present here (Figure 5) a black and white substitute and attempt to use line patterns as a surrogate for color. The patterns highlight spectral bands where common minerals, if present, would display

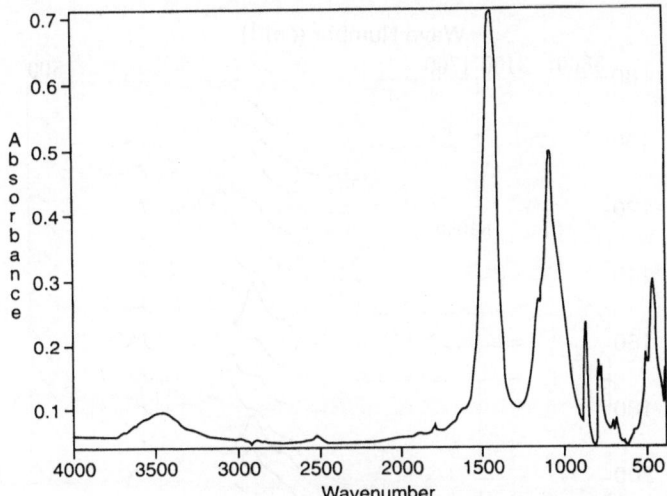

Figure 3

A typical FTIR absorption spectrum for a calcite-rich sandstone. Note the dominant peaks at 1500 and 1000 wave number. These are calcite and quartz, respectively. The peak at 3500 wave number (λ) is due to clays.

Mineral Spectra

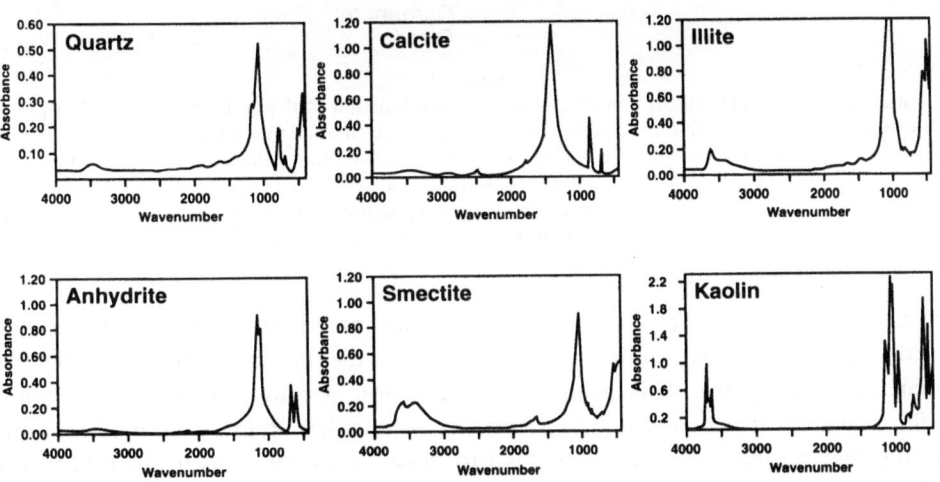

Figure 4

Examples of absorption spectra for some common minerals. Comparison of an unknown spectrum (Figure 3) to these reference spectra allows qualitative mineralogy to be determined.

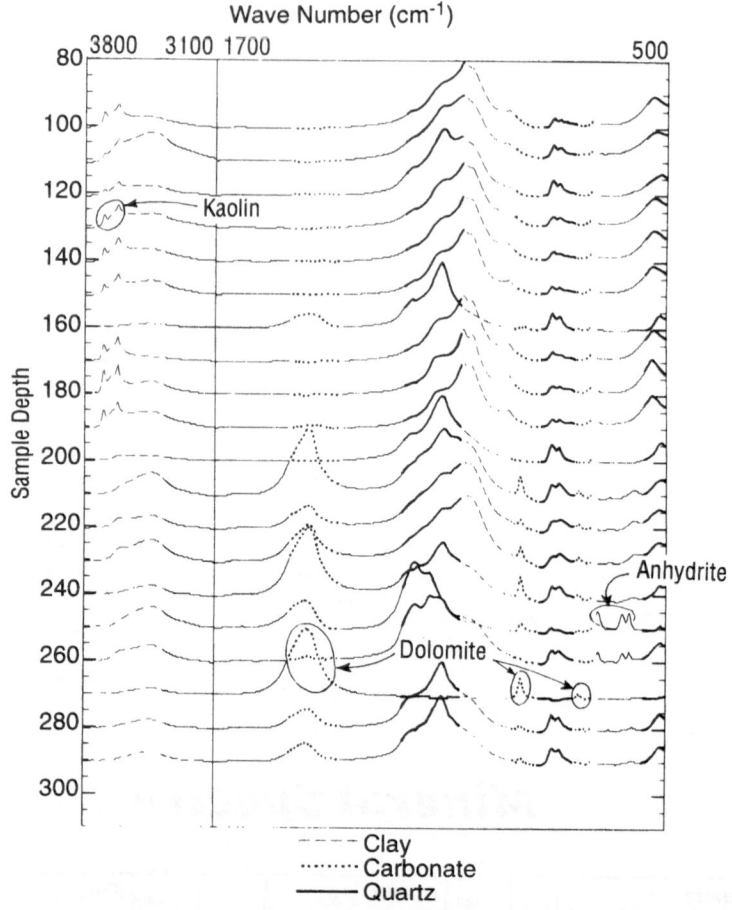

Figure 5

A waterfall plot of FTIR absorption spectra plotted as a function of depth. Each spectrum is a plot of absorbance against wave number. Spectra are normalized and line texture highlighted over a range of wave numbers (cm⁻¹) associated with a particular mineral. Clays, for example, have dominant peaks in the wave number range of 3800–3100 (cm⁻¹). Kaolin is identified. Carbonates have three dominant peaks (dotted). The region of the third peak (lowest wave number) is diagnostic for calcite and dolomite. Anhydrite displays strong peaks close to a wave number of 500.

prominent absorption peaks. The rock in question (Figure 3) can be classified crudely as a limestone, sandstone or shale, based on the dominance of carbonate, quartz or clay peaks. This primitive classification permits one to select a more refined quantitative model for the respective lithologies. Regression analysis, comparing spectral features of the unknown to those of the minerals in the lithological mineral suite, yields a quantitative estimate of the weight percentage of the minerals present. Generic to any such quantification is a model of mineralogical constituents. Thus, the quantitative analysis will only find the relative percentages of the minerals defined to constitute a given lithology. At present, a typical mineralogical suite for sandstones is composed of about 17 minerals. Regression analysis, comparing spectral features of the unknown to those of the minerals in the lithological mineral suite, yields a quantitative estimate of the weight percentage of the minerals present.

In summary, we can routinely determine the bulk dry, ρ_d, wet, ρ_s, and grain densities, ρ_g, Boyle's law, ϕ_B, and effective porosity, ϕ_E, magnetic susceptibility, χ, TOC, qualitative and quantitative mineralogy, compressional (V_p), and shear wave (V_s) velocities as a function of effective pressure at two saturations, birefringence (also as a function of effective stress), compressive failure strength, σ_f, under triaxial stress, static Young's modulus, E, and Poisson's ratio, v, and air permeability, k_{air}, etc. Obviously one can also look at time-dependent phenomena such as stress relaxation (TEUFEL, 1983; WOLTER and BERCKHEMER, 1990). When fresh core is supplied, it is also quite apparent that other measurement capabilities are a mere matter of software modifications. With on-board coring and machining capabilities, it is not only practical to take horizontal and vertical cores but also possible to core at any orientation for anisotropic elastic constant determination.

The portability of GEM provides a unique opportunity to move the laboratory to the drill site. The need for this capability was explicitly stated by DÉLIAC et al. (1991), in reviewing Texaco's drilling in Paraguay (GUNN, 1991). This has tremendous advantages when dealing with friable core and unstable shales or when one is studying any time-dependent phenomena such as stress relaxation. The characterization of sealing shales requires on-site deployment. Figure 6 shows two shales recovered at a well site. These friable and chemically reactive samples often fall apart before they can be shipped back to a laboratory for analysis. On-site deployment also provides an opportunity to measure samples with 100% native saturation and to sample pore fluids.

Applications

The applications of data generated by GEM impact every aspect of oil and gas exploration and exploitation. Since the measurements made in the GEM system are performed on a small well-defined volume of rock, relationships among these measured properties are not plagued by typical depth shifts, depth smearing, rugosity corrections, etc., which affect log-based interpretations. Neither logs nor core are unequivocal truth. However, one must naturally ask how the two compare. Let us first examine the agreement between V_p and V_s determined from a full waveform log in a "good hole" and pressure-corrected core measurements (Figure 7). Pressure-corrected implies that measurements of V_p and V_s are performed or extrapolated to an effective pressure equivalent to that which the rock would experience at the depth from which it was recovered. Note that in general the core values (symbols) and log values (solid curves) agree very well. Keep in mind that the velocity measurements were performed at a frequency of several hundred

Figure 6
Examples of friable shales soon after recovery. These samples would not survive the trip back to a laboratory for measurement. However, one can make meaningful measurements on these delicate samples at the well site. Six months after these photographs were taken the samples were reduced to powder.

kilohertz while the logs were obtained at tens of kilohertz. This figure clearly indicates that core- and log-derived data are equally valid, especially in the absence of independent direct evidence. Let us now examine a different portion of the hole (Figure 8) which was subjected to a different drilling history. This section of the hole is shallower and was first drilled at high rotary speeds with a diamond coring bit and a mining rig (WALKER and MILHEIM, 1990; RANDOLPH et al., 1991); The core was recovered and measured in GEM as in the lower section (Figure 7). The hole was then drilled with a conventional tricone bit to a depth of 2,000 feet and logged with a full waveform tool. The core lithology and velocity values indicated that a converted shear wave should have been detected extending to the surface.

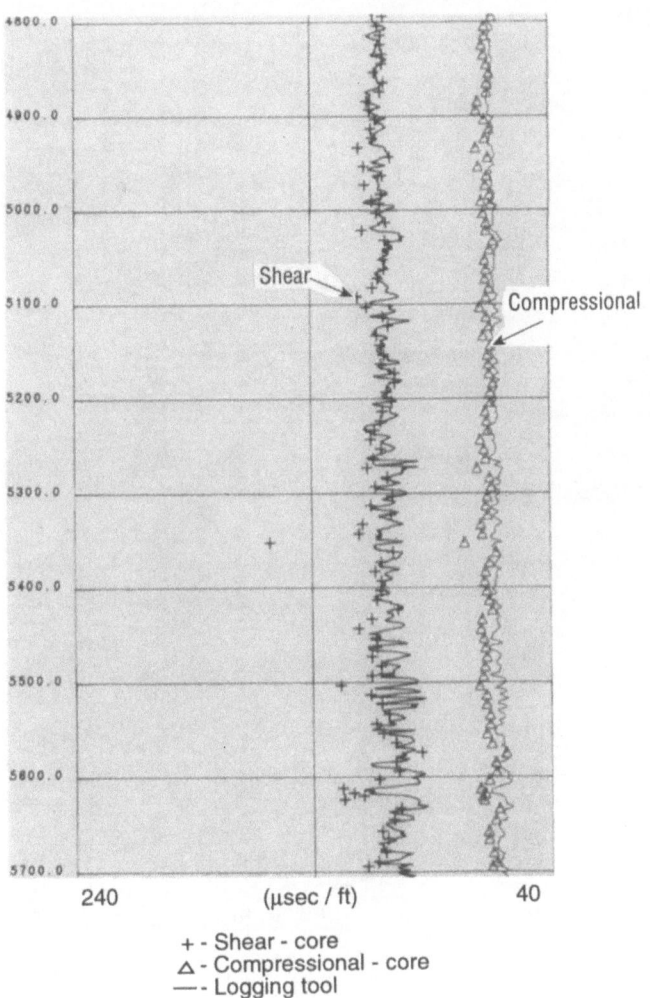

Figure 7

Plot of log-derived interval transit times (μsec/ft) (solid curve) and pressure-corrected core values (symbols). The triangles are compressional and crosses shear interval transit times. The agreement over the 1500 ft of well is excellent. This clearly indicates that the magnitude of velocity dispersion is small.

The full waveform logging tool radiates axisymmetric compressional waves into the annulus of borehole fluid. These waves convert to critically refracted waves at the interface between the fluid and formation, provided the fluid velocity is less than the corresponding compressional or shear velocity of the formation. The near-surface layer in this well is a dolomite. All the values of V_p and V_s determined on the core are consistently faster than the *in situ* values. There is no way one can remove a core from depth and measure a velocity greater than the *in situ* value. It is quite clear from this comparison that the hole was damaged during the drilling process

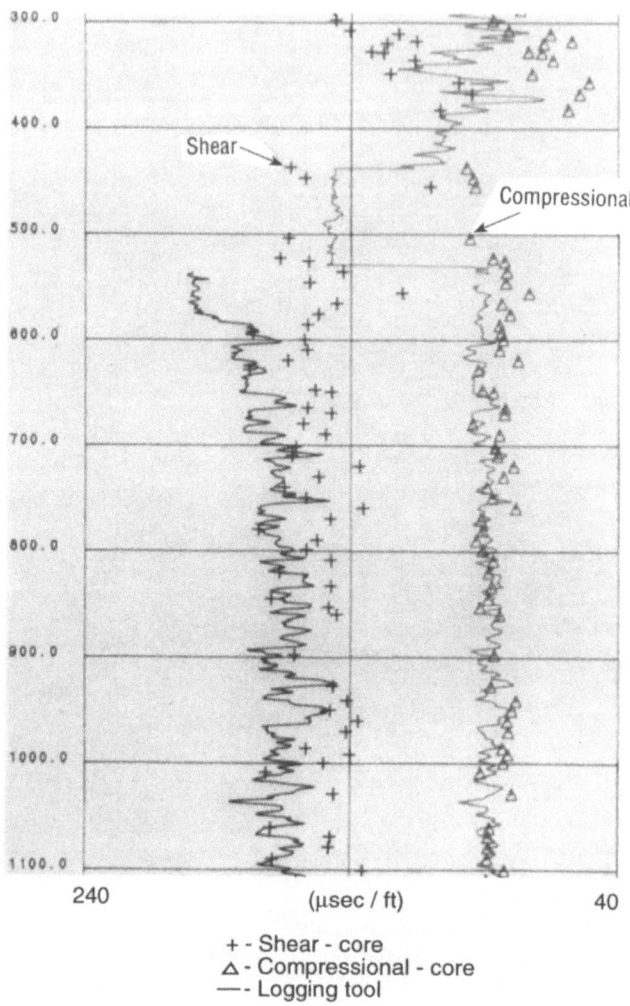

Figure 8

Same plot as Figure 7 except over a different depth range. This section of the wellbore was subjected to a different drilling history and has quite obviously suffered extensive formation damage due to drilling. Both V_p and V_s measured on cores are considerably faster than those deduced from the logs.

with the tricone bit. The damage appears more intense in the dolomite than the underlying sandstones. The fact that the logging tool was unable to measure a shear velocity in the dolomite layer indicates that the drilling damage reduced the formation shear velocity below the fluid velocity. The specter raised by these observations should at least make us cautious in accepting at face value the veracity of log-derived information. It is also commonly observed that fracture attribute logs commonly display more features above and below cored intervals. There is no uncertainty in the ability of the logging tools to reflect the nature of what they sense. It is truly a matter of that which they measure being altered by the drilling process.

Another example of the uniqueness and utility of GEM data lies in the measurement of grain densities. GEM is capable of measuring grain densities over the entire depth of the well. There is no logging tool capable of determining grain density. However, grain density is a required input for porosity calculations from log data. This proves to be a minor problem in mature exploitation scenarios but quite the contrary in analyzing the potential and risk in frontier plays. Certainly, accurate porosity values are of vital importance in equity decisions in mature areas. The data presented in Figure 9 indicate that classical assumption of $\rho_g = 2.65$ gm/cc is not valid for the interval in question between about 4,000 and 5,000 ft as this zone is rich in feldspars. The grain densities of feldspars are substantially less than quartz and thus lead to erroneous porosity estimates through the interval. Feldspars also wreak havoc with gamma ray logs because they are rich in potassium. A direct knowledge of their existence also connotes a proximity to sediment sources.

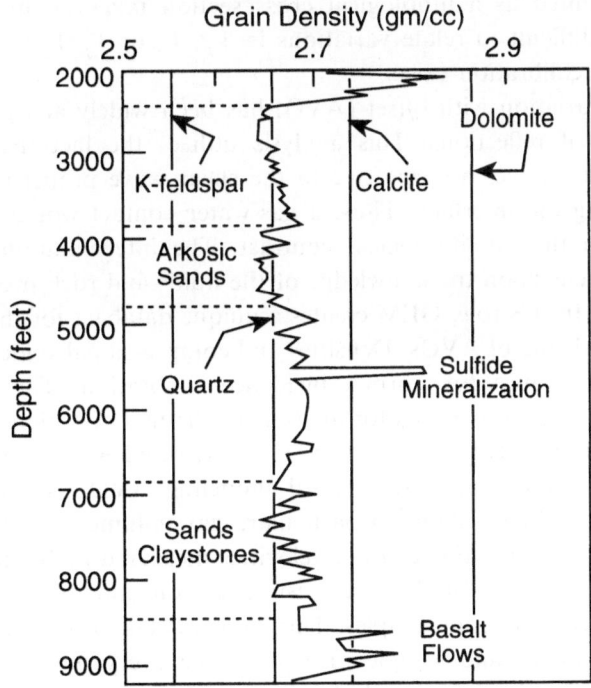

Figure 9

Plot of grain density measured on core as a function of depth in a well. Note the variability. The major mineralogical influences are annotated. K feldspar is extremely problematical in both grain density determination and in influencing standard gamma ray logs. Porosity logs require grain densities as input. It is clear that log-derived porosities can only be believed when there is no mineralogical variation.

Mineralogy as shown in Figures 3 and 5 gives instantaneous feedback regarding geologic environments and/or facies. Clearly evident in the waterfall presentation of absorption spectra is the presence of secondary but important mineralization. Anhydrite, for example, is clearly shown in Figure 5. The geological implications of anhydrite require more information as to whether it is primary or secondary in origin. The simple presence of anhydrite requires that extreme care be taken in determining porosities and permeabilities in these samples. Cores cleaned in toluene and methanol mixtures will likely dissolve salts associated with some anhydrites, leading to optimistic values for porosities.

Using a subset of the data generated in GEM, one can easily determine if relationships exist between seismically mappable parameters and mineralogy, porosity or permeability, for example. Having found a dependency, it is a simple matter to develop functional forms of the relations and employ them to map physical properties and their variations in the subsurface. The same information can be used to determine the feasibility of 3D surveys to map subsurface properties. A velocity tomogram can be transformed into a porosity or lithology time or depth section. From such a map the structure and integrity of seals can easily be ascertained. The tomogram presented as a lithological cross section takes on new meaning to a geologist. It is difficult to relate variations in V_p, V_s or V_p/V_s with their geologic causes when no calibration exists.

Amplitude variation with offset (AVO) has been widely analyzed to refine the possible causes of reflections. This analysis utilizes the fact that amplitudes of obliquely reflected waves are sensitive to the shear wave properties of the formations constituting the interface. Thus, a gas/water contact would have a different AVO expression than a lithological contrast. The interpretation of an AVO is critically dependent upon the knowledge of the fluids and rock properties comprising an interface. In this role, GEM creates a unique database for the interpretation, and forward modeling of AVOs. Densities and compressional velocities provide the basic impedance values for normal incidence calibration. The shear velocities provide the necessary constraints for analysis of the first and higher order terms in the AVO formulations (e.g., BORTFELD, 1961; AKI and RICHARDS, 1980; SHUEY, 1985). More recent attempts at forward modeling AVO, WARD et al. (1991), require estimates of mineralogy, in particular, clay volumes. To use core data for this purpose requires that one can use velocities measured at 100 KHz on core for seismic problems. Figure 10 shows a seismic section and the calculated acoustic impedance derived from core samples. This data was taken from a well cored from surface to final depth and sampled roughly once every ten feet. The sampling interval was designed to be much finer than the seismic wavelengths passing through these formations. Since the depth of each sample is known, one has to simply identify the rock properties with a prominent reflection amplitude. A strong correlation can be found between the changes in acoustic impedance and reflection events. The ability to tie the two sets of data allows us to convert the seismic time

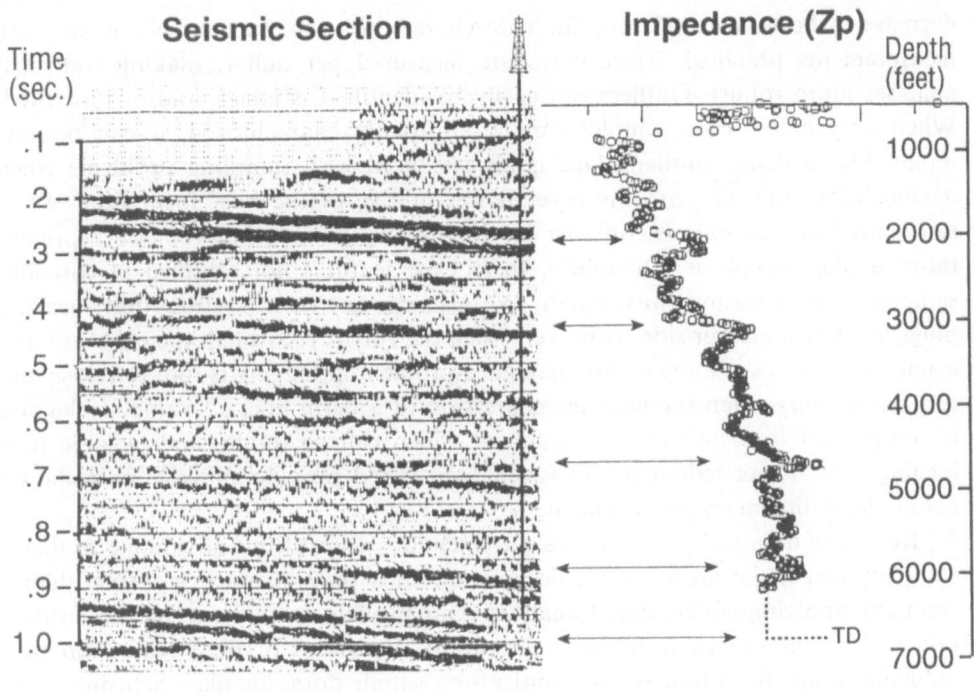

Figure 10

A seismic section with major events highlighted. Plotted alongside are smoothed core-derived values for acoustic impedance ($Z_p = \rho V_p$). Since Z_p controls reflectivity, the two should be strongly correlated and they are. Core measurements done properly can be used to address seismic problems of depth conversion and lithological interpretation. Again reinforcing the observations that velocity dispersion for properly measured core is not a major obstacle in applying core-derived velocities to field problems.

scales to a true depth scale and confirm seismic targets in real time. After gaining confidence in matching seismic horizons, prediction ahead of the drill bit in real-time manner becomes possible. In this deployment the target prediction depth based upon core data was within 15 ft of the depth at which the drill bit penetrated it. An error, even an order of magnitude, would have been quite acceptable. The important finding is that core measurements made at a well site can be applied to seismic problems at low frequency.

Real-time deployment provides a host of additional (unforeseen) capabilities, it becomes possible to have instantaneous feedback to drilling problems. Real-time mineralogy allows us to identify unstable clays in the formation and hydrates precipitating in drilling muds. Real-time measures of velocities and densities give us the capability to analyze unpenetrated horizons and determine distance to them as well as their relative penetrability. Such information permits optimization of bit trips and facilitates scheduling of consumables.

Critical elements in the deployment of a technology in these cost conscious times are the cost and time involved in its use. GEM provides an order of magnitude decrease in both cost and time for core characterization. In fact this makes core measurements practical. More cores are measured per dollar, making statistical analyses more robust. Outliers can easily be identified, studied, and/or discarded. When only a few core samples can be affordably characterized, it can become impossible to detect outliers. One is always faced with sampling problems when dealing with core, i.e., are the cores representative of the host formation. When sampling from continuous core one develops a very good feeling for how representative a plug sample is. Obviously, there exist features not sampled in core but sampled over a seismic wavelength. We also do not core fractures but measure plugs taken from either side. However, when the core is recovered, the existence and extent of fractures is known. An analogous problem in logging is tool response and resolution. Very often the geologic expression of a berm line, a swamp, or lagoon is compressed to a half inch of sediment which can be totally undetectable to a logging tool. These sediments, however, once cored provide valuable information about the sedimentary environment.

Real-time data analysis provides a tremendous competitive advantage in that a company can make an informed decision about a play in much less time than a company applying conventional analysis. This provides confidence in play evaluation and permits efficient resource allocation. Estimates of play evaluation time reduction range from four to six months for a simple domestic play. Scheduling the deployment of GEM so that it is in a foreign country prior to core retrieval can substantially reduce exploration play evaluation time. This is shown schematically in Figure 11. A possible exploration time line is given in Figure 11. The upper half of the figure depicts a conventional scenario and the bottom is a conservative estimate of time saved when GEM is used. Both scenarios involve the same basic evaluation steps, beginning with geological and geophysical surveys to define the location of a test well. At this point the models diverge, the lower time line capitalizes on being able to perform many operations in parallel, i.e., drilling, core evaluation, data evaluation and integration. This is a direct consequence of real-time analysis performed in GEM. With core literally in hand, one can also reduce the number of logging runs and save additional rig time. These processes must be carried out serially in conventional scenarios. The absolute magnitude of the time saved can vary substantially but the estimate provided in Figure 11 is meant to be conservative. More important perhaps than the time saved is the information derived from the core. For example, core can reveal whether a repeated sampling of a formation within a well bore results from simple folding or faulting. Very often a feature such as graded bedding, which is diagnostic in these situations, is not detectable via wireline logging tools. Many wells are kept active for some time after drilling to final depth simply to perform a Vertical Seismic Profile (VSP) survey. Often the purpose of this VSP is to confirm that the seismic horizon imaged, for example in Figure 10, was actually penetrated by the well. This is obviously no longer necessary when GEM data are available.

A Possible Exploration Time Line

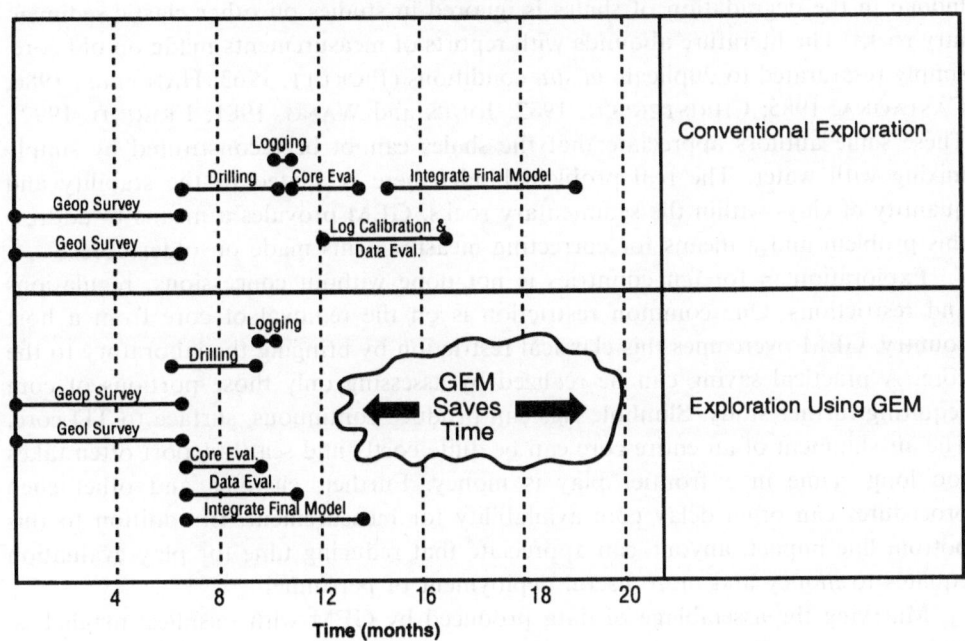

Figure 11

A possible exploration time line comparison between a conventional approach and one which incorporates GEM. Savings in time are realized when GEM is utilized because it permits a number of evaluation processes to be carried out simultaneously.

Summary

The benefits of GEM are many. The two most important are technological and scientific. Technologically, GEM is a quantum jump in the state-of-the-art core characterization making real-time, cost effective, remote analyses possible. Scientifically, the book remains to be written, but this chapter clearly indicates that when core measurements are performed properly, the values can be used to solve exploration problems. Among the obvious findings are the documentation of borehole damage, the definition of true shale properties, the interdependence of elasticity, mineralogy, porosity and TOC in a local sense. The unique data set provided through GEM has been used to address fundamental rock physics problems, including evaluation of constitutive equations for the elastic behavior of sedimentary rock (e.g., BIOT, 1956a,b) and estimation of microstructural parameters (e.g., O'CONNELL and BUDIANSKY, 1974).

GEM provides an opportunity to evaluate the worth of measurements on existing core. For those who have visited core warehouses with the hope of retrieving shales for physical property measurements and have been disappointed, GEM provides a rare chance to actually characterize such lithologies at the well site. Shales are perhaps the most dramatic example of the problem. The message hidden in the degradation of shales is ignored in studies on other clastic sedimentary rocks. The literature abounds with reports of measurements made on old core, simply resaturated to duplicate *in situ* conditions (PICKETT, 1963; HAN *et al.*, 1986; CASTAGNA, 1985; CHRISTENSEN, 1982; JONES and WANG, 1981; FREUND, 1992). These same authors appreciate that the shales cannot be reconstituted by simply mixing with water. The real problem hidden here is related to the stability and quantity of clays within the sedimentary rocks. GEM provides a means to address this problem and a means for correcting measurements made on older cores.

Exploration in foreign countries is not done without concessions, regulations and restrictions. One common restriction is on the removal of core from a host country. GEM overcomes this classical restriction by bringing the laboratory to the core. A practical saving can be realized by assessing only those portions of core requiring further study. Slimhole rigs can produce continuous, surface to TD core. The air shipment of an entire core can be quite costly and sea transport often takes too long. Time in a frontier play is money. Further, customs and other such procedures can often delay core availability for measurement. In addition to this bottom line impact, anyone can appreciate that reducing time for play evaluation equates to money and more useful deployment of personnel.

Marrying the assemblage of data produced by GEM with seismics, magnetics, and/or gravity allows a more accurate picture of the play to be developed. Contained within the core, of course, are geological features which alter or reinforce play concepts, suggesting alternate, deeper or different targets. These assessments can be performed as the well is drilling, permitting optimal utilization of the drill rig and optimal placement of subsequent seismic surveys. Interactive seismic processing and migration becomes a reality.

While this paper is oriented towards the application of this technology to oil exploration and exploitation, however, such technology with little or no modification can serve the scientific community in general. A major frontier left for discovery is the third dimension of the earth's crust, OLIVER (1991). In the last decade we have witnessed a greater number of continental scientific drilling programs designed to sample this frontier, for example, the deep Soviet drilling (KOZLOVSKY, 1986), Cajon Pass (ZOBACK *et al.*, 1988), the German deep borehole (KTB) (BHER *et al.*, 1988) and the Newark Basin project (KENT, 1991). The experience documented in this paper requires that careful consideration be given to performing real-time measurements at the well site in order to maximize the scientific value of information derived from these costly efforts.

Acknowledgements

We would like to thank Amoco for allowing us to make a contribution to this volume. We acknowledge the efforts and suggestions of the reviewers (H. Wang and

B. C. Schreiber). The efforts of Jim Denton and Bruce Spears are recognized as instrumental in the fabrication and deployment of the measurement system. We are also fortunate to have worked with a programmer with the skills of Bruce Blackwell when we converted the original real-time data collection system to a UNIX environment. The encouragement and support of Mr. M. Waller turned vision into reality. One of us, Carl H. Sondergeld, would like to acknowledge the early contributions of Professor E. Schreiber in nurturing and shaping his experimental and scientific capabilities.

REFERENCES

AKI, K., and RICHARDS, P. G., Quantitative Seismology Theory and Methods, vol. I (W. H. Freeman and Co., San Francisco 1980) 557 pp.

API (1960), Recommended Practice for Core Analysis Procedure, American Petroleum Institute RP-40.

BHER, H. J., KEHRER, P., and RISCHMULLER, H., The German continental deep drilling program, objectives and state of work. In Deep Drilling in Crystalline Bedrock, vol. 2 (eds. Boden, A. and Erikson, K. G.) (Springer-Verlag, New York 1988) pp. 64–81.

BIOT, M. M. (1956a), Theory of Propagation of Elastic Waves in a Fluid-saturated Porous Solid. I. Low-frequency Range, J. Acoust. Soc. Am. 28, 168–178.

BIOT, M. M. (1956b), Theory of Propagation of Elastic Waves in a Fluid-saturated Porous Solid. II. Higher-frequency Range, J. Acoust. Soc. Am. 28, 179–191.

BORTFELD, R. (1961), Approximation to the Reflection and Transmission Coefficients of Plane Longitudinal and Transverse Waves, Geophys. Prospect 9, 485–502.

CASTAGNA, S. P., BATZLE, M. L., and EASTWOOD, R. L. (1985), Relationships between Compressional-wave and Shear-wave Velocities in Clastic Silicate Rocks, Geophysics 50 (4), 571–581.

CHRISTENSEN, N. J., Seismic velocities. In Handbook of Physical Properties for Rocks, II (ed. Carmichael, R. S.) (CRC Press Inc. 1982).

DÉLIAC, E. P., MESSINES, J. P., and THIERRÉE, B. A. (1991), Mining Technique Finds Applications in Oil Exploration, Oil and Gas J., May 6, 85–90.

ENDERLIN, M. B., HANSON, D. K. T., and HOYT, B. R. (1991), Rock Volumes: Considerations for Relating Well Log and Core Data in Reservoir Characterization. II (eds. Lake, L. W., Carroll, H. B. Jr., and Wesson, T. C.) (Academic Press, New York 1991) pp. 277–288.

FREUND, D. (1992), Ultrasonic Compressional and Shear Velocities in Dry Elastic Rocks as a Function of Porosity, Clay, Content, and Confining Pressure, Geophys. J. Int. 108, 125–135.

GRIFFITHS, P. R., and DE HASETH, J. A., Fourier Transform Infrared Spectrometry (John Wiley, New York 1986).

GUNN, K. B. (1991), Well Cored to 9,800 ft in Paraguay, Oil and Gas J., 51–55, May 13.

HAN, D., NUR, A., and MORGAN, D. (1986), Effects of Porosity and Clay Content on Wave Velocities in Sandstones, Geophys. 51, 2093–2107.

JONES, L. E. A., and WANG, H. F. (1981), Ultrasonic Velocities in Cretaceous Shales from the Williston Basin, Geophys. 46, 288–297.

JUSTICE, J. H., VASSILIOU, A. A., SINGH, S., LOGEL, J. D., HANSEN, P. A., HALL, B. R., HUTT, P. R., and SOLANKI, J. J. (1989), Geophysics: The Leading Edge, 12–19, Feb.

KENT, D. V. (1991), Continental Drilling Program to Establish Fundamental Data Base for Rift Basin Evolution Models, CSD News 2, 1–2.

KOZLOVSKY, Y. A. (editor), The Super Deep Well of the Koal Peninsula (Springer-Verlag, New York 1986) 558 pp.

LINES, L. (1991), Applications of Tomography to Borehole and Reflection Seismology, Geophysics: The Leading Edge, 11–17.

MAVKO, G., and JIZBA, D. (1991), Estimating Grain-scale Fluid Effects on Velocity Dispersion in Rocks, Geophys. 56, 1940–1949.

MUELLER, M. C. (1991), *Prediction of Lateral Variability in Fracture Intensity Using Multicomponent Shear-wave Surface Seismic as a Precursor to Horizontal Drilling in the Austin Chalk*, Geophys. J. Int. *107* (3), 409–415.

O'CONNELL, R. J., and BUDIANSKY, B. (1974), *Seismic Velocities in Dry and Saturated Cracked Solids*, J. Geophys. Res. *7935*, 5417–5426.

OLIVER, J. E., *The Incomplete Guide to the Art of Discovery* (Columbia Univ. Press 1991) 208 pp.

OSTRANDER, W. J. (1984), *Plane-wave Reflection Coefficients for Gas Sands at Non-normal Angle of Incidences*, Geophys. *49*, 1637–1648.

PICKETT, G. R. (1963), *Acoustic Character Logs and their Applications in Formation Evaluation*, J. Petr. Tech. *15*, 650–667.

RANDOLPH, S. B., and JOURDAN, A. P. (1991), *Slimhole Continuous Coring and Drilling in Tertiary Sediments*, Paper SPE/IADC 21906, SPE/IADC Drilling Conference, Amsterdam, March 11–14.

ROBERTSON, J. D. (1989), *Reservoir Management Using 3-D Seismic Data*, Geophysics: The Leading Edge, 25–31, Feb.

SCHREIBER, E., ANDERSON, O. L., and SOGA, N., *Elastic Constants and their Measurement* (McGraw-Hill Book Co., NY 1973) 196 pp.

SHUEY, R. T. (1985), *A Simplification of the Zoeppritz Equations*, Geophys. *50*, 609–614.

SONDERGELD, C. H., and RAI, C. S. (1992), *Laboratory Observations of Shear-wave Propagation in Anisotropic Media*, Geophysics: The Leading Edge *11*, 38–43, Feb.

TEUFEL, L. W. (1983), *Determination of in situ Stress from Anelastic Strain Recovery Measurements of Oriented Core*, SPE/DOE Symp, 11649, Denver, CO, 421–430.

WALKER, S. H., and MILHEIM, K. K. (1990), *An Innovative Approach to Exploration and Exploitation Drilling: The Slim-hole, High-speed, Drilling System*, J. Petrol. Tech. *42* (9), 1184–1191.

WARD, J., RANDAZZO, S., LIEBER, R., and SMITH, R. (1991), *Synthetic Shear Wave Sonic Logs for Seismic Models*, poster presented at SEG Research Workshop on Lithology: Relating elastic properties to Lithology at all scales, July 28, St. Louis, MO.

WINKLER, K. W. (1986), *Estimates of Velocity Dispersion between Seismic and Ultrasonic Frequencies*, Geophys. *51* (1), 183–189.

WOLTER, K. E., and BERCKHEMER, H. (1990), *Estimation of in situ Stresses by Evaluation of Time-dependent Strain Recovery of KTB Drill Cores*, Tectonophys. *178*, 255–259.

ZOBACK, M. A., SILVER, L. T., HENYEY, T., and THATCHER, W. (1988), *The Cajon Pass Scientific Drilling Experiment: Overview of Phase 1*, Geophys. Res. Lett. *15*, 933 pp.

(Received March 23, 1993, revised September 7, 1993, accepted October 6, 1993)

PAGEOPH, Vol. 141, No. 2/3/4 (1993)   0033-4553/93/040269-18$1.50 + 0.20/0

Quasi-static Poroelastic Parameters in Rock and Their Geophysical Applications

HERBERT F. WANG[1]

Abstract—The constitutive equations of poroelasticity contain four static moduli. Different sets of moduli are reviewed in the context of their laboratory measurement and their geophysical applications. One complete set consists of the drained bulk modulus and Poisson's ratio, and their undrained counterparts. Skempton's coefficient (ratio of pore pressure increment to mean stress increment under undrained conditions) and the Biot-Willis parameter serve equally well for the undrained bulk modulus and Poisson's ratio, because they permit the drained and undrained moduli to be related to each other. Time dependence is introduced into poroelastic behavior through Darcy's law. Geophysical applications that can be approximated by undrained conditions (fast loading) include seismicity, tidal and barometric loading, and tectonic compression. Several of these problems are most directly formulated in terms of Skempton's coefficient, undrained Poisson's ratio, and hydraulic diffusivity.

Key words: Poroelasticity, geophysics, hydrogeology, seismicity, Biot theory, barometric loading, tidal loading, undrained compression.

Introduction

The standard isotropic, homogeneous elastic solid is characterized by two independent elastic moduli. Any pair among Young's modulus, bulk modulus, Poisson's ratio, shear modulus, Lamé's constant, etc. suffice to characterize the linear stress-strain behavior of a rock and any constant can be expressed in terms of any other pair. Each of these moduli is defined in terms of some specific stress or strain boundary conditions, and for certain problems only one elastic modulus may be necessary, e.g., bulk modulus is sufficient to obtain volumetric strain for a rock in a hydrostatic stress field. If the void space is fluid-filled, then the number of elastic moduli required to characterize the stress-strain response increases to four (BIOT, 1941; RICE and CLEARY, 1976; DETOURNAY and CHENG, 1993). As in standard elasticity, many choices can be made for the set of four independent

---

[1] Department of Geology and Geophysics, 1215 W. Dayton Street, Universtiy of Wisconsin–Madison, Madison, Wisconsin 53706, U.S.A.

poroelastic moduli and the right choice can sometimes reduce the number of moduli needed to characterize poroelastic behavior for particular problems.

Poroelastic behavior is time-dependent as a result of fluid flow in response to pressure gradients (Darcy's law). Therefore, the hydrogeologic parameters of permeability and specific storage are additional poroelastic parameters. The poroelastic phenomena considered in this paper are quasi-static because of the time-dependent fluid flow.

Dynamic or wave propagation phenomena also depend on the static poroelastic moduli, but display a complex frequency dependence (BIOT, 1956a,b) that is beyond the scope of the present paper. This paper reviews experimental techniques for measuring the static poroelastic moduli and hydrogeologic parameters with particular emphasis on those constants that are useful for solving typical geophysical problems. A similar review of the poroelastic parameters has been published recently by KÜMPEL (1991).

Static Poroelastic Moduli

The constitutive equations that extend standard linear elasticity to poroelastic materials were originally presented by BIOT (1941). The constitutive equations were reformulated by RICE and CLEARY (1976), and their equations, or close variants thereof, are most frequently used in the geophysical literature. Stress applied to a small volume element of a fluid-filled porous material produces the usual strains, but in addition changes of fluid pressure in the volume element induce volumetric strains as well.

$$2Ge_{ij} = \sigma_{ij} - \frac{1}{3}\left(1 - \frac{2G}{3K}\right)\sigma_{kk}\delta_{ij} + \frac{2G}{3}\left(\frac{1}{K} - \frac{1}{K_s}\right)P\delta_{ij} \tag{1}$$

where e_{ij} and σ_{ij} are the tensor strains and stresses of the bulk solid, respectively, G is the shear modulus, K is the drained bulk modulus, K_s is the unjacketed solid frame modulus ($\beta_s = 1/K_s$ is defined in equation (12) below), and P is fluid pressure. The summation convention is implied for the repeated subscripts kk, and δ_{ij} is the Kronecker delta. The linear approximation can be considered valid if the strains, stresses and fluid pressure are considered to be incremental quantities relative to some reference state. The sign convention used in (1) is the extensional stresses and strains are positive. This sign convention appears to be used most commonly in poroelasticity literature, including rock mechanics and geophysics. It is important to emphasize that the strains, stresses, and fluid pressure in (1) are considered to be *incremental* changes relative to a reference state. The material constants are anticipated to be functions of the stress and fluid pressure of the reference state.

Variants of the stress-strain relation (1) are obtained using substitutions based on the standard isotropic elastic relationship among the shear modulus, bulk modulus, and Poisson's ratio.

$$K = G \frac{2(1 + v)}{3(1 - 2v)}. \tag{2}$$

Also, the pore pressure term may be expressed in terms of the Biot-Willis parameter α.

$$\alpha = 1 - \frac{K}{K_s}. \tag{3}$$

Summing the normal strains in (1) yields an expression for volumetric strain in terms of the changes in mean stress and pore pressure.

$$e_{kk} = \frac{1}{K}\left(\frac{\sigma_{kk}}{3} + \alpha P\right). \tag{4}$$

An additional constitutive equation relates the change in fluid mass per unit volume to both mean stress and pore pressure changes.

$$\Delta m_f = m_f - m_0 = \frac{\rho_0 \alpha}{K}\left[\frac{\sigma_{kk}}{3} + \frac{1}{B}P\right] \tag{5}$$

where m_f is the fluid mass per unit bulk volume, m_0 is the fluid mass per unit bulk volume in the reference state, ρ_0 is the fluid density in the reference state, and B is Skempton's coefficient (or the undrained pore pressure buildup coefficient), which describes the undrained pore pressure increase due to an increase in mean stress (Figure 1).

$$B \equiv -3(dP/d\sigma_{kk})_{dm_f = 0}. \tag{6}$$

The value of B ranges between zero and one.

Drained and Undrained Conditions

Two limiting situations occur in terms of the fluid flow rate relative to external stress changes. Undrained conditions ($\Delta m_f = 0$) are satisfied for times short relative to applied stress changes, and drained conditions ($P = 0$) are satisfied for times long relative to applied stress changes. Undrained and drained conditions are completely analogous to the thermoelastic equivalents of adiabatic (no heat flow) and isothermal (no temperature change) conditions, respectively (RICE and CLEARY, 1976).

Several useful relations can be derived to show the physical significance behind the drained and undrained constants. Solving (5) for P and substituting into (4) gives

$$e_{kk} = \frac{1}{3K}(1 - \alpha B)\sigma_{kk} + \frac{B}{\rho_0}\Delta m_f. \tag{7}$$

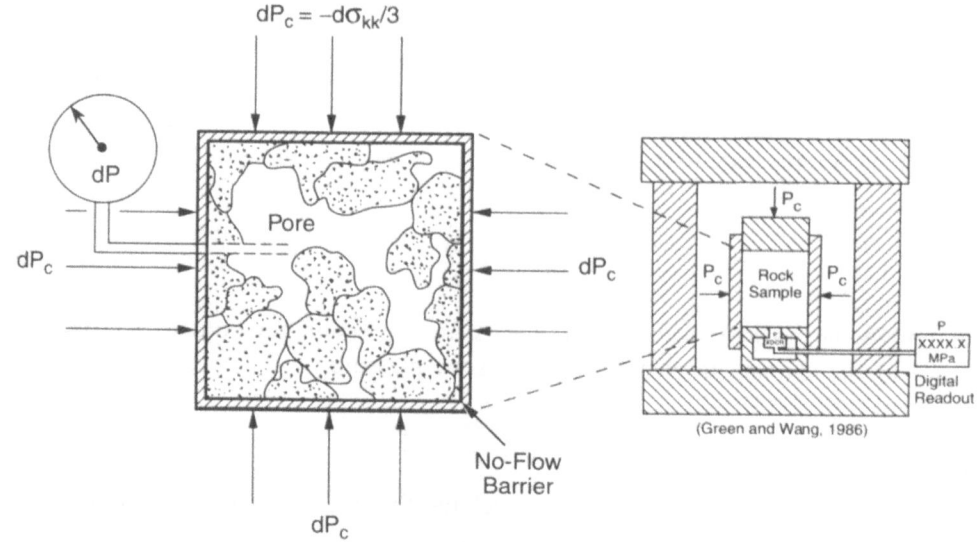

SKEMPTON'S COEFFICIENT

Figure 1
Undrained test configuration to measure Skempton's coefficient (GREEN and WANG, 1986). Detail of rock microstructure from Green (personal communication).

Undrained conditions mean $\Delta m_f = 0$, so that the factor multiplying $\sigma_{kk}/3$ can be identified as the undrained compressibility $\beta_u = 1/K_u$, i.e.,

$$\frac{1}{K_u} = \frac{1}{K}(1 - \alpha B). \tag{8}$$

The undrained compressibility is smaller than the drained compressibility. From (8) equation (7) can then be written simply as

$$e_{kk} = \frac{\sigma_{kk}}{3K_u} + B\frac{\Delta m_f}{\rho_0}. \tag{9}$$

Equation (9) is used for identifying separately the contribution to volumetric strain due to the change in mean stress and that due to the change in fluid mass per unit volume (SEGALL, 1989).

Poisson's ratio can also be defined in the usual way but for undrained conditions. Solving (5) as before for undrained conditions, i.e., $P = -B\sigma_{kk}/3$, and putting the result in (1) for the strain ratio $v_u = -e_{22}^u/e_{11}^u$, for a longitudinal stress σ_{11}, and $\sigma_{22} = \sigma_{33} = 0$ give

$$v_u = \frac{3v + \alpha B(1 - 2v)}{3 - \alpha B(1 - 2v)}. \tag{10}$$

The undrained Poisson's ratio is larger than the drained value. The undrained Poisson's ratio (10) and undrained compressibility (8) are sufficient to describe the response to stress for undrained conditions. These two undrained constants and their two drained counterparts form a complete set of four poroelastic constants. Alternatively, the Biot-Willis parameter α and Skempton's coefficient B can be the additional two poroelastic constants in addition to the drained constants K and v because they allow the transformations (8) and (10) between drained and undrained constants. More complete tables of relations between the various poroelastic moduli have been presented by GREEN and WANG (1986) and KÜMPEL (1991).

Skempton's Coefficient

Undrained conditions are achieved in a pressure vessel by placing solid endcaps on either side of a jacketed, saturated sample. Skempton's coefficient can be measured by incorporating a small, semiconductor pressure transducer within one endcap to measure the pore pressure increase as the confining pressure is increased (GREEN and WANG, 1986; BERGE et al., 1993). Locating the pressure transducer with no intervening tubing volume is important for achieving high accuracy, because no correction is needed for the fluid compressibility or tubing flexure.

Measured values of B are approximately unity at low effective pressures (confining pressure minus pore pressure) for both natural and synthetic sandstones. The effective pressure remains zero with increasing confining pressure when $B = 1$ because pore pressure builds up equally. Higher effective pressures can be obtained by undersaturating the sample prior to an experiment. The pore pressure remains zero until the pore volume is reduced sufficiently by confining pressure to bring the saturation to one. Sandstone values of B decrease to 0.5–0.6 at effective pressures of 20 MPa (DROPEK et al., 1978; GREEN and WANG, 1986; BERGE et al., 1993).

A theoretical equation for Skempton's coefficient has been derived expressing B in terms of the bulk drained compressibility β, the fluid compressibility β_f, the unjacketed solid frame compressibility β_s, and the unjacketed pore compressibility β_ϕ (BROWN and KORRINGA, 1975; ZIMMERMAN et al., 1986; BERRYMAN, 1992).

$$B = \frac{\beta - \beta_s}{\beta - \beta_s + \phi(\beta_f - \beta_\phi)} \tag{11}$$

where

$$\beta_s = \frac{-1}{V}\left(\frac{\partial V}{\partial P}\right)_{dP = dP_c} \tag{12}$$

and

$$\beta_\phi = \frac{-1}{V_\phi}\left(\frac{\partial V_\phi}{\partial P}\right)_{dP = dP_c}. \tag{13}$$

The symbol V_ϕ is the pore volume and P_c is the confining pressure. An alternative expression for B is obtained from solving (8) and using (3):

$$B = \frac{\dfrac{1}{K} - \dfrac{1}{K_u}}{\dfrac{1}{K} - \dfrac{1}{K_s}} = \frac{\beta - \beta_u}{\beta - \beta_s}. \tag{14}$$

A common assumption in (11) is that $\beta_s = \beta_\phi$. Although no completely consistent set of measurements have been made of all the parameters in (11), the unjacketed pore compressibility for sandstones appears to be closer to the fluid compressibility than the unjacketed solid frame compressibility at low effective pressures but it approaches the solid compressibility at higher effective pressures (GREEN and WANG, 1986; BERGE et al., 1993).

Biot-Willis Parameter

The physical significance of the Biot-Willis parameter, $\alpha = 1 - K/K_s$ is that it appears in the effective stress law for volumetric strain (NUR and BYERLEE, 1971; BERRYMAN, 1992). The effective stress for volumetric strain is the linear combination of confining pressure P_c and pore pressure that leaves volumetric strain in (4) unchanged. Measurements of α are obtained from measurements of jacketed and unjacketed bulk moduli (e.g., NUR and BYERLEE, 1971; COYNER, 1984). Coyner's values of the Biot-Willis parameter range between 0.55 and 0.89 at 10 MPa confining pressure for Navajo sandstone, Weber sandstone, Berea sandstone, and Westerly granite (BERRYMAN, 1992). The values of α decrease with confining pressure so that the range is 0.37 to 0.74 at 20 MPa confining pressure.

Fluid Flow Parameters

Time dependence enters poroelastic phenomena because fluid flow occurs in response to pore pressure gradients according to Darcy's law.

$$q_i = -\rho_0 \kappa \frac{\partial P}{\partial x_i} \tag{15}$$

where q_i is the *mass flux* in the x_i direction and $\kappa = k/\mu$, where k is the permeability (units of area), and μ is the fluid viscosity. The symbol κ is called *coefficient of permeability* by RICE and CLEARY (1976), but is also known as the *mobility coefficient* in reservoir engineering literature. The ratio of permeability to viscosity in Darcy's law (15) is, therefore, an essential poroelastic parameter in addition to the four poroelastic moduli appearing in the constitutive equations (1) and (5). Quasi-static conditions are satisfied when the poroelastic moduli are independent of frequency, which is the case for frequencies lower than a few hundred cycles per second.

The continuity condition is expressed as

$$\frac{\partial q_i}{\partial x_i} + \frac{\partial m_f}{\partial t} = 0. \tag{16}$$

Combining (5), (15) and (16) gives one of two coupled equations relating fluid pressure and mean stress (VAN DER KAMP and GALE, 1983).

$$\kappa \, \nabla^2 P = \frac{S'}{\rho_0 g} \frac{\partial}{\partial t} \left(P + \frac{B}{3} \sigma_{kk} \right) \tag{17}$$

where

$$S' = \rho_0 g \left[\left(\frac{1}{K} - \frac{1}{K_s} \right) + \phi \left(\frac{1}{K_f} - \frac{1}{K_\phi} \right) \right] \tag{18}$$

and g and ϕ are the acceleration of gravity and porosity, respectively. The parameter S' is a storage coefficient similar in concept to the specific storage coefficient employed in hydrogeology. It is defined rigorously in hydrogeologic fashion (NARASIMHAN and KANEHIRO, 1980) as

$$S' = g \left(\frac{dm_f}{dP} \right)_{\sigma_{kk} = 0}. \tag{19}$$

The parameter S' represents the change in fluid weight per unit volume due to a change in fluid pressure *under conditions of no change in mean stress*. This storage coefficient is different from the usual specific storage in hydrogeology, because the defining boundary conditions are different. It has been called the "three-dimensional" storage coefficient by KÜMPEL (1991), as the usual specific storage is sometimes called the "one-dimensional" storage coefficient in reference to the condition of uniaxial strain.

The strain compatibility equations yield a second equation between fluid pressure and mean stress (RICE and CLEARY, 1976; VAN DER KAMP and GALE, 1983; DETOURNAY and CHENG, 1993).

$$\nabla^2 \sigma_{kk} + 4\eta \, \nabla^2 P = 0 \tag{20}$$

where

$$\eta = \frac{\alpha(1 - 2v)}{2(1 - v)}. \tag{21}$$

Finally, combining (17) and (20) leads to the result that the fluid mass content satisfies the diffusion equation (RICE and CLEARY, 1976).

$$\nabla^2 m_f = \frac{S_s}{\rho_0 g \kappa} \frac{\partial m_f}{\partial t} \tag{22}$$

where

$$S_s = S'(1 - 4\eta B/3). \tag{23}$$

The diffusivity is defined to be (RICE and CLEARY, 1976; VAN DER KAMP and GALE, 1983; GREEN and WANG, 1990):

$$c = \frac{\rho_0 g \kappa}{S_s}. \tag{24}$$

The diffusion equation (22) for fluid mass content incorporates stress coupling to the fluid pressure and is a generalization of the usual diffusion equation used in hydrogeology in which pressure alone takes the place of m_f.

$$\nabla^2 P = \frac{S_s}{\rho_0 g \kappa} \frac{\partial P}{\partial t}. \tag{25}$$

The decoupling of the stress from the pore pressure is exact for boundary conditions of zero lateral strain and constant vertical stress. For that case, the specific storage in (25) is identical to that in (22), and therefore, specific storage is defined explicitly (GREEN and WANG, 1990) as

$$S_s = g \left(\frac{dm_f}{dP} \right)_{\sigma_{11} = 0; \, e_{22} = e_{33} = 0}. \tag{26}$$

Comparison of (26) with the three-dimensional storage coefficient S' in (19) shows that they have different stress/strain boundary conditions while undergoing a change in fluid pressure. The reason for giving special status to S_s is that the stress and strain boundary conditions reduce the diffusion equation (22) for m_f to the standard hydrogeologic one (25) for P. The importance of stress and strain boundary conditions in the definition of a poroelastic modulus such as storage coefficient is entirely analogous to the standard elastic moduli (Figure 2). For example, Young's modulus is the ratio of longitudinal stress to longitudinal strain for *zero change in lateral stress*. A different uniaxial strain modulus would describe the same ratio but under conditions of zero change in lateral strain.

Permeability and Specific Storage

Pressure Injection Methods

The technique developed by BRACE et al. (1968), or variants thereof, is to observe the pressure decay in a rock-reservoir system after a sudden injection of a fluid into one reservoir. Transient pulse decay methods are often thought of as a means to measure permeability in low permeability samples, but transience necessarily involves the specific storage as well. In fact, the hydraulic diffusivity appears alone in the transient flow equation but the permeability enters as a flow boundary condition into the upper and lower reservoirs. Thus, any two of the three parame-

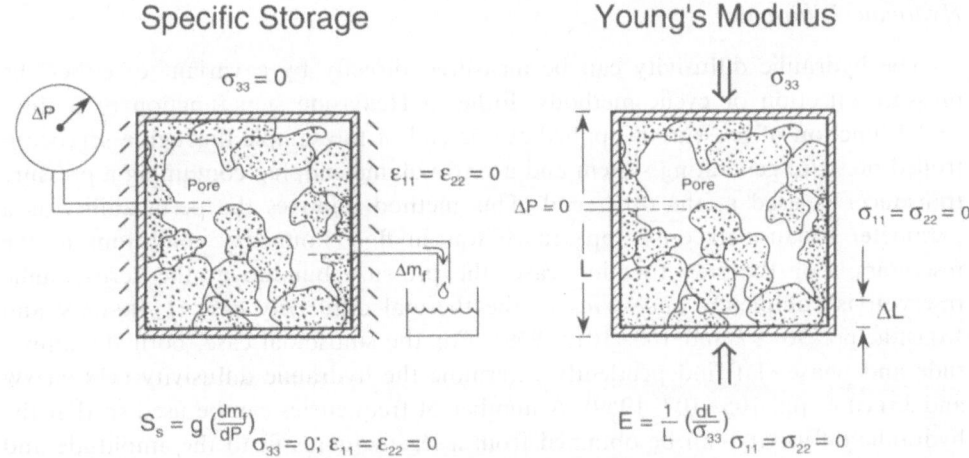

Figure 2

Stress strain boundary conditions inherent in definitions of specific storage and drained Young's modulus.

ters among hydraulic diffusivity, permeability, and specific storage characterize the rock-reservoir flow response. HSIEH *et al.* (1981) have written the equations for pressure in the upper and lower reservoirs on either end of the rock sample in terms of permeability and specific storage, as well as experimental design parameters such as the volume storage of the reservoirs, including tube flexure effects. Therefore, fitting the pressure response can yield both hydrogeologic parameters.

An error analysis of the pressure decay/buildup equations has been made recently by WANG and HART (1993). The procedure is to calculate the partial derivatives of the pressure decay with respect to permeability and specific storage (sensitivity coefficients) from the analytical equations of HSIEH *et al.* (1981). The covariance matrix is calculated from the sensitivity coefficient for a particular time series of pressure data points. The variance of each model parameter normalized by the variance of the data is its diagonal value in the covariance matrix. The variances depend on the sampling interval and total time span of the experiment.

Cyclic Methods

A cyclic method to measure permeability and hydraulic diffusivity has been developed by KRANZ *et al.* (1990) analogous with methods for measuring thermal diffusivity. Forced, periodic pore pressure variations in one reservoir are attenuated and shifted in phase at the opposite reservoir. The amplitude ratio and phase angle are directly related to the permeability and hydraulic diffusivity, respectively. The driving frequency needs to be adjusted for the diffusivity of the sample, and low permeability samples can require extremely low oscillation frequencies.

Hydraulic Diffusivity

The hydraulic diffusivity can be measured directly by a variant of either the pressure injection or cyclic methods. Either a Heaviside step function or a sinusoidal function of pressure is applied at one end of the sample through a servocontrolled pressure generating system and a zero-volume endplug containing a pressure transducer is used at the other end. This method removes the permeability as a parameter because its only appearance was in flow boundary conditions at the reservoirs. For the step function case, the pressure buildup at the zero-volume reservoir is completely analogous to the thermal diffusivity case (CARSLAW and JAEGER, pp. 96–97 and 100–101, 1959). For the sinusoidal case, both the amplitude and phase shift independently determine the hydraulic diffusivity (CARSLAW and JAEGER, pp. 105–107, 1959). A number of frequencies can be used so that the hydraulic diffusivity can be obtained from a least-squares fit to the amplitude and phase shift versus frequency. The hydraulic diffusivity for Berea sandstone was determined by this method (REN, 1992) using the endplug employed by GREEN and WANG (1986) in the measurement of Skempton's coefficient.

Geophysical Applications

The application of quasi-static poroelastic theory to geophysical problems can be divided into three broad categories. First, seismicity phenomena involve poroelasticity, either because fault movement couples to the fluid pressure field, or conversely. Second, the hydrogeologic parameters can be obtained literally by "stressing" the aquifer via atmospheric and tidal loading rather than pumping. The parameters inferred from the well response to such stresses may differ significantly from laboratory measured values, in part because larger scale features such as fractures are not represented by small samples, in part because the parameters may not be particularly sensitive to the loads, and in part because the loads may not be well-known. The third category of geophysical application is tectonic modeling in which poroelasticity is used to account for effects such as the generation of high-fluid pressure during tectonic compression. The examples presented will also illustrate how the different poroelastic constants already discussed plus several others arise naturally in the sense of being particularly suited to the particular boundary conditions of a problem.

Seismicity

Fault slip in the earth's crust creates a strain field that induces changes in the pore pressure distribution. An example of this phenomenon is water well level changes due to propagating creep events (ROELOFFS and RUDNICKI, 1984/85). They model the event in plane strain as a propagating edge dislocation (Figure 3).

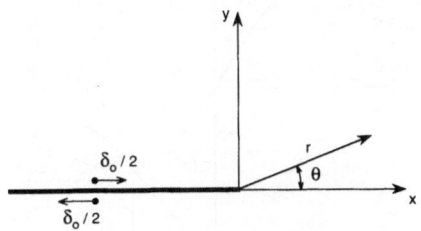

Shear Dislocation
(Roeloffs and Rudnicki, 1984/85)

Figure 3
Edge dislocation geometry used in poroelastic formulation of induced pore pressure changes (ROELOFFS and RUDNICKI, 1984/1985).

If the propagation speed is slow relative to the rate of fluid diffusion, then the strain field is not coupled to the pore pressure and the normal elastic solution using drained elastic moduli will apply. If, however, the propagation speed is fast relative to pore fluid flow, then undrained conditions apply and the poroelastic problem is solved for this limiting case by using the undrained Poisson's ratio in place of the drained and by using Skempton's coefficient to calculate the induced pore pressure field from the mean stress (ROELOFFS and RUDNICKI, 1984/1985).

$$P(x, y) = \frac{B(1 + v_u)}{3(1 - v_u)} \left(\frac{G\delta_0}{\pi}\right) \left(\frac{y}{r^2}\right) \tag{27}$$

where δ_0 is the relative slip on the negative x axis. This induced pore pressure field is the initial condition for a standard hydrogeologic boundary value problem, i.e., the diffusion equation (25) is solved for pressure. The same procedure was applied to calculate the water level excursion due to an earthquake beneath Yucca mountain (BREDEHOEFT, 1992; CARRIGAN et al., 1991).

The converse phenomenon is seismicity in hydrocarbon reservoirs as the result of large-scale pore pressure reduction inducing changes in the stress field (SEGALL, 1989). The extraction of fluid in a hydrocarbon reservoir produces measurable strains at the ground surface. SEGALL (1989) used equation (7) to calculate the stress changes from Δm_f and the strains. He presented the analytical expressions for the horizontal and vertical displacements at the surface for an idealized plane-strain model in which fluid is extracted uniformly from a strip of thickness T and halfwidth a buried at a depth D (Figure 4).

$$u_x = \frac{2(1 + v_u)BT \, \Delta m_f}{6\pi\rho_0} \log\left[\frac{1 + \xi_+^2}{1 + \xi_-^2}\right] \tag{28}$$

$$u_y = \frac{2(1 + v_u)BT \, \Delta m_f}{3\pi\rho_0} [\tan^{-1}(\xi_-) - \tan^{-1}(\xi_+)] \tag{29}$$

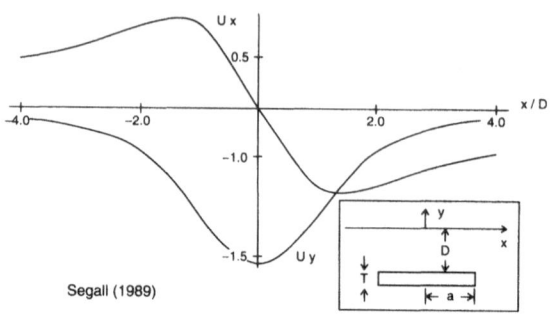

Figure 4
Surface displacements resulting from fluid extraction from finite strip at depth (SEGALL, 1989).

where $\xi_- = (x - a)/D$ and $\xi_+ = (x + a)/D$. (Note that the x and y axes are interchanged from those used by SEGALL.) The calculated stress changes are consistent with faulting within the producing fields. The problem was formulated most efficiently using Skempton's coefficient, undrained Poisson's ratio, and undrained bulk modulus as the poroelastic parameters.

Induced seismicity is often associated with the impoundment of deep reservoir lakes behind high dams. The location of earthquakes, their fault mechanism, and their timing with respect to reservoir height may be caused by the direct stress load imposed by the water elevation, the induced pore pressure due to the direct stress load, or the pore pressure diffusion in response to the stress and pore pressure boundary condition at the base of the reservoir lake (ROELOFFS, 1988). These factors have been examined with an idealized plane strain model in one dimension. The surface boundary conditions for a Heaviside step function change of reservoir level (ROELOFFS, 1988) are

$$\sigma_{zz}(0, t) = - H(t) \tag{30}$$

and

$$P(0, t) = H(t). \tag{31}$$

The solution is

$$P(z, t) = (1 - \gamma) \text{ erf } c(z/\sqrt{4ct}) + \gamma H(t) \tag{32}$$

where

$$\gamma = \frac{B(1 + v_u)}{3(1 - v_u)} \tag{33}$$

and c, the diffusivity is defined by equation (24). Similarly, the boundary conditions for a cyclic variation (angular frequency ω) of reservoir level are

$$\sigma_{zz}(0, t) = - P_0 \exp(i\omega t) \tag{34}$$

and

$$P(0, t) = P_0 \exp(i\omega t). \tag{35}$$

The solution has also been presented by ROELOFFS.

$$P(z, t) = [(1 - \gamma) \exp\{-(1 + i)\sqrt{\omega/2c}\ z\} + \gamma]P_0 \exp(i\omega t). \tag{36}$$

The step function and cyclic pressure variations with depth are plotted in Figures 5 and 6. For both (32) and (36), the pore pressure response with depth for the uncoupled case ($B = 0$) is the same as the solution of (25). For nonzero values of B, the pore pressure response consists of an instantaneous undrained response due to the Skempton coefficient effect and a pore pressure diffusion term. The parameter γ is the ratio of the instantaneous induced pore pressure to the mean stress, where the horizontal stresses are $v_u/(1 - v_u)$ as a result of the zero horizontal strain assumption. The diffusion term in the step function case is the same as if the pressure diffusion equation were solved for a step of $(1 - \gamma)$.

Tidal and Barometric Loading

Aquifer properties are normally obtained by "stressing" the aquifer through pump tests. Poroelastic coupling means that aquifer properties can be obtained theoretically from well level response to the more literal case of stressing the aquifer

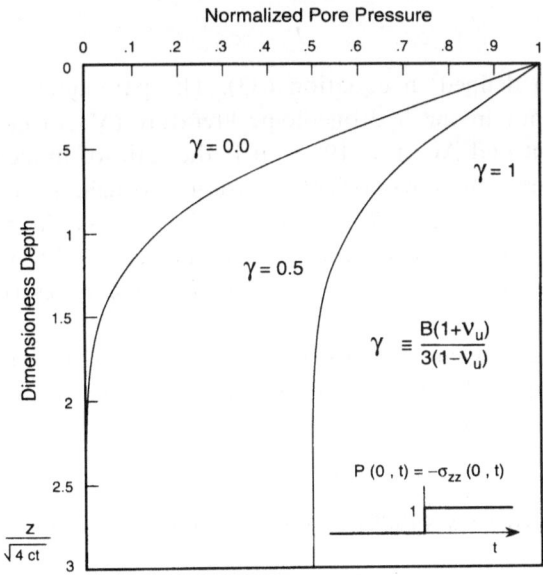

Figure 5
Relative pore pressure amplitude versus dimensionless depth for step change of reservoir level
(ROELOFFS, 1988).

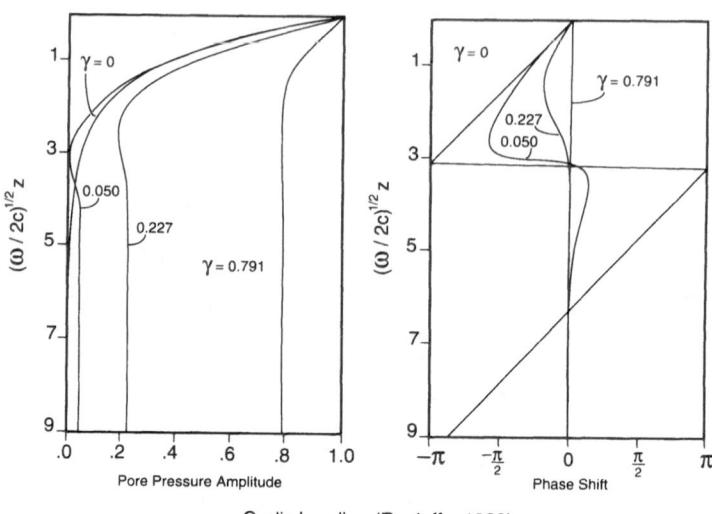

Cyclic Loading (Roeloffs, 1988)

Figure 6

Relative pore pressure amplitude and phase lag versus dimensionless depth for cyclic change of reservoir level (ROELOFFS, 1988).

via tidal and barometric loading. The barometric load $\sigma_{zz} = -\sigma_B$ is considered to be large in areal extent and therefore the assumption is made that zero lateral strain boundary conditions apply (VAN DER KAMP and GALE, 1983). Then

$$P = \gamma\sigma_B \tag{37}$$

where γ has been defined in equation (33). The parameter γ is often called the barometric efficiency in the hydrogeologic literature (VAN DER KAMP and GALE, 1983; ROJSTACZER and AGNEW, 1989); it is the ratio of induced pore pressure to applied vertical stress for zero horizontal strain. As such, a more descriptive name for γ might be *uniaxial strain Skempton's coefficient*. It could be measured directly in the laboratory under zero lateral strain boundary conditions while monitoring the pore pressure buildup using the zero-volume endcap described in GREEN and WANG (1986).

For earth tides, the areal strain $e_A = e_{xx} + e_{yy}$ is assumed known, and plane stress ($\sigma_{zz} = 0$) and undrained conditions are also assumed. VAN DER KAMP and GALE (1983) express the pore pressure change as

$$P = -2G\gamma e_A. \tag{38}$$

ROJSTACZER and AGNEW (1989) express the pore pressure change as

$$P = -\frac{B}{\beta_u}\frac{1 - 2v_u}{1 - v_u} e_A \tag{39}$$

which is obtained by substituting (33) and the undrained equivalent of (2) into (38).

Tectonic Loading

OLIVER (1986) has proposed that both vertical and horizontal compressive loading by thrust sheets act as "giant squeegees" driving pore fluids for long distances from foreland basins. The transient hydrogeological effects have been examined with a poroelastic model by GE and GARVEN (1992). GE and GARVEN use (17) with the assumption of incompressible solid grains, although they use the notation S_s for S' and call it specific storage (their equations (10) and (12)). Except for terminology, they correctly express their "specific storage" in terms of the bulk compressibility rather than the vertical compressibility, i.e., their expression for S_s is (18) for incompressible solid grains.

The degree of coupling depends on the value of Skempton's coefficient and the rate of loading relative to the rate of pore pressure diffusion. For incompressible grains, Skempton's coefficient (11) can be expressed as

$$B = \frac{\beta}{\beta + \phi \beta_f}. \tag{40}$$

From this expression, GE and GARVEN (1992) emphasize the role of the bulk compressibility β in the tectonic loading problem. However, the poroelastic nature of the problem is probably illuminated better by emphasizing the key role of Skempton's coefficient directly.

Pore pressure response to gravitational loading is similar to tectonic loading and can also be treated as a poroelastic problem (GREEN and WANG, 1986). However, nonlinear compaction effects can be significant and they are not incorporated in the linear theory presented here. The converse problem of erosional unloading has been treated poroelastically by NEUZIL and POLLOCK (1983).

Summary

Linear poroelasticity is a useful constitutive theory for analyzing coupled stress and pore pressure effects in geophysical problems for which loading rates are comparable to pore fluid diffusion rates. Foremost among the applications is fault behavior, but tectonic and gravitational loading/unloading can also exhibit important poroelastic effects. In the fast loading limit, undrained conditions apply. The induced pore pressure is then described by Skempton's coefficient for hydrostatic loading or by barometric efficiency for uniaxial strain loading. The induced pore pressure stiffens the elastic response, i.e., both the undrained bulk modulus and undrained Poisson's ratio are greater than their drained values. Stress and fluid pressure coupling is intrinsic to aquifer and reservoir behavior, so that the hydrogeologic parameters of permeability and specific storage also are key poroelastic parameters. Tidal and barometric loading of the aquifers can be used to estimate these parameters. The diffusivity that controls diffusion of fluid mass per unit volume is the same as the hydraulic diffusivity that controls pressure diffusion when the pressure is assumed to be decoupled from the stress field.

Symbols

B Skempton's coefficient [dimensionless]

c Hydraulic diffusivity [m²/s]

e_{ij} Strain tensor component [dimensionless]

g Acceleration of gravity [m/s²]

G Shear modulus [Pa]

K Drained bulk modulus [Pa]

K_f Fluid bulk modulus [Pa]

K_s Solid frame modulus [Pa]

K_u Undrained bulk modulus [Pa]

K_ϕ Unjacketed pore bulk modulus [Pa]

m_f Fluid mass per unit bulk volume [kg/m³]

P Fluid pressure [Pa]

P_c Confining pressure [Pa]

q_i Mass flux in x_i direction [kg/m² · s]

S' Three-dimensional storage coefficient

S_s Specific storage coefficient

α Biot-Willis parameter [dimensionless]

β Drained compressibility [Pa⁻¹]

β_f Fluid compressibility [Pa⁻¹]

β_s Solid frame compressibility [Pa⁻¹]

β_u Undrained bulk compressibility [Pa⁻¹]

β_ϕ Unjacketed pore compressibility [Pa⁻¹]

δ_{ij} Kronecker delta [dimensionless]

γ Barometric efficiency [dimensionless]

κ Coefficient of permeability or mobility coefficient [m²/Pa · s]

μ Fluid viscosity [Pa · s]

ν Drained Poisson's ratio [dimensionless]

ν_u Undrained Poisson's ratio [dimensionless]

ϕ Porosity [dimensionless]

ρ_0 Fluid density in reference state [kg/m³]

σ_{ij} Stress tensor component [Pa]

Acknowledgments

Ed Schreiber was very kind in assisting me twenty years ago during the halycon days of single crystal elastic constant and lunar rock measurements, even though I was a graduate student at a different institution. Based on that experience, I proudly take credit for bringing Ed to the attention of George Kolstad at DOE in 1981 to succeed me as university rotator in the Geosciences program. He served this program in residence on two separate occasions. I had the enjoyment of working with Ed in the context of the DOE program from 1981 until his untimely death. Ed Schreiber was a direct inspiration for writing a review of poroelastic measurement techniques. He was constantly in my mind as the audience. My poroelasticity research is supported by the Geosciences program of DOE under grant DE-FG02–91ER14194.

REFERENCES

BERGE, P. A., WANG, H. F., and BONNER, B. P. (1993), *Pore Pressure Buildup Coefficient in Synthetic and Natural Sandstones*, Int. J. Rock. Mech. Min. Sci. and Geomech. Abstr., in press.

BERRYMAN, J. G. (1992), *Effective Stress for Transport Properties of Inhomogeneous Porous Rock*, J. Geophys. Res. 97, 17,409–17,424.

BIOT, M. A. (1941), *General Theory of Three-dimensional Consolidation*, J. Appl. Physics 12, 155–164.

BIOT, M. A. (1956a), *Theory of Propagation of Elastic Waves in a Fluid-saturated Porous Solid, I, Low-frequency Range*, J. Acoust. Soc. Am. 28, 168–178.

BIOT, M. A. (1956b), *Theory of Propagation of Elastic Waves in a Fluid-saturated Porous Solid, II, Higher-frequency Range*, J. Acoust. Soc. Am. 28, 179–191.

BRACE, W. F., WALSH, J. B., and FRANGOS, W. T. (1968), *Permeability of Granite under High Pressure*, J. Geophys. Res. 73, 2225–2236.

BREDEHOEFT, J. D., *Response of the ground-water system at Yucca Mountain to an earthquake*. In *Groundwater at Yucca Mountain* (National Academy Press, Washington, D.C. 1992) pp. 212–222.

BROWN, R. J. S., and KORRINGA, J. (1975), *On the Dependence of the Elastic Properties of a Porous Rock on the Compressibility of the Pore Fluid*, Geophysics 40, 608–616.

CARRIGAN, C., KING, G. P., BARR, G. E., and BIXLER, N. E. (1991), *Potential for Water Table Excursions Induced by Seismic Events at Yucca Mountain, Navada*, Geology 19, 1157–1160.

CARSLAW, H. S., and JAEGER, J. C., *Conduction of Heat in Solids*, 2nd ed. (Oxford University, Oxford, 1959).

COYNER, K. B., *Effects of Stress, Pore Pressure, and Pore Fluids on Bulk Strain, Velocity, and Permeability of Rocks* (Ph.D. thesis, MIT 1984).

DETOURNAY, E., and CHENG, A. H-D., *Fundamentals of poroelasticity*. In *Comprehensive Rock Engineering: Principles, Practice and Projects*, Vol. 2, Chap. 5 (ed. Hudson, J. A.) (Pergamon Press, Oxford 1993) pp. 113–171.

DROPEK, R. K., JOHNSON, J. N., and WALSH, J. B. (1978), *The Influence of Pore Pressure on the Mechanical Properties of Kayenta Sandstone*, J. Geophys. Res 83, 2817–2824.

GE, S., and GARVEN, G. (1992), *Hydromechanical Modeling of Tectonically Driven Groundwater Flow with Application to the Arkoma Foreland Basin*, J. Geophys. Res. 97, 9119–9144.

GREEN, D. H., and WANG, H. F. (1986), *Fluid Pressure Response to Undrained Compression in Saturated Sedimentary Rock*, Geophysics 51, 948–956.

GREEN, D. H., and WANG, H. F. (1990), *Specific Storage as a Poroelastic Constant*, Water Resources Res. 26, 1631–1637.

HSIEH, P. A., TRACY, J. V., NEUZIL, C. E., BREDEHOEFT, J. D., and SILLIMAN, S. E. (1981), *A Transient Laboratory Method for Determining the Hydraulic Properties of Tight Rocks, I. Theory*, Int. J. Rock Mech. Min. Sci. and Geomech. Abstr. 18, 245–252.

KRANZ, R. L., SALTZMAN, J. S., and BLACIC, J. D. (1990), *Hydraulic Diffusivity Measurements on Laboratory Rock Samples Using an Oscillating Pore Pressure Method*, Int. J. Rock. Mech. Min. Sci. and Geomech. Abstr. 27, 345–352.

KÜMPEL, H.-J. (1991), *Poroelasticity: Parameters Reviewed*, Geophys. J. In. *105*, 783–799.

NARASIMHAN, T. N., and KANEHIRO, B. Y. (1980), *A Note on the Meaning of Storage Coefficient*, Water Resour. Res. *16*, 423–429.

NEUZIL, C. E., and POLLOCK, P. W. (1983), *Erosional Unloading and Fluid Pressures in Hydraulically "Tight" Rocks*, J. Geology *91*, 179–193.

NUR, A., and BYERLEE, J. D. (1971), *An Effective Stress Law for Elastic Deformation of Rock with Fluids*, J. Geophys. Res. *76*, 6414–6419.

OLIVER, J. (1986), *Fluids Expelled Tectonically from Orogenic Belts: Their Role in Hydrocarbon Migration and Other Geological Phenomena*, Geology *14*, 99–102.

REN, X., *Experimental Determination of Hydraulic Diffusivity of Berea Sandstone* (M.S. thesis, University of Wisconsin–Madison 1992).

RICE, J. R., and CLEARY. M. P. (1976), *Some Basic Stress-diffusion Solutions for Fluid-saturated Elastic Porous Media with Compressible Constituents*, Rev. Geophys. Space. Phys. *14*, 227–241.

ROELOFFS, E. (1988), *Fault Stability Changes Induced Beneath a Reservoir with Cyclic Variations in Water Level*, J. Geophys. Res. *93*, 2107–2124.

ROELOFFS, E., and RUDNICKI, J. W. (1984/85), *Coupled Deformation-diffusion Effects on Water-level Changes due to Propagating Creep Events*, Pure Appl. Geophys. *122*, 560–582.

ROJSTACZER, S., and AGNEW, D. C. (1989), *The Influence of Formation Material Properties on the Response of Water Levels in Wells to Earth Tides and Atmospheric Loading*, J. Geophys. Res. *94*, 12,403–12,411.

SEGALL, P. (1989), *Earthquakes Triggered by Fluid-extraction*, Geology *17*, 942–946.

VAN DER KAMP, G., and GALE, J. E. (1983), *Theory of Earth Tide and Barometric Effects in Porous Formations with Compressible Grains*, Water Resources Res. *19*, 538–544.

WANG, H. F., and HART, D. J. (1993), *Experimental Error for Permeability and Specific Storage from Pulse Decay Measurements*, Int. J. Rock Mech. Min. Sci. and Geomech. Abstr., in press.

ZIMMERMAN, R. W., SOMERTON, W. H., and KING, M. S. (1986), *Compressibility of Porous Rocks*, J. Geophys. Res. *91*, 12,765–12,777.

(Received April 7, 1993, revised/accepted October 19, 1993)

PAGEOPH, Vol. 141, No. 2/3/4 (1993)

0033-4553/93/040287-37$1.50 + 0.20/0
© 1993 Birkhäuser Verlag, Basel

Controls on Sonic Velocity in Carbonates

FLAVIO S. ANSELMETTI[1] and GREGOR P. EBERLI[2]

Abstract — Compressional and shear-wave velocities (V_p and V_s) of 210 minicores of carbonates from different areas and ages were measured under variable confining and pore-fluid pressures. The lithologies of the samples range from unconsolidated carbonate mud to completely lithified limestones. The velocity measurements enable us to relate velocity variations in carbonates to factors such as mineralogy, porosity, pore types and density and to quantify the velocity effects of compaction and other diagenetic alterations.

Pure carbonate rocks show, unlike siliciclastic or shaly sediments, little direct correlation between acoustic properties (V_p and V_s) with age or burial depth of the sediment so that velocity inversions with increasing depth are common. Rather, sonic velocity in carbonates is controlled by the combined effect of depositional lithology and several post-depositional processes, such as cementation or dissolution, which results in fabrics specific to carbonates. These diagenetic fabrics can be directly correlated to the sonic velocity of the rocks.

At 8 MPa effective pressure V_p ranges from 1700 to 6500 m/s, and V_s ranges from 800 to 3400 m/s. This range is mainly caused by variations in the amount and type of porosity and not by variations in mineralogy. In general, the measured velocities show a positive correlation with density and an inverse correlation with porosity, but departures from the general trends of correlation can be as high as 2500 m/s. These deviations can be explained by the occurrence of different pore types that form during specific diagenetic phases. Our data set further suggests that commonly used correlations like "Gardner's Law" (V_p-density) or the "time-average-equation" (V_p-porosity) should be significantly modified towards higher velocities before being applied to carbonates.

The velocity measurements of unconsolidated carbonate mud at different stages of experimental compaction show that the velocity increase due to compaction is lower than the observed velocity increase at decreasing porosities in natural rocks. This discrepancy shows that diagenetic changes that accompany compaction influence velocity more than solely compaction at increasing overburden pressure.

The susceptibility of carbonates to diagenetic changes, that occur far more quickly than compaction, causes a special velocity distribution in carbonates and complicates velocity estimations. By assigning characteristic velocity patterns to the observed diagenetic processes, we are able to link sonic velocity to the diagenetic stage of the rock.

Key words: Sonic velocity, carbonates, physical properties, porosity, diagenesis, compaction.

[1] Swiss Federal Institute of Technology ETH, Geologisches Institut, Sonneggstr. 5, CH-8092 Zürich, Switzerland.

[2] University of Miami, Rosenstiel School of Marine and Atmospheric Science, MGG, 4600 Rickenbacker Causeway, Miami, FL 33149, U.S.A.

1. Introduction

Knowledge of the relation between sonic velocity in sediments and rock lithology is one of the keys to interpreting data from seismic sections or from acoustic logs of sedimentary sequences. Reliable correlations of rock velocity with other petrophysical parameters, such as porosity or density, are essential for calculating impedance models for synthetic seismic sections (BIDDLE et al., 1992; CAMPBELL and STAFLEU, 1992) or identifying the origin of reflectivity on seismic lines (SELLAMI et al., 1990; CHRISTENSEN and SZYMANSKI, 1991). Velocity is thus an important parameter for correlating lithological with geophysical data.

Recent studies have increased our understanding of elastic rock properties in siliciclastic or shaly sediments. The causes for variations in velocity have been investigated for siliciclastic rocks (VERNIK and NUR, 1992), mixed carbonate siliciclastic sediments (CHRISTENSEN and SZYMANSKI, 1991), synthetic sand-clay mixtures (MARION et al., 1992) or claystones (JAPSEN, 1993). The concepts derived from these studies are however only partly applicable in pure carbonates. Carbonates do not have large compositional variations that are, as is the case in the other sedimentary rocks, responsible for velocity contrasts. Pure carbonates are characterized by the lack of any clay or siliciclastic content, but are mostly produced and deposited on the top or on the slope of isolated or detached carbonate platforms, that have no hinterland as a source of terrigeneous material (WILSON, 1975; EBERLI, 1991). They consist of over 95% of the carbonate minerals calcite (low- and high-Mg), dolomite and aragonite. These minerals have very similar physical properties, which excludes compositional variation as a major reason for the large variability in velocity of carbonates.

Theories that describe sonic wave propagation in porous media (GASSMAN, 1951; BIOT, 1956) are hard to apply in the complex system of pure carbonates because they form a variety of unique diagenetic rock fabrics with specific elastic properties. In order to quantify the physical properties, sonic velocity in pure carbonate samples from three different areas that cover a wide range of depositional environments and lithologies have been measured. Measurements of compressional-wave velocity (V_p) and shear-wave velocity (V_s) were performed under confining and pore-fluid pressures, which accurately simulate in situ subsurface conditions. Our study includes carbonates at all stages of diagenetic alteration and complements studies on the velocity of carbonates which were limited to highly lithified, low porosity carbonate rocks (RAFAVICH et al., 1984; WANG et al., 1991) or to pelagic, deep water carbonates (SCHLANGER and DOUGLAS, 1974; MILHOLLAND et al., 1980; URMOS and WILKENS, 1993).

Sonic velocity measurements were done in combination with a thorough lithologic and diagenetic examination of thin sections and XRD analysis. Porosity in the samples ranges from 0 to 60% and the depositional environment varies from the protected shallow water platform over reefal and platform-marginal sediments to deeper slope deposits. The correlation of the velocity measurements with the lithology and the mineralogy data enables us to assign depositional and diagenetic stages to characteristic velocities. Furthermore it allows tracing of diagenetic evolution and velocity development from the time of deposition through different burial stages, recognizing that each diagenetic process alters velocity in its characteristic way.

2. Sample Areas

This study presents the correlation of physical rock properties with rock lithology based on velocity analyses of 210 discrete samples from three different areas; (1) modern carbonate mud from Florida Bay, (2) two deep drill holes in Great Bahama Bank and (3) the Maiella, an exhumed carbonate platform in Central Italy. An understanding of the geological setting of the three areas is essential in order to relate the physical properties of the carbonates to the rock lithology.

A. Velocity Samples from Modern and Unconsolidated Carbonate Sediments: Artificially Compacted Carbonate Mud from Cluett Key, Florida Bay (South Florida)

The velocities of 20 carbonate mud samples were measured at various stages of artificial compaction in order to determine the increase of velocity caused by the porosity reduction during pure mechanical compaction. The mud was collected with push cores of approximately 70 cm length from the interior pond of Cluett Key (Figure 1), a mangrove island in Florida Bay.

Florida Bay is a triangular shaped shallow lagoon on the southern part of the Florida peninsula. This protected bay is subdivided by a series of mudbanks with several mangrove-fringed islands (ENOS and PERKINS, 1979). The Holocene sediments on the islands overlie Pleistocene bedrock and are up to 4 m thick. The base of the Holocene is often marked by a peat layer which is overlain by a succession of dominantly mud to wackestones with few intercalations of shell-rich storm layers. Unconsolidated carbonate mud of the upper part of the Holocene section was used for the compaction-velocity experiments. The samples were taken from parts of the cores in which no roots or shell fragments disturb the homogeneous mud.

Mud from the islands and the mudbanks in Florida Bay have porosities that range from 61 to 78% (ENOS and SAWATSKY, 1981). Gamma ray attenuation measurements with cores from Cluett Key gave an average porosity for the Holocene sediments of 63% (VIDLOCK, 1983). The mineralogical composition, determined on carbonate mud from Jimmy Key, an adjacent island (BURNS and

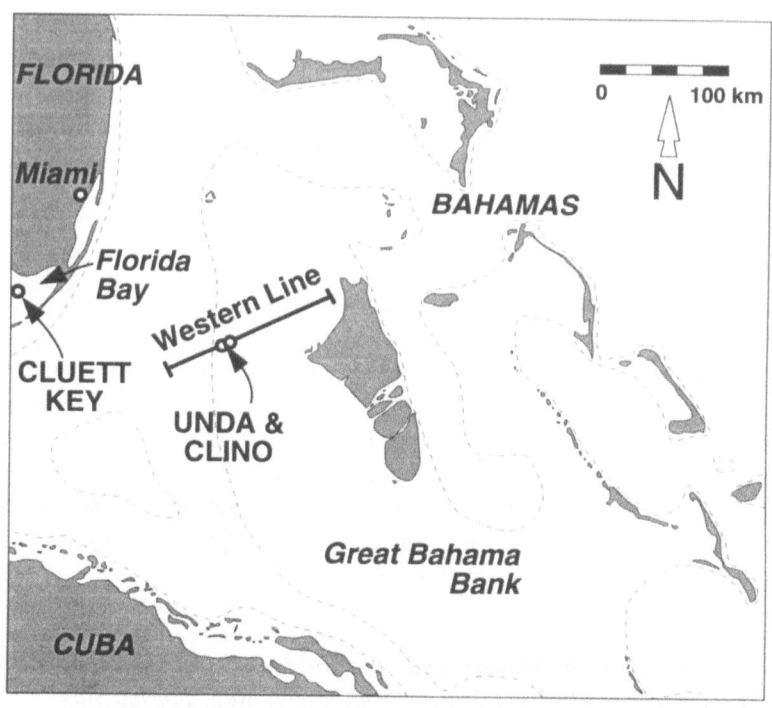

Figure 1
Location map showing the positions of the deep core borings Unda and Clino along the Western seismic line on Great Bahama Bank and the location of Cluett Key in Florida Bay.

SWART, 1992) averages 65% aragonite, 20% high Mg-calcite and 15% low Mg-calcite. These values are stable for the whole Holocene section and only traces ($<5\%$) of dolomite are observed. Between the surface and 70 cm depth, no detectable diagenetic alterations occur, although variations in pore water chemistry indicate that early diagenetic processes such as dolomitization have already started (BURNS and SWART, 1992).

B. *Velocity Samples from Cores of Deep Drillholes: Pleistocene to Miocene Carbonates from Core Borings in Great Bahama Bank (Bahamas Drilling Project)*

Two continuous core borings from the Bahamas Drilling Project, located on a multi-channel seismic line on Great Bahama Bank (Figures 1 and 2), provide an excellent opportunity to correlate the physical properties of Miocene to Pleistocene carbonate sediments with their depositional lithologies and diagenetic stages. Eighty-nine samples from both drill holes were analyzed. Unlike older and exhumed outcrop samples, the young age of the drilled sediments allows measurement of sonic velocities of carbonates that are partly unconsolidated and that are not in

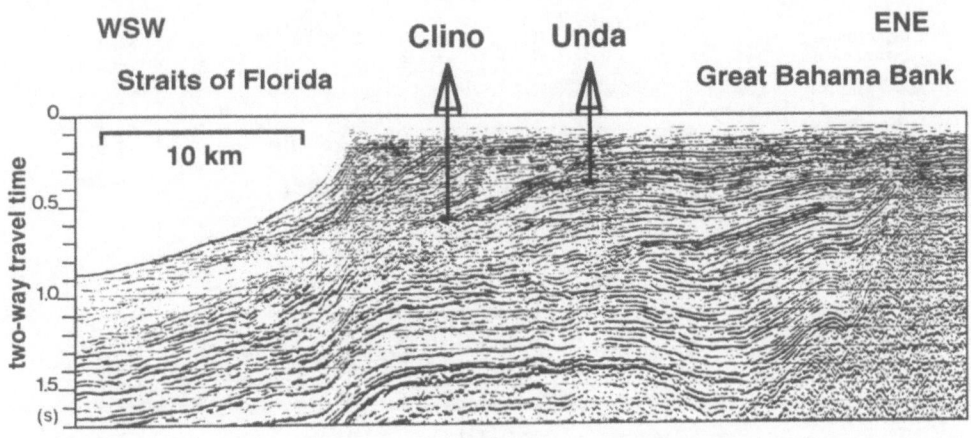

Figure 2

Part of Western line displaying modern platform margin and drill sites Unda and Clino. The succession of inclined reflectors below the modern shallow water platform document the progradation of the platform edge over inclined slope sediments for a distance of over 25 km.

their final stage of post-depositional alteration. The variety of diagenetic processes encountered enabled us to trace the velocity evolution of different carbonate sediments under different diagenetic conditions through burial history and time.

The two holes (Unda and Clino) penetrated to depths of 442 and 662 m below seafloor, respectively. The continuous cores had an average recovery of over 80%. The top of the rock section in both holes is of Pleistocene age. The oldest drilled sediments are dated as Middle Miocene at the bottom of Unda, whereas the bottom of Clino reaches an age of Late Miocene (Figure 3). The retrieved lithologies range from platform-interior to platform-margin and slope carbonates; there is no silici-clastic sediment on this isolated carbonate platform (KENTER et al., 1991).

Hole Unda, located 10 km from the modern platform edge, is characterized by three successions of shallow-water platform sands and reefal deposits that alternate with fine-sand and silt-sized deeper marginal deposits. The two intervals of deeper-water sediments record periods of rapid rise of sea level and probable backstepping of the platform and reefal units. Hole Clino, 7 km closer to the modern platform edge, penetrated a single interval of shallow platform and reefal sediments overlying a thick succession of slope sediments. The nearly 500 m of fine-sand to silt-sized slope sediments below 200 m have a variable amount of planktonic foraminiferas and are, except for some intervals with coarse-grained, mainly skeletal sands, remarkably poor in coarser material.

This succession of depositional environments (Figure 3) shows the progradation of the whole platform over the underlying slope sediments. The platform rim prograded over 25 km to the west into the Straits of Florida (EBERLI and GINSBURG, 1989).

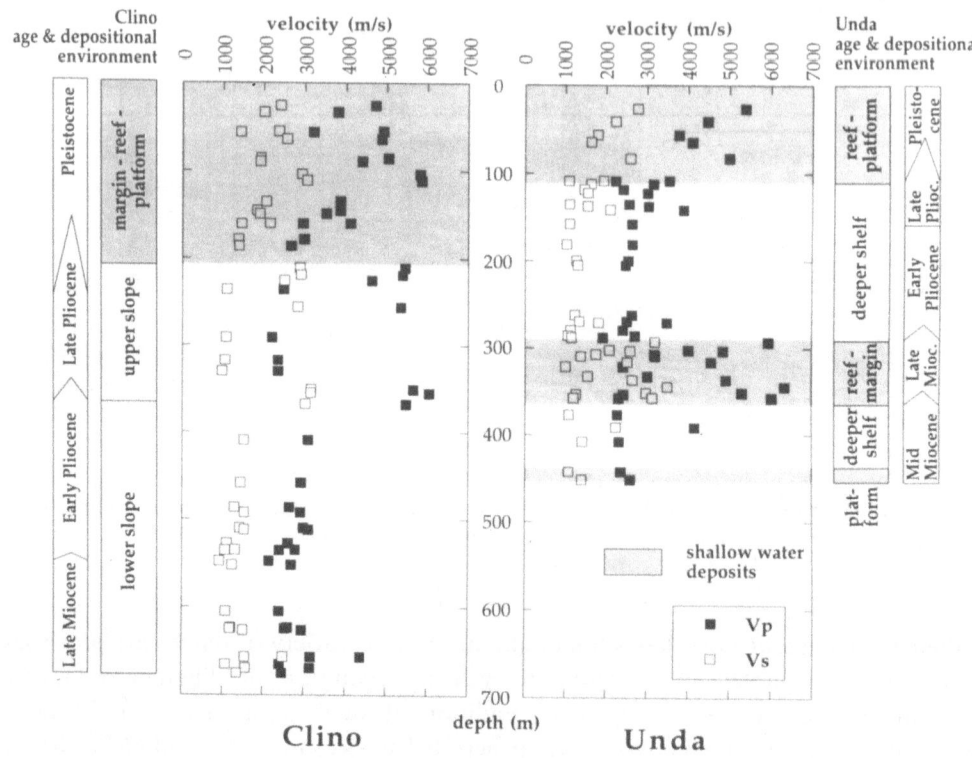

Figure 3

Correlation of V_p and V_s (at 8 MPa effective pressure) with depth, depositional environment and age of the drilled sediments from the two drill holes Unda and Clino on Great Bahama Bank. Velocity inversions are common in both holes and show that the effect of diagenetic alterations and sediment type dominate over the velocity effect of depth. Velocities of carbonates that were deposited on the shallow water platform (shaded areas in graph) have larger variability and higher velocities than velocities from deeper water samples.

Not only the depositional lithology, but also the diagenetic overprinting changes several times downhole. The upper parts of both holes are characterized by early marine and subsequent intense freshwater diagenesis. Many samples show intense dissolution features, as well as extensive cementation, which led to specific rock fabrics with characteristic elastic properties. In the lower part of Clino, the periplatform slope sediments show no major alterations and only the platform derived turbidite layers are more cemented. Little dolomite occurs below a hard-ground at 536 m. Dolomitization in Unda is considerably more pervasive and forms either a fabric destructive sucrosic dolomite or a crystalline mimetic dolomite (DAWANS and SWART, 1988), depending on the precursor. In the lowest part of Unda, dolomite disappears and the rocks again show intense dissolution features.

C. *Velocity Samples from Outcrops: Montagna della Maiella (Abruzzi, Italy)*

The Maiella is an uplifted and exhumed carbonate platform in Central Italy (Figure 4) that is exposed in several valley flanks. The exposed platform and slope carbonates range in age from the Lower Cretaceous to the Upper Miocene. One hundred and one samples were collected and velocities were determined. The knowledge of the physical properties in combination with the assessment of the large-scale geometrical pattern of the outcropping rock formations enables us to calculate synthetic seismic sections using computer simulations in order to see the seismic response of a particular geological setting (ANSELMETTI and EBERLI, 1991).

The margin of the Maiella carbonate platform is characterized by a steep escarpment during Early Cretaceous time that became buried during the Late Cretaceous and developed into a low-angle ramp in the Paleogene. The sediments of the external platform are mostly rudist biostromes and carbonate sandbodies whereas the platform interior is mainly made of limestones deposited in a shallow subtidal to supratidal environment, such as wackestones and fenestreal mudstones (CRESCENTI et al., 1969; SANDERS, 1994). A distinct mid-Cretaceous, karstic unconformity separates the Cretaceous platform section into an upper and a lower unit. On the adjacent slope, several mega-breccias onlap this platform margin (EBERLI et al., 1993; VECSEI, 1991). They were deposited during sea-level lowstands that caused the exposure of the platform top and the erosion and downslope

Figure 4
Map showing the location of the Montagna della Maiella, in the Abruzzi, Central Italy.

transport of platform fragments. These breccias are intercalated with calcareous turbidites and pelagic carbonates that form the normal background sedimentation. In the lower Paleogene, a relative deepening of the platform resulted in a backstepping of the platform margin and the steep escarpment was slowly infilled. Finally, during the Oligocene, reefal units of the platform margin prograded over the former deeper shelf and slope deposits and formed a wide and shallow shelf. This general evolution of backstepping and prograding of an isolated carbonate platform has striking similarities with the evolution of the modern Great Bahama Bank.

Unlike the Great Bahama Bank, the Maiella platform shows almost no signs of dolomitization. This explains why, despite their older age, most samples are better preserved than many dolomitized Bahamas carbonates. Some of the bioclastic sands of the Upper Cretaceous still have porosities of over 30%, whereas most of the platform deposits are densely cemented.

3. Methods

A. Sampling Technique

The samples used for velocity determinations are cylindrical miniplugs 2.5 or 3.8 cm in diameter. The miniplugs of unconsolidated mud from Florida Bay were sampled from short push cores 7.6 cm in diameter. So as to avoid compaction and fabric destruction during sampling, a thin-wall tube with a diameter of 3.8 cm was used to cut the miniplugs vertically out of the cores. The 2–4 cm long cylinders were compacted longitudinally by a hydraulic press with a uniaxial pressure of up to 170 MPa. The velocities of the mud samples were measured at variable degrees of compaction. The maximum compaction reached was approximately 50% so that the initial porosity of 63% was reduced to 26%.

The samples from the Bahamas deep drillings were cored horizontally, or occasionally vertically, into the 7.6 cm diameter cores. Plugs from the Maiella were cored from hand samples collected in outcrops. All rock cores were trimmed to a length between 1.5 and 5 cm. The end surfaces were polished to make them flat and parallel in order to allow a good transmission of the acoustic signal.

B. Velocity Measurements

The velocities from all Bahamas samples and from 29 of the Maiella samples were measured applying a pulse transmission technique (BIRCH, 1960) with an apparatus shown in Figure 5. The velocity samples were water saturated and jacketed in rubber or heat shrink tubing which seals the pore fluid from the confining oil in the pressure vessel. Confining and pore-fluid pressures are chosen

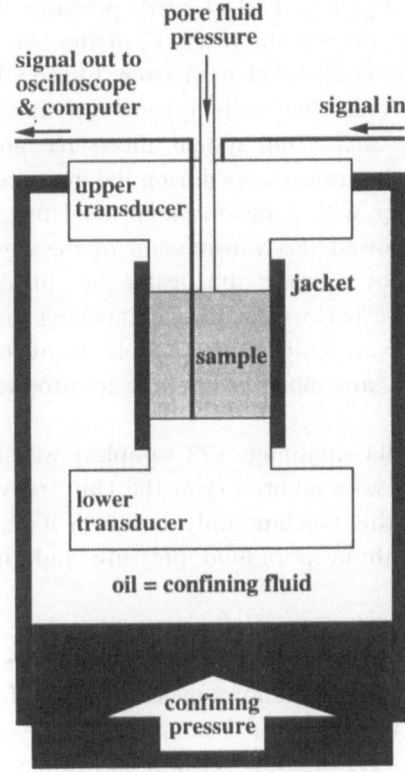

Figure 5
Schematic cross section of ultrasonic velocity meter.

independently to simulate most accurately *in situ* conditions of a buried sediment. Pore-fluid pressures as high as 50 MPa can be obtained but most experiments were run at 2 MPa. Confining pressure is varied between 3 and 100 MPa, resulting in an effective maximum pressure (confining pressure minus pore-fluid pressure) of up to 98 MPa. However, many samples collapsed and failed at pressures below 100 MPa.

Within the end caps, piezoelectric crystals create a signal with a center frequency of 0.6 to 1.2 MHz. The same pair of transducers generates one compressional-wave signal (V_p) and two orthogonally polarized shear-wave signals (V_{s1}, V_{s2}). The transducers are arranged so that the waves propagate along the core axis. The electrical signal produced by the receiver crystal is amplified, filtered, and fed into a digital oscilloscope. The oscilloscope digitizes the ultrasonic signals and transfers the digitized waveforms to a Macintosh Quadra computer for display and time series analysis. A customized analysis software package collects the data as a function of effective pressure, and calculates the travel times of the signals as well as the three velocities (V_p, V_{s1}, V_{s2}) at every pressure step. The V_s used for the calculation of the V_p/V_s ratio is the mean V_s of the two measurements.

The velocities of the compacted mud from Florida Bay were measured with the same set of transducers but with a benchtop measuring system not under confining pressure. This measuring system allows recognition of compaction due to the axial pressure of the transducers during the measurement. The two transducers were pressed together with a piston at an axial pressure of 0.1–1 MPa. This relative low pressure allowed the transmission of the signal from the transducers into the mud but did not compact drastically the still deformable mud samples. Some measurements were performed on uncompacted mud, however the minimal required transducer pressure reduced the sample length by a few percent. Corrections for length change are made so as not to produce errors in the velocity determination.

A part of the Maiella miniplugs (72 samples) was measured with a similar apparatus in the petrophysics laboratory at the University of Geneva, Switzerland. The transducer pair of this machine only creates a p-wave signal. Measurements were performed dry without pore-fluid pressure and under confining pressures varying up to 400 MPa.

The precision of the velocity measurements is mainly a function of the quality of the sample. In well cemented, high velocity samples, the lower transducer receives a clear peak as first break which allows measurements of velocity with an error of less than 1%. Friable, unconsolidated samples tend to compact and reduce their sample length by up to 5%. In addition, they sometimes produce only a moderate first break signal, especially for V_s, so that the error of velocity determinations in these difficult samples probably amounts to approximately 5%.

C. Additional Properties

In addition to the velocity determinations, several other analyses were performed. Dry bulk densities were calculated by weighing the oven dried rock plugs and calculating the volume by measuring diameter and length. XRD analyses were performed on the cut-offs of drilled plugs from the Bahamas samples. Calibration with carbonate-standards allowed determination of the percentage of calcite, dolomite and aragonite. Because these almost pure carbonates consist dominantly ($>95\%$) of three carbonate minerals, the percentage of the minerals determines a theoretical grain density:

$$\rho_{grain} = \frac{\%\text{calcite} \cdot \rho_{calcite} + \%\text{dolomite} \cdot \rho_{dolomite} + \%\text{aragonite} \cdot \rho_{aragonite}}{100}$$

$$\rho_{calcite} = 2.71 \text{ g/cm}^3; \quad \rho_{dolomite} = 2.87 \text{ g/cm}^3; \quad \rho_{aragonite} = 2.93 \text{ g/cm}^3.$$

The rock porosity is calculated by comparing the calculated grain density with the measured dry or saturated bulk density

$$\frac{\%\text{porosity}}{100} = \frac{\rho_{\text{grain}} - \rho_{\text{bulk}}}{\rho_{\text{grain}} - \rho_{\text{fluid}}}.$$

This easy way to determine rock porosity in pure carbonates was compared with the results from other techniques such as helium densitometry and Archimedes principle. Our porosity values are systematically 0–3% higher than porosities obtained by the other methods. The difference is caused by the fact that the helium densitometry as well as the Archimedes method are based on penetration of a pore fluid or gas (water, and helium respectively) into the pore space and therefore are a function of permeability. In addition, isolated and closed porosity is not penetrated by the pore fluid and is therefore not detected, whereas our method based on the density and X-ray analyses also considers this closed porosity.

Cut-offs of the mini-plugs were also used to make thin sections from most velocity samples. Thin sections were examined in order to determine factors such as sediment type, composition, grain size, porosity type and diagenetic alterations. These examinations enable us to correlate the physical properties with the lithological parameters.

4. Velocity Data

A. V_p and V_s

In the following descriptions and correlations, velocities at a confining pressure of 10 MPa and a pore-fluid pressure of 2 MPa are discussed. The resulting effective pressure of 8 MPa (10 MPa for dry samples) is high enough to allow a good signal transmission but does not cause significant fracturing in the high porous samples. The V_p measurements on dry samples are also presented here because major differences between dry and saturated V_p only have to be expected in rocks with a dominant crack porosity (NUR and SIMMONS, 1969), whereas the saturation of round-shaped pores, abundant in our samples, does not influence V_p drastically.

Despite the limited variability in mineralogy, the measured carbonates have an extraordinarily wide range in velocities. V_p varies between 1700 and 6500 m/s, V_s between 700 and 3400 m/s. The three different data sets have different ranges in V_p and V_s (Figure 6). The unconsolidated mud samples from Florida Bay have the lowest velocities with a minimum V_p and V_s of 1700 and 700 m/s, respectively. The Bahamas and the Maiella samples reach velocities of up to 6500 m/s (V_p) and 3400 m/s (V_s). Unlike siliciclastic sediments, where variations in mineralogy (e.g., clay-content) can cause large velocity contrasts, the different carbonate minerals, calcite, dolomite and aragonite, have very similar physical properties so that

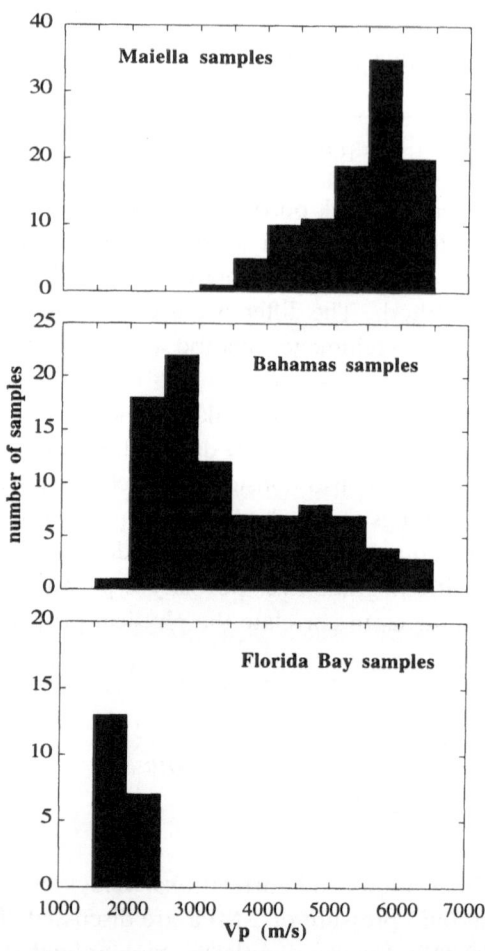

Figure 6

Range of V_p for the three investigated areas (at 8 MPa effective pressure). The Maiella samples have a
higher average velocity than the Bahamas samples. The artificially compacted mud samples from Florida
Bay have, despite compaction of up to 50%, only low velocities. The large range of velocity in all
samples is remarkable for the restricted mineralogy of the pure carbonates.

differences between them cannot be responsible for the large variability in velocities.
Consequently, the wide range of V_p and V_s in carbonates has to be explained with
different fabrics and textures and not with the different minerals of the rocks.

B. *Acoustic Impedance*

The observed range in V_p and V_s and therefore the large contrasts in acoustic
impedance (Figure 7) can explain the excellent seismic reflectivity of many seismic

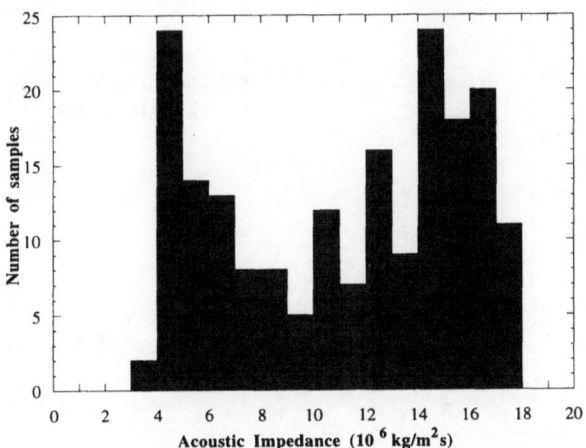

Figure 7

Range of acoustic impedance of the Bahamas and Maiella samples (at 8 MPa effective pressure). Impedances of high velocity and dense rocks are over five times higher than impedances of low velocity rocks. These impedance variations in the pure carbonates are caused by differences in fabric and not by differences in composition. The large range explains the good reflectivity observed in many seismic sections of carbonates.

sections in pure carbonates. The two drillholes in the Bahamas, for instance, have acoustic impedance values that range from $3.8-17.4 \cdot 10^6 \, kg/m^2 \, s$. The observed good reflectors on seismic sections, often believed to be caused by intercalations of noncarbonate sediments, can in fact be explained by variations in the fabric of the carbonates.

C. V_p/V_s

Similar to V_p and V_s, the V_p/V_s, which was only measured under saturated conditions, also has a wide range. The cross plot of V_p/V_s with V_p (Figure 8) shows that the ratio in indurated rocks with high V_p normally falls between 1.8 and 2. At lower V_p, the V_p/V_s ratio can reach substantially higher values of up to 2.6. These higher V_p/V_s ratios reflect the fact that in general V_s is more affected by the highly porous fabric of the low-velocity carbonates than V_p. It must be taken into account that some readings of the shear wave velocity and thus the V_p/V_s ratio might have a large error, due to a bad V_s-signal quality, e.g., when a low shear wave amplitude is combined with a high background noise. Therefore some of the extreme V_p/V_s values might be unreliable; however, these few values do not change the general pattern of the V_p/V_s range. The artificially compacted samples from Florida Bay have an extreme range of V_p/V_s from 1.7 to 2.8 within a narrow range of V_p from 1700 to 2300 m/s. The uncompacted mud-fabric with a porosity of approximately

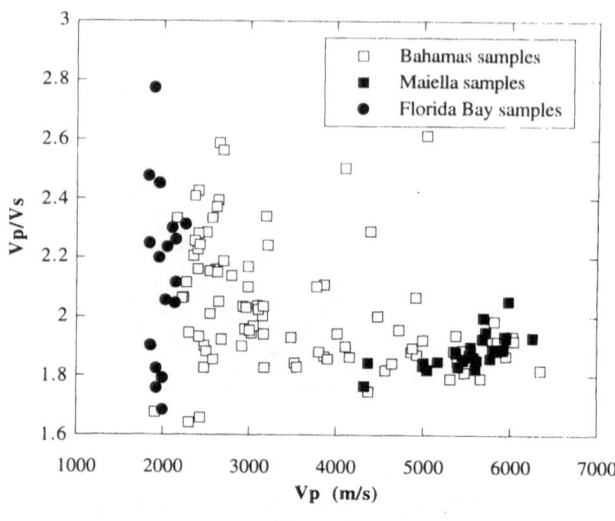

Figure 8

V_p/V_s as a function of V_p (at 8 MPa effective pressure). The wider range of V_p/V_s towards higher values in low velocity rocks shows that shear waves are more affected by high porosity fabrics than compressional waves. Some of the extreme low and high values might be caused by a questionable registration of the arriving V_s signal, in particular in high porosity rocks.

63% is not strong enough to sustain shear stresses (LAUGHTON, 1957) and inhibits transmission of the shear-wave signal. The little compacted mud samples have high V_p/V_s of up to 2.8. With increasing compaction, the V_p/V_s of these Florida Bay samples approach more "normal" values around 2.2.

5. Factors Affecting Velocity

In many sedimentary rocks, the concept of grain and matrix supported fabric with a critical porosity is able to explain the variations in velocity (NUR *et al.*, 1991) and to relate them to differential composition. The high susceptibility of carbonates towards diagenetic changes however, causes cementation, dissolution and recrystallization processes that form rock fabrics unique to carbonates with velocity patterns that do not simply reflect the compositional variations of the sediment.

Acoustic velocity in carbonates is a complex funciton of several factors. We can distinguish between rock-intrinsic and rock-extrinsic parameters. Intrinsic parameters, such as porosity, pore type, composition or grain size, are factors that are connected with the lithology and thus, with the physical properties of the rock fabric. Rock-extrinsic parameters are factors that are not physically connected to the rock fabric, but are determined by external boundary conditions. Examples of rock-extrinsic parameters are burial depth, confining pressure and age of the sediment. It will be shown that in carbonates the rock-intrinsic factors are more important than the extrinsic ones.

A. *Velocity as a Function of Rock-extrinsic Parameters*

The effect of mechanical compaction

The compaction experiments and the velocity measurements on modern carbonate mud from Florida Bay were performed to determine the change in velocity due to a porosity reduction from solely mechanical compaction. The samples are pure carbonate mud and have a special lithology that is rarely encountered in the other measured carbonate samples. However, most of the measured samples have, together with the coarser grain fraction, a large amount of micrite in the matrix. Therefore, we suspect that compaction in our other samples would have a similar effect on velocity as in the compacted mud.

It is known that porosity has a major influence on velocity and a porosity decrease usually produces a velocity increase. The Florida Bay samples show a relatively subtle increase in velocity with increasing compaction or decreasing porosity (Figure 9). At porosities close to 60%, poorly compacted samples have a V_p of 1700 m/s and no measurable V_s. The samples had to be compacted by 10–15% in order to receive a V_s signal. This corresponds well with the measurements of LAUGHTON (1957) who only detected shear waves in unconsolidated

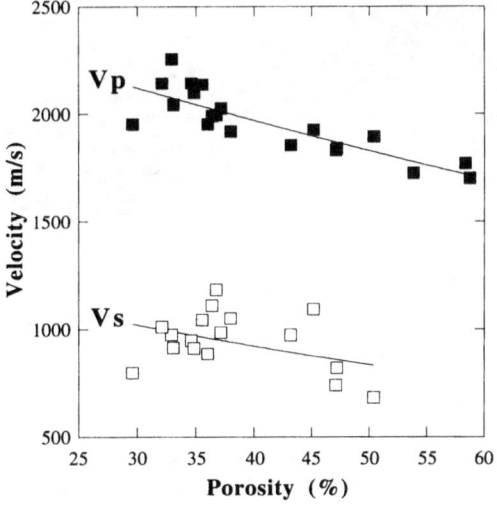

Figure 9

Increase of V_p and V_s at decreasing porosities in the differently compacted Florida Bay samples. The velocity increase is the effect of pure compaction. Initial porosity of the carbonate mud is on average 63%. The little compacted samples with porosities above 50–55% could not transmit a V_s signal.

ocean sediments above a compacting pressure of 5 MPa. At maximum compaction with porosities of 29%, V_p increases to 2200 m/s and V_s lies around 900–1100 m/s. The gradient of the measured increase in velocity of the compacted mud is significantly lower than the observed gradients in the other Bahamas and Maiella samples (Figure 10). The mud samples display, due to their low shear modulus, a behavior similar to material that has no rigidity as suggested by HAMILTON (1971). He showed that, unlike liquids, most unconsolidated marine sediments do possess rigidity (shear modulus > 0) and have a definite structure. The Wood equation (WOOD, 1941), valid for mediums without shear modulus or rigidity, can be used to compare the observed porosity-velocity relation of the artificially compacted sediments.

$$V_p = [(\Phi\beta_{\text{fluid}} + (1 - \Phi)\beta_{\text{solid}})(\Phi\rho_{\text{fluid}} + (1 - \Phi)\rho_{\text{solid}})]^{-1/2}$$

Φ = fractional porosity; β = compressibility; ρ = density.

The Wood equation was calculated using values for water ($4.06 \cdot 10^{-10}$ m²/N) and for calcite ($1.34 \cdot 10^{-11}$ m²/N) compressibility, because the elastic properties of aragonite are not well-known. The comparison of the calculated with the measured velocities reveals that the measured velocities of the mud samples are in fact only slightly above the values predicted by the Wood equation, whereas all other samples that were altered during diagenesis show much higher velocities (Figure 10). The

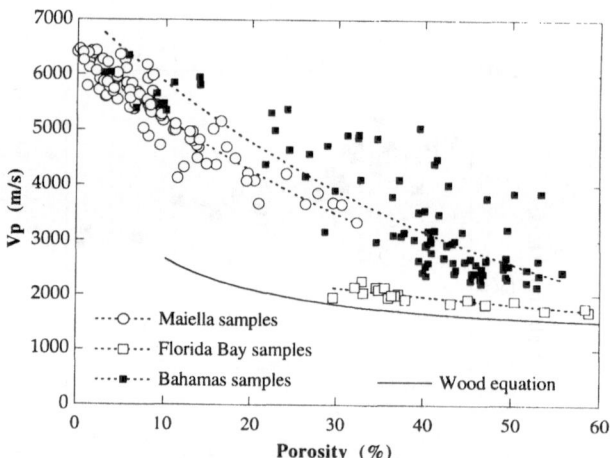

Figure 10
Velocity as a function of porosity for the three data sets. V_p and porosity show clearly an inverse correlation, but the gradient of increasing V_p at decreasing porosity is much lower in the compacted Florida Bay samples than in the natural rocks of the Bahamas and the Maiella. Velocities of the mud samples are only slightly higher than velocities of mediums without rigidity, calculated using the Wood equation (WOOD, 1941). This shows that porosity reduction due to mechanical compaction has only a minor effect on V_p, whereas porosity reduction due to diagenetic processes (e.g., cementation) can increase rigidity which results in higher velocities.

lower gradient for the mud samples can thus be explained by the absence of additional processes that increase the rigidity of the rock and normally accompany the compaction of sediments. The artificial compaction experiments happen so fast that no diagenetic alterations are initiated. Normally, the effects of diagenetic processes, such as recrystallization or cementation, are superimposed on the effect of porosity reduction. The velocity, therefore, represents the combined effect of these different processes. The difference between the velocity-porosity correlation in natural rocks and in the artificial compacted rocks demonstrates how much diagenesis contributes to the observed velocity increase. In fact, our samples document that diagenetic alterations are more effective in increasing velocity than compaction, because they significantly increase the rigidity of the rock.

With a uniaxial pressure of 170 MPa, the porosity could not be reduced to under 29%, indicating that the microfabric of the rock, consisting of 65% aragonite needles, is close to the densest packing that can be reached just by mechanical compaction. This experimental compaction also shows that pure mechanical compaction only plays a minor role in carbonates. Samples with porosities between 0 and 25 percent can only reach their actual porosity with the aid of diagenetic closing (cementation) of part of the pore space.

Burial depth and age of the sediment

The measured samples taken from the two core borings of the Bahamas Drilling Project clearly show that in these carbonates, velocity is neither primarily a function of the sediment age, nor of the burial depth. In contrast to the usual assumption that velocity increases with depth (HAMILTON, 1980; JAPSEN, 1993), the depth plots of V_p and V_s (Figure 3) in the two drillholes display velocity inversions that make velocity predictions, based only on depth, impossible. Both holes display a pattern of high variability of velocity in the carbonates that were deposited in a shallow-water environment like sediments from the reef, platform margin or platform interior. These high velocity zones, in Clino above 220 m and in Unda between 290 and 370 m and above 120 m, overlie zones of low velocities that cause the observed velocity inversions. The distinct jump to higher velocities at Unda 293 m, for instance, marks the transition from a fully dolomitized carbonate sand to a dolomitized reefal unit. Below the reefal unit, both V_p and V_s decrease again, resulting in a velocity inversion. Rocks of the low velocity zones are mostly carbonates that were deposited in deeper water and underwent less diagenetic alteration than the shallow-water carbonates. The inverse trend with decreasing velocities at greater depths indicates that diagenetic processes other than simple compaction substantially control the velocity evolution. In the young sediments of shallow Clino and Unda, high velocities are attained due to intense diagenetic alterations which occur much faster than compaction due to an increased overburden.

The Maiella carbonates also demonstrate that depth, or in this case age, does not necessarily influence velocity. Some of the Upper Cretaceous rudist sands are among the oldest but also have the slowest velocities of the measured samples from this data set. The V_p of 3300 m/s is remarkably low for Cretaceous carbonates, documenting again the insignificance of absolute age for velocity evolution in these samples. The reason for this low velocity is the preserved high porosity and the associated interparticle pore type.

As a consequence, the depositional environment and the diagenetic alteration is much more important for velocity evolution than age or depth. Velocity predictions cannot be made solely with the knowledge that a carbonate sediment has a certain age and/or is at a certain depth, rather the acquisition of some additional, intrinsic rock parameters is necessary to produce a reliable velocity estimation.

Effective pressure

To investigate the pressure dependence of V_p and V_s, sonic velocities of the Bahamas and Maiella samples were measured under varying confining and pore-fluid pressures. Sonic velocity in rocks is a function of the differential pressure, or effective pressure, which is the difference between the confining and the pore-fluid pressure. Minor departures from this relation are caused by changing pore-fluid properties at different pressures (COYNER, 1984).

At low pressures, all samples show an increase in velocity with increasing effective pressure due to better grain contacts, changing pore shapes and closing of microcracks (GARDNER *et al.*, 1974). This increase is large for slow, unconsolidated samples, whereas the velocities of indurated, dense samples are usually less affected by higher pressures. All velocity-pressure traces of the Bahamas samples plotted in the same graph (Figure 11) display a systematic pattern with higher gradients for low-velocity samples and lower gradients for fast samples. A minor part of the observed velocity increase is an artifact because compaction at increasing pressures reduces the sample length which is used to calculate the velocities.

A characteristic of many samples is that both V_p and V_s reach a maximum during increasing pressure and suddenly begin to decrease above a critical pressure. This velocity decrease is caused by a continuous disintegration and collapse of the sample in the pressure vessel, which progressively destroys the partly cemented grain contacts that supported the transmission of the acoustic signal. Eventually, velocity can increase again (e.g., sample Unda 141 m, Figure 12d) because the newly formed fractures that reduced the velocity can close by a further increase in confining pressure. In these samples, velocities that are measured under decreasing pressure at the end of a hysteresis loop (Figures 12d,e) are much lower than at the same pressures in the first part of the loop, because the fractured plugs completely fall apart and form an unconsolidated fabric with loose fragments. V_p/V_s increases dramatically above the critical pressure (Figure 12f) and continues to increase with decreasing pressure in the hysteresis loop. Shear-wave velocity is extremely affected

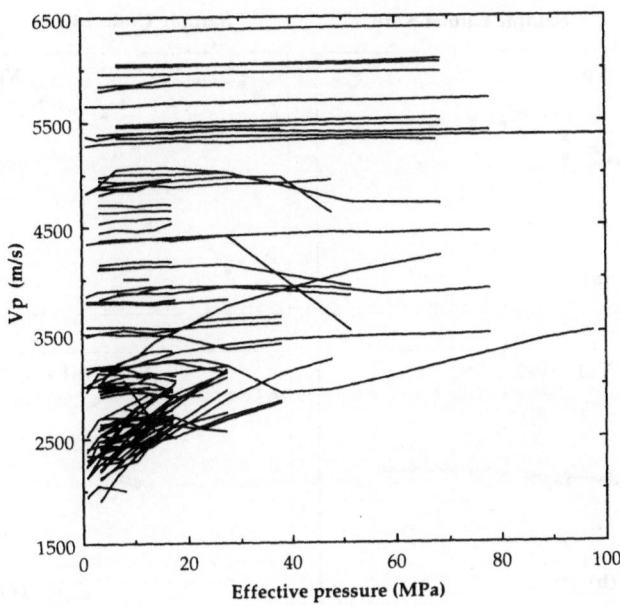

Figure 11

Velocity evolution of the Bahamas samples at increasing effective pressure. Each trace represents the velocities for one sample at different pressures. The gradient of low-velocity samples is higher than the increase of velocities in samples with high velocities. Decreasing velocities at higher pressures mark a critical pressure at which the samples are intensely fractured and collapse. A minor part of the velocity increase is an artifact caused by compacted sample length, that is used for velocity calculation.

by the destroyed fabric that results in a nonelastic behavior. A similar behavior with decreasing velocity during increasing pressure was also observed in some Cretaceous samples from the Maiella, indicating that age or burial depth is not a guarantee for consolidation and lithification of a carbonate sediment.

In contrast to the nonelastic behavior, the hysteresis loops of fast, more lithified samples (e.g., sample, Clino 657 m (Figures 12a–c) show a gentle increase in velocities with increasing pressure. V_p, V_s and also V_p/V_s eventually reach a plateau at high pressures and they nearly reach the former velocities at decreasing pressures. In these cases, the plugs are perfectly intact when they are removed from the pressure vessel, documenting the elastic behavior of the high-velocity rocks.

The critical pressure at which the first velocity decrease occurs varies with the different lithologies. Dense, indurated rocks display no evidence of fabric destruction up to the highest measured pressures of 100 MPa, whereas soft, unconsolidated samples show signs of velocity decrease already at 5 MPa. These low critical pressures demonstrate that some carbonates, especially most slope deposits or sucrosic dolomites, became buried without being progressively indurated. Under hydrostatic conditions, an effective pressure of 5 MPa is equivalent to a burial of

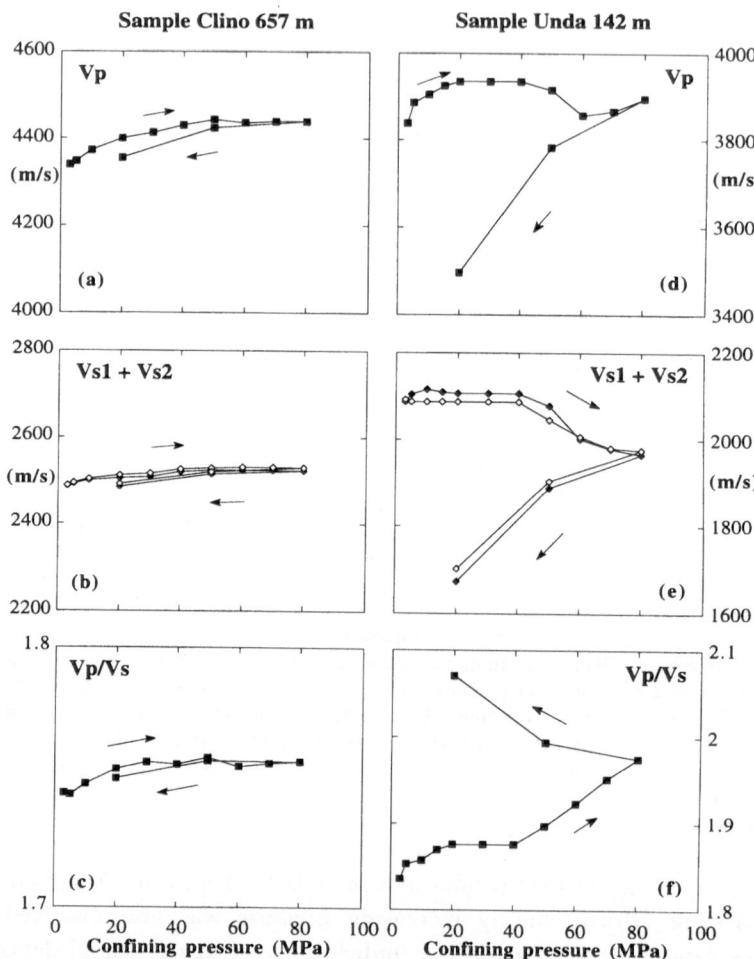

Figure 12
Two examples for elastic and nonelastic behavior: Sample Clino 657 m shows a steady increase in V_p (a) and V_s (b); this increase is mainly caused by the closing of microcracks at elevated pressures. The increasing V_p/V_s (c) shows that V_p increases more than V_s. The values reach a plateau at high pressures and approach starting conditions at the end of the hysteresis-loop. After the experiment, the plug shows no signs of damage. Sample Unda 142 m shows a similar increase in V_p at low pressures but at a critical pressure of 40 MPa, V_p and V_s start to decrease (d and e). This decrease is the result of fabric destruction in the pressure vessel. The plug is fractured and cemented grain contacts are destroyed so that velocities decrease. Above 60 MPa, V_p starts to increase again (d) because the newly formed fractures are progressively closed. With the release of pressure, V_p and V_s decrease dramatically because the fractured plug disintegrates. The V_p/V_s increases remarkably above the critical pressure (40 MPa) and becomes even higher at the end of the hysteresis-loop (f). Pore-fluid pressure equals 2 MPa for all samples.

less than 500 meters. Therefore, some parts of the drilled cores must have *in situ* conditions that cause development of cracks and fractures. This observation coincides well with open or partly cemented fractures that are visible in part of the cores. Also remarkable is that many samples show no signs of fabric destruction up to high pressures, demonstrating that porosity within partly cemented rocks can be preserved, even at pressures of 100 MPa or at depth of approximately 5 km.

B. *Velocity as a Function of Rock-intrinsic Parameters*

Depositional lithology

The lithology of a carbonate sediment at the time of deposition has strong influence on the evolution of velocity, in that it controls future alterations of the rock. At the time of deposition all unconsolidated sediments have similar velocities between 1550 and 1800 m/s. The different lithologies have, despite their similar velocities, different susceptibilities to diagenetic alteration that will change the physical properties and thus the velocities. The diagenetic susceptibility of special sediment types causes fast or slow alterations of the rock fabric, depending on the diagenetic potential of the sediment (SCHLANGER and DOUGLAS, 1974) and on the diagenetic regime.

The diagenetic potential in carbonates is mainly a factor of the grain size and the amount of metastable minerals. A high content of fine-grained micritic material, as found in mud or wackestones, results in a low permeability. The resulting low fluid flow inhibits or slows diagenetic alterations that rely on transport of chemical components in the water. In contrast to fine-grained rocks, sediments with a grain-supported fabric and a low content of micrite (grain- to packstone) have a higher permeability and, as a consequence, a higher fluid flow. This accelerates diagenetic processes and the sediment is quickly altered and consolidated. Thus, original coarse-grained rocks can reach higher velocities after a short time of burial, whereas fine-grained rocks tend to preserve their unaltered fabric and their slow velocity for longer burial durations.

In addition to grain size, the amount of metastable minerals, such as aragonite or high-magnesium calcite controls the diagenetic potential (SCHLANGER and DOUGLAS, 1974). A high amount in metastable components causes a high diagenetic potential and leads to a rapid dissolution or recrystallization of the sediment. This alteration can enhance the elastic properties and increase permeability, resulting in accelerated lithification.

The comparison of the velocity range with the depositional environment confirms these relationships. For example, if the Maiella samples are grouped into two categories of depositional environment, (1) platform deposits and (2) basin, slope and deeper shelf deposits (Figure 13), the velocity range of the two categories form two different clusters that only overlap in the high velocity area. The platform

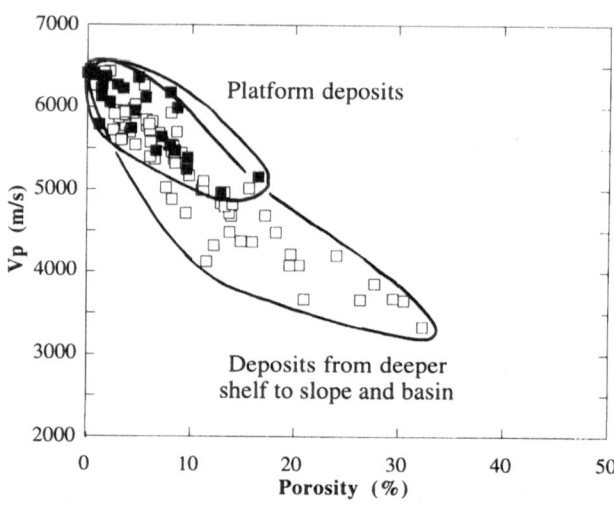

Figure 13
Velocity vs. porosity (at 8 MPa effective pressure) compared with depositional environments from the
Maiella samples. Carbonates deposited on the shallow-water platform have a narrow range with high V_p
and low porosities. Sediments from the deeper shelf, slope or basin show a higher variability towards
lower V_p, but maximal V_p are the same as in the platform rocks.

carbonates have a significantly higher average velocity than the basin and slope
carbonates. In the Bahamas cores, the relation between shallow water deposits and
deeper water deposits is slightly different with an overlap of velocities of the two
categories in the low velocity area (Figure 3). The majority of the Bahamian slope
carbonates are unconsolidated and thus slow, whereas the platform carbonates
show a larger range towards higher velocities, but have similar minimal velocities.

The explanation for these observed velocity patterns is that the platform
carbonates are usually high in coarse-skeletal grains or non-skeletal grains (ooids,
peloids). They consist predominantly of aragonite which is metastable in sea water.
Slope or deeper water deposits are normally characterized by a high micritic grain
fraction and by a higher content in pelagic calcareous organism (foraminiferas,
coccolithes) and consist mainly of more stable low-Mg calcite shells. Therefore,
shallow-water carbonates fulfill both conditions for fast diagenetic alterations:
coarse grain size and a high amount of metastable minerals. Some turbidites
deposited on the slopes, that contain many skeletal and aragonitic fragments from
the platform top, are similar to platform deposits (EBERLI, 1988), and thus different
from the normal background slope deposits. The differential depositional lithologies
explain the different ranges of the velocity measurements in the different sediment
types. In the older Maiella samples, most of the platform carbonates had enough
time for diagenetic alterations to reach their high, final velocities, and as a
consequence cluster at higher values than the slope sediments (Figure 13). In the

Bahamas cores, which are younger than the Maiella samples, not all the platform deposits reach high velocities, but both the average velocity and overall velocity range are much higher than in most of the slope samples. Only the few turbidites in the slope section containing platform derived material have high velocities resulting in some velocity variability (Figure 3).

The data strongly suggest that the depositional environment of a carbonate sediment affects the starting conditions under which a sediment undergoes diagenetic alterations. This indirect influence controls direct, rock-intrinsic parameters such as porosity, pore type, density and mineralogy.

Mineralogy

In siliciclastic rocks, the physical variety of minerals is large (e.g., quartz and clay), and mineralogy has more influence on sonic velocity than in carbonates (CHRISTENSEN and SZYMANSKI, 1991). The minimal influence of mineralogy on velocity in carbonates can be partially explained by the small velocity contrasts of the two dominant carbonate minerals calcite (6500 m/s) and dolomite (6900 m/s). Pure carbonates have little initial velocity differences due to mineralogy. The measured samples from the Bahamas cores are comprised >95% of minerals calcite, dolomite and aragonite and the Maiella carbonates consist almost purely of calcite.

Our data suggest that changes of this mineralogical composition have no major influence on velocity. A plot of the dolomite content versus velocity of the Bahamas samples clearly shows that there is no correlation between dolomite content and velocity (Figure 14). This lack of correlation is also shown by two measured plugs of Unda, only 7 m apart and both made of 100% dolomite: sample Unda 286 m has a V_p of 2697 m/s and a V_s of 1052 m/s, whereas sample Unda 293 m has a V_p and V_s of 5953 m/s and 3187 m/s, respectively. The high-velocity sample is a "reefal"-dolomite with a fabric preserving dolomitic cementation resulting in a total porosity of 14%, whereas the low-velocity sample is a sucrosic dolomite with high interparticle porosity of 49%. This example demonstrates that velocity depends on the type of dolomite and thus the associated porosity and pore type, and that mineralogy alone is not a characteristic parameter for determination of velocity in carbonates.

While the mineralogical composition in carbonates has little influence on velocity, the processes that alter mineralogy, such as sucrosic dolomitization or dolomitic cementation, have a strong influence on velocity. These processes also alter, in concert with changing mineralogy, porosity and porosity type. For example, fabric destructive dolomitization also destroys most of the earlier cementation, creating an undercemented and loose dolostone with petrophysical characteristics similar to a semi-lithified carbonate sand.

Porosity and pore types

Velocity is strongly dependent on the rock porosity (WANG et al., 1991; RAFAVICH et al., 1984). A plot of porosity versus velocity displays a clear inverse

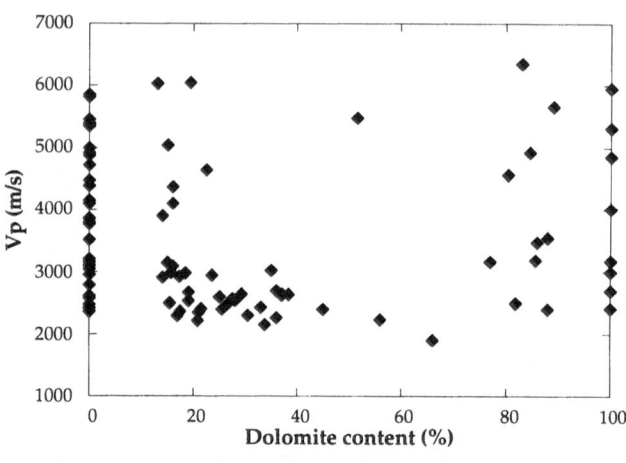

Figure 14

Velocity at 8 MPa effective pressure as a function of dolomite content in the Bahamas samples. There
is no correlation between these two factors. Different dolomite types, such as sucrosic dolomite or
dolomitic cement have totally different effects on velocity, therefore dolomite content alone cannot be
used as an indicator for velocity.

trend; an increase in porosity produces a decrease in velocity (Figure 15). The
general trend of the Bahamas and Maiella samples has correlation coefficients of
0.94 for V_p and 0.92 for V_s. Nevertheless, the measured values display a large
scatter around this inverse correlation in the velocity-porosity diagram. Velocity
differences at equal porosities can be over 2500 m/s, in particular at higher
porosities. For example, rocks with porosities of 40% can have velocities between
2100 m/s and 5000 m/s, which is an extraordinary range for rocks with the same
chemical composition and the same amount of porosity. This discrepancy is caused
by the ability of carbonates to form cements and special fabrics with pore types that
can enhance the elastic properties of the rock without filling all the pore space. The
high elastic moduli result in velocities that are higher than velocities predicted by
theoretical equations, such as the time average equation (WYLLIE *et al.*, 1956), as
shown in Figure 15.

In other data sets, such as synthetic sand-clay mixtures (MARION *et al.*, 1992)
or siliciclastic sediments (VERNIK and NUR, 1992), a similar scattering in the
velocity-porosity diagram is observed. But unlike carbonates, the scattering in these
rocks can be explained by compositional variations, in particular by changes in clay
content. WILKENS *et al.* (1991) noticed that velocities of low-porous basalts are
very dependent on the pore shapes. Samples containing pores with low aspect ratios
(cracks) are associated with lower velocities, compared to samples with round pores
or high aspect ratios. As a result, high velocity contrasts are observed between
rocks without large variations in total porosity. The pores in our high porosity

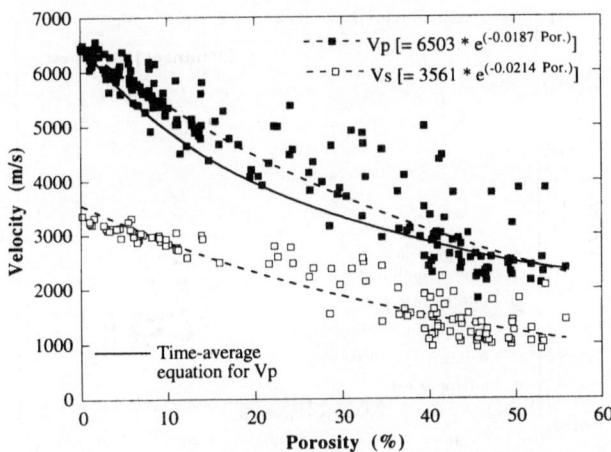

Figure 15

V_p and V_s from the Bahamas and the Maiella samples at 8 MPa effective pressure as a function of porosity with exponential best fit equations (dashed lines). Both V_p and V_s demonstrate the trend of decreasing velocities with increasing porosities, but scatter, especially at higher porosities, around the dotted best fit lines. The scattering is a result of special fabrics and pore types that enhance the elastic moduli of the rock without filling all the pore space. As a consequence, measured V_p are higher than V_p predicted by theoretical equations such as time-average-equation (WYLLIE *et al.*, 1956).

carbonates generally have high aspect ratios. In this case, the high velocity contrasts between rocks with similar total porosity can be related to specific pore types resulting in characteristic and very different elastic properties. Based on thin section observations, the Bahamas samples can be grouped into five categories of predominant pore types which all have characteristic clusters in the velocity-porosity diagram (Figure 16). The five dominant pore types that can be distinguished are:

(1) *Interparticle and intercrystalline porosity (Figures 17a,b):* The porosity between the components of a sediment is the interparticle porosity. This porosity predominates after deposition of a sediment when grains form a loose package with little cementation. Intercrystalline porosity develops at a later stage during diagenesis, when newly crystallized minerals such as dolomite rhombohedra form a loose aggregate. It has a similar petrophysical behavior as interparticle porosity. The accumulation of unconnected grains without cement or matrix results in a low velocity because the rock has low elastic moduli due to the lack of a rigid framework. Most of these samples therefore show a negative departure from the average velocity-porosity correlation (Figure 17c).

(2) *Micro-porosity:* Micro-pores ($< 10 \, \mu m$) are abundant in carbonate mud, either in a micritic grain or in the micritic matrix. High micro-porosity is thus expected in carbonates with a high micritic content. Due to the lack of cementation that results in an unconnected grain fabric, micro-porosity has a similar effect on

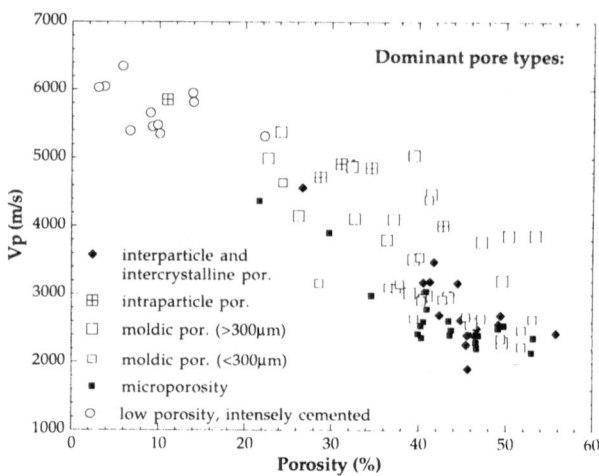

Figure 16

Velocity vs. porosity diagram from the Bahamas samples, with observed categories of different pore types. The large scattering, e.g. velocities from 2200 to 5000 m/s at porosities of 40%, are a result of different predominant pore types in the analyzed samples. Rocks with moldic or intraparticle porosity have positive departures from the general trend whereas rocks with interparticle, intercrystalline or micro-porosity have relatively low velocities and thus show negative departures. Velocities are taken at an effective pressure of 8 MPa.

velocity as fine-grained, interparticle porosity and also shows a negative departure from the average velocity-porosity trend.

(3) *Moldic porosity (Figures 17d,e):* Moldic porosity develops by dissolution of grains with a metastable mineralogy (e.g., grains of aragonite and high Mg-calcite). Selective dissolution can occur before, after or during cementation of the interparticle pore space. After dissolution, the rock consists mainly of molds and the partially cemented former interparticle pore space which is a fabric type with high elastic properties. Samples in which moldic porosity predominates, have higher

---

Figure 17

Two examples for pore types with characteristic clusters in the V_p-porosity diagram: a) Photomicrograph of sample Unda 286 m as an example for a rock with intercrystalline porosity. The rock consists completely of micro-sucrosic dolomite. Plug-porosity is 49%. b) Computer scan of photomicrograph above (porosity black, particles white) with characteristic pattern of loose particles (dolomite rhombohedra) surrounded by connected pore space. c) V_p-porosity diagram of all samples with dominant interparticle and intercrystalline porosity. V_p are in general below average trend due to the lack of connections between the grains. d) Photomicrograph of sample Unda 65 m as an example for a rock with coarse moldic porosity. All grains were dissolved after cementation of the interparticle pore space. Plug-porosity is 37%. e) Computer scan of photomicrograph above (porosity black, particles white). The nonconnected molds are integrated in a framework of sparry cement. f) V_p-porosity diagram of all samples with dominant coarse moldic porosity. V_p are significantly above average V_p-porosity trend due to the rigid framework of the rocks.

velocities than expected from their total porosities, and therefore a positive departure from the best fit curve (Figure 17f). These high velocities are caused by the self-supporting framework made of cement and micrite surrounding the molds. The travel time through this framework is faster than through grains that are only connected by point contacts, as found in rocks with interparticle porosity. In

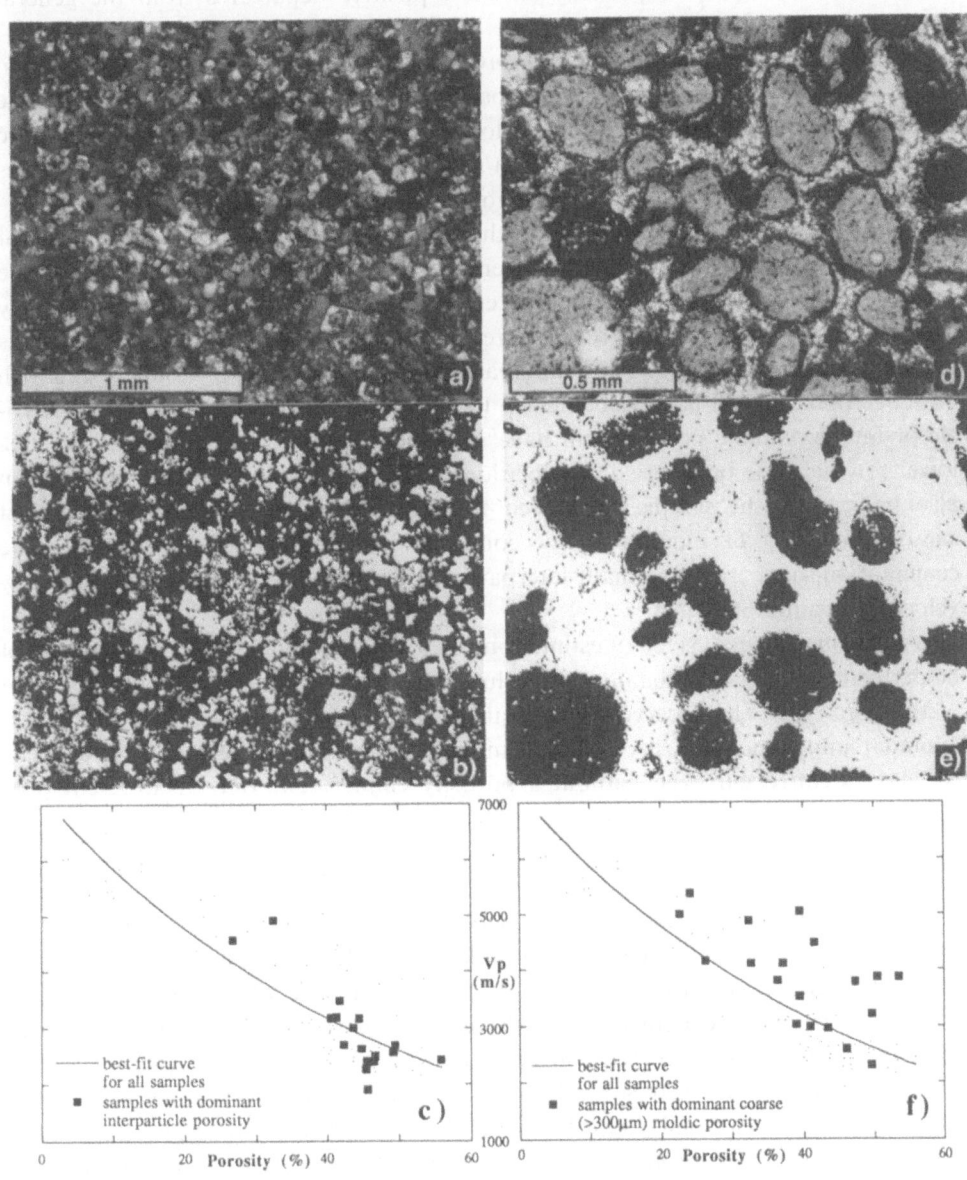

addition, velocity is dependent on the diameter of the molds. Velocities are higher in coarse moldic rocks, whereas fine-moldic samples are relatively slower.

(4) *Intraparticle porosity:* Framestones and boundstones, formed by organisms such as corals or bryozoans, consist of a constructional framework with a porosity that is embedded in the solid frame. Therefore, these samples show a similar velocity-porosity pattern to rocks with coarse moldic porosity that also have a framework with high elastic rigidity, resulting in high velocities. The samples with predominant intraparticle porosity all show positive departures from the general trend in the velocity-porosity diagram.

(5) *Low porosity samples with dense cementation:* These samples show an extensive, blocky cementation with porosities of 20% or less. They are close to the final stage of diagenetic evolution. Velocities are high and close to the intrinsic velocities of the minerals calcite (6500 m/s) and dolomite (6900 m/s). These samples form the upper part of the velocity-porosity correlation line.

As discussed above for the case of the Bahamas samples, the specific effects of the various pore types on elastic properties of rocks explain why rocks with the same porosity can have extremely different velocities. The most significant velocity contrasts at equal porosities are measured between coarse moldic rocks and rocks in which interparticle porosity predominates (Figures 17 and 19). Moldic rocks with 40–50% porosity can have velocities up to 5000 m/s, whereas rocks with interparticle or intercrystalline porosity can have velocities that are up to 2500 m/s or 50% lower for the same porosities. This relationship between pore type and velocity can also be seen in the samples measured from the Maiella. The Cretaceous rudist sands, consisting of individual, not connected rudist fragments with only little cementation, have a predominant interparticle porosity and have therefore very low velocities around 3000 m/s.

As a consequence, velocity estimation for a given carbonate sample should not be performed using only the porosity values, but in combination with an assessment of the pore type. The observed complicated velocity-porosity pattern, which causes a similar impedance-porosity pattern, implies that an impedance contrast between two layers can occur even without a porosity change, due only to different pore types.

Density

Seismic reflection patterns are a function of acoustic impedance and therefore the combined products of velocity and bulk density. In many case studies, only one parameter, either velocity or density, is known and the other factor has to be estimated with empirical correlations. Because density is closely related to porosity, velocity shows a good correlation with density (Figure 18). Despite the good correlation coefficients of 0.94 for V_p and 0.93 for V_s, the data in a plot of velocity vs. density scatters around a best-fit curve which also reflects the carbonate specific pore types. Using the general velocity-density trend in our data set, we can improve

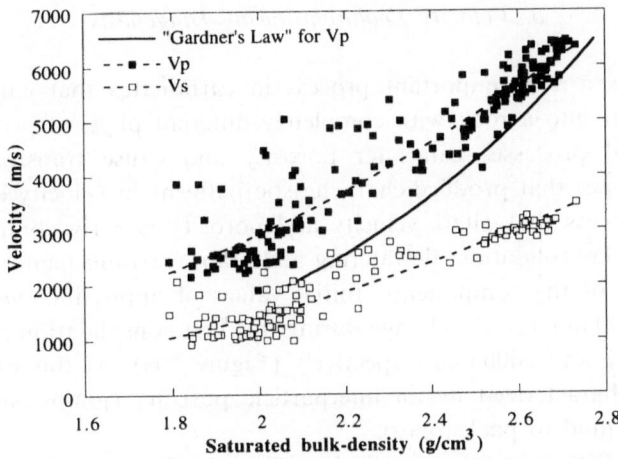

Figure 18

Velocity as a function of density in the Bahamas and Maiella samples at an effective pressure of 8 MPa. The solid line represents V_p calculated by "Gardner's Law" [density(g/cm^3) = 0.23 · V_p(ft/sec)$^{1/4}$], an empirical formula for all sedimentary rocks, which is often used to calculate impedance values only from velocity or density data (GARDNER et al., 1974). The velocities of the measured carbonates are all higher than the Gardner velocities. The equation has thus to be modified towards higher velocities in order to produce more reliable velocity-density pairs in carbonates (for suggested equations see text).

the velocity-density correlations for pure carbonate rocks because most empirical formulas, such as Gardner's Law (GARDNER et al., 1974), are mainly valid in siliciclastic rocks. Gardner's Law is an empirical equation for sedimentary rocks relating V_p to density.

$$\text{density(g/cm}^3) = 0.23 \cdot [V_p(\text{ft/sec})]^{1/4} \quad \text{or} \quad V_p(\text{m/s}) = 108.9 \cdot [\text{density(g/cm}^3)]^4.$$

This formula is mainly used to calculate impedance values from either density or velocity data so as to make impedance estimations for seismic models in sedimentary sequences. However, all our measured velocities are higher than the Gardner equation predicts. This implies that Gardner's equation, which is an average formula for all sedimentary rocks, requires a modification for carbonates towards higher velocities to predict reliable velocity-density pairs. Based on the data from the Bahamas and the Maiella samples, we suggest these empirical correlations

$$V_p(\text{m/s}) = 524 \cdot [\text{density(g/cm}^3)]^{2.48} \quad V_s(\text{m/s}) = 199 \cdot [\text{density(g/cm}^3)]^{2.84}.$$

These correlations, specific for carbonates, show a better fit and describe more accurately the velocity-density relation, but should not be applied in siliciclastic or mixed carbonate-siliciclastic rocks.

6. Velocity Evolution during Diagenesis

Diagenesis is a very important process in carbonates that can transform a lithified sediment into a rock with completely different physical properties. These post-depositional processes can alter porosity and cause transformations into different pore types that produce characteristic patterns in velocity evolution.

The first process that alters velocity and porosity is early compaction of the sediment: initial consolidation, dewatering and grain rearrangement with no cracking or breaking of the components. Initial values of approximately 50–60% for porosity and 1600 m/s for V_p, change during this first consolidation stage to values close to 40–50% and 2000 m/s respectively (Figure 20a). At this early stage, the sediments are characterized by an interparticle porosity (grainstones) or a high micro-porosity (mud to packstones).

Different diagenetic processes will also affect the future evolution of porosity. The velocity effect of this evolution, in particular the effect of the transformation of pore types, can be described by a velocity-porosity path (Figures 19 and 20). During its burial history, every sediment undergoes such a specific velocity-porosity path, which starts at deposition and ends at the measured velocity-porosity values of the last diagenetic stage. This path is not necessarily a straight line because different pore types are created and eventually destroyed during diagenesis. To pass the different clusters in the velocity-porosity diagram caused by the specific pore types, the shape of the path is rather a curved line which depends on the timing of the diagenetic events.

A good example for a loop during the velocity-porosity path is the fabric inversion of a grainstone to a coarse moldic rock. A clean unconsolidated ooid sand from Cat Cay (Bahamas), that has a depositional, mainly interparticle porosity of 40–50% (ENOS and SAWATSKY, 1981), has a V_p of 1779 m/s at 8 MPa effective pressure (Figure 19a). The "same" rock (sample Unda 65 m), but after cementation of the interparticle pore space and after the dissolution of the ooids and peloids (Figure 19b), has a completely inverted fabric with a porosity of 37% (mainly moldic) and a V_p of 4105 m/s. During the fabric inversion, the rock must have

Figure 19

Example for reconstruction of a porosity-velocity path. Figures (a) and (b) are porosity scans (porosity = black; particles = white) of photomicrographs, short side equals 1 mm. a) Ooid-grainstone at time of deposition (Cat Cay, Bahamas) with interparticle porosity, representing starting conditions of the path; por. = 45%, V_p = 1779 m/s. b) Sample Unda 65 m (also shown in Fig. 17d), por. = 37%, V_p = 4105 m/s, V_s = 1640 m/s. Former ooid(?)-grainstone with coarse moldic porosity. After a few marine alterations and an intense blocky cementation, all grains were dissolved and left molds behind (black). c) Reconstructed velocity-porosity path for a nonskeletal grainstone from deposition (a) to the diagenetic stage observed in (b). The transformation of pore types together with the dissolution of grains that took place after cementation led to a fabric inversion, which resulted in a characteristic loop of the path. The moldic framework (b) provides high elastic rigidity and thus high velocity at high porosity.

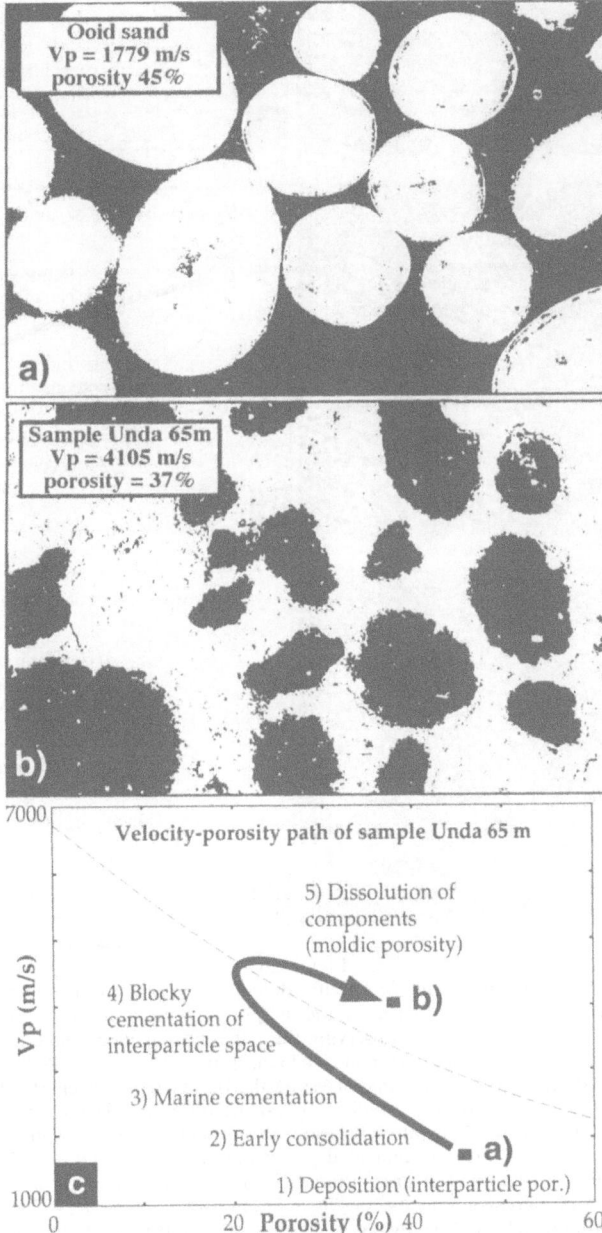

Ooid sand
Vp = 1779 m/s
porosity 45%

a)

Sample Unda 65m
Vp = 4105 m/s
porosity = 37%

b)

7000

Velocity-porosity path of sample Unda 65 m

5) Dissolution of
components
(moldic porosity)

Vp (m/s)

4) Blocky
cementation of
interparticle space

3) Marine cementation

2) Early consolidation

c

1000

1) Deposition (interparticle por.)

0   20 **Porosity (%)** 40   60

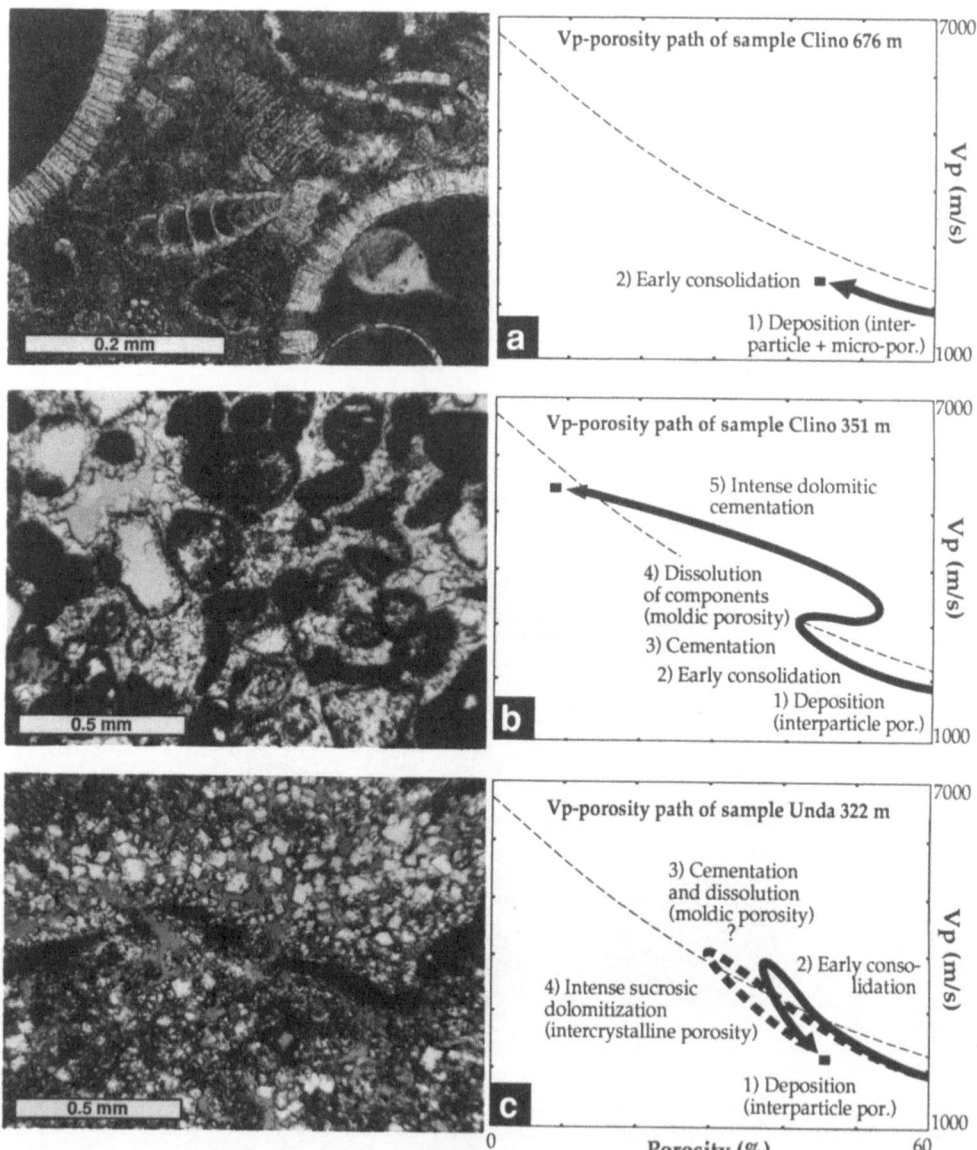

Figure 20

Examples of inferred V_p-porosity paths for specific Bahamas samples from different depths, shown in photomicrographs (left). The black square in the V_p-porosity diagram (right) marks the measured values. The arrow path is constructed by observing the different diagenetic stages of the sample and by relating them to V_p-porosity values. a) Clino 676 m, por. = 44%, V_p = 2478 m/s, V_s = 1356 m/s. Periplatform slope sediment with globigerinas. Despite the burial depth (deepest sample of both cores) only minor compaction and matrix recrystallization can be observed. The V_p-porosity path is thus a short, straight line from conditions at deposition to present times. b) Clino 351 m, por. = 9%, V_p = 5661 m/s, V_s = 3158 m/s. Densely cemented grainstone. Similar evolution as in sample of Figure 19, but an additional cementation after dissolution filled the earlier created moldic pore space, resulting in reduced porosity and increased velocity. c) Unda 322 m, por. = 46%, V_p = 2405 m/s, V_s = 991 m/s. Sucrosic dolomite with few relicts of redalgaes. The sedimentary fabric has been completely destroyed by sucrosic dolomitization that created a dominant intercrystalline porosity.

undergone a stage with a considerably lower porosity, because cementation occurred before dissolution. This stage marks the turning point of the loop in the velocity-porosity path (Figure 19c). This evolution clearly documents how diagenesis can invert the fabric and change the elastic properties of the rocks, even without changing total porosity.

Some samples undergo, in addition to a dissolution stage, a subsequent intense cementation of the newly created moldic pore space (Figure 20b). The beginning of the velocity-porosity path is, in this example, the same as in the example described above, but the last stage of cementation reduces porosity, increases velocity and creates a much denser rock fabric. In contrast to cementation, fabric destructive processes, as shown with the example of sucrosic dolomization (Figure 20c), can decrease velocity and increase porosity of an already altered sediment.

In siliciclastic sediments, the diagenetic potential is much lower than in carbonates, and increasing burial pressure leads to an increase in velocity at greater depth (JAPSEN, 1993). All the described diagenetic processes occur much faster than compaction and the carbonate sediments are quickly dissolved, cemented and recrystallized. These processes control and alter the velocity before compaction can play a significant role. The dominance of lithification by diagenesis over lithification by compaction is the reason why the velocities of carbonates show no clear correlation with increasing depth. Before burial pressure compacts the rock fabric, the sediment is already altered and the cemented fabric, as well as part of the porosity, survives the increasing overburden.

7. Summary and Conclusions

The measured velocities in carbonate rocks have a remarkably wide range of over 4500 m/s for V_p and over 2500 m/s for V_s. The maximum velocities (V_p 6500 m/s, V_s 3400 m/s) are four times higher than the minimum velocities (V_p 1700 m/s, V_s 800 m/s). These velocity contrasts cause, together with the density variations, large impedance differences that explain the excellent seismic reflectivity of pure carbonates observed in seismic sections.

The performed analyses document that the variability in velocity of carbonates is a product of several factors that have different relevances and effects:

(1) Changes in mineralogy are not a reason for the large variability in velocities, because all measured samples contain only carbonate minerals that have very similar physical properties. Fully dolomitized rocks can be extremely fast but also extremely slow, demonstrating the insignificance of mineralogical composition.

(2) Carbonates that are deposited in shallow water generally have a higher average velocity than carbonates from the deeper shelf, slope or basin. This relation can be explained by the higher diagenetic potential of shallow-water carbonates.

(3) The comparison of the velocity with burial depth or age shows that neither one has a major control on the velocity evolution and that velocity inversions with increasing depth or age are common. In fact, a velocity increase, caused only by pure mechanical compaction, as measured in compacted carbonate mud, is lower than the observed velocity increase with decreasing porosities in natural rocks. This difference in velocity increase shows clearly that velocity is mainly influenced by other post-depositional, diagenetic processes and not just by pure compaction at increasing burial.

(4) Increasing effective pressure can lead to a fracturing of the rock and thus to a dramatic decrease in velocity. The critical pressure, at which fractures are first formed, varies substantially and can be as low as 5 MPa.

(5) Porosity is the most important physical factor that influences velocity. V_p and V_s increase with decreasing porosity, but there are large departures from this general trend.

(6) Different velocities in rocks with equal porosities are the result of different pore types. Rocks with frame-forming pore types, such as moldic or intraparticle porosity, can have very high velocities even at high-porosity fabrics, whereas rocks with interparticle, intercrystalline or high micro-porosity have, at the same porosities, much lower velocities,

(7) The high elastic properties of self-supporting framework fabrics enable the rock to maintain its extensive porosity even at a high overburden pressure. In particular the cementation-dissolution processes, that are in this extent unique in carbonates, create during early stages very stable rock fabrics with high velocities. As a consequence, commonly used velocity-density or velocity-porosity correlations, such as Gardner equation or time-average equation, have to be modified towards higher velocities in order to produce reliable velocity data in carbonates.

In summary, it is impossible to make a velocity prediction for carbonates based solely on depth or age data. High susceptibility towards diagenetic alterations distinguishes carbonates from other sedimentary rocks. Diagenetic processes, such as dissolution, cementation or recrystallization, take place so fast and in such shallow depth of burial, that the rock fabric is completely altered before compaction becomes effective. The sediment is mainly lithified by diagenesis and to a lesser degree by increased overburden, which explains the lack of correlation between velocity and age or depth.

Most of the physical properties are a combined result of (1) the initial sediment type and (2) the diagenetic alterations. The initial lithology determines the diagenetic potential and controls, together with the succession of diagenetic processes, the post-depositional history of the carbonate sediment. The timing of the different diagenetic events controls the porosity evolution and thus the velocity development. Our analyses suggest characteristic signatures in the porosity-velocity evolution to the specific diagenetic processes.

Acknowledgments

Many of the samples are from the cores of the Bahamas Drilling Project which was sponsored by the American National Science Foundation grant OCE– 89117295 and by contributions from several industrial sponsors and the Swiss National Science Foundation. The geologic background information was produced by and with our collaborators Robert Ginsburg, Peter Swart, Jeroen Kenter, Donald McNeill and Leslie Melim. Flavio Anselmetti thanks Jaap Focke and Volker Vahrenkamp from the production research laboratory at Shell Holland (K.S.E.P.L.) for financial support. Gregor Eberli was supported by DOE grant DE–FG05–92ER14253. Collection and initial measurements of Italian samples were supported by Daniel Bernoulli through Swiss National Science Foundation grant 20–27457.91. Part of these samples were measured at the petrophysics laboratory at the University of Geneva, Switzerland; we thank Souad Sellami and Jean-Jacques Wagner for access and help. Karl Coyner from Verde GeoScience manufactured and installed the ultrasonic velocity meter in Miami. Phil Kramer provided the boat and helped us to take samples from Florida Bay for the compaction experiments. Lisa Dell'Angelo improved the English and Urs Gerber made the photographic figures. Roy Wilkens and an anonymous reviewer made useful suggestions and important comments.

REFERENCES

ANSELMETTI, F. S., EBERLI, G. P., SELLAMI, S., and BERNOULLI, D., *From outcrops to seismic profiles: An attempt to model the carbonate platform margin of the Maiella, Italy.* In *Abstract with Programs* (Geol. Society of America, Annual Meeting, San Diego 1991).

BIDDLE, K. V., SCHLAGER, W., RUDOLPH, K. W., and BUSH, T. L. (1992), *Seismic Model of a Progradational Carbonate Platform, Picco di Vallandro, the Dolomites, Northern Italy*, American Association of Petroleum Geologists Bull. *76*, 14–30.

BIOT, M. A. (1956), *Theory of Propagation of Elastic Waves in a Fluid-saturated Porous Solid, I. Low Frequency Range, II. Higher Frequency Range*, J. Acoust. Soc. Am. *28*, 168–191.

BIRCH, F. (1960), *The Velocity of Compressional Waves in Rocks to 10 Kilobars*, Part 1, J. Geophys. Res. *65*, 1083–1102.

BURNS, S. J., and SWART, P. K. (1992) *Diagenetic Processes in Holocene Carbonate Sediments: Florida Bay Mudbanks and Islands*, Sedimentology *39*, 285–304.

CAMPBELL, A. E., and STAFLEU, J. (1992), *Seismic Modelling of an Early Jurassic, Drowned Platform: The Djebel Bou Dahar, High Atlas, Morocco*, American Association of Petroleum Geologists Bull. *76*, 1760–1777.

CHRISTENSEN, N. I., and SZYMANSKI, D. L. (1991), *Seismic Properties and the Origin of Reflectivity from a Classic Paleozoic Sedimentary Sequence, Valley and Ridge Province, Southern Appalachians*, Geol. Soc. Am. Bull. *103*, 277–289.

COYNER, K. B. (1984), *Effects of Stress, Pore Pressure, and Pore-fluids on Bulk Strain, Velocity and Permeability in Rocks* (Ph.D. Thesis, Massachusetts Institute of Technology).

CRESCENTI, U., CROSTELLA, A., DONZELLI, G., and RAFFI, G. (1969), *Stratigrafia della serie calcarea dal Lias al Miocene nella regione Marchigiano-Abruzzese, Parte II—Litostratigrafia, Biostratigrafia, Paleogeografia*, Mem. Soc. Geol. It. *8*, 343–420.

DAWANS, J. M., and SWART, P. K. (1988), *Textural and Geochemical Alterations in Late Cenozoic Bahamian Dolomites*, Sedimentology 35, 385–403.

EBERLI, G. P., *Physical properties of carbonate turbidite sequences surrounding the Bahamas: Implications for slope stability and fluid movements*. In *Proceedings of the Ocean Drilling Program, Scientific Results* 101 (eds. Austin, J. A., Jr., and Schlager, W.) (1988) pp. 305–314.

EBERLI, G. P., and GINSBURG, R. N., *Cenozoic progradation of Northwestern Great Bahama Bank, a record of lateral platform growth and sea-level fluctuations*. In *Controls on Carbonate Platform and Basin Development* (SEPM Special Publication No. 44 1989) pp. 339 351.

EBERLI, G. P., *Growth and demise of isolated carbonate platforms: Bahamian controversies*. In *Controversies in Modern Geology* (Academic Press Limited 1991) pp. 231 248.

EBERLI, G. P., BERNOULLI, D., SANDERS, D., and VECSEI, A. (1993), *From aggradation to progradation: The Maiella platform (Abruzzi, Italy)*. In *Atlas of Cretaceous Carbonate Platforms* (eds. Simo, J. T., Scott, R. W., and Masse, J.-P.) Amer. Assoc. of Petroleum Geologist Memoir 56, 213 232.

ENOS, P., and SAWATSKY, L. H. (1981), *Pore Networks in Holocene Carbonate Sediments*, J. Sed. Petrol. 51, 961–985.

ENOS, P., and PERKINS, R. D. (1979), *Evolution of Florida Bay from Island Stratigraphy*, Geol. Soc. Am. Bull. 90, 59–83.

Gardner, G. H. F., GARDNER, L. W., and GREGORY, A. R. (1974), *Formation Velocity and Density: The Diagnostic Basics for Stratigraphic Traps*, Geophysics 39, 770 780.

GASSMANN, F. (1951), *Elastic Waves through a Packing of Spheres*, Geophysics 16, 673 685.

HAMILTON, E. L. (1971), *Elastic Properties of Marine Sediments*, J. Geophys. Res. 76/2, 579–604.

HAMILTON, E. L. (1980), *Geoacoustic Modeling of the Sea-floor*, J. Acoust. Soc. Am. 68, 1313–1340.

JAPSEN, P. (1993), *Influence of Lithology and Neogene Uplift on Seismic Velocities in Denmark: Implications for Depth Conversion of Maps*, American Association of Petroleum Geologists Bull. 77, 194–211.

KENTER, J. A. M., GINSBURG, R. N., EBERLI, G. P., MCNEILL, D. F., and LIDZ, B. H. (1991), *Mio-Pliocene Sea-level Fluctuations Recorded in Core Borings from the Western Margin of Great Bahama Bank*, Abstract, GSA Annual Meeting, San Diego, California.

LAUGHTON, A. S. (1957), *Sound Propagation in Compacted Ocean Sediments*, Geophysics 22, 233 260.

MARION, D., NUR, A., YIN, H., and HAN, D. (1992), *Compressional Velocity and Porosity in Sand-clay Mixtures*, Geophysics 57, 554–563.

MILHOLLAND, P., MANGHANI, M. H., SCHLANGER, S. O., and SUTTON, G. H. (1980), *Geoacoustic Modeling of Deep-sea Carbonate Sediments*, J. Acoust. Soc. Am. 68/5, 1351–1360.

NUR, A., and SIMMONS, G. (1969), *The Effect of Saturation on Velocity in Low Porosity Rocks*, Earth and Planet. Sci. Lett. 7, 183–193.

NUR, A., MARION. D., and YIN, H., *Wave velocities in sediments*. In *Shear Waves in Marine Sediments* (Kluwer Academic Publishers 1991) pp. 131–140.

RAFAVICH, F., KENDALL, C. H. St. C., and TODD, T. P. (1984), *The Relationship between Acoustic Properties and the Petrographic Character of Carbonate Rocks*, Geophysics 49, 1622–1636.

SANDERS, D. G. K. (1994), *The Cenomanian to Miocene Evolution of a Carbonate Platform to Basin Transition: Montagna della Maiella Abruzzi, Italy* (unpubl. Diss. ETH Zürich, Switzerland).

SCHLANGER, S. O., and DOUGLAS, R. G., *The pelagic ooze-chalk-limestone transition and its implications for marine stratigraphy*. In *Pelagic Sediments* (eds. Hsu, K. J., and Jenkyns, H. C.) (Special Publication Int. Assoc. of Sedimentologists 1 1974) pp. 117–148.

SELLAMI, S., BARBLAN, F., MAYERAT, A.-M., PFIFFNER, O. A., RISNES, K., and WAGNER, J.-J. (1990), *Compressional Wave Velocities of Samples from the NFP-20 East Seismic Reflection Profile*, Mém. Soc. Géol. Suisse 1, 77–84.

URMOS, J., and WILKENS, R. H. (1993), *In situ Velocities in Pelagic Carbonates: New Insights from Ocean Drilling Program Leg 130, Ontong Java Plateau*, J. Geophys. Res. 98/B5, 7903–7920.

VECSEI, A. (1991), *Aggradation und Progradation eines Karbonatplattform-Randes: Kreide bis Mittleres Tertiär der Montagna della Maiella, Abruzzen*, Mitteilungen aus dem Geologischen Institut der Eigdenössischen Technischen Hochschule und der Universität Zürich, 294.

VERNIK, L., and NUR, A. (1992), *Petrophysical Classification of Siliciclastics for Lithology and Porosity Prediction from Seismic Velocities*, American Association of Petroleum Geologists Bull. 76, 1295–1309.

VIDLOCK, S. (1983), *The Stratigraphy and Sedimentation of Cluett Key, Florida Bay*, M.S. Thesis, University of Connecticut.

WANG, Z., HIRSCHE, W. K., and SEDGWICK, G. (1991), *Seismic Velocities in Carbonate Rocks*, J. Can. Petr. Tech. *30*, 112–122.

WILKENS, R. H., FRYER, G. F., and KARSTEN, J. (1991), *Evolution of Porosity and Seismic Structure of Upper Oceanic Crust: Importance of Aspect Ratios*, J. Geophys. Res. *96*, 17981–17995.

WILSON, J. L., *Carbonate Facies in Geologic History* (Springer, New York 1975).

WOOD, A. B. (1941), *A Textbook of Sound* (Macmillan, New York 1941).

WYLLIE, M. R., GREGORY, A. R., and GARDNER, G. H. F. (1956), *Elastic Wave Velocities in Heterogeneous and Porous Media*, Geophysics *21*/1, 41–70.

(Received April 14, 1993, revised October 1993, accepted November 1993)

Acoustic Studies of the Elasticity and
Equation of State of Minerals

PAGEOPH, Vol. 141, No. 2/3/4 (1993)

0033–4553/93/040327–13$1.50 + 0.20/0

Accuracy in Measurements and the Temperature and Volume Dependence of Thermoelastic Parameters

ORSON L. ANDERSON[1] and DONALD G. ISAAK[1,2]

Abstract — To obtain the temperature T and volume V (or pressure P) dependence of the Anderson-Grüneisen parameter δ_T, measurements with high sensitivity are required. We show two examples: P, V, T measurements of NaCl done with the piston cylinder and elasticity measurements of MgO using a resonance method. In both cases, the sensitivity of the measurements leads to results that provide information about $\delta_T(\eta, T)$, where $\eta \equiv V/V_0$ and V_0 is the volume at zero pressure. We demonstrate that determination of δ_T leads to understanding of the volume and temperature dependence of $q = (\partial \ln \gamma / \partial \ln V)_T$ over a broad V, T range, where γ is the Grüneisen ratio.

Key words: Elasticity, thermoelasticity, piston-cylinder, rectangular parallepiped resonance, MgO, NaCl, Anderson-Grüneisen parameter, equation of state.

1. Introduction

Information pertaining to the volume V and temperature T dependence of physical quantities such as the Anderson-Grüneisen parameters δ_T and q given by

$$\delta_T = -\frac{1}{\alpha K_T}\left(\frac{\partial K_T}{\partial T}\right)_P = -\left(\frac{\partial \ln K_T}{\partial \ln \eta}\right)_P = \left(\frac{\partial \ln \alpha}{\partial \ln \eta}\right)_T \tag{1}$$

and

$$q = \left(\frac{\partial \ln \gamma}{\partial \ln \eta}\right)_P \tag{2}$$

where

$$\gamma = \frac{\alpha K_T}{\rho C_V} \tag{3}$$

[1] Center for Chemistry and Physics of Earth and Planets, Institute of Geophysics and Planetary Physics, UCLA, Los Angeles, CA 90024-1567, U.S.A.
[2] Also at Azusa Pacific University.

requires accurate measurements of thermoelastic and material equation of state properties over a range of pressures and temperatures. In (1)–(3) the appropriate quantities are: the volume coefficient of thermal expansivity α; the isothermal bulk modulus K_T; the constant volume heat capacity C_V; the density ρ; the ratio of the volume V to the zero pressure volume V_0 given by η, i.e., $\eta = V/V_0$; and the pressure P. This paper is dedicated to the principle of striving for the highest possible accuracy in measurements in order to obtain a database from which information on higher order physical properties such as δ_T and q, and how they vary throughout P, V, T (or P, η, T) space, can be obtained. We provide an example based on work done by Reinhard Boehler in the laboratory of George Kennedy, using the piston-cylinder method to measure the P, V, T values of NaCl (BOEHLER and KENNEDY, 1980). In a later section, we give an example using the rectangular parallelepiped resonance (RPR) method of measuring elastic constants. In both cases we illustrate how information pertaining to higher order physical parameters is obtained.

2. NaCl and Piston Cylinder Measurements

We wish to demonstrate that precise static measurements for P, V, T can be used to provide information on the thermal pressure ΔP_{TH} and the Anderson-Grüneisen parameter δ_T. The results of BOEHLER and KENNEDY (B&K) (1980) are used to represent the P, η, T values of NaCl at regular increments in P and T (see Table 1). B&K (1980) used a piston cylinder apparatus to obtain P, η curves at six different temperatures ranging from 298–773 K. The pressure range of the B&K

Table 1

V/V₀ for NaCl as a function of P and T (from Table 3 of BOEHLER and KENNEDY (1980)). A small extrapolation beyond the highest measured pressure of 3.2 GPa is required to represent the 3.5 GPa values

P GPa	$\eta = V/V_0$					
	298 K	373 K	473 K	573 K	673 K	773 K
0	1.0000	1.0093	1.0225	1.0368	1.0523	1.0691
0.5	0.9802	0.9883	0.9999	1.0125	1.0258	1.0401
1.0	0.9627	0.9700	0.9802	0.9912	1.0229	1.0154
1.5	0.9469	0.9536	0.9628	0.9725	0.9829	0.9938
2.0	0.9325	0.9386	0.9470	0.9557	0.9651	0.9748
2.5	0.9191	0.9247	0.9325	0.9406	0.9491	0.9578
3.0	0.9067	0.9117	0.9191	0.9268	0.9347	0.9424
3.5	0.8949	0.8994	0.9065	0.9142	0.9215	0.9285

(1980) experiments was 0–3.2 GPa. Since

$$K_T \equiv -\eta \left(\frac{\partial P}{\partial \eta}\right)_T, \tag{4}$$

differentiation of the P, η curves provides $K_T(\eta, T)$ (or, alternatively, $K_T(P, T)$). Isotherms of K_T are used by B&K (1980) to represent $K_T(P)$ according to the modified Murnaghan equation

$$K_T = K_{T_0} + K'_{T_0}P + \frac{1}{2}K''_{T_0}P^2, \tag{5}$$

where K_{T_0} is the zero pressure isothermal bulk modulus (temperature arbitrary) and primes indicate pressure derivatives. A least squares fit of the K_T, P results then yields the following values for parameters in (5) at 298 K (B&K, 1980): $K_{T_0} = 23.60$ GPa; $K'_{T_0} = 5.85$; and $K''_{T_0} = -0.008$ GPa^{-1}. At 773 K, the respective values are 15.89 GPa, 4.66, and 0.000 GPa^{-1}. The P, η, T values shown in Table 1 result from integrating (5) using the definition of K_T provided in (4).

It should be noted that the values for K_{T_0}, K'_{T_0}, and K''_{T_0}, at both low and high temperature, are in good agreement with the dynamic measurements (ultrasonic interferometry) of SPETZLER et al. (1972), who obtained their data over the ranges in P and T of 0–0.8 GPa and 300–800 K, respectively. An exception is that SPETZLER et al. (1972) find $K'_{T_0} = 5.35$ at ambient temperature, compared to 5.85 from B&K (1980). SPETZLER et al. (1972) also find that K''_{T_0} at $T = 800$ K has nearly the same value as at $T = 300$ K, whereas B&K (1980) find K''_{T_0} at 773 K negligible. These differences are small but measurable. We cannot say with certainty why some values differ when comparing the static and dynamic equation of state results, but note: (1) the difference in pressure ranges used by the two experiments, 3.2 GPa (B&K, 1980) and 0.8 GPa (SPETZLER et al., 1972), and (2) the numerical results are likely to vary somewhat due to differences in data reduction techniques and the use of different equations of state. The difference in data reduction is necessary due to the different experimental techniques that were used. In any case, the differences between these two data sets produce very little effect on our derived results that follow.

The pressure $P(\eta, T)$ can be expressed by

$$P(\eta, T) = P_{298}(\eta) + \Delta P_{TH}(T), \tag{6}$$

where

$$\Delta P_{TH}(T) = P_{TH}(T) - P_{TH}(298), \tag{7}$$

and P_{TH} is referred to as the thermal pressure. In (6), $P_{298}(\eta)$ represents the pressure due to volume compression at some constant reference temperature, which is assumed to be 298 K. The change in thermal pressure $\Delta P_{TH}(T)$ represents the pressure required to maintain the compression at η as temperature increases from

298 K to the new temperature T. Using the data in Table 1, we can compute the value of $\Delta P_{TH}(T)$ for a given η along different isotherms as seen in Table 2.

From Table 2 we note that for NaCl, ΔP_{TH} is substantially independent of volume; that is, along isotherms

$$\left(\frac{\partial P_{TH}}{\partial \eta}\right)_T \cong 0. \tag{8}$$

It is also apparent from Table 2 that along isochores ΔP_{TH} is proportional to T; that is,

$$\left(\frac{\partial P_{TH}}{\partial T}\right)_\eta \cong \text{constant.} \tag{9}$$

A number of results follow immediately from (8) and (9) by using thermodynamic identities. First we note that

$$\alpha K_T \equiv \left(\frac{\partial P}{\partial T}\right)_\eta, \tag{10}$$

so from (8) and (10) we see that for NaCl

$$\left[\frac{\partial(\alpha K_T)}{\partial \eta}\right]_T \cong 0, \tag{11}$$

Table 2

Change of thermal pressure ΔP_{TH} from room temperature (298 K) to elevated temperature at various compressions for NaCl along isotherms

	$\Delta P_{TH} = P_{TH}(T) - P_{TH}(298 \text{ K})$ (GPa)					
η	298 K	373 K	473 K	573 K	673 K	773 K
1.00	0.0	0.216	0.501	0.785	1.067	1.349
0.99	0.0	0.220	0.500	0.784	1.067	1.348
0.98	0.0	0.223	0.510	0.783	1.068	1.348
0.97	0.0	0.215	0.499	0.783	1.069	1.348
0.96	0.0	0.214	0.499	0.784	1.071	1.348
0.95	0.0	0.214	0.499	0.786	1.071	1.349
0.94	0.0	0.214	0.502	0.789	1.078	1.351
0.93	0.0	0.213	0.502	0.792	1.084	–
0.92	0.0	0.213	0.500	–	–	–
0.91	0.0	0.213	–	–	–	–

Note that ΔP_{TH} is substantially independent of volume. Data measured by BOEHLER and KENNEDY (1980) as reported by ANDERSON *et al.* (1982).

over a wide range of P and T. However, from the identity (ANDERSON et al., 1992b; ANDERSON and YAMAMOTO, 1987)

$$\left[\frac{\partial(aK_T)}{\partial\eta}\right]_T \equiv -\frac{1}{\eta}\left(\frac{\partial K_T}{\partial T}\right)_\eta, \tag{12}$$

it follows that

$$\left(\frac{\partial K_T}{\partial T}\right)_\eta \cong 0. \tag{13}$$

We also have the identity (ANDERSON et al., 1993),

$$\left(\frac{\partial K_T}{\partial T}\right)_\eta \equiv \left(\frac{\partial K_T}{\partial T}\right)_P + \alpha K_T \left(\frac{\partial K_T}{\partial P}\right)_T, \tag{14}$$

which implies that

$$-\left(\frac{\partial K_T}{\partial T}\right)_P \cong \alpha K_T \left(\frac{\partial K_T}{\partial P}\right)_T \tag{15}$$

when (11) (or (8)) is true. From the definition of δ_T given by (1), with (15), we have

$$\delta_T \cong \left(\frac{\partial K_T}{\partial P}\right)_T \equiv K'_T. \tag{16}$$

As shown by BIRCH (1986), $\delta_T \approx K'_T$ over a wide range of pressure for NaCl (see Table 3). The BIRCH (1986) values in Table 3 are average values for the T range 298–773 K. The values shown in Table 3 use the B&K (1980) and SPETZLER et al.

Table 3

NaCl values of K'_T, δ_T, γ found by BIRCH (1986) using the data of BOEHLER and KENNEDY (1980), SPETZLER et al. (1972), and FRITZ et al. (1971). Mean values for the T range 298–773 K are shown

P GPa	K'_T	δ_T	γ	$\eta = V/V_0$
0	5.5	5.3	1.62	1.000
1	5.0	4.9	1.55	0.963
2	4.8	4.7	1.51	0.932
3	4.6	4.5	1.46	0.907
4	4.4	4.3	1.43	0.885
5	4.2	4.2	1.40	0.865
10	3.8	3.8	1.27	0.791
15	3.4	3.5	1.19	0.740
20	3.2	3.1	1.12	0.700
25	3.0	2.9	1.07	0.669
30	2.8	2.7	1.03	0.642

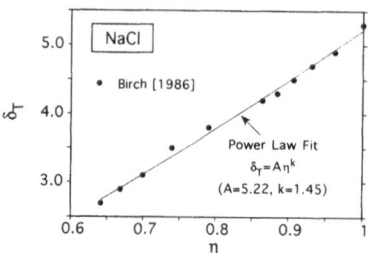

Figure 1

δ_T versus $\eta = V/V_0$ for NaCl. The data points (shown as circles) are from BIRCH (1986). Linear or a power law fits ($\delta_T = 5.22\eta^{1.45}$) (solid line) represent the data equally well.

(1972) data, together with the shock data of FRITZ et al. (1971), to fit to a higher order equation of state (EoS) than is represented by (5). (See the EoS labelled BE$_2$ in BIRCH (1978, 1986).)

From Table 3, we find that both δ_T and K'_T for NaCl decrease with decreasing η. Figure 1 illustrates the η dependence of δ_T at ambient T. The data in Table 3 and Figure 1 can be represented equally well by either a linear relationship between δ_T and η

$$\delta_T = -1.696 + 6.88\eta \tag{17}$$

or by an exponential equation

$$\delta_T = 5.22\eta^{1.45}. \tag{18}$$

For a determination of the dependence of δ_T upon η using the linear relationship for several other solids, the reader is referred to CHOPELAS and BOEHLER (1992).

3. Determination of Thermoelastic Constants in MgO

The fact that δ_T is close in value to K'_T for NaCl follows from the empirical result represented by (8) as described above. But for other materials it is not necessarily true that δ_T is as close to K'_T as found for NaCl. ANDERSON and ISAAK (1993) determined $\delta_T(\eta)$ and $K'_T(\eta)$ of MgO for various isotherms, using results from an *ab initio* calculation (potential induced breathing electron gas model) (ISAAK et al., 1990).

Consider the identity given by (12). Expanding the left side of (12) we find the following identity (ANDERSON and ISAAK, 1993)

$$-\left(\frac{\partial K_T}{\partial T}\right)_\eta \equiv \alpha K_T(\delta_T - K'_T). \tag{19}$$

The negative sign on the left side of our equation (19) is incorrectly omitted in equation (7) in ANDERSON and ISAAK (1993). Equation (19) again shows that $\delta_T = K'$ requires $(\partial K_T / \partial T)_\eta = 0$. We can find $(\partial K_T / \partial T)_\eta$ from measurements of K_T vs. T at $P = 0$ (YAMAMOTO and ANDERSON, 1987) provided that K'_T is known, as is seen in (14) and (19). Both terms on the right side of (14) can be evaluated from experiments. It turns out that few minerals have the property $(\partial K_T / \partial T)_\eta = 0$ found in NaCl. For example, MgO (ANDERSON and ZOU, 1989), KCl (YAMAMOTO and ANDERSON, 1987), olivine (ISAAK, 1992) and corundum (GOTO et al., 1989) do not have this property. ANDERSON and ISAAK (1993) show how one could find the trace in P, η space where $\delta_T = K'$. For NaCl $\delta_T \cong K'$ at all T at $\eta = 1$. But for MgO, results of calculations (ISAAK et al., 1990) suggest that $\delta_T = K'_T$ only along the trace of the curve shown in Figure 2. Using the data behind the Figure 2 and evaluating $K'(\eta)$ along isotherms from an equation of state, we can evaluate $\delta_T(\eta, T)$ (ANDERSON and ISAAK, 1993) from (14). The results for MgO are shown in Figure 3.

In order to calculate the data represented in Figures 2 and 3, we need the values of δ_T at $P = 0$ and high temperatures from experiments (ANDERSON and ISAAK,

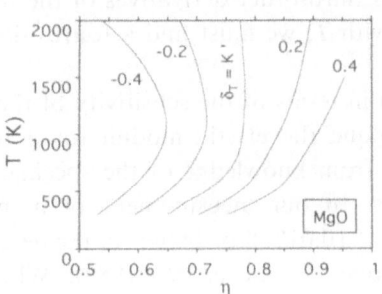

Figure 2

The trace (η, T) of constant values of x, where $x \equiv \delta_T - K'_T$. Along the curve where $\delta_T = K'_T$ (or $x = 0$), it also holds that $(\partial K_T / \partial T)_\eta = 0$.

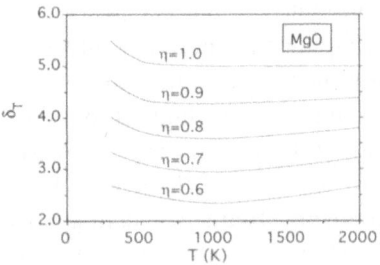

Figure 3
Isochores of δ_T versus T for MgO.

1993). The extension to elevated pressure is then done using model calculations. It is important to know whether $\delta_T(T)$ is independent of T since $\delta_T(\eta, T) - K'(\eta, T)$ controls $(\partial K_T/\partial T)_\eta$ along an isochore. The solution for $(\partial K_T/\partial T)_\eta$ is not reliable unless $\delta_T(T)$ at $P = 0$ is accurately known.

4. The RPR Method for Determining $\delta_T(T)$ at High T

To define thermoelastic properties as a function of temperature, we need an overall accuracy to within a few percent. For example, if we wish to know how the Anderson-Grüneisen constant, δ_T, varies with temperature at high T, we must determine whether δ_T is increasing or decreasing with T. In terms of the Helmholtz free energy, F, we have

$$\delta_T = -V \frac{\left(\dfrac{\partial^3 F}{\partial^2 V \partial T}\right)}{\left(\dfrac{\partial^2 F}{\partial V \partial T}\right)}. \tag{20}$$

It is clear that δ_T involves third-order derivatives of the free energy, and if we wish to find how δ_T changes with T, we must find a fourth derivative of the Helmholtz free energy.

We view this problem in terms of the sensitivity of the measurements. With the RPR experimental technique the elastic moduli for a rectangular parallelepiped specimen are determined from knowledge of the specimen's edge lengths, density, and resonant frequencies. In our measurement of a modal frequency, we can achieve a sensitivity of 10^{-4} (0.01%), or better, in the determination and reproducibility of a particular mode (ISAAK et al., 1989). While the sensitivity of our measurements is very high, the accuracies with which the elastic moduli are determined are considerably less (typically 0.1–1.0%). Since frequencies are combined with information on specimen density and edge lengths when reducing the data, the precision with which the moduli are known is limited by the precision of the density and length measurements ($\sim 0.1\%$). Further limitations on the accuracy of determining the elastic moduli are due to uncertainties in crystal orientation ($\sim 1/2°$), imperfections in shaping the specimen into a rectangular parallelepiped, and the presence of inhomogeneities (chemical and structural) within a crystalline specimen. Nevertheless, we aim for the highest possible sensitivity in frequency measurements since *changes* in frequency with increasing temperature are what provide the essential data to determine $(\partial K_T/\partial T)_P$. With a sensitivity of 10^{-4} in absolute frequency measurements, frequency changes over 100 K are accurately measured to 10^{-2} (1%), or better. The other experimental factors mentioned above (density, length, specimen geometry, and inhomogeneities) that reduce the accuracy of the measured moduli have much less effect on reducing the accuracy of the

Table 4

Variation of δ_T with T at P = 0 for six minerals (from ANDERSON et al. 1992a). Note that for T above the Debye temperature (θ_D), δ_T tends to be constant

T	MgO	Al_2O_3	Mg_2SiO_4	$MgAl_2O_4$*	Olivine	Gross. Garnet	Pyrope
300	5.26	5.71	5.94	7.73	6.59	6.30	6.27
400	4.83	5.16	5.58	7.36	5.95	5.36	5.70
500	4.69	5.03	5.49	7.07	5.65	4.98	5.46
600	4.67	5.08	5.49	6.82	5.51	4.80	5.35
700	4.70	5.17	5.47	6.62	5.44	4.70	5.30
800	4.74	5.29	5.46	6.47	5.40	4.64	5.28
900	4.78	5.37	5.47	6.35	5.36	4.60	5.30
1000	4.84	5.42	5.46	6.24	5.36	4.58	5.32
1100	4.92	5.42	5.49	–	5.37	4.57	–
1200	4.99	5.39	5.44	–	5.38	4.57	–
1300	5.18	5.32	5.37	–	5.35	4.58	–
1400	5.12	5.22	5.38	–	5.41	–	–
1500	5.13	5.08	5.40	–	5.43	–	–
1600	5.07	4.92	5.42	–	–	–	–
1700	4.95	4.73	–	–	–	–	–

$\theta_D(K)$: MgO, 945; Al_2O_3, 1034; Mg_2SiO_4, 763; $MgAl_2O_4$, 863; Olivine, 731; Gross. Garnet, 824; Pyrope, 785.
*CYNN, H. (1992).

temperature variation of the moduli, since these factors remain essentially constant with changes in temperature.

The point is that we could not reasonably hope to achieve an accurate experimental determination of how δ_T varies with T without a beginning point in our experiment representing 10^{-4}, or better, sensitivity in measurement. For our best determinations of δ_T versus T at P = 0 for a group of important minerals see Table 4.

5. Determining How $(\delta_T - K')$ and q Vary with η and T

A determination of $\delta_T(T)$ is useful for many applications. We present one example here. We intend to show that $\delta_T(T)$ affects the behavior of q over a wide temperature and pressure range, where q is defined as

$$q \equiv \left(\frac{\partial \ln \gamma}{\partial \ln \eta} \right)_T, \tag{21}$$

and γ is the Grüneisen constant defined in (3). The parameter q is an essential quantity in the reduction of shock data. Equation (21) can be written in the form

$$\gamma = \gamma_0 \eta^q. \tag{22}$$

From the definition of γ given by (3) and by taking appropriate derivatives as indicated in (21) and (3), q can be expressed as (ANDERSON et al., 1993)

$$q \equiv \delta_T - K' + 1 - \left(\frac{\partial \ln C_V}{\partial \ln \eta}\right)_T. \tag{23}$$

Consider first the properties in the high temperature domain where $(\partial \ln C_V / \partial \ln \eta)_T$ is approximately zero, so that (23) becomes

$$q \cong \delta_T - K' + 1. \tag{24}$$

From (19) we then obtain

$$q \cong \frac{-1}{\alpha K_T}\left(\frac{\partial K_T}{\partial T}\right)_\eta + 1. \tag{25}$$

Here we see that when $(\partial K_T / \partial T)_\eta$ is zero, then $q = 1$. From the data on NaCl (Table 3), $\delta_T \cong K'_T$, and therefore $(\partial K_T / \partial T)_V$ must be zero. For the case of NaCl we expect $q = 1$ to be a good solution for large ranges of pressure and temperature. But for MgO, as shown in Figure 4, and possibly other minerals, $\delta_T - K'$ decreases with increasing compression.

From Figure 4 we see that there is a range of high compression and high T where $\delta_T - K'$ approaches zero, but at low compression and low T we find $\delta_T - K'$ is definitely greater than zero. It is a simple matter to determine from Figure 4 the behavior of q versus η and T using (25). This is shown in Figure 5.

In general, the higher the temperature, the closer q approaches unity at high compression. This is especially true for temperatures in excess of 2000 K. Therefore we expect q to be close to unity for shock compression conditions. As seen from (22), if $q = 1$, then $\rho\gamma$ is constant. However, at low temperatures, in order to

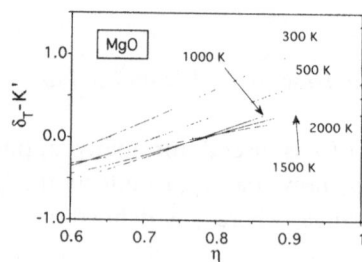

Figure 4
Isotherms of $\delta_T - K'$ versus η determined from Figures 2 and 3 (ANDERSON and ISAAK, 1993).

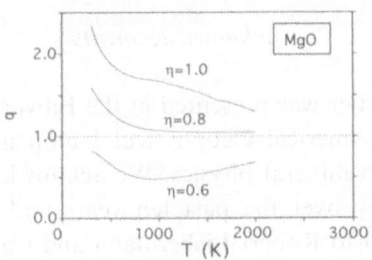

Figure 5
The variation of q versus T along isochores for MgO (after ANDERSON *et al.*, 1993). An important difference between Figures 4 and 5 is that in Figure 5, the effect of the $(\partial \ln C_V/\partial \ln \eta)_T$ in (24) becomes significant for isotherms where $T < \theta$ ($\theta = 950$ K). It is clear that q converges to unity as $T \to 3000$ K and higher.

determine q one must have information pertaining to $(\partial C_V/\partial \eta)_T$ (ANDERSON *et al.*, 1992b, 1993) and consequently for some arbitrary η, q is not necessarily close to unity at low T. A higher value of q, i.e., $q > 1$ is determined from static high pressure measurements at room temperature.

6. Conclusions

We discuss the importance of accurate measurements that are required to determine the P (or η) and T dependence of essential geophysical parameters such as δ_T and q. In elasticity measurements using the resonance technique, modal frequencies are measured with a sensitivity of 10^{-4} or better. This level of sensitivity typically leads to T derivatives of elastic moduli that are accurate to a percent or so. We also discuss important thermodynamic relationships, which, when coupled with measured quantities, elucidate the pressure (or volume) and temperature dependence of δ_T and q. Two examples, NaCl (static piston-cylinder data) and MgO (dynamic resonance data), are described. For NaCl, the Anderson-Grüneisen parameter δ_T is very close in value to K'_T throughout a wide range of P and T. This result follows from the empirical observation that for NaCl $(\partial P_{TH}/\partial \eta)_T \cong 0$. Although both δ_T and K'_T decrease with increasing compression, the fact that $\delta_T - K'_T \approx 0$ implies that $(\partial K_T/\partial T)_\eta \approx 0$ and that $q \approx 1$ throughout P, T space. In the case of MgO, $\delta_T - K'_T$ apparently varies with increasing compression. Thus the approximation $q \approx 1$ is not generally valid. However, at high T ($T > 2000$ K), $\delta_T - K'_T$ tends to approach 0 (implying $q \approx 1$) for a wide range of compressions.

These examples illustrate that when the fundamental data are obtained with great care, some understanding of the higher order thermoelastic parameters important to descriptions of lower mantle conditions is gained.

Acknowledgements

The essence of this paper was presented at the Edward Schreiber Symposium at the Fall meeting of the Americal Geophysical Union held in December 1992, to honor his contribution to mineral physics. We acknowledge many colleagues who worked in our laboratory over the past ten years and who helped us build our database. We are grateful to Robert Liebermann and Carl Sondergeld for organizing the Schreiber Symposium. This paper is dedicated to the memory of Ed Schreiber, who was a close colleague of one of us (OLA) for ten years. We acknowledge financial support from NSF Grant EAR 91-17280 and support from the Office of Naval Research through Grant No. 014-92-5-1354. IGPP contribution no. 3849.

REFERENCES

ANDERSON, O. L., BOEHLER, R., and SUMINO, Y., *Anharmonicity in the equation of state at high temperature for some geophysically important solids.* In *High-pressure Research in Geophysics* (eds. Akimoto, S., and Manghnani, M. H.) (Center for Academic Publications, Tokyo, Japan 1982) pp. 93–113.

ANDERSON, O. L., and ISAAK, D. G. (1983), *The Dependence of the Anderson-Grüneisen Parameter δ_T Upon Compression at Extreme Conditions,* J. Phys. Chem. Solids *54*, 221–227.

ANDERSON, O. L., ISAAK, D., and ODA, H. (1992a), *High Temperature Elastic Constant Data on Minerals Relevant to Geophysics,* Rev. Geophys. *30*, 57–90.

ANDERSON, O. L., ODA, H., and ISAAK, D. (1992b), *A Model for Computation of Thermal Expansivity at High Compression and High Temperatures: MgO as an Example,* Phys. Chem. Minerals *19*, 1987–1990.

ANDERSON, O. L., ODA, H., CHOPELAS, A., and ISAAK, D. (1993), *A Thermodynamic Theory of the Grüneisen Ratio at Extreme Conditions: MgO as an Example,* Phys. Chem. Minerals *19*, 369–380.

ANDERSON, O. L., and YAMAMOTO, S., *The interrelationship of thermodynamic properties obtained by the piston-cylinder high pressure experiments and RPR high temperature experiments for NaCl.* In *High Pressure Research in Mineral Physics,* Geophys. Monogr. Ser., Vol. 39 (eds. Manghnani, M. H., and Syono, Y.) (AGU, Washington, D.C. 1987) pp. 289–298.

ANDERSON, O. L., and ZOU, K. (1989), *Formulation of the Thermodynamic Functions for Mantle Minerals: MgO as an Example,* Phys. Chem. Minerals *16*, 642–648.

BIRCH, F. (1978), *Finite Strain Isotherm and Velocities for Single-crystal and Polycrystalline NaCl at High Pressures and 300°K,* J. Geophys. Res. *83*, 1257–1268.

BIRCH, F. (1986), *Equation of State and Thermodynamic Parameters of NaCl to 300 kbar in the High-temperature Domain,* J. Geophys. Res. *91*, 4949–4954.

BOEHLER, R., and KENNEDY, G. C. (1980), *Equation of State of Sodium Chloride up to 32 Kbar and 400°C,* J. Phys. Chem. Solids *41*, 517–523.

CHOPELAS, A., and BOEHLER, R. (1992), *Thermal Expansivity of the Lower Mantle,* Geophys. Res. Lett. *19*, 1983–1986.

CYNN, H., *Effects of Cation Disordering in $MgAl_2O_4$ Spinel on the Rectangular Parallelepiped Resonance and Raman Measurements of Vibrational Spectra,* Ph.D. thesis, University of California at Los Angeles, 1992.

FRITZ, J. N., MARSH, S. P., CARTER, W. J., and McQUEEN, R. G., *The Hugoniot equation of state of sodium chloride in the sodium chloride structure.* In *Accurate Characterization of the High Pressure Environment,* N.B.S. Spec. Publ. *326* (ed. Lloyd, E. C.) (National Bureau of Standards, Washington, DC. 1971) pp. 201–208.

GOTO, T., YAMAMOTO, S., OHNO, I., and ANDERSON, O. L. (1989), *Elastic Constants of Corundum up to 1825 K*, J. Geophys. Res. *94*, 7588–7602.

ISAAK, D. G. (1992), *High Temperature Elasticity of Iron-bearing Olivines*, J. Geophys. Res. *97*, 1871–1885.

ISAAK, D. G., ANDERSON, O. L., and GOTO, T. (1989), *Measured Elastic Moduli of Single-crystal MgO up to 1800 K*, Phys. Chem. Minerals *16*, 704–713.

ISAAK, D. G., COHEN, R. E., and MEHL, M. J. (1990), *Calculated Elastic and Thermal Properties of MgO at High Pressures and Temperatures*, J. Geophys. Res. *95*, 7055–7067.

SPETZLER, H., SAMMIS, C. G., and O'CONNELL, R. J. (1972), *Equation of State of NaCl: Ultrasonic Measurements to 8 Kbar and 800°C and Static Lattice Theory*, J. Phys. Chem. Solids *33*, 1727–1750.

YAMAMOTO, S., and ANDERSON, O. L. (1987), *Elasticity and Anharmonicity of Potassium Chloride at High Temperature*, Phys. Chem. Minerals *14*, 332–340.

(Received April 13, 1993, revised August 17, 1993, accepted October 6, 1993)

PAGEOPH, Vol. 141, No. 2/3/4 (1993)

0033-4553/93/040341-37$1.50 + 0.20/0

A New Ultrasonic Interferometer for the Determination of Equation of State Parameters of Sub-millimeter Single Crystals

HARTMUT A. SPETZLER,[1] GANGLIN CHEN,[1] SCOTT WHITEHEAD,[1]
and IVAN C. GETTING[1]

Abstract — A new giga-Hertz ultrasonic interferometer has been developed, based on ultrasonic microscopy technology. The interferometer operates from 0.3 GHz to 1.5 GHz. The high frequency and associated small wavelengths together with the large bandwidth make it possible to measure travel times in samples with thicknesses of several microns and allow for unprecedented accuracy in bond corrections. An absolute accuracy of 1 part in 10^5 in travel time measurements is achievable in single crystals (thickness of ~200 microns) or glasses of interest to the earth sciences. The high precision travel time data, combining with sample length measurements using a laser interferometer built in our laboratory, yield very high precision ultrasonic velocities.

The interferometer is intended for use in conjunction with a newly developed 4 GPa gas piston cylinder apparatus (GETTING and SPETZLER, 1993) for equation of state measurements under simultaneous pressure and temperature. A separate correction for the bond will be made for each datum at every point in temperature pressure space.

Key words: GHz ultrasonic interferometry, diffraction, acoustic wave, bond effect, equation of state, elastic constant, acoustic velocity.

Introduction

Ultrasonic interferometry for the determination of equation of state parameters has been introduced to the geophysical community by O. L. Anderson, E. Schreiber, N. Soga (ANDERSON and SCHREIBER, 1965; SCHREIBER and ANDERSON, 1966; SOGA et al., 1966) in the mid 1960s. Today ultrasonics (JACKSON and NIESLER, 1982; ANDERSON et al., 1991) along with optical (WEIDNER et al., 1975; BASS, 1989; BASSETT et al., 1982; CHOPELAS and NICOL, 1982; BROWN et al., 1989) and X-ray techniques (AKAOGI et al., 1987; OLIJNYK et al., 1991; WANG et al., 1991; SMYTH and SWOPE, 1992; MAO et al., 1992) are important in deciphering the composition and phase of earth's interior (ANDERSON and BASS, 1984; WEIDNER and ITO, 1987; DUFFY and ANDERSON, 1989; ITA and STIXRUDE, 1992).

---

[1] CIRES and Department of Geological Sciences, University of Colorado, Boulder, CO 80309, U.S.A.

In this paper we describe the start of a new generation of ultrasonic interferome-ters which employ technology developed in ultrasonic microscopy (BRIGGS, 1992). The major difference between the interferometer described here and others is the high frequency (from 0.3 GHz up to 1.5 GHz) and consequently small elastic wavelengths (1 to 10 μm) at which it operates. The small wavelengths allow the use of small samples (thicknesses as small as 100 μm have been used) and thus extend the available sample pool of synthetic high-pressure phases and of many small naturally occurring minerals. With this interferometer, we have measured the acoustic wave travel time through a material of several microns thick to a precision of 0.3%.

Here we describe the physical interferometer and the data reduction which is necessary to convert the measured interferences into travel times. In particular, we describe the means of correcting for undesirable amplitude modulation of the measurement system and the phase effect introduced by bonding the sample to the buffer rod. A laser interferometer has been built to measure the sample length. The high precision travel time data, combining with the sample length measurements, yield the ultrasonic velocities and allow us to study the equations of state of candidate mantle minerals (CHEN *et al.*, 1994). In a companion paper (SPETZLER and YONEDA, this issue) we present a complete travel time equation of state (CT-EOS) and show how we can obtain the complete set of the equations of state parameters from the ultrasonic travel time data. Combining the CT-EOS with independent measurements of the lattice spacings by X-ray methods offers the potential of a new improved pressure scale (YONEDA *et al.*, 1993).

The Interferometer Instrumentation

1. Description of the Interferometer

The microwave acoustic interferometer used in this work was patterned after the ultrasonic instrumentation described by SPETZLER (1970). This modern version of the acoustic interferometer is capable of measurements from 300 MHz up to 1.5 GHz. The high frequency enables travel-time measurements with high accuracy on very small high Q samples because of the short wavelengths in relation to the sample size.

Figure 1 shows a block diagram of the interferometer. A microwave radio frequency (RF) synthesizer is used as the signal source. This signal generator is capable of a 16 dBm output and is frequency stable to better than 10 parts per million with an aging rate of less than $\pm 5 \times 10^{-7}$/month. The continuous output from the synthesizer is gated by a pulse generator via an external modulation port on the synthesizer. The pulse generator is capable of sending out two gating pulses with variable pulse widths and spacings accurate to one picosecond. The resulting signals from the synthesizer are phase coherent sinusoidal tone bursts. The pulse generator, the synthesizer, and a sampling oscilloscope share the 10 MHz clock

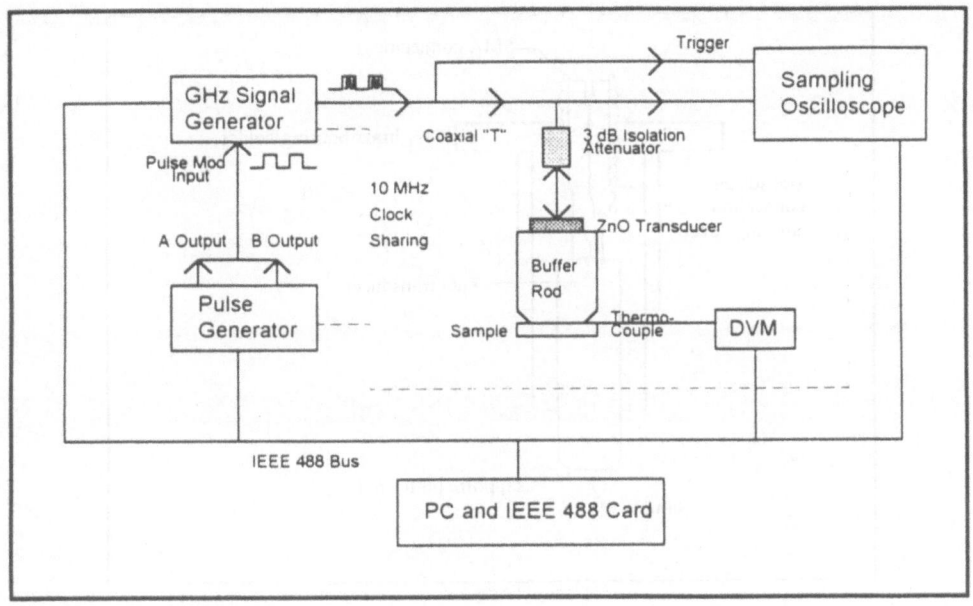

Figure 1

Schematic of the GHz ultrasonic interferometer. Details are explained in the text.

frequency of the synthesizer. This interfacing between the pulse generator and the synthesizer allows the oscilloscope to trigger on the same phase of the two tone bursts sent out of the synthesizer.

As a sinusoidal tone burst leaves the signal generator, it is split at a coaxial "T". A portion of the signal energy travels to the transducer/buffer-rod/sample assembly. This assembly is shown in Figure 2 and will be described later. The remaining portion of the split sinusoidal wave goes to the sampling oscilloscope trigger input. As the tone burst reaches the transducer, a stress wave is sent into the buffer rod. Reflections of the stress wave from the buffer rod-sample interface and occur from the free end of the sample. To simplify the following discussion, the reflection from the buffer-rod and sample interface will be called the *buffer rod echo* (BR); the first reflection of the transmitted wave from the free end of the sample will be called the *first sample echo* (S1); and the second reflection of the transmitted wave from the free end of the sample will be called the *second sample echo* (S2); etc. The various echoes travel back to the transducer, where they are converted to electrical signals and are recorded by the oscilloscope. A typical echo train is shown in Figure 3a. Figure 3b shows the expanded S1 echo. By simply measuring the time between the beginning of the BR echo and the S1 echo, a rough two-way travel time through the sample can be estimated.

2. Interferometry Practice

The instrument is put into an interferometric mode by sending out two sinusoidal tone bursts from the signal generator. The second tone burst is time-de-

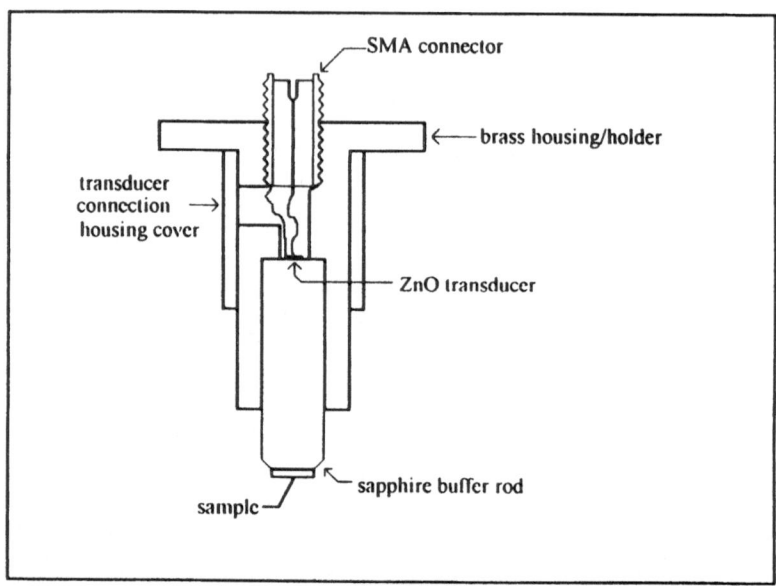

Figure 2
Schematic of the cross section of transducer/buffer-rod/sample assembly (Not to scale). The termination
of the coaxial cable transducer is detuned to achieve the broad bandwidth.

layed from the first tone burst such that the BR echo from the second tone burst
overlaps with the S1 echo of the first tone burst (see Figure 4). Because the
sinusoidal tone bursts originate from one continuous RF signal source, the two
tone bursts are phase coherent. At this interferometric mode of the system, the S1
echo of the second tone burst also overlaps the S2 echo of the first tone burst. In
this section, we will illustrate the principles of our interferometer. We assume that
the sample is perfectly coupled to the buffer rod with no bonding agent. The term
"perfectly coupled" means both stress and displacement are continuous across the
contact. The situations where nonperfect contact occurs and where bonding agents
are needed will be discussed in later sections. In the perfect coupling situation,
constructive or destructive interference between the echoes of the first tone burst
and the echoes of the second tone burst occurs when the RF of the sinusoid is
adjusted such that the number of wavelengths in twice the sample length corresponds to an integer or half integer number. The condition of integer or half integer
numbers needed for constructive or destructive interference is determined by the
relative acoustic impedances of the buffer rod and the sample. We also assume that
the Q of the sample is high so that the attenuation is negligible (the velocity of the
sample v is a real value).

The acoustic impedance of a material Z is the product of the density of the
material ρ and the sound velocity within the material v: $Z = \rho v$. To generalize the
condition of phase shift due to reflections, consider an interface between medium 1

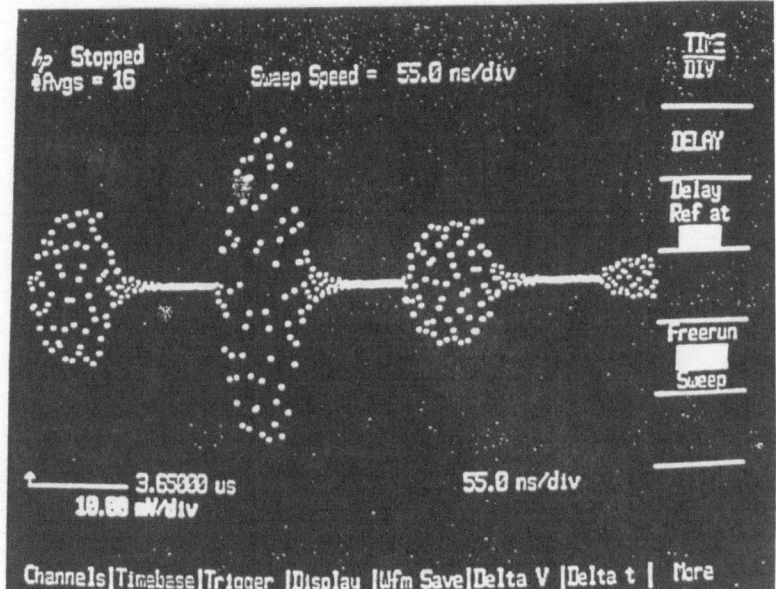

(a)

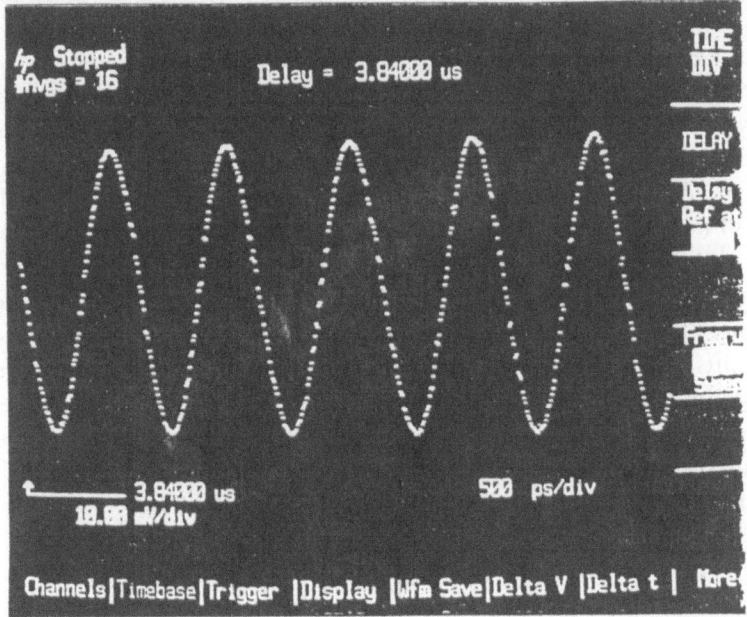

(b)

Figure 3

a. A typical echo train from the ultrasonic interferometer as observed on the oscilloscope screen. The first tone burst is the buffer rod echo. The subsequent tone bursts are echoes from the free end of the sample (S1, S2, and S3—partially seen in the figure). The dots result from the aliasing between the digitization rate of the oscilloscope and the high frequency in the tone burst. There are approximately 100 full cycles of sinusoid in one tone burst. An expanded view of the sample echo is shown in Figure 3b.

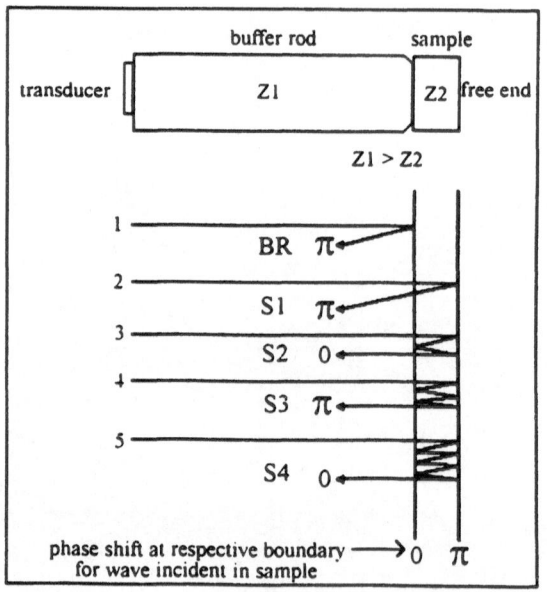

Figure 4

Schematic showing the phase shifts of the stress wave echoes of the buffer rod/sample assembly for the case of $Z_1 > Z_2$.

with acoustic impedance Z_1 and medium 2 with acoustic impedance Z_2. A plane stress wave traveling in medium 1 is normally incident in the interface. The reflection coefficient R for the stress wave at the interface is $(Z_2 - Z_1)/(Z_2 + Z_1)$ (equation (A1.4a) in Appendix 1). If $Z_1 < Z_2$, then $R > 0$ and the reflected stress wave will not change its phase. If $Z_1 > Z_2$, then $R < 0$, and the reflected wave will exhibit a phase change of π radians.

For the case of our interferometer, most materials under study have acoustic impedances less than that of the sapphire buffer rod ($Z_1 > Z_2$). Figure 4 shows the phase shifts of the first few reflected stress waves within the sample when there are an integral number of wavelengths in twice the sample length, $2l$. At the free end of the sample, a π phase shift will always occur. The general equation that is used to calculate the total phase shift in the nth echo for the case of the buffer rod impedance higher than the sample impedance is given by

$$\Phi_n = n\pi\left[1 + 2\left(\frac{2l}{\lambda} - m\right)\right] \qquad (1)$$

where n is the echo number, l is the sample length, λ is the acoustic wavelength in the sample, and m is the integer number of whole wavelengths in $2l$. Maximum or minimum amplitude between two interfering tone burst echoes occurs if $[(2l/\lambda) - m]$ is an integer or half integer. If $[(2l/\lambda) - m] = 0$, then in $2l$ sample length there is an integral number of wavelengths, and the phase shift between two consecutive echoes

(nth and $(n + 1)$th) is π radians. A destructive interference occurs and a minimum amplitude condition will be attained. If $[(2l/\lambda) - m] = 1/2$ then $\Phi_n = 2n\pi$ and consecutive echoes are in phase. A constructive interference occurs and a maximum amplitude condition is present.

For the case of $Z_1 < Z_2$, Figure 4 differs in that the BR echo will show a phase shift of zero and each sample echo will have a phase shift of π radians. The equation governing the total phase shift in the nth echo is given by

$$\Phi_n = \pi\left[1 + 2n\left(\frac{2l}{\lambda} - m\right)\right]. \tag{2}$$

In this case, if $[(2l/\lambda) - m]$ is zero, the phase shift between two consecutive echoes is zero and a maximum amplitude will be realized. If this quantity is $1/2$, then the phase shift between consecutive echoes will be π and a minimum in amplitude is seen (SPETZLER, 1970).

The two-way travel time through the sample is computed from the interference data. Minima and maxima of the two interfering tone burst echoes are found by measuring the amplitudes of the overlapping echoes as the frequency of the synthesizer is swept in small steps. The frequency difference at which two consecutive minima or maxima occur yields the two-way travel time through the sample. In practice, the two-way sample travel time calculated from this method is too crude and more sophisticated data reduction algorithms are required. In a later section we describe the data reduction procedures for our GHz interferometer. These procedures remove the travel time trends due to the system response. The effect of the wave field diffraction and the effect of the bond on the sample travel time data are examined.

3. Transducer Assembly and Sample Preparation

The transducer/buffer-rod assembly was constructed by Professor Pierre Khuri-Yakub at Stanford University (Figure 2; KHURI-YAKUB, personal communication, 1990). It consists of a thin film zinc-oxide transducer vapor-deposited onto a single crystal sapphire buffer rod whose long dimension is oriented to the C axis. Zinc-oxide transducers are commercially used in bulk acoustic wave delay devices for aviation (e.g., Teledyne Microwave). The transducer is detuned to achieve a large bandwidth. The assembly has a nominal impedance around 50 ohms. The buffer rod and the transducer are housed in a brass enclosure with an SMA connector for interfacing to the test equipment. A 3 dB attenuator is connected at the transducer connector to partially isolate the transducer from the electronic system. It was discovered that if the attenuator/isolator was omitted, system nonlinearities would not allow for continuous measurements throughout the usable bandwidth of the interferometer (approximately 0.3 to 1.5 GHz).

Because the interferometer works at frequencies that correspond to wavelengths in the range of 1 to 10 μm in the sample and the buffer rod, the surfaces of these components must be very flat and parallel. Without a flat and smooth polish on these surfaces, wave propagation through these boundaries will be poor at best. Sample surface preparation must be of optical quality.

The following sample preparation procedures are currently employed. Disks of single crystal samples are carefully polished successively with 14 μ and 8 μ silicon carbide, 3 μ alumina and finally finished with 0.05 μ alumina powder. The samples are then dried at room conditions for at least 24 hours. Before bonding the sample onto the buffer rod, the sample faces and the free end of the buffer rod are cleaned with acetone and then isopropyl alcohol. Cyanoacrylate (superglue) is used as the bonding agent to acoustically couple the sample to the buffer rod. Mercury and gallium bonds have also been used successfully. After the bonding, the cyanoacrylate bond is allowed to cure for 24 hours. The curing is taking place under a normal pressure of $\sim$0.6 MPa applied by a dead weight on the free surface of the sample. The 24 hours curing time was determined from a series of exploratory experiments. In the exploratory experiments, the frequency at a minimum of the interference echo was continuously monitored as the bond was curing, until the frequency at which the minimum occurred was stable at a frequency interval of several parts in 10^5. After an experiment, the sample is removed by dipping the buffer-rod/sample assembly into acetone. After removal of the sample, the free end of the buffer rod is cleaned with acetone. The travel-time measurements reported in this paper were made on commercial compressional wave transducers of X-cut single quartz crystals which were polished by the manufacturer.

Travel-time measurements are made under a carefully controlled temperature environment. The temperature of the buffer rod and sample are continuously monitored with a chromel-alumel thermocouple and recorded. Sample temperature control of 0.1 K is accomplished with a closed loop temperature controller.

4. System Control and Data Collection

The instrumentation is interconnected via an IEEE 488 bus and controlled with an IBM compatible personal computer (PC) (Figure 1). The PC thus allows for automatic control of the interferometer. With the current configuration, the controller steps the RF generator through a particular frequency range at predetermined steps. The step size depends on the travel-time resolution desired and on the travel time through the sample. The amplitudes of the overlapping tone burst echoes are recorded at each frequency step. The data reduction described in a later section demands high quality signal amplitude data. We circumvent the limit of the measurement precision posed by the commercial oscilloscope by downloading the digitized waveform of each echo to the computer. A least square fit is performed on the waveform while the oscilloscope is setting up for the next measurement. For each echo, only the central portion of the waveform of constant amplitude is used to compute the signal amplitude. At each frequency, the echo amplitudes and the sample temperature are recorded for further data reduction. At present a complete frequency sweep for a sample at one temperature takes several hours. We expect to reduce this time period of sampling to about one hour for high pressure and high temperature experiments.

Diffraction of the Acoustic Wave

One of the basic assumptions for most acoustic measurement systems is the plane wave propagation assumption. This assumption holds approximately where the wavelength of the acoustic wave is much smaller than the propagation path length, and the region in which the wave propagates can be approximated as an infinite medium compared to the dimensions of the source and the wavelength. For high precision measurements, the plane wave assumption must be carefully examined. To study this problem, we adopted the method of calculating the average response to the acoustic excitation of the transducer at positions or planes of interest (YONEDA, 1990). This average response will be called the average plane wave in the following discussion. The amplitude and the phase of the average plane wave was then compared with the results from the ideal plane wave assumption. The Fresnel parameter, S, is useful in describing the wave propagation field (KINO, 1987):

$$S = \frac{z\lambda}{a^2} \tag{3}$$

where a is the transducer radius; λ is the wavelength; and z is the spatial distance from the transducer plane to the position under consideration (along the normal of the transducer plane).

Plotted in Figure 5 are the phase difference and the log of the amplitude ratios between the average plane wave and the ideal plane wave. The following properties characterize the so-called Fresnel zone where $S < 1$ and the Fraunhofer zone where $S > 1$. They are summarized from KINO (1987). In the Fresnel zone:

(1) the differences of both the amplitude and the phase between the average plane wave and the ideal plane wave vary steadily and slowly with increasing S;

(2) the displacements of the acoustic wave vary rapidly with radius in the plane normal to the propagation direction; and

(3) the acoustic beam is very well collimated and most of the energy is channeled within the radius a of the transducer.

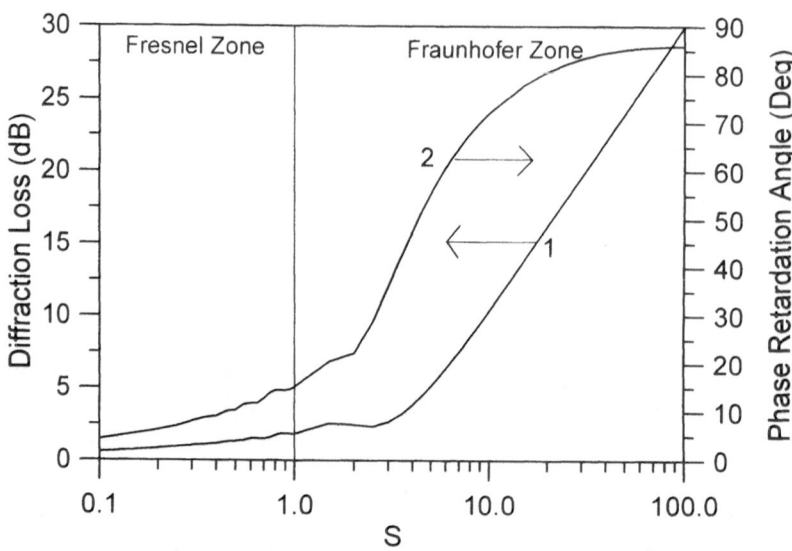

Figure 5
Propagation effect of the acoustic wave on the average plane wave amplitudes and phases as compared with the ideal plane wave. Plotted in this figure are two curves: the diffraction loss curve (Curve 1) and the phase retardation curve (Curve 2). The S values (abscissa) are the so-called Fresnel parameter. The region where $S < 1$ is called the near-field zone or Fresnel zone. The region where $S > 1$ is called the far-field zone or Fraunhofer zone. The diffraction loss curve (Curve 1) is calculated as the log of the amplitude ratios between the average plane wave and the ideal plane wave ($20 \log_{10}(A_{avg}/A_{ideal})$, A_{avg} is the amplitude of the average plane wave and A_{ideal} is the amplitude of the ideal plane wave). The phase retardation curve (Curve 2) is calculated by taking the difference between the phase of the average plane wave and that of the ideal plane wave.

In the Fraunhofer zone:

(1) the differences of both the amplitude and the phase between the average plane wave and the ideal plane wave vary far more rapidly than in the Fresnel zone;

(2) the displacements of the acoustic wave vary smoothly with radius in the plane normal to the propagation direction; the wavefront over the area defined by the transducer is very close to being planar; and

(3) the acoustic beam is broadened over an area larger than that defined by the transducer; the intensity of the acoustic beam at the axis falls off with distance Z monotonically as $1/Z^2$.

For our ultrasonic interferometer, the transducer radius a is 0.285 mm; the length of the buffer rod is 20 mm; over the frequency range from 300 MHz to 1.5 GHz S varies from 3.3 (at 1.5 GHz) to 16.4 (at 300 MHz). For echoes which have traversed a quartz sample (0.57 mm thick) for one round trip S changes by 0.06 at $S = 3.3$ for the 1.5 GHz case and by 0.27 at $S = 16.4$ for the 300 MHz case. Thus our experiments are performed in the Fraunhofer zone. Taking the phase retardation dispersion, we find that at 300 MHz and 1.5 GHz the retardations correspond to 0.3° and 0.7° or to 2.8 and 1.3 picoseconds, respectively. The sample travel times are corrected for this diffraction effect in our experiments.

In summary, the plane wave assumption should be carefully studied where high precision acoustic measurement data are sought. In particular, in the case where the transducer is directly bonded to the sample and the interference of the first reflection and second reflection of the acoustic wave inside the sample is considered (e.g., McSKIMIN, 1950; JACKSON et al., 1981), the diffraction effect may be significant. In this situation, the S value of the second sample reflection is twice that of the first sample reflection. In another case the sample length is comparable to the buffer rod length (e.g., WINKLER and PLONA, 1982; NIESLER, 1986). A third situation employs a long buffer rod (e.g., KATAHARA et al., 1981). In this case, the plane wave assumption may not be valid if the radius of the transducer is close to that of the buffer rod and the acoustic beam is not highly focused. It may correspond to a cylindrical wave in this situation (MEEKER and MEITZLER, 1964). In general an interferometer should be designed such that (i) the difference in S for the two interfering signals ΔS is small; (ii) the S value should not be too large or the signal amplitude would be too small to be resolved with high precision due to the diffraction loss.

Data Reduction

In an earlier section we illustrated the basics of the ultrasonic interferometer under ideal situations (no bond and no system response effect). Here we will describe procedures developed to reduce the interference data in the real ultrasonic interferometry. Two effects on the sample travel time are significant; the system response and the bond effect. In this section we discuss the logistics necessary to remove the system response from the interference data. We devote a separate section to a discussion of the bond effect on the sample travel time.

1. Reducing the Interference Signal Data

In Appendix 2, we show that the amplitude of the interference signal A_{int} between the buffer rod echo and the sample echo recorded by the oscilloscope is given by:

$$A_{int} = \sqrt{(A')^2 + (A'')^2 + 2A'A'' \cos(\theta_s + \theta_t - \theta_r + \pi)} \qquad (4)$$

where A' and A'' are the amplitudes of the uninterfered BR and S1 echoes; θ_r and θ_t are the phase shifts of the acoustic wave due to reflection and transmission across the composite buffer rod-bond-sample interfaces; θ_s is the phase shift inside the sample; the π phase shift is due to the reflection of the S1 echo from the free end of the sample. Simple arithmetic yields the reduced interference amplitude:

$$A_i = 2\cos(\theta_s + \theta_t - \theta_r + \pi) = \frac{A_{int}^2 - (A')^2 - (A'')^2}{A'A''}. \tag{5}$$

The raw interference data A_{int} are modulated by the system (Fig. 6a). The frequencies at which maxima and minima occur are thus biased by the system modulation (Fig. 6b). Equation (5) shows a means to remove the system response from the interference amplitude data. This is done by simple arithmetic on the interference amplitude data A_{int} (the amplitude of the interference between BR and S1) and its separately measured component amplitude data A' and A'' (the amplitudes of the BR echo and S1 echo). Ideally, the reduced interference amplitudes A_i are perfect sinusoids, whose maxima and minima are entirely determined by the phase shifts inside the sample (θ_s) and across the buffer rod-bond-sample interfaces ($\theta_t - \theta_r$) (see Figure 7a).

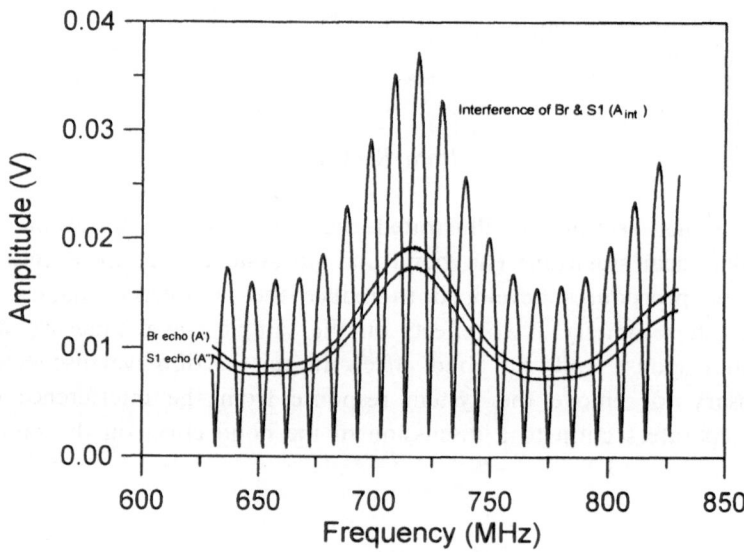

Figure 6a

Raw amplitude data over a frequency range from 630 MHz to 830 MHz. The interference pattern of the small frequency interval (~ 10 MHz) results from the interference of the buffer rod echo (BR echo, A' in equation (5)) and the first sample echo (S1 echo, A'' in equation (5)). The frequency spacing of the interference between the buffer rod echo and the first sample echo yields the 2-way travel time through the sample. This interference (A_{int}) between the buffer rod echo and the first sample echo is modulated by the interference caused by the internal reflections in the coaxial cable (frequency spacing ~ 125 MHz). Superimposed upon the cable modulations is the transducer response. These amplitude modulations caused by the coaxial cables and the transducer bias the interpretation of the sample travel time from the interference patterns (Fig. 6b). Also plotted are the amplitude data for the BR echo (A') and S1 echo (A'') which are necessary to compute the reduced interference data (A_i) from equation (5).

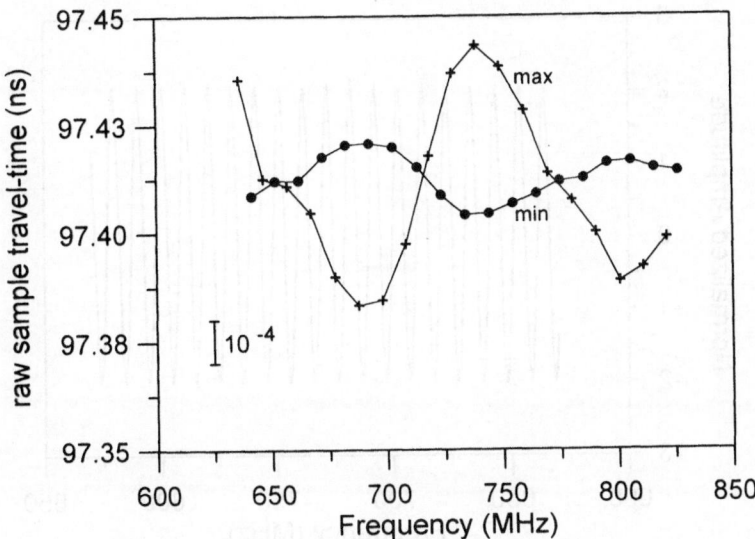

Figure 6b

Sample travel-time data from the raw interference amplitude data are shown in Figure 6a. The plus symbols are from the maxima and the filled circles are from the minima. The curve connecting the maxima crosses that connecting the minima where the system modulation shows zero first derivatives. The travel times from the maxima and minima differ the most where the system modulation to the interference amplitudes has the largest first derivatives (in absolute values).

The travel times determined from the maximum and minimum frequencies of the reduced interference amplitude data (A_i in equation (5)) are shown in Figure 7b as the diamonds and stars. To emphasize the effect of this procedure the travel times determined from the raw data are included in Figure 7b. The data reduction clearly removes the effect of the system modulation from the maximum and minimum frequency data.

2. Obtaining Raw Sample Travel Time

Referring to equation (5), the frequencies at which the maxima and minima occur depend on both the sample phase shift (θ_s) and the phase shift across the composite buffer rod-bond-sample interfaces ($\theta_t - \theta_r$). The maxima occur when

$$\theta_s + \theta_t - \theta_r + \pi = 2n\pi$$

and the minima occur when

$$\theta_s + \theta_t - \theta_r + \pi = (2n + 1)\pi$$

where n are integers in both cases. With $\theta_s = \omega t_{2s} = 2\pi f t_{2s}$, where t_{2s} is the two-way travel time through the sample and f is the frequency. The above two

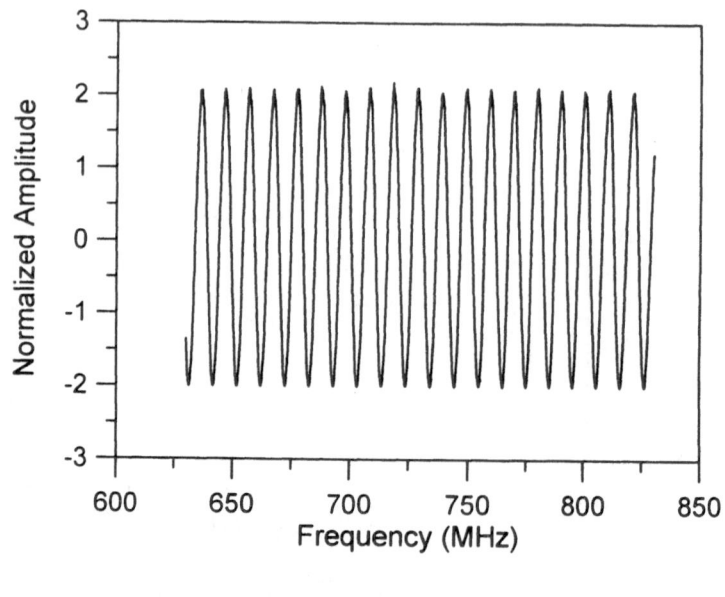

(a)

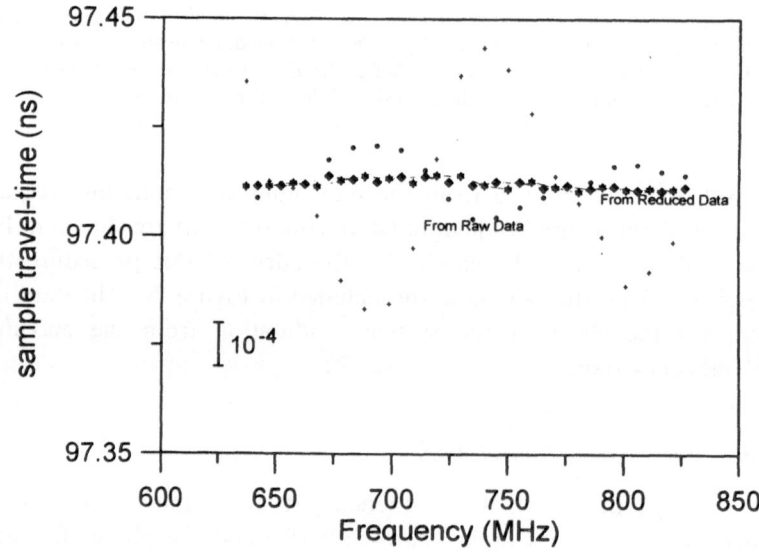

(b)

Figure 7
Reduced interference amplitude according to equation (5) and the travel-time data. The raw interference amplitude data are shown in Figure 6a. Figure 7a shows the reduced interference amplitude data. The data reduction process removes the system modulation of the interference amplitudes and gives a clean sinusoidal curve. Plotted in Figure 7b are the sample travel-time data computed from the reduced amplitude data shown in Figure 7a. For comparison, the sample travel-time data computed from the raw interference amplitude data are also plotted as connected by two background ghost dashed curves. All the data cross each other where the system modulation to the interference amplitudes has zero first derivatives. The sample travel-time data computed from the reduced interference minima (filled diamonds) and maxima (stars) trace each other and do not follow the system amplitude modulation.

equations for the maximum and minimum conditions can be combined and re-written as:

$$t'_{2s} = t_{2s} + \frac{\theta_t - \theta_r + \pi}{2\pi f_m} \equiv t_{2s} + t_{bc} = \frac{m}{f_m} \tag{6}$$

where values of m are either integers (for maxima) or half integers (for minima) and f_m are the frequencies at which the maxima or minima occur. θ_t and θ_r represent the bond effect and are frequency dependent. We call t'_{2s} the raw travel time. If the bond effect could be neglected, $\theta_t = 0$, $\theta_r = \pi$ (for sample impedance less than sapphire impedance); and thus $t'_{2s} = t_{2s} = m/f_m$ would be frequency independent. Another situation in which t_{2s} would be equal to t'_{2s} or m/f_m is where the sample were perfectly coupled to the buffer rod without any bonding material (both stress and displacement are continuous across the contact). However, even with a dry lapped contact, perfect couple may be difficult to achieve (see the discussion in next section). Instead, one may achieve a situation where across the contact, the stresses are continuous while the displacements are not (SCHOEN-BERG, 1980). The discontinuity in the displacement field can result in phase shifts to the reflected and transmitted signals. Thus the measured sample travel time t'_{2s} still needs to be corrected for the phase shifts.

Computer algorithms have been developed to obtain t'_{2s} as a function of frequency based on the following logic. The integer (or half integer) between the consecutive f_m can be obtained experimentally. Assume that the integer (or half integer) corresponding to the first extremum frequency f_{m0} is m_0. Consequently, all the integers (or half integers) corresponding to the first extremum frequencies $f_{m0}, f_{m1}, f_{m2}, \ldots, f_{mn}$ can be represented by $m_0, m_0 + m_1, m_0 + m_2, \ldots, m_0 + m_n$, etc.; where $m_1, m_2, \ldots, m_n$ have been obtained experimentally. The task now is to find the index m_0. To the first order, t'_{2s} can be taken as a constant. Equation (6) becomes:

$$C \equiv t'_{2s} = \frac{m_0 + m_i}{f_{mi}}$$

or

$$f_{mi} = m_i/C + m_0/C \equiv am_i + b. \tag{7}$$

Equation (7) states that if we fit the experimental data f_{mi} vs. m_i with a straight line, the slope of that line is $1/C$ or $1/t'_{2s}$ and the intercept of that line is m_0/C or m_0/t'_{2s}. The m_0 is obtained by $m_0 = b/a$. Because of the bond effect, which is small in comparison to uncertainties in m_0, the m_0 obtained by this method is not an integer (or a half integer). We then adjust m_0 to be an integer (or a half integer) such that we derive minimum frequency dispersion for t'_{2s} ($= (m_0 + m_i)/f_{mi}$). The large range in frequency allows us to determine m_0 unambiguously and with ease.

The following section discusses the experiments performed to study the nature of the bond and the effect of the bond on the sample travel-time data (t_{bc} term in equation (6)).

The Bond Effect

Bonds are used in ultrasonic measurements to couple acoustic energy to the sample. The effect of the bond has been one of the most abominable problems plaguing high precision ultrasonic measurements. There has been a long history in studying bond effects (e.g., MCSKIMIN, 1950, 1961; DAVIES and O'CONNELL, 1977; JACKSON *et al.*, 1981; NIESLER, 1986; VINCENT, 1987). However, conclusive results on the nature and the effect of the bond are still lacking. The high precision of the data, the high frequency (short wavelength), and the large frequency range of our interferometer enables us to study the bond effects better than was previously possible.

DAVIES and O'CONNELL (1977) systematically examined the bond phase shifts for the two different setups of ultrasonic interferometers (transducer/sample, MC-SKIMIN (1961); transducer/buffer-rod/sample, SPETZLER (1970)). Under ambient conditions, the bond was estimated to be 0.2 to 0.4 μm thick from the comparison between the theoretical and measured phase shifts as functions of carrier frequency over a frequency range from 25 MHz to 70 MHz. For the transducer/bond/sample (no buffer rod) configuration, the travel-time data and their pressure derivatives can be significantly biased by the bond effect if the interferometer is operated at frequencies away from the free resonant frequency of the transducer. For the transducer/buffer-rod/sample configuration, dry lapped buffer-rod/sample contact was chosen as the standard (no bond and thus no phase shift). The observed phase shifts for shear waves were found to be consistent with the theoretical estimate with a bond 1 μm thick.

DAVIES and O'CONNELL (1977) showed that the *P*-wave data cannot be explained by the bond phase shift alone. They argued that the lapped buffer-rod/sample contact may not be perfect and should not serve as a standard for *P* waves. As suggested by the work of SCHOENBERG (1980) and PYRAK-NOLTE (1990), and pointed out by NIESLER (1986), the nonperfect contact of lapped buffer-rod/sample may result from the displacement discontinuity of the acoustic waves at the contact interface. A perfect contact interface is one across which both stress and displacement are continuous. Real contacts between solids usually occur at distributed spots (BOWDEN and TABOR, 1950). Consequently, although the overall macroscopic stress across the interface is continuous, the displacement field is not (SCHOENBERG, 1980). The discontinuity of the displacement results in phase shifts when an acoustic wave reflects from and transmits through the interface (SCHOENBERG, 1980).

JACKSON *et al.* (1981) developed techniques to correct for the bond effect from comparison experiments instead of using theoretical models. The transducer/bond/sample setup was used (no buffer rod). Two types of comparison experiments were performed: (i) adding another 'identical' transducer to the opposite face of the sample; (ii) using another 'identical' sample of doubled thickness. While the first technique depends on the physical identity of the transducers and the bonds, the second depends on the physical identity of the samples, e.g., bond corrections are limited to the resolution to which the sample lengths can be measured. Both techniques depend on the reproducibility of the bond. Accuracies of several parts in 10^4 were reported in their travel-time measurements (similar to those achieved by MCSKIMIN (1961) who theoretically corrected the travel-time data for the bond effect). JACKSON *et al.*'s interferometer has been used in pressure derivative measurements of mantle minerals at pressures up to 3 GPa (room temperature) (JACKSON and NIESLER, 1982; GWANMESIA *et al.*, 1990).

Our ultrasonic interferometer system uses the transducer/buffer-rod/sample setup. Bonding material is used to couple the acoustic signals to the sample. The signals reflected from and transmitted through the buffer rod-bond-sample interfaces actually contain the reverberation of the signal inside the thin film of the bonding material (Figure 8). The resultant signals are thus phase shifted, compared to the perfect contact situation (no bond).

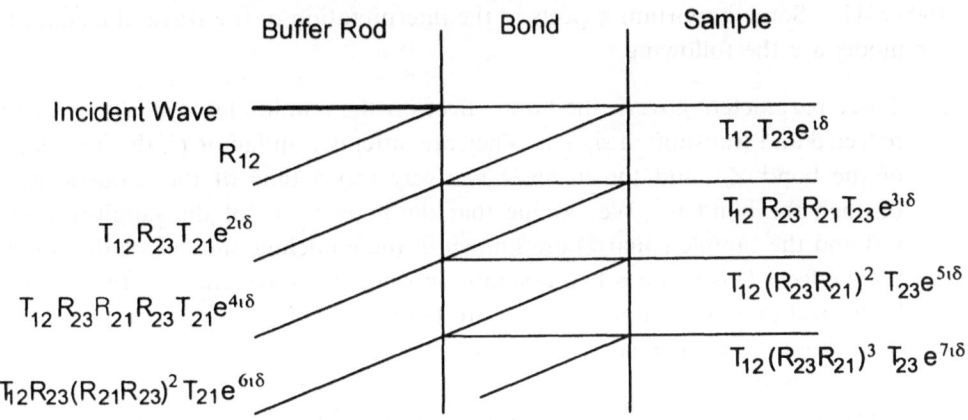

Figure 8

Schematic of the reflection and transmission of the acoustic wave at the composite buffer rod-bond-sample interfaces. For a normally incident wave, both the reflected and transmitted waves align with the incident wave. For illustration, the reflected waves have been drawn to the direction different from the incident wave. The resultant reflected and transmitted waves at the buffer rod-bond-sample interfaces are actually composite interference signals due to the reverberation inside the thin bond film. This reverberation results in phase shifts and amplitude modulation to the incident wave.

R_{21}

In our interferometer, the bond correction becomes especially important because of its relatively large dimension in comparison to the small sample dimensions. A one way trip of a 1 GHz acoustic signal through a bond of 250 nm thickness (typical bond thickness reported by DAVIES and O'CONNELL (1977)) results in a phase shift of $\pi/2$ (assuming sound velocity of the bond is 1 μm/ns). Reverberation of the acoustic signals inside the bond complicates the considerations. We derive formulas based on plane wave propagation theory in Appendix 1. KATAHARA *et al.* (1981) modeled the acoustic wave transmission through thin melts in the MHz frequency range. Our development is similar to that of KATAHARA *et al.* (1981) with further extension to the wave reflection problem. In the following, we will first discuss some of the important features of the theoretical model detailed in Appendix 1. Then we will describe and discuss two types of experiments that were performed to examine the nature of the bond and the bond effect on the sample travel-time data. The sample used in these two experiments is the same and is a commercial single crystal quartz transducer of a resonant frequency of 5 MHz (thickness = 570 μm).

1. About the Theoretical Model on Thin Bonds (Appendix 1)

We have carried out numerical analyses to study the features of the thin bond (superglue) model in the frequency range of interest to us (from 200 MHz to 1600 MHz). Some important aspects in the interpretation of the travel-time data by the model are the following.

(1) Three parameters govern the bond effect on the amplitudes and phases of the reflected and transmitted signals. They are: attenuation factor Q, the impedance of the bond Z_b, and the intrinsic two-way travel time of the acoustic wave through the bond t_{2b}. We assume that the impedances of the sapphire buffer rod and the sample (quartz) are known in the modeling and calculation of the bond effect. This is not a bad assumption since the impedances of the sapphire buffer rod and the sample (quartz) are known to precision considerably better than those of the acoustic properties of the bond.

(2) The effects of Z_b and t_{2b} on the measured sample travel time from the BRS1 interference are coupled. When the Q of the bond is fixed, a group of paired Z_b and t_{2b} values yields the same frequency dispersion trend of bond travel-time shift to the BRS1 interference data (frequency dependence of t_{bc} term in equation (6)). Those paired Z_b and t_{2b} also yield the same values for the bond correction to the sample travel-time data (the absolute value of t_{bc} in equation (6)). This is a remarkable feature of the model since it actually reduces the number of the controlling parameters of the bond model to two. In fitting the experimental data to the model, we fixed Z_b to a value of 2.5×10^6 kg m^{-2} s^{-1} which we constrained experimentally. The values of the Q and t_{2b} are allowed to vary to fit the data.

(3) The reflection coefficient of the bond from the model (amplitude of equation (A1.8) of Appendix 1) exhibits periodicity as a function of the frequency of the acoustic wave. The periodicity originates from the interference of the reverberation signals within the thin bond. The frequency interval (periodicity) is determined by t_{b2}. The smaller the t_{2b}, the larger the interval. In the frequency range from 300 MHz to 1500 MHz, the lower limit of t_{2b} at which a complete period in the reflection coefficient can be observed is ~ 0.7 ns. For values of t_{2b} smaller than this value, only a portion of the period in the reflection coefficient is spanned and can be measured experimentally in the frequency range from 300 MHz to 1500 MHz.

(4) When Q and Z_b are fixed and t_{2b} is smaller than ~ 0.7 ns, larger values of t_{2b} also result in larger corrections to the travel-time data. In other words, larger values of t_{2b} give rise to larger values of t_{bc} (equation (6)). Larger values of t_{2b} also result in steeper slopes of t_{bc} vs. frequency (stronger frequency dependence in t_{bc}; Figure 9).

(5) For a bond of several tens of nanometers thick (t_{2b} = several tens of picoseconds), the upper limit of Q that affects the t_{bc} value is at about 30. Further increase in Q value gives the same t_{bc} value as $Q = 30$. When Z_b and t_{2b} are fixed, smaller Q values give smaller values for t_{bc} (smaller correction to the

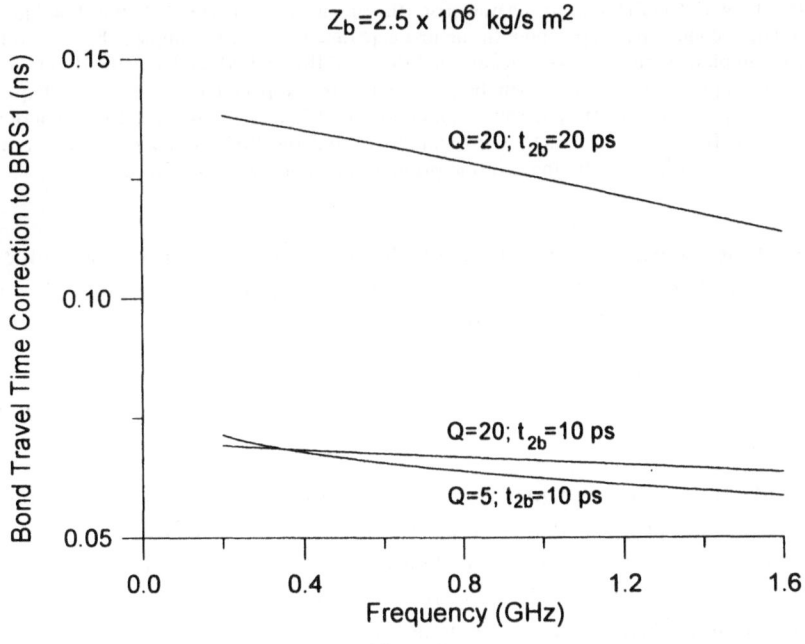

Figure 9

Theoretical calculations of the bond travel-time correction to BRS1 for different values of Q (inverse of attenuation) and t_{2b} (intrinsic bond two-way travel time). The bond impedance is constrained experimentally to be 2.5×10^6 kg m^{-2} s^{-1}.

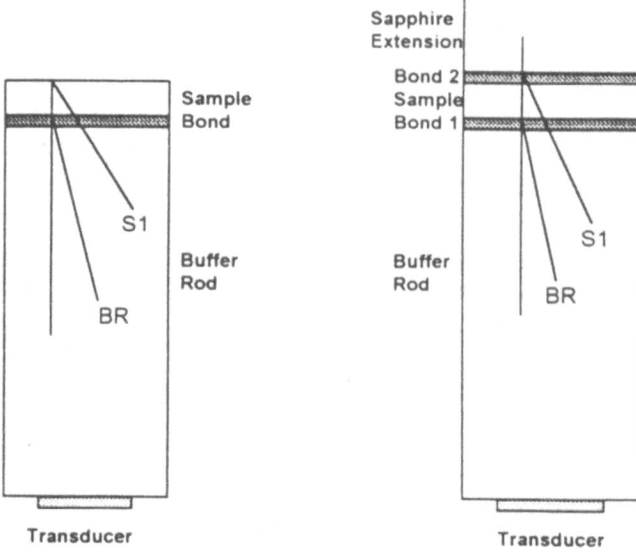

Experiment 1: Naked Sample Experiment   Experiment 2: Sandwiched Sample Experiment

Figure 10

Schematic of two experiments to characterize the acoustic properties of the bonding material (cyanoacrylate—commercial superglue). In the first experiment, a quartz sample is bonded to the buffer rod (naked sample experiment). The signal amplitudes of BR and S1 and the BRS1 interference are measured. A sapphire extension is then bonded to the free top end of the sample (Experiment 2: sandwiched sample experiment). The signal amplitudes of BR and S1 and the BRS1 interference are again measured. By comparing the S1 echo amplitudes and the BRS1 interferences between the two experiments, the acoustic properties of bond 2 are obtained.

travel time measured from BRS1 interference) and result in steeper slopes in the frequency dispersion trend in the travel-time data (steeper slope for t_{bc} as a function of frequency; Figure 9).

2. *Acoustic Properties (Impedance, Two-way Travel Time, Attenuation) of the Bonding Material*

To study the acoustic properties of the bonding material, the following experiment is performed (Figure 10). A single crystal sample of X-cut quartz was bonded onto the free end of the buffer rod. The echo amplitudes of BR and S1 and the interference between BR and S1 were measured over a frequency range from 300 MHz to 1400 MHz. After these measurements were completed, a single crystal sapphire (the buffer rod is also made of single crystal sapphire) extension was bonded onto the top of the quartz sample. The quartz sample is now sandwiched by two pieces of single crystal sapphire (Figure 10). The echo amplitude of S1 and the interference between BR and S1 for the quartz sample were again measured over the same wide frequency range. The sample temperatures for both experiments were controlled at 302.4 ± 0.1 K. The results are shown in Figures 11a and 11b. For convenience in the discussion of the results, we call the former experiment the naked sample experiment and the latter the sandwiched sample experiment. In the sandwiched sample experiment, two bonds are introduced: one is between the sample and the buffer rod and the other is between the sample and the sapphire extension. To avoid confusion, we will only discuss the bond between the sample and the sapphire extension throughout this section (bond 2 in Figure 10).

Referring to Figure 10, the amplitude ratios of the S1 echoes between the naked sample experiment and the sandwiched sample experiment are plotted as a function of frequency in Figure 11a. This ratio is the reflection coefficient of the bond between the quartz sample and the sapphire extension. The theoretical reflection coefficient is given by equation (A1.8) in Appendix 1. If we take the amplitude of equation (A1.8), we obtain:

$$|A_r|^2 = \frac{R_{12}^2 + R_{32}^2 e^{-2\delta/Q} - 2R_{12}R_{32} e^{-\delta/Q} \cos(2\delta)}{1 + R_{12}^2 R_{32}^2 e^{-2\delta/Q} - 2R_{12}R_{32} e^{-\delta/Q} \cos(2\delta)}. \tag{8}$$

The meaning of each symbol in equation (8) is explained in Appendix 1 and will not be repeated here. It becomes self-evident from equation (8) that the periodicity of $|A_r|$ is determined by δ ($=$ angular frequency $\omega \times$ bond 2-way travel time t_{2b}). Through numerical calculations, it is found that the top levels of $|A_r|$ are determined by the impedance of the bond, while the depth of the spikes is controlled by the value of the intrinsic attenuation of the bond Q (Figure 11a). We obtain the following constraints on the acoustic properties of the bond through this amplitude data set: Attenuation factor $Q = 40 \pm 5$; Impedance $Z_b = (2.5 \pm 0.5) \times 10^6$ kg m^{-2} s^{-1}; Two-way travel time through the bond $t_{2b} = 3.8 \pm 0.1$ ns. For comparison, CHALLIS et al. (1992) measured the attenuation and acoustic velocity of thin layers of adhesives (120–170 μm thick). They obtained a value of $\alpha\lambda$ from less than 0.1 to $\sim$0.5 over a frequency range from 1 MHz to 40 MHz. The corresponding Q ($=\pi/\alpha\lambda$) ranges from $\sim$6 to greater than 30. The acoustic velocities of different adhesives they measured range from 1600 m/s to 2400 m/s (1.6 to 2.4 μm/ns).

The interference data between BR and S1 in the naked sample and sandwich sample experiments give us the phase shift of the bond to the reflected signal from the bond (the S1 echo in the sandwiched sample experiment). In Figure 11b we plot the difference of the sample travel time calculated from the BRS1 interference between the sandwiched sample and naked sample experiments (the "+" symbols). The abscissa is the frequency in GHz. The solid curve is from a theoretical calculation based on equation (A1.8) of Appendix 1 for the following acoustic properties for the bond: $Q = 40$, $Z_b = 2.5 \times 10^6$ kg m^{-2} s^{-1}, and $t_{2b} = 3.73$ ns. Given the Q and Z_b determined from the echo amplitude data, the phase shifts of

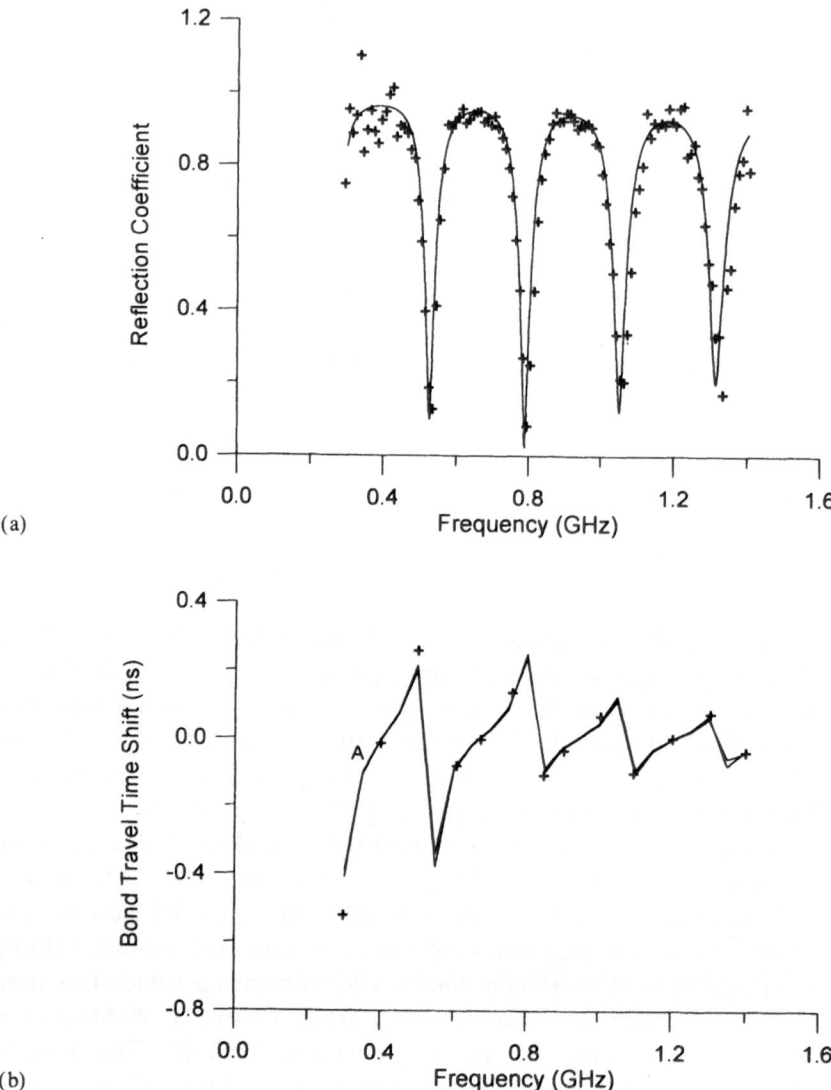

(a)

(b)

Figure 11

a. Amplitude ratio of S1 echoes between the naked sample experiment and the sandwiched sample experiment (see Figure 10). This amplitude ratio gives the reflection coefficient of the bond (bond 2 in Figure 10). The "+" symbols are measured experimentally. The solid curve is calculated from the thin bond model (Appendix 1). The following acoustic properties for the bond are obtained as constrained by the experimental data: Attenuation factor $Q = 40 \pm 5$; Impedance $Z_b = (2.5 \pm 5) \times 10^6 \, \text{kg m}^{-2} \, \text{s}^{-1}$; and two-way travel time through the bond $t_{2b} = 3.8 \pm 0.1$ ns. b. Bond travel-time shift for the sandwiched sample experiment (bond 2 of Figure 10). The "+" symbols are measured experimentally. The solid curves are calculated from the thin bond model (Appendix 1). The data plotted as "+" symbols are obtained by taking the difference of the sample travel times calculated from the BRS1 interference between the sandwiched sample experiment and the naked sample experiment. For thin bonds ($t_{2b} < 0.7$ ns), the bond travel-time shift always corresponds to the first increasing branch (A) of the curve.

the bond between the sapphire extension and the quartz sample provide a tight constraint on the bond two-way travel time.

This set of experiments reveals that the bond properties can be adequately described by the theoretical derivations outlined in Appendix 1. The attenuation and the impedance of the bond and two-way travel time of the acoustic wave through the bond can be constrained experimentally. Moreover, this set of experiments implies that with the GHz interferometer, we can measure the acoustic wave travel time through materials of very small thickness (on the order of 5 μm) with high precision ($\sim 1/400$ or $\sim 0.3\%$). The development of this GHz interferometer thus may open a new arena for high precision determination of acoustic properties of materials only available in micron size or of thin films sputtered on a buffer rod.

3. Effect of Thin Bonds on the Sample Travel-time Measurements

The results from the previous experiment indicate that a bond several microns thick was produced in that experiment. With such a "thick" bond, the bond properties are easily constrained utilizing the high frequency and wide bandwidth of the interferometer. However, such a thick bond between the sample and the buffer rod is not usable in the sample travel-time measurement experiments. The amplitude of the sample echo would be too small to be resolved with high accuracy. With our current bonding technique, bonds with thicknesses less than 1 μm, probably on the order of several tens of nanometers, are commonly produced. When the bond is so thin, the periodicity in the reflection coefficient, as shown in Figure 11a, results in much larger frequency intervals and can no longer be observed directly. Only a portion of a period in the reflection coefficient curve is spanned and measured. It is considerably more difficult to constrain the acoustic properties of a thin bond (several tens of nanometers) than to constrain the acoustic properties of a thick bond (several microns).

To investigate the acoustic behavior of thin bonds, we performed the following experiment. First measurements were performed on the buffer rod without the sample bonded to it. The amplitudes of two consecutive buffer rod echoes and their interference were measured. One buffer rod echo travels one round trip through the buffer rod, and the other travels two round trips. We called these measurements the naked buffer rod experiment (Figure 12). Subsequently the sample was bonded to the buffer rod. The amplitudes of the two buffer rod echoes and their interference were again measured. Meanwhile, the interference between the one round trip buffer rod echo and the S1 echo for the quartz sample is also measured. We called the second type of measurements the bonded buffer rod experiment (Figure 12).

We took the amplitude ratio of the one round trip buffer rod echoes between the naked buffer rod experiment and the bonded buffer rod experiment. This amplitude ratio is the reflection coefficient of the thin bond between the sample and the buffer rod. Theoretical calculations based on equation (A1.8) of Appendix 1 were per-

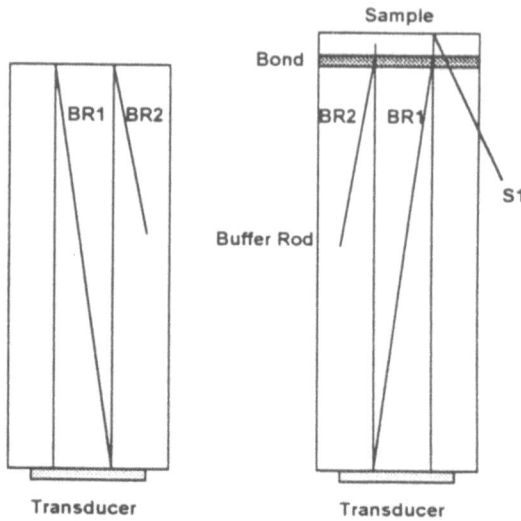

Experiment 1: Naked Buffer Rod   Experiment 2: Bonded Buffer Rod

Figure 12

Schematic of experiments correcting the measured sample travel time for the bond effect. In Experiment 1, no sample is bonded to the buffer rod (naked buffer rod experiment). The amplitudes of two consecutive BR echoes and their interference are measured. In Experiment 2, a sample is bonded to the buffer rod (bonded buffer rod experiment). The measurements of Experiment 1 are repeated. In addition, the BRS1 interference through the sample is measured. By comparing the amplitudes of the two consecutive BR echoes and their interference between the naked and bonded buffer rod experiments, the bond acoustic properties are constrained. Given these bond acoustic properties, bond travel-time shift to the measured sample travel time can be calculated. The bond effect on the sample travel time is therefore resolved.

formed to fit the measured reflection coefficient. Another constraint on the fit is the interference between the two consecutive buffer rod echoes. From this interference we obtain the raw travel time of the acoustic wave through the buffer rod. By taking the difference between the two sets of raw travel-time data (naked buffer rod and bonded buffer rod), we obtained the travel-time shift due to the phase shift of the bond to the reflected buffer rod echoes.

The bond phase shifts obtained from the interference data provide very tight constraints on the acoustic properties of the bond when fit to the thin bond model detailed in Appendix 1. This is because the absolute value of the phase shift as well as its relative variation over the frequency range (frequency dispersion trend) must be fit by the model. The amplitude ratio of the buffer rod echoes between the two types of the experiments (naked buffer rod and bonded buffer rod) only gives loose constraints on the acoustic properties of the bond. We obtained the following fit parameters for the bond from the buffer rod interference data: $Q = 15 \pm 2$; $Z_b = 2.5 \times 10^6 \text{ kg m}^{-2} \text{ s}^{-1}$; $t_{2b} = 31 \pm 2$ ps. The experimental data and the model fitting curve are shown in Figure 13a.

In the bonded buffer rod experiment we also measured the interference between the one round trip buffer rod echo and the S1 echo for the quartz sample. From equation (6) (Data Reduction section), we observe that the raw sample travel-time data t'_{2s} contain the travel-time shifts of the bond to the buffer rod echo ($\theta_r/2\pi f_m$) and to the S1 echo ($\theta_t/2\pi f_m$). We just obtained the travel-time shift of the bond to the buffer rod echo ($\theta_r/2\pi f_m$) experimentally, which is the difference between the two travel-time data sets from the buffer rod interference (naked buffer rod experiment and bonded buffer rod experiment). Thus this experiment provides the means to remove the bond effect on the buffer rod echo of the sample travel-time data. The results before and after the removal of the bond effect on the BR are shown in Figure 13b. The raw sample travel times ("+") decrease ~ 50 ps over a frequency range from 0.3 GHz to 1.4 GHz. After being corrected for the bond effect on the BR echo, the frequency dispersion in the sample travel time ("*"-stars) is greatly reduced (less than 20 ps over the same frequency range). The correction amounts to 70 ps which is on the order of 3.5×10^{-4} of the two-way travel time (~ 200 ns) of the acoustic wave through the sample.

The travel-time shift of the bond to the S1 echo cannot be measured physically by experiments. We must resort to the model. The acoustic properties of the bond have been obtained earlier by fitting the thin bond model to the travel-time data of the buffer rod (naked buffer rod and bonded buffer rod experiments). When these bond parameters (Q, t_{2b}, Z_b) were used to calculate the travel-time shift of the bond to the sample travel time, the model fits the measurements nicely (Figure 13c). We obtained the travel time for the quartz sample 196.465 ± 0.002 ns. Over the frequency range of measurement (0.3 to 1.4 GHz), there is no obvious frequency dispersion for the sample travel time (Figure 13d).

4. Repeated Thin Bond Experiment on the Same Sample as 3 with a Separately Prepared Bond

In the above experiments, we demonstrated that (i) the bond model described in Appendix 1 has remarkable success in predicting the acoustic properties of bonds several microns thick; and (ii) the bond model can be applied to correct for the frequency dispersion trend in the measured sample travel-time data from BRS1 interference. At this point, we carried out a further and crucial experiment to validate the applicability of the bond model to thin bonds.

This experiment duplicates the experiment described in 3 (*Effect of Thin Bonds on the Sample Travel-time Measurements*). The only difference is that the bond is separately prepared. After the experiment described in 3, we removed the sample from the buffer rod and then rebonded the sample to the buffer rod again. We will show in the following that a different bond has been produced in this process and a different bond correction to the measured sample travel-time data is required.

After the correction, we obtained a sample travel time which has no frequency dispersion trend. The sample travel time corrected for the bond effect from the two separate experiments (two physically different bonds) overlaps at the 10^{-5} level.

We will not repeat the details of the experiment here. However, we would like to point out that the bond acoustic properties are again constrained tightly by the

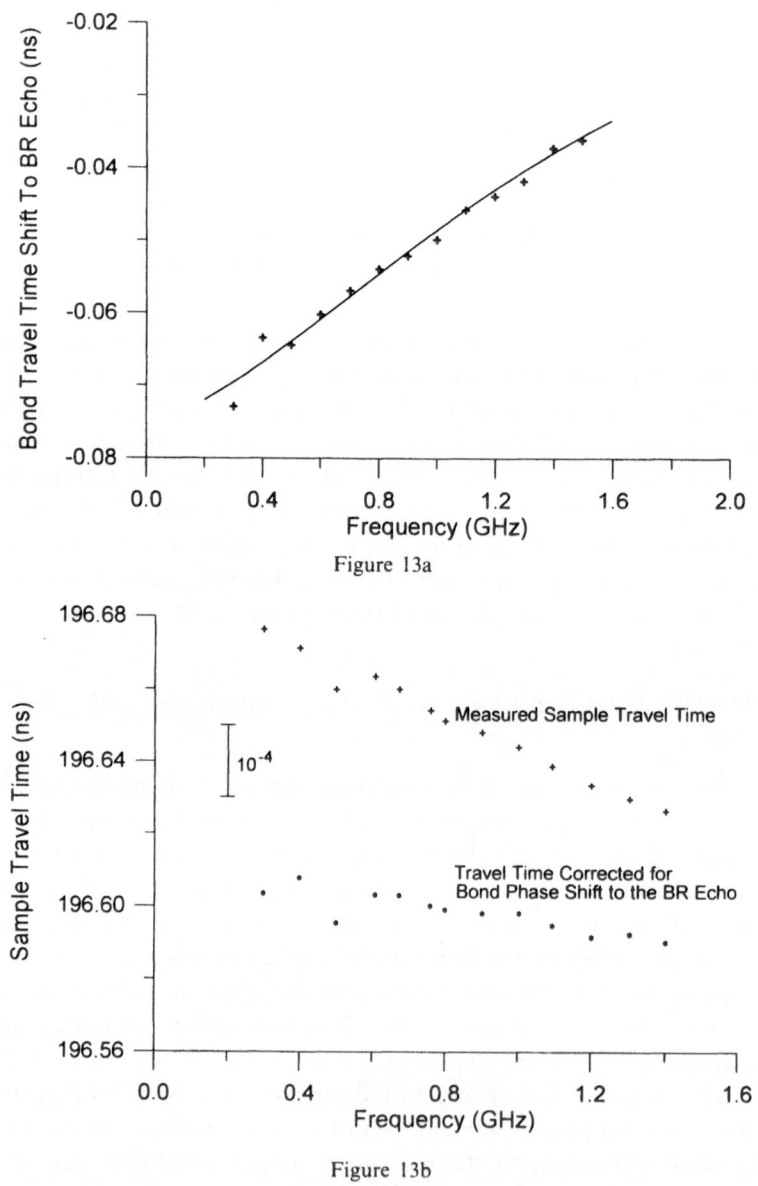

Figure 13a

Figure 13b

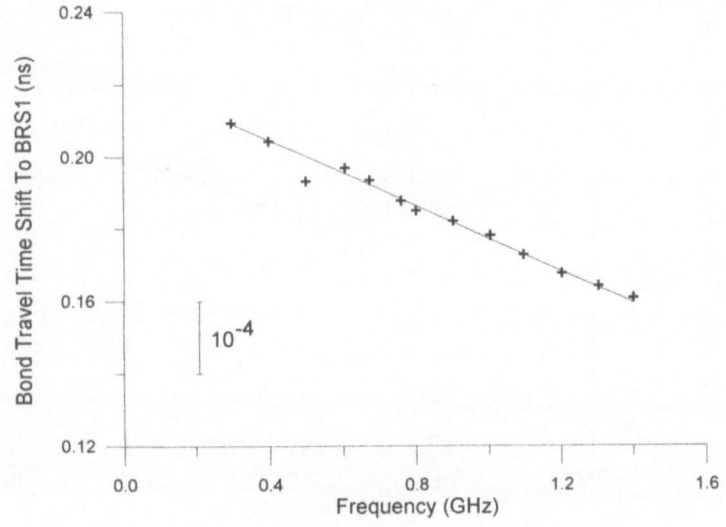

Figure 13c

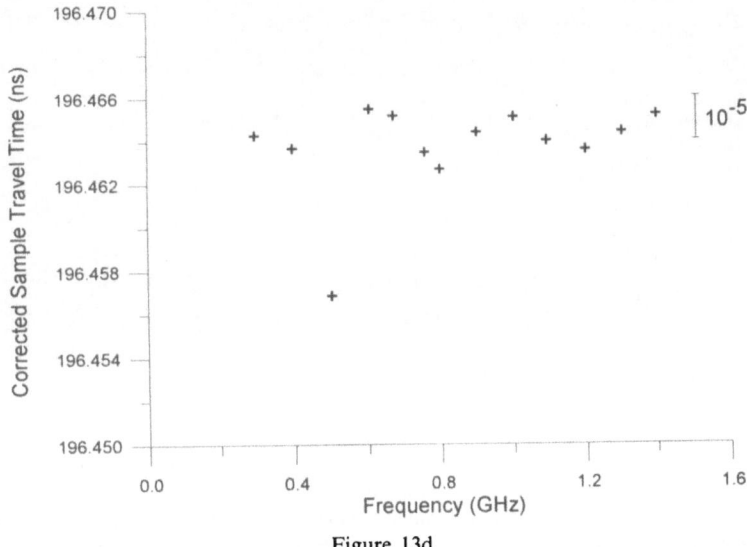

Figure 13d

Figure 13

Procedures in solving the bond acoustic properties and the bond effect on the sample travel time from the experiments shown in Figure 12. Details of these figures are explained in the text.

interference of the two consecutive **BR** echoes. In Figure 14a we plot the bond travel-time phase shift to the **BR** echo. The solid lines are from the model calculation. The "+" symbols are from the difference of the measured buffer rod travel time between the naked buffer rod experiment and the bonded buffer rod

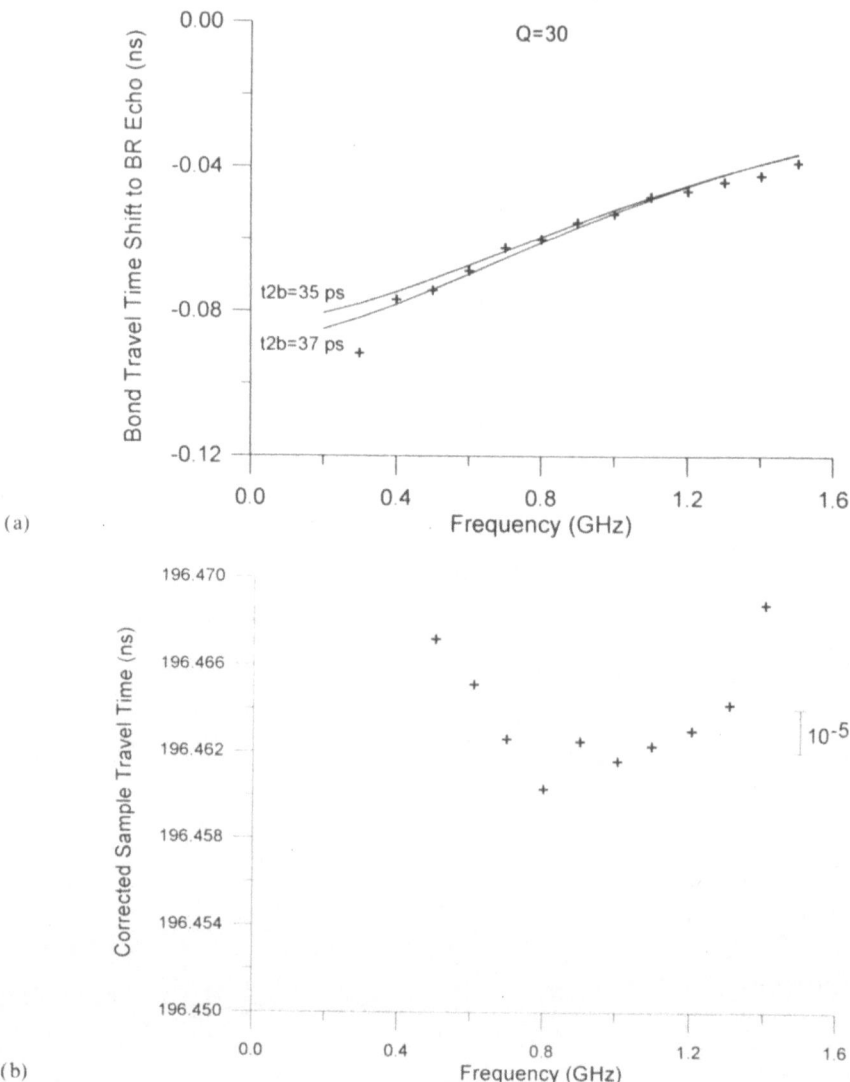

Figure 14
Results from the repeated naked and bonded buffer rod experiments. The experimental setup is the same as those shown in Figure 12. The sample is removed from the previous experiment and rebonded onto the buffer rod. A physically different bond is prepared in this process. The figures are explained in the text.

experiment. We obtained the following bond acoustic properties: $Q = 30$; $t_{2b} = 36 \pm 1$ ps; $Z_b = 2.5 \times 10^6$ kg m^{-2} s^{-1}.

The bond acoustic properties constrained by the internal buffer rod interference are used to calculate the bond travel-time shift to the BRS1 interference of the sample. The corrected sample travel-time data are shown in Figure 14b with the

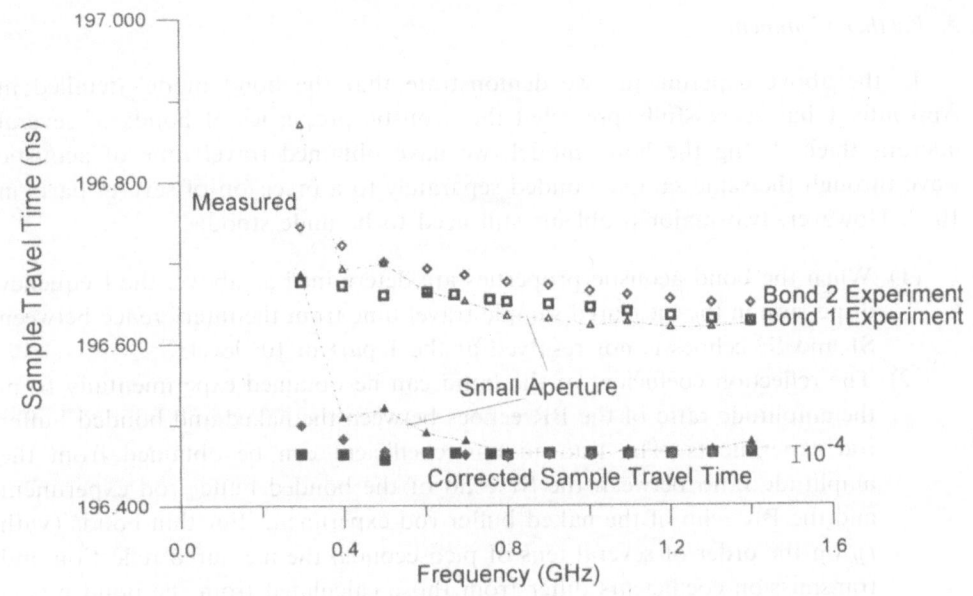

Figure 15
Summary of the experimental results with three separately prepared bonds on the same sample (5P quartz sample). The measured sample travel times (thin open diamonds, squares, triangles) differ from each other by ~20 ps between experiments. The bond travel-time shift amounts to 200 ps or on the order of 10^{-3} of the total sample travel time. After correcting for the Fraunhofer dispersion and the bond effect, the three sets of sample travel time (filled diamonds, squares, triangles) overlap at the 10^{-5} level. Two data sets (diamonds and squares) are from measurements with a buffer rod aperture of ~2 mm, while the small aperture (triangles) is for a buffer rod aperture of ~400 microns.

bond acoustic properties: $Q = 30$, $t_{2b} = 36$ ps, $Z_b = 2.5 \times 10^6$ kg m^{-2} s^{-1}. To emphasize the importance of the correction by the model, the results from the two experiments with separately prepared bonds are summarized in Figure 15. The difference of the uncorrected sample travel-time data between the two bond experiments indicates that two physically different bonds have been made. The difference in the bonds results in more than 20 ps difference in the apparent sample travel times. The bond travel-time shift amounts to more than 0.2 ns or on the order of 10^{-3} of the total sample travel time. After correcting for the bond effects, both experiments yield sample travel times overlapping at the 10^{-5} level (Figure 15). At the low frequency end ($\lesssim 600$ MHz), the data are spurious, probably due to two reasons: (i) the number of wavelengths in the sample is smaller, resulting in less accuracy in the measured sample travel time; and (ii) the amplitude data become considerably more noisy. At this stage, we do not concern ourselves with the differences of the corrected sample travel times at the low frequency end.

5. Further Comments

In the above experiments, we demonstrate that the bond model detailed in Appendix 1 has successfully predicted the acoustic properties of bonds of several microns thick. Using the bond model, we have obtained travel time of acoustic wave through the same sample bonded separately to a precision of several parts in 10^{-5}. However, two major problems still need to be understood:

(1) When the bond acoustic properties are determined as above, the frequency dispersion in the measured sample travel time from the interference between S1 and S2 echoes is not resolved at the 1 part in 10^5 level.

(2) The reflection coefficient of the bond can be obtained experimentally from the amplitude ratio of the BR echoes between the naked and bonded buffer rod experiments. The transmission coefficient can be obtained from the amplitude ratio between the S1 echo of the bonded buffer rod experiment and the BR echo of the naked buffer rod experiment. For thin bonds (with t_{2b} on the order of several tens of picoseconds) the measured reflection and transmission coefficients differ from those calculated from the bond model when the bond acoustic properties are determined from the travel-time data from the interference of internal buffer rod echoes (naked and bonded buffer rod experiments) and from the interference of BRS1 of the sample.

We examined the possible frequency shift of the signals by an edge wave, a possible secondary effect of wave field diffraction (CHALLIS, 1982). The entire digitized waveforms at different frequencies are fit by the least square method to solve for the frequency and the amplitude of the signal. No difference is found at the 10^{-6} level between the computed frequency and the frequency read from the signal generator. At this time, the source of the above problems is not clear to us. Possible sources may include misalignment between the buffer rod and the sample, surface roughness resulting in uneven bond properties, and nonparallelism of the sample.

The bond correction algorithms have been applied to a set of measurements on the temperature derivatives of the acoustic velocities of *b*-axis olivines with different Mg/Fe ratios. The travel-time data, corrected for the bond effect, show that the temperature derivatives of the acoustic velocities can discriminate between different composition of olivines over a slightly elevated temperature range, from 293 K to 323 K (WHITEHEAD, 1993). At present, measurements of the P and T derivatives of the acoustic velocities on the same suite of olivines along the other orientations (*a* and *c* axes) are underway. The temperature derivative of the acoustic velocity of the *c* axis FO67 is found to differ from that of the *b* axis FO67 by about 5%. These high precision acoustic velocities and their P and T data on mantle candidate minerals will enable us to distinguish different equations of state and to extrapolate to deep mantle conditions with confidence (SPETZLER and YONEDA, 1993).

Acknowledgment

Professor Edward Schreiber played a very important role, both as valued colleague and as dear friend in my (H.S) life and career. While searching for a Ph.D. thesis I met Ed at Lamont-Doherty Lab. of Columbia University. We quickly became colleagues and friends and Ed came to Cal-Tech in Pasadena for an extended period. There we developed our lasting friendship and an ultrasonic interferometer which then advanced the state-of-the-art (see references including E. Schreiber) for equation of state measurements at simultaneous high pressure and temperature. Our present effort, once again using the most modern technology available, is a return to what Ed and I pursued with great vigor almost three decades ago. We thus dedicate this paper to the memory of Ed Schreiber.

The development of this new interferometer was only possible with a rapid education of one of us (H.S.) by experts in ultrasonic microscopy. Professors Andrew Briggs from Oxford University, Pierre Khuri-Yakub from Stanford University, Leonard J. Bond from the University of Colorado and Drs. Chris Fortunko and Fred Wahls from the National Institute of Standards and Technology were the teachers, and all of us are grateful. Without their assistance this project would have been far more difficult and slower in progressing from idea to working instrument. Discussions with Professor Ian Jackson at the Australian National University, Professor Gordon Kino at Stanford University, and Professor R. E. Challis at the University of Keele have been very helpful in clarifying certain problems. The patience and technical expertise of Mr. Tom Weiss from Tektronix and Mr. Ray Kagiama from Hewlett Packard were much appreciated in the difficult decisions in acquiring specialized instruments. Dr. Ernst Kirchner, Dr. Lane Gore, and Mr. Lawrence Novak of Teledyne Microwave are helping us in new transducer and buffer-rod designs. In addition, financial support was provided by the National Science Foundation through grant EAR–8916327 and a matching grant from the University of Colorado.

Appendix 1. Reverberation of GHz Acoustic Signal inside a Thin Bond Film

The following theoretical development is generally applicable to plane wave propagation through an isotropic thin film or layer whose thickness is less than the wavelength.

Consider a thin bond film sandwiched by a buffer rod on one side and a sample on the other side (Figure 8). An acoustic sinusoidal wave of frequency f comes from the buffer rod side. This acoustic wave impinges on the buffer rod and bond interface. Reflection and transmission of the acoustic wave at the buffer rod and bond interface occur. The transmitted wave arrives at the bond and sample interface at a slightly delayed time. Reflection and transmission of the wave also occur at the bond-sample interface. For a thin bond, the reflected signals at the buffer rod-bond-sample composite interfaces superimpose to yield a composite signal and similarly for the signals transmitted through the buffer rod-bond-sample interface (Figure 8). If a stress wave of unity amplitude and phase of ωt ($\omega = 2\pi f$) is incident on the buffer rod-bond interface, which we will represent by the wave function ϕ_{inc}:

$$\phi_{\text{inc}} = e^{i\omega t} \tag{A1.1}$$

the reflected and transmitted stress waves can be represented by wave functions ϕ_r and ϕ_t, respectively:

$$\phi_r = A_r e^{i\omega t} \tag{A1.2}$$

$$\phi_t = A_t e^{i\omega t} \tag{A1.3}$$

where A_r and A_t are complex quantities representing the amplitude and phase modulations due to the reverberation of the acoustic wave inside the thin bond film. A_r and A_t will be derived as follows. For convenience, we will call A_r and A_t the composite reflection coefficient and composite transmission coefficient due to the bond modulation.

The reflection coefficient and transmission coefficient for the amplitude of a stress wave at the interface between medium i and medium j are:

$$R_{ij} = \frac{Z_j - Z_i}{Z_j + Z_i} \quad \text{and} \quad T_{ij} = \frac{2Z_j}{Z_i + Z_j}. \tag{A1.4a}$$

For a displacement wave,

$$R_{ij} = \frac{Z_i - Z_j}{Z_i + Z_j} \quad \text{and} \quad T_{ij} = \frac{2Z_i}{Z_i + Z_j}. \tag{A1.4b}$$

Z_i and Z_j are the acoustic impedances of medium i and medium j, respectively. $Z_i = \rho_i V_i$ with ρ_i and V_i are the density and acoustic velocity of medium i. It should be pointed out that the phase of the interference echo (BRS1) is governed by the same formula (equation (6) of Data Reduction Section) for displacement waves or stress waves if the appropriate reflection and transmission coefficients are used. The same argument is true for the amplitudes of the composite reflection and transmission coefficients at the buffer rod-bond-sample interfaces. Throughout this paper, we choose to discuss stress waves.

The composite reflection coefficient A_r is (Figure 9, L_2 and V_2 are the bond thickness and bond acoustic velocity, respectively):

$$\begin{aligned}
A_r &= R_{12} + T_{12} R_{23} T_{21} e^{2i\omega L_2 / V_2} + T_{12} R_{23} (R_{21} R_{23}) T_{21} (e^{2i\omega L_2 / V_2})^2 \\
&\quad + T_{12} R_{23} (R_{21} R_{23})^2 T_{21} (e^{2i\omega L_2 / V_2})^3 + \cdots \\
&= R_{12} + (T_{12} R_{23} T_{21} e^{2i\omega L_2 / V_2})[1 + (R_{21} R_{23} e^{2i\omega L_2 / V_2}) \\
&\quad + (R_{21} R_{23} e^{2i\omega L_2 / V_2})^2 + (R_{21} R_{23} e^{2i\omega L_2 / V_2})^3 + \cdots].
\end{aligned}$$

Since

$$\left| R_{21} R_{23}\, e^{2i\omega L_2 / V_2} \right| = \left| \frac{Z_2 - Z_1}{Z_2 + Z_1} \frac{Z_2 - Z_3}{Z_2 + Z_3} \right| < 1$$

the relation

$$1 + q + q^2 + q^3 + \cdots = 1/(1 - q) \quad \text{for} \quad |q| < 1$$

applies to the above expression for A_r

$$A_r = R_{12} + T_{12} R_{23} T_{21}\, e^{2i\delta}\, \frac{1}{1 - R_{21} R_{23}\, e^{2i\delta}} = \frac{R_{12} - R_{32}\, e^{2i\delta}}{1 - R_{12} R_{23}\, e^{2i\delta}} \tag{A1.5}$$

where $\delta = \omega L_2 / V_2$ is the one-way phase shift through the bond. In deriving the last expression, the following relationships are useful:

$$T_{12} T_{21} - R_{12} R_{21} = 1 \quad \text{and} \quad R_{ij} = -R_{ji}.$$

Similar derivation yields

$$A_t = \frac{T_{12} T_{23}\, e^{i\delta}}{1 - R_{12} R_{32}\, e^{2i\delta}}. \tag{A1.6}$$

Setting the subscript 3 equal to 1 (media 3 and 1 are the same) in equation (A1.5) gives the expression derived by KATAHARA et al. (1981).

The attenuation of the bond can be easily incorporated into the above formulation by allowing for the impedance of the bond to be frequency dependent and the attenuation loss of the acoustic wave through the bond (AKI and RICHARDS, pp. 167–185, 1980). Assuming that the bond attenuation factor Q is effectively constant over the frequency range under consideration. In this case, the elastic velocity of the bond V_2 in the above formulation needs to be replaced by:

$$V_2 \rightarrow V_2' \left[1 + \frac{1}{\pi Q} \ln\left(\frac{\omega}{\omega'}\right) - \frac{i}{2Q} \right] \tag{A1.7}$$

where V_2' is the acoustic velocity of the bond at some reference frequency ω'. This substitution includes the effect of the attenuation on both the frequency dependence of the acoustic velocity and the attenuation loss of the acoustic wave transmitted through the bond. The modification to the equations (A1.4) and (A1.5) in the presence of attenuation in the bonding material results in:

$$A_r = \frac{R_{12}' - R_{32}'\, e^{-\delta/Q}\, e^{2i\delta}}{1 - R_{12}' R_{32}'\, e^{-\delta/Q}\, e^{2i\delta}} \tag{A1.8}$$

$$A_t = \frac{T_{12}' T_{23}'\, e^{-\delta/2Q}\, e^{i\delta}}{1 - R_{12}' R_{32}'\, e^{-\delta/Q}\, e^{2i\delta}} \tag{A1.9}$$

where the superscript prime of the simple reflection coefficients R'_{12} and R'_{32} and transmission coefficients T'_{12} and T'_{23} denotes that the bond impedance is frequency dependent, given by the formula (constant Q):

$$Z_2 = Z'_2\left[1 + \frac{1}{\pi Q} \ln\left(\frac{\omega}{\omega'}\right)\right].$$ (A1.10)

Equations (A1.4a, b) still apply in calculating these reflection and transmission coefficients with the substitution of equation (A1.10) for the acoustic impedance of the bond.

Appendix 2. Buffer Rod Echo, Sample Echo, and the Interference Signal

Following the previous sections, we will call the stress wave reflected from the composite buffer rod-bond-sample interfaces the *buffer rod echo*; the stress wave transmitted through the composite buffer rod-bond-sample interface and reflected back from the free end face of the sample will be called the *sample echo*.

Assuming the incident stress wave is represented by equation (A1.1)

$$\phi_{\text{inc}} = e^{i\omega t}.$$

The buffer rod echo received by the oscilloscope can be represented by

$$\phi_{br} = a_r A_r e^{i\omega t}$$

where a_r is a complex quantity representing the system (transducer, buffer rod, and electronics, etc.) modulation to the amplitude and the phase of the acoustic wave. A_r is the composite reflection coefficient at the buffer rod-bond-sample interfaces and is given by equation (A1.8) in the case of a lossy bond:

$$A_r = \frac{R'_{12} - R'_{32} e^{-\delta/Q} e^{2i\delta}}{1 - R'_{12} R'_{32} e^{-\delta/Q} e^{2i\delta}}.$$

The sample echo received by the oscilloscope can be represented by

$$\phi_s = a_t A_t A'_t e^{i(\theta_s + \pi)} e^{i\omega t}$$

where a_t is again the system modulation (transducer, buffer rod, and electronics, etc.) to the amplitude and phase of the acoustic wave. A_t and A'_t are the composite transmission coefficients at the buffer rod-bond-sample interfaces: A_t is for wave incident from the buffer rod side and A'_t is for wave incident from the sample side. θ_s is the phase shift of the acoustic wave due to 2-way traveling through the sample. The phase shift π is caused by reflection of the stress wave from the free end of the sample. A_t is given by equation (A1.9)

$$A_t = \frac{T'_{12} T'_{23} e^{-\delta/2Q} e^{i\delta}}{1 - R'_{12} R'_{32} e^{-\delta/Q} e^{2i\delta}}$$

and A_t' is given by interchanging the subscripts of 1 and 3 in the expression for A_t:

$$A_t' = \frac{T_{32}' T_{21}' e^{-\delta/2Q} e^{i\delta}}{1 - R_{12}' R_{32}' e^{-\delta/Q} e^{2i\delta}}.$$

The interference between the buffer rod echo and the sample echo received at the oscilloscope can be represented by:

$$\phi_{int} = \phi_{br} + \phi_s = a_r A_r e^{i\omega t} + a_t A_t A_t' e^{i(\omega t + \theta_s + \pi)}$$

$$= [a_r A_r + a_t A_t A_t' e^{i(\theta_s + \pi)}] e^{i\omega t}$$

$$= [A' e^{i\theta_r} + A'' e^{i(\theta_t + \theta_s + \pi)}] e^{i(\omega t + \theta_{sys})} \qquad (A2.1)$$

where $A' = |a_r A_r|$; $A'' = |a_t A_t A_t'|$; θ_r and θ_t are the phases of A_r and $A_t A_t'$, respectively; θ_{sys} is the phase modulation of the system, which is assumed to be the same for both the buffer rod echo and the sample echo.

Without loss of generality, consider the sine component of the above expression for the interference signal in equation (A2.1):

$$\varphi_{int} = A' \sin(\omega t + \theta_{sys} + \theta_r) + A'' \sin(\omega t + \theta_{sys} + \theta_t + \theta_s + \pi)$$

$$= \sqrt{(A')^2 + (A'')^2 + 2A'A'' \cos(\theta_s + \theta_t - \theta_r + \pi)} \; \sin(\omega t + \theta_{sys} + \vartheta) \qquad (A2.2)$$

where

$$\vartheta = \arctan\left(\frac{A' \sin \theta_r + A'' \sin(\theta_t + \theta_s)}{A' \cos \theta_r + A'' \cos(\theta_t + \theta_s)}\right).$$

The effect of the composite buffer rod-bond-sample interfaces is introducing a phase shift ϑ and amplitude modulation to the interference signal.

The amplitudes of the interference signal, recorded through the oscilloscope, are given by

$$A_{int} = \sqrt{(A')^2 + (A'')^2 + 2A'A'' \cos(\theta_s + \theta_t - \theta_r + \pi)}. \qquad (A2.3)$$

References

Akaogi, M., Navrotsky, A., Yagi, T., and Akimoto, S., *Pyroxene-garnet transformation: Thermochemistry and elasticity of garnet solid solutions, and application to a pyrolite mantle.* In *High-pressure Research in Mineral Physics*, Geophys. Mono. 39 (eds. Manghnani, M. H., and Syono, Y.) (Terra Scientific Publishing Co., Tokyo 1987) pp. 251–260.

Aki, K., and Richards, P. G., *Quantitative Seismology, Theory and Methods*, Vol. 1 (W. H. Freeman and Company, New York 1980) pp. 167–185.

Anderson, D. L., and Bass, J. D. (1984), *Mineralogy and Composition of the Upper Mantle*, Geophys. Res. Lett. *11*, 637–640.

Anderson, O. L., and Schreiber, E. (1965), *The Pressure Derivatives of the Sound Velocities of Polycrystalline Magnesia*, J. Geophys. Res. *70* (20), 5,241–5,248.

Anderson, O. L., Isaak, D. L., and Oda, H. (1991), *Thermoelastic Parameters for Six Minerals at High Temperature*, J. Geophys. Res. *96* (B10), 18,037–18,046.

BASS, J. D. (1989), *Elasticity of Grossular and Spessartite Garnets by Brillouin Spectroscopy*, J. Geophys. Res. *94*, 7,621–7,628.

BASSETT, W. A., SHIMIZU, H., and BRODY, E. M., *Pressure dependence of elastic moduli of forsterite by Brillouin scattering in diamond cell*. In *High-pressure Research in Geophysics* (eds. Akimoto, S., and Manghnani, M. H.) (Center for Academic Publications, Tokyo 1982) pp. 115–124.

BOWDEN, F. P., and TABOR, D., *The Friction and Lubrication of Solids. Part I* (Clarenden Press, Oxford 1950).

BRIGGS, A., *Acoustic Microscopy* (Clarenden Press, Oxford 1992).

CHALLIS, R. E. (1982), *Diffraction Phenomena in the Field of a Circular Piezoelectric Transducer Viewed in the Time Domain*, Ultrasonics *20*, 168–172.

CHALLIS, R. E., COCKER, R. P., HOLMES, A. K., and ALPER, T. (1992), *Viscoelasticity of Thin Adhesive Layers as a Function of Cure and Service Temperature Measured by a Novel Technique*, J. Appl. Polymer Sci. *44*, 65–81.

CHEN, G., SPETZLER, H., WHITEHEAD, S., and GETTING, I. C. (1994), *Characterizing sub-millimeter crystals with a wide-band GHz interferometer*. In Proc. *Ultrasonics International '93*, in press.

CHOPELAS, A., and NICOL, M. F. (1982), *Pressure Dependence to 100 Kbar of the Phonons of MgO at 90 and 295 K*, J. Geophys. Res. *87*, 8,591–8,597.

DAVIES, G. F., and O'CONNELL, R. J. (1977), *Transducer and bond phase shifts in ultrasonics, and their effects on measured pressure derivatives of elastic moduli*. In *High-pressure Research, Application in Geophysics* (eds. Manghnani, M. H., and Akimoto, S.) (Academic Press, Inc. 1977) pp. 533–562.

DUFFY, T. S., and ANDERSON, D. L. (1989), *Seismic Velocity in Mantle Minerals and the Mineralogy of the Upper Mantle*, J. Geophys. Res. *94* (B2), 1,895–1,912.

GETTING, I. C., and SPETZLER, H. A. (1993), *Gas-charged Piston-cylinder Apparatus for Pressures to 4 GPa*, EOS, Trans. Am. Geophys. Union *74* (16), 163.

GWANMESIA, G. D., RIGDEN, S., JACKSON, I., and LIEBERMANN, R. C. (1990), *Pressure Dependence of Elastic Wave Velocity for β-Mg$_2$SiO$_4$ and the Composition of the Earth's Mantle*, Sci. *250*, 794–797.

ITA, J., and STIXRUDE, L. (1992), *Petrology, Elasticity, and Composition of the Mantle Transition Zone*, J. Geophys. Res. *97* (B5), 6,849–6,866.

JACKSON, I., NIESLER, H., and WEIDNER, D. J. (1981), *Explicit Correction of Ultrasonically Determined Elastic Wave Velocities for Transducer-bond Phase Shifts*, J. Geophys. Res. *86B*, 3,736–3,748.

JACKSON, I., and NIESLER, H., *The elasticity of periclase to 3 GPa and some geophysical implications*. In *High-pressure Research, Application in Geophysics* (eds. Manghnani, M. H., and Akimoto, S.) (Academic Press, Inc. 1982) pp. 93–113.

KATAHARA, K. W., RAI, C. S., MANGHNANI, M. H., and BALOGH, J. (1981), *An Interferometric Technique for Measuring Velocity and Attenuation in Molten Rocks*, J. Geophys. Res. *86B*, 11,779–11,786.

KINO, G. S., *Acoustic Waves, Devices, Imaging, and Analog Signal Processing* (Prentice-Hall, Inc. 1987) pp. 164–175.

MAO, H. K., HEMLEY, R. J., FEI, Y., SHU, J. F., CHEN, L. C., and BASSETT, W. A. (1991), *Effect of Pressure, Temperature, and Composition on Lattice Parameters and Density of (Fe, Mg)SiO$_3$-perovskites to 30 GPa*, J. Geophys. Res. *96*, 8,069–8,079.

McSKIMIN, H. J. (1950), *Ultrasonic Measurement Techniques Applicable to Small Solid Specimens*, J. Acoust. Soc. Am. *22*, 413–418.

McSKIMIN, H. J. (1961), *Pulse Superposition Method for Measuring Ultrasonic Wave Velocities in Solids*, J. Acoust. Soc. Am. *33*, 12–16.

MEEKER, T. R., and MEITZLER, A. H., *Guided wave propagation in elongated cylinders and plates*. In *Physical Acoustics*, Vol. I (Part A) (ed. Mason, W. P.) (Academic Press Inc., New York 1964) pp. 111–167.

NIESLER, H., *Measurement of Elastic Wave Velocities on Jacketed Polycrystals at High Pressure*, M.S. Thesis (Australian National University, 1986).

OLIJNYK, H., PARIS, E., GEIGER, C. A., and LAGER, G. A. (1991), *Compressional Study of Katoite [Ca$_3$Al$_2$(O$_4$H$_4$)$_3$] and Grossular Garnet*, J. Geophys. Res. *96* (B9), 14,313–14,318.

PYRAK-NOLTE, L. J., MYER, L. R., and COOK, N. G. W. (1990), *Transmission of Seismic Waves across Single Natural Fractures*, J. Geophys. Res. *95B*, 8,617–8,638.

SCHOENBERG, M. (1980), *Elastic Wave Behavior across Linear Slip Interfaces*, J. Acoust. Soc. Am. *68* (5), 1,516–1,521.

SCHREIBER, E., and ANDERSON, O. L. (1966), *Temperature Dependence of the Velocity Derivatives of Periclase*, J. Geophys. Res. *71* (12), 3,007–3,012.

SMYTH, J. R., and SWOPE, R. J. (1992), *Crystal Chemistry of Mantle Eclogite Garnets*, Geol. Soc. Am. 1992 Ann. Meeting, Abstracts with Programs *24* (7), A129.

SOGA, N. E., SCHREIBER, E., and ANDERSON, O. L. (1966), *Estimation of Bulk Modulus and Sound Velocities of Oxides at Very High Temperatures*, J. Geophys. Res. *71*, 1,320–1,336.

SPETZLER, H. (1970), *Equation of State of Polycrystalline and Single-crystal MgO to 8 Kilobars and 800 K*, J. Geophys. Res. *75*, 2,073–2,087.

SPETZLER, H., SCHREIBER, E., and NEWBIGGING, D. (1969a), *Leak Detection in High Pressure Gas System*, Rev. Sci. Instr. *40*, 179.

SPETZLER, H., SCHREIBER, E., and NEWBIGGING, D. (1969b), *Coupling of Ultrasonic Energy through Lapped Surfaces at High Temperature and Pressure*, J. Acoust. Soc. Am. *45*, 1057.

SPETZLER, H., SCHREIBER, E., and PESELNICK, L. (1969c), *Coupling of Ultrasonic Energy through Lapped Surfaces: Application to High Temperatures*, J. Acoust. Soc. Am. *45*, 520.

SPETZLER, H., SCHREIBER, E., and O'CONNELL, R. (1972), *Effect of Stress-induced Anisotropy and Porosity on Elastic Properties of Polycrystals*, J. Geophys. Res. *77*, 4938–4944.

SPETZLER, H. A., and YONEDA, A. (1993), *Performance of the Complete Travel-time Equation of State at Simultaneous High Pressure and Temperature*, Pure and Appl. Geophys., this issue.

VINCENT, A. (1987), *Influence of Wearplate and Coupling Layer Thickness on Ultrasonic Velocity Measurement*, Ultrasonics *25*, 237–243.

WANG, Y., WEIDNER, D. J., LIEBERMANN, R. C., LIU, X., KO, J., VAUGHAN, M. T., ZHAO, Y., YEGANEN-HAERI, A., and PACALO, R. E. G. (1991), *Phase Transition and Thermal Expansion of $MgSiO_3$ Perovskite*, Science *251*, 410–413.

WEIDNER, D. J., and ITO, E., *Mineral physics constraints on a uniform mantle composition*. In *High-pressure Research in Mineral Physics*, Geophys. Mono. 39 (eds. Manghnani, M. H., and Syono, Y.) (Terra Scientific Publishing Co., Tokyo 1987) pp. 439–446.

WEIDNER, D. J., SWYLER, K., and CARLETON, H. R. (1975), *Elasticity of Microcrystals*, Geophys. Res. Lett. *2*, 189–192.

WHITEHEAD, S. M., *Microwave Ultrasonic Travel-time Measurements on an Olivine Series*, M.S. Thesis (University of Colorado at Boulder, 1993).

WINKLER, K. W., and PLONA, T. J. (1982), *Technique for Measuring Ultrasonic Velocity and Attenuation Spectra in Rocks under Pressure*, J. Geophys. Res. *87* (B13), 10,776–10,780.

YONEDA, A. (1990), *Pressure Derivatives of Elastic Constants of Single Crystal MgO and $MgAl_2O_4$*, J. Phys. Earth *38*, 19–55.

YONEDA, A., SPETZLER, H., and GETTING, I. (1994), *Implication of the complete travel time equation of state for a new pressure scale*, In *High-pressure Science and Technology* (eds. Schmidt, S. C., Shaner, J. W., Samara, G. A., and Ross, M.) (American Institute of Physics, Woodbury, New York 1993) pp. 1609.

(Received April 23, 1993, revised October, 1993, accepted November, 1993)

PAGEOPH, Vol. 141, No. 2/3/4 (1993)

0033-4553/93/040379-14$1.50 + 0.20/0

Performance of the Complete Travel-time Equation of State at Simultaneous High Pressure and Temperature

Hartmut A. Spetzler[1] and Akira Yoneda[2,3]

Abstract — The complete travel-time equation of state (CT-EOS) is presented by utilizing thermodynamics relations, such as;

$$K_T = K_S(1 + \alpha\gamma T)^{-1}, \quad \gamma = \frac{\alpha K_S}{\rho C_P}, \quad \left(\frac{\partial C_P}{\partial P}\right)_T = -\frac{T}{\rho}\left[\alpha^2 + \frac{\partial \alpha}{\partial T}\right)_P\right], \quad \text{etc.}$$

The CT-EOS enables us to analyze ultrasonic experimental data under simultaneous high pressure and high temperature without introducing any assumption, as long as the density, or thermal expansivity, and heat capacity are also available as functions of temperature at zero pressure. The performance of the CT-EOS was examined by using synthesized travel-time data with random noise of 10^{-5} and 10^{-4} amplitude up to 4 GPa and 1500 K. Those test conditions are to be met with the newly developed GHz interferometry in a gas medium piston cylinder apparatus. The results suggest that the combination of the CT-EOS and accurate experimental data (10^{-4} in travel time) can determine thermodynamic and elastic parameters, as well as their derivatives with unprecedented accuracy, yielding second-order pressure derivatives ($\partial^2 M/\partial P^2$) of the elastic moduli as well as the temperature derivatives of their first-order pressure derivatives ($\partial^2 M/\partial P\,\partial T$). The completeness of the CT-EOS provides an unambiguous criterion to evaluate the compatibility of empirical EOS with experimental data. Furthermore because of this completeness, it offers the possibility of a new and absolute pressure calibration when X-ray (i.e., volume) measurements are made simultaneously with the travel-time measurements.

Key words: Elastic constant, thermodynamics, equation of state, acoustic velocity, thermal expansivity, heat capacity, Grüneisen constant, high pressure, high temperature.

Introduction

Equation of state measurements in the laboratory on pertinent mantle minerals are necessary for the interpretation of seismic data in terms of the composition of earth's interior. Seismic data consist of travel times of various strain waves which are subsequently inverted to yield compressional and shear wave velocities for

[1] Cooperative Institute for Research in Environmental Sciences and Department of Geological Sciences, University of Colorado, Boulder, Colorado 80309-0216, U.S.A.

[2] Cooperative Institute for Research in Environmental Sciences, University of Colorado, Boulder, Colorado 80309-0216, U.S.A.

[3] Permanent Address: Department of Earth and Planetary Sciences, Nagoya University, Chikusa, Nagoya 464-04, Japan.

earth. The most direct comparison between laboratory results and field observations (seismic data) comes from experiments in which the appropriate velocities are
measured under the conditions of pressure and temperature as they exist in earth.

Of the many equation of state measurement techniques, Brillouin scattering,
laser induced phonon scattering (LIPS), and the various ultrasonic measurements
can yield compressional and shear velocities with no need for thermodynamic
assumptions (DANDEKAR, 1970; SPETZLER et al., 1972). In the optical techniques
(e.g., WEIDNER, 1975; BASSETT et al., 1982; BROWN et al., 1989), the measured
frequency shifts are directly related to the velocities, while in ultrasonics the
measurements correspond to either travel times through a sample or some resonance frequency of the sample. We will use the term 'travel time' in a general sense,
such that a resonance frequency could be viewed as an inverse travel time. Thus the
sample dimensions must be obtained in addition to the directly measured quantities.

In this paper we show:

1) When measurements at simultaneous high pressure and temperature by
optical methods (Brillouin scattering and LIPS), or ultrasonic methods are combined with thermal expansivity and heat capacity data as a function of temperature
at zero pressure, a complete equation of state is obtained. In this paper, we assume
an elastically isotropic mineral for the sake of simplicity. The extension to any
lower symmetry is possible in principle, but requires at least as many measurement
sets as there are elastic constants and the thermal expansivity tensor at zero
pressure. By complete equation of state we mean that the following parameters are
unambiguously determined over the range in pressure and temperature for which
the travel-time or velocity measurements were made: thermal expansivity α; heat
capacity C_P; elastic wave velocities v_p, v_s; adiabatic and isothermal bulk moduli K_S,
K_T; shear modulus μ, density ρ, and any combination of these such as the
Grüneisen parameter γ.

2) How the precision of the measurements will translate into the precision with
which the various thermodynamic parameters are determined within the range over
which the measurements were taken.

3) How the uncertainty increases when the data in the measured range are
extrapolated by using certain empirical equations of state.

The Nature of the Data and their Conversion to Equation of State Parameters

At least in part, this paper is motivated by the recent development of an
ultrasonic interferometer (SPETZLER et al., this issue) which enables us to make
high precision (1 part in 10^5) travel-time measurements at GHz (10^9 Hz) frequencies. At such high frequencies the wavelengths (λ) of the ultrasonic waves are near
optical wavelengths, i.e., on the order of microns. Since the phonon wavelengths
used in the optical methods are also quite short in comparison to those at seismic

frequencies, i.e., on the order of kilometers, we compare the strain waves in terms of their adiabatic or isothermal nature.

Seismic waves are adiabatic; i.e., the heat generated in the compressional part of a cycle does not have time for significant diffusion into the cooler dilatational part. The thermal relaxation time τ_{th} for one wavelength of the seismic wave is much greater than the period of the seismic wave. From the heat diffusion equation, the thermal relaxation time is

$$\tau_{th} = \lambda^2/\chi \qquad (1)$$

where χ is the thermal diffusivity and has a typical value of 10^{-6} m^2/s for rocks. The period of an elastic wave is

$$\tau_{ew} = \lambda/v. \qquad (2)$$

For a strain wave to be isothermal requires that the thermal relaxation time be shorter than the period of the wave, i.e.,

$$\tau_{th} < \tau_{ew} \quad \text{or} \quad \lambda < \chi/v. \qquad (3)$$

For a typical seismic velocity of 10 km/s, Eq. (3) requires the wavelengths of strain waves to be shorter than 10^{-10} m. This corresponds to a frequency of $\sim 10^{14}$ Hz. Thus at frequencies substantially below 10^{14} Hz (corresponds to near-infrared optical frequencies) elastic waves are adiabatic. Furthermore, phonon dispersion curves (e.g., KITTEL, 1986) for most minerals are straight for $\lambda > 10^{-7}$ m, i.e., for frequencies $< 10^{11}$ Hz. Phonon experiments with wavelengths on the order of microns are therefore well in the adiabatic range and should be directly comparable in terms of velocities with seismic velocities (~ 1 Hz). If there is significant anelasticity between the experimental and the seismic frequency range the resulting frequency dispersion must be considered.

For this discussion we assume that the ultrasonic and optical phonon measurements determine adiabatic travel times and velocities. While the adiabatic measurements are made, the sample is in isothermal equilibrium with its environment. As mentioned above, the optical techniques yield the velocities directly while the ultrasonic measurements provide only the travel times. This analysis is therefore somewhat more complicated for the ultrasonic case and will be shown here. However for the optical data to yield density and heat capacities the conversion from adiabatic to isothermal parameters is also needed and thus follows the analysis below. The isotropic mineral has two elastic velocities v_p and v_s (t_p and t_s are the travel times through the sample for compressional and shear waves respectively) and a scalar thermal expansivity coefficient. The following analysis is similar to that given in SPETZLER et al. (1972).

Assume that C_P and α are given as functions of temperature at zero pressure ($T, 0$) and a data set $t_p(T, P)$ and $t_s(T, P)$ have been obtained at simultaneous temperature and pressure. Furthermore, the sample dimensions l_0 and density ρ_0 are known at ambient conditions $(0, 0)$. From $\rho(0, 0)$ and $l(0, 0)$, and $\alpha(T, 0)$ we calculate $\rho(T, 0)$ and $l(T, 0)$.

$$\rho = \rho_0 \exp\left(-\int_{T_0}^{T} \alpha \, dT\right), \quad l = l_0\left(\frac{\rho_0}{\rho}\right)^{1/3}. \tag{4}$$

The velocities and the adiabatic bulk modulus are readily calculated at $(T, 0)$ as

$$v_p = \frac{l}{t_p}, \quad v_s = \frac{l}{t_s} \tag{5}$$

$$K_S = \rho(v_p^2 - \tfrac{4}{3}v_s^2). \tag{6}$$

Using the thermodynamic Grüneisen parameter

$$\gamma = \frac{\alpha K_S}{\rho C_P} = \frac{\alpha K_T}{\rho C_V} \tag{7}$$

and the thermodynamic identity

$$K_T = K_S(1 + \alpha\gamma T)^{-1} \tag{8}$$

allows the calculation of all required parameters at $(T, 0)$. We now extend the calculation into the pressure regime by an increment ΔP. Since we have K_T along the temperature axis we can calculate ρ and l at $(T, \Delta P)$ and thus the velocities and K_S as above.

$$\rho = \rho_0 \exp\left(\int_{P_0}^{P} \frac{dP}{K_T}\right), \quad l = l_0\left(\frac{\rho_0}{\rho}\right)^{1/3}. \tag{9}$$

To extend the calculation by the next pressure interval requires C_P at $(T, \Delta P)$; i.e., we need $\partial C_P/\partial P)_T$ in order to find K_T from K_S. From the definition of $C_P = T\,\partial S/\partial T)_P$ and differentiating with respect to P we get

$$\left.\frac{\partial C_P}{\partial P}\right)_T = \left.\frac{\partial T}{\partial P}\right)_T \left.\frac{\partial S}{\partial T}\right)_P + \left.\frac{\partial\left[\left.\frac{\partial S}{\partial T}\right)_P\right]}{\partial P}\right)_T T. \tag{10}$$

The first term vanishes since $\partial T/\partial P)_T \equiv 0$. We introduce the Gibbs free energy, G, and reverse the order of differentiation on the second term to obtain

$$\left.\frac{\partial C_P}{\partial P}\right)_T = \left.\frac{\partial\left[\left.\frac{\partial S}{\partial P}\right)_T\right]}{\partial T}\right)_P T = \left.\frac{\partial\left[-\frac{\partial^2 G}{\partial P\,\partial T}\right)\right]}{\partial T}\right)_P T = \left.\frac{\partial\left[-\left.\frac{\partial V}{\partial T}\right)_P\right]}{\partial T}\right)_P T \tag{11}$$

$$= \left.\frac{\partial(-\alpha V)}{\partial T}\right)_P T = -T\left[\left.\frac{\partial\alpha}{\partial T}\right)_P V + \alpha\left.\frac{\partial V}{\partial T}\right)_P\right] \tag{12}$$

$$= -\frac{T}{\rho}\left[\alpha^2 + \left.\frac{\partial\alpha}{\partial T}\right)_P\right]. \tag{13}$$

Thus all parameters can be calculated for every new interval in pressure and therefore all thermodynamic parameters can be calculated over the entire range in temperature and pressure over which the travel times were measured. It is perhaps interesting to note that when travel times are measured along the temperature axis at one pressure, all adiabatic and all isothermal thermodynamic variables considered here can be calculated along that temperature axis. This does not hold for travel-time measurements along a pressure axis at one temperature. Since the sample length must be determined through K_T and the conversion of adiabatic to isothermal parameters requires α and C_P, assumptions must be made about the pressure dependence of α and C_P. If in addition to the travel-time measurements, the sample volume were to be measured simultaneously (by X-ray method perhaps) as a function of temperature and pressure, a redundant equation of state would be determined. This redundancy could be used to backcalculate the pressure and then determine an *absolute pressure scale* which depends only on our ability to measure time (travel time) and angle (X-ray diffraction). We anticipate an accuracy in pressure of $\sim 0.01\%$ up to 4 GPa by means of the new absolute pressure scale (YONEDA *et al.*, 1993).

In the following section we will examine the effect of the integration in pressure and temperature space on the precision of the data.

Performance of the CT-EOS Using Synthesized Travel-time Data

The recently achieved accuracy in travel-time measurements (JACKSON *et al.*, 1981; SPETZLER *et al.*, this issue) and the extension to high pressures (3 GPa for JACKSON and NIESLER, 1982; 8 GPa for YONEDA, 1990) and temperatures (~ 1800 K; GOTO and ANDERSON, 1988; ISAAK *et al.*, 1989) warrant a close look at the implied accuracy with which the pertinent thermodynamic parameters can be determined. SPETZLER *et al.* (this issue) have achieved a precision in travel-time measurements of 1 part in 10^5 in the newly developed GHz interferometer; we anticipate using it for measurements at simultaneous high pressure and high temperature up to 4 GPa and 1500 K in a gas medium piston cylinder apparatus (GETTING and SPETZLER, 1993). Here we investigate the expected accuracy of the thermodynamic and elastic parameters up to lower mantle conditions (25 GPa and 1500 K) by assuming uncertainties of acoustic travel times and other input parameters; the noise amplitudes were realistically chosen as 10^{-5} and 10^{-4} for travel times, 10^{-2} for α and C_P (e.g., SUZUKI, 1975; WATANABE, 1982), and 10^{-4} for ρ_0 and l_0. In this analysis, the uncertainty in pressure and temperature measurements was not considered directly, because uncertainties of ~ 0.012 GPa and ~ 1.9 K are equivalent to 10^{-4} uncertainty of travel time in the present model data. Therefore those values are the target accuracy of pressure and temperature control and measurement (at least as to detectivity, those are realistic targets within reach of experimental techniques; GETTING and KENNEDY, 1970; NIESLER *et al.*, 1988). If

Table 1

List of input parameters. These values are quite similar to those for MgO (periclase). Note that $T' = (T - 300 \text{ K})/1200 \text{ K}$ ($T' = 0$ at $T = 300 \text{ K}$; $T' = 1$ at $T = 1500 \text{ K}$)

$$\rho_0 = 3585 \text{ kg/m}^3$$

$$\alpha(T') = a_1 + a_2 T'$$

$$a_1 = 3.0 \times 10^{-5} \text{ K}^{-1}, \quad a_2 = 2.0 \times 10^{-5} \text{ K}^{-1}$$

$$C_P(T') = b_1 + b_2 T'$$

$$b_1 = 900 \text{ J/kgK}, \quad b_2 = 300 \text{ J/kgK}$$

$$K_{S_0}(T') = c_1 - c_2 T' - c_3 T'^2$$

$$c_1 = 160 \text{ GPa}, \quad c_2 = 12.0 \text{ GPa}, \quad c_3 = 2.5 \text{ GPa}$$

$$\mu_0(T') = d_1 - d_2 T' - d_3 T'^2$$

$$d_1 = 130 \text{ GPa}, \quad d_2 = 18.0 \text{ GPa}, \quad d_3 = 2.5 \text{ GPa}$$

$$\xi_{K_S}(T') = \frac{K_{T_0}}{K_{S_0}} (e_1 + e_2 T' + e_3 T'^2) - 4.0$$

$$e_1 = 4.3, \quad e_2 = 0.25, \quad e_3 = 0.15$$

$$\xi_\mu(T') = \frac{K_{T_0}}{\mu_0} (f_1 + f_2 T') - 4.0$$

$$f_1 = 2.0, \quad f_2 = 0.25$$

$$\eta_{K_S}(T') = 0$$

$$\eta_\mu(T') = 0$$

we establish the new pressure scale mentioned in the former section, the target accuracy in pressure determination can be achieved not only in detectivity but also in absolute value.

Table 1 gives the input parameters used for the model travel-time data (see Appendix for the details of synthesizing the model travel-time data and the procedure of the analysis and extrapolation to 25 GPa).

Summary of the Uncertainties Involved in the CT-EOS

Table 2 gives the fractional variation in the parameters (e.g., the density ρ changes by 2.4% at 300 K and by 2.9% at 1500 K when the pressure is changed

Table 2

Comparison between normalized changes in selected parameters and their standard deviations along the pressure and temperature axis. The first and fifth numerical columns are normalized standard deviations at 4 GPa, 300 K and 4 GPa, 1500 K, respectively. The second and fourth columns are normalized changes of the corresponding parameter over 4 GPa at 300 K and 1500 K, respectively. The third (middle) column shows the changes of the parameters from 300 K to 1500 K at 4 GPa. Note that the noise amplitudes are 10^{-4} for travel times, 10^{-2} for α and C_P, and 10^{-4} for ρ_0 and l_0. The '−' signs before values for std(X)/(X) indicate that the X's are negative, and those before $\Delta X/|X|$ indicate that X decreases with increasing temperature or pressure

	Parameter	std(X)/X at 4 GPa 300 K	$\Delta X_P/|X|$ at 300 K	$\Delta X_T/|X|$ at 4 GPa	$\Delta X_P/|X|$ at 1500 K	std(X)/X at 4 GPa 1500 K
	ρ	1.0E−4	2.4E−2	−4.1E−2	2.9E−2	2.6E−4
	K_S	2.9E−4	1.0E−1	−7.0E−2	1.2E−1	3.3E−4
M	K_T	4.4E−4	1.0E−1	−1.5E−1	1.4E−1	1.4E−3
	μ	2.0E−4	5.8E−2	−1.3E−1	7.6E−2	2.6E−4
	α	1.4E−2	−8.4E−2	5.9E−1	−1.1E−1	8.2E−3
Φ	C_P	8.4E−3	−5.8E−3	3.0E−1	−2.4E−2	7.7E−3
	γ	1.6E−2	−8.1E−3	1.8E−1	−1.1E−2	9.5E−3
	v_p	1.0E−4	2.7E−2	−3.4E−2	3.4E−2	1.0E−4
	v_s	1.0E−4	1.6E−2	−5.2E−2	2.2E−2	1.1E−4
	K'_S	6.8E−3	−2.3E−2	6.9E−2	−4.0E−2	1.1E−2
M'	K'_T	6.8E−3	−2.3E−2	7.0E−2	−4.3E−2	9.7E−3
	μ'	7.1E−3	−8.7E−2	1.0E−1	−9.9E−2	4.6E−3
	K''_S	−5.4E−1	1.7E−1	−7.8E−1	2.4E−1	−3.9E−1
M''	K''_T	−4.8E−1	1.7E−1	−9.5E−1	2.0E−1	−3.8E−1
	μ''	−1.4E−1	1.4E−1	−2.1E−1	2.0E−1	−8.6E−2
	$(\partial K_S/\partial T)_P$	−3.8E−1	7.3E−2	−3.5E−1	1.0E−1	−1.9E−1
$(\partial M/\partial T)_P$	$(\partial K_T/\partial T)_P$	−3.3E−1	4.1E−2	−5.4E−1	4.8E−2	−1.9E−1
	$(\partial\mu/\partial T)_P$	−8.0E−2	5.2E−2	−3.1E−1	3.3E−2	−3.5E−2
	$(\partial\alpha/\partial P)_T$	−3.3E−1	2.0E−1	−1.1E−0	2.5E−1	−1.9E−1
Φ'	$(\partial C_P/\partial P)_T$	−7.8E−1	1.5E−1	−4.2E−0	2.0E−1	−4.9E−1
	$(\partial\gamma/\partial P)_T$	−4.5E−0	2.9E−1	−6.7E−1	2.6E−1	−4.6E−0

from 0 to 4 GPa; it changes by 4.1% when the temperature is changed from 300 K to 1500 K while the sample is at a pressure of 4 GPa) as well as the uncertainties with which they can be determined (e.g., the fractional uncertainty with which the density is determined at 4 GPa and 300 K is 0.01%, the corresponding value at 4 GPa and 1500 K is 0.026%) over the range from 0 to 4 GPa and 300 K to 1500 K.

(1) The unexpected accuracy of 2.6×10^{-4} for ρ at 4 GPa and 1500 K is due to the uncertainty of ρ_0 and α at 0 GPa, rather than due to the uncertainty of travel time.

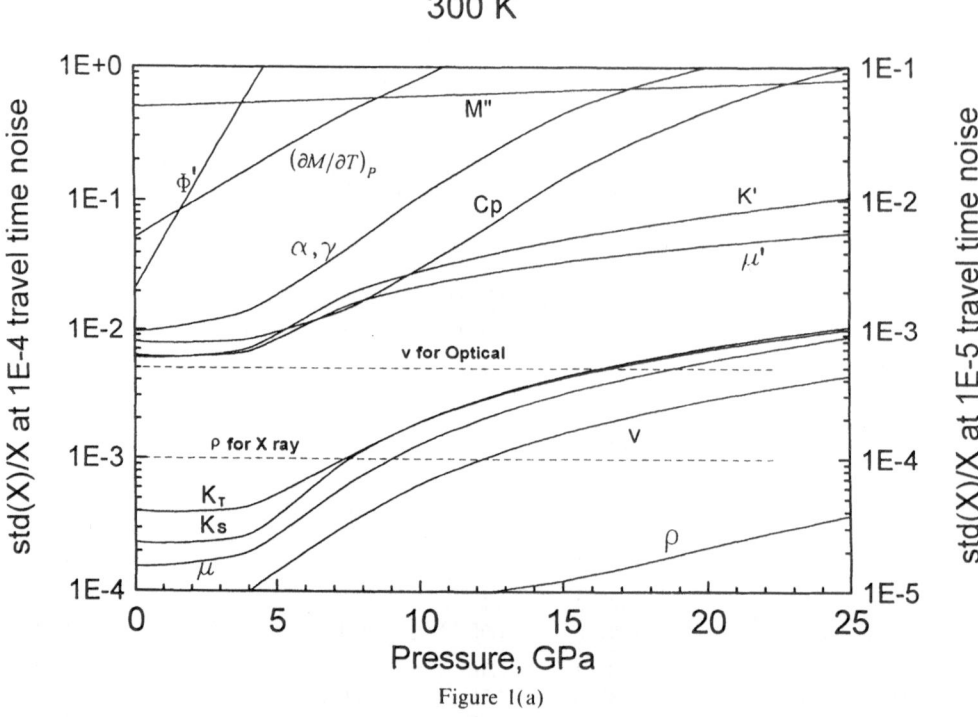

Figure 1(a)

(2) The M (set of K_S, K_T and μ) are determined as accurate as $10^{-4} \sim 10^{-3}$ even at 1500 K.

(3) The Φ (set of α, C_P, and γ) can be determined as accurate at $\sim 10^{-2}$ even at 1500 K; this implies that the present method is most useful to measure Φ under high pressure.

(4) The uncertainty of v_p and v_s, $\sim 10^{-4}$, is essentially due to the uncertainty of the initial sample length.

(5) The uncertainties of M' (the first-order pressure derivatives of M) are smaller than their variations along the pressure axis ($\Delta M'_P$) at 300 K and 1500 K, and along the temperature axis ($\Delta M'_T$) at 4 GPa; this implies that it is possible to obtain M'' (the second-order pressure derivatives of M), and $(\partial M'/\partial T)_P$ up to 4 GPa.

(6) It is not possible to obtain the $(\partial M''/\partial T)_P$ and $M^{(3)}$ (the third-order pressure derivatives of M); this can be seen by comparing the uncertainty of M'' with $\Delta M''_T$ and $\Delta M''_P$.

(7) We may obtain some information for Φ' (pressure derivative of Φ) except for γ.

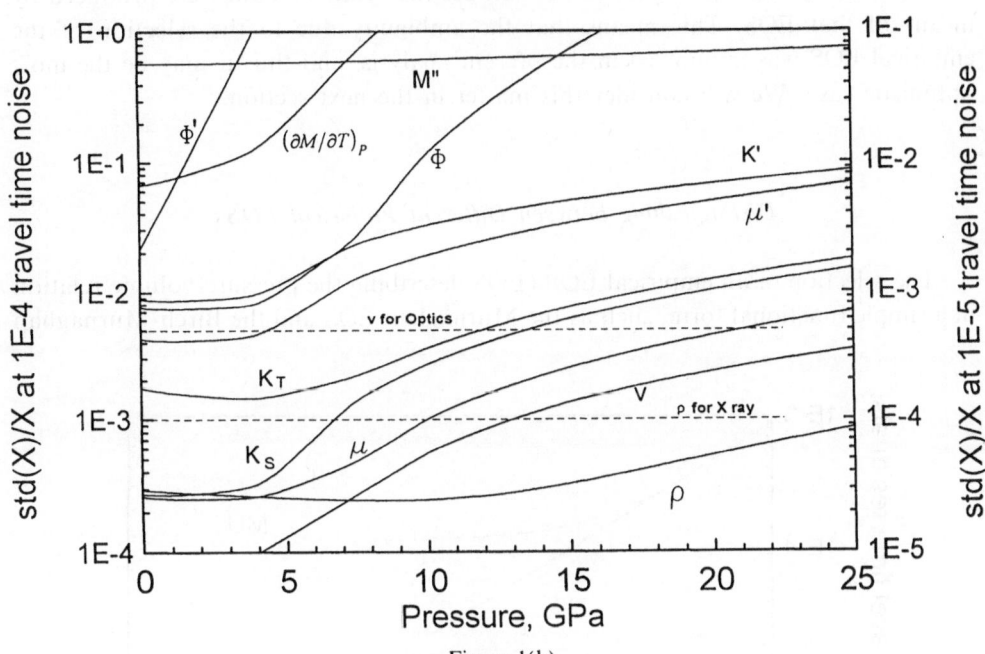

Figure 1(b)

Figure 1

Results of the error analysis up to 25 GPa and 1500 K. The left abscissa corresponds to the results for 10^{-4} noise amplitude, and the right for 10^{-5}. M, v, and Φ, shown in the figure, are abbreviations for the set of K_S, K_T, μ, for the set of v_p, v_s, and for the set of α, C_P, γ, respectively. Note that M'', $(\partial M/\partial T)_P$, and Φ' are shown schematically. Figure 1(a) normalized errors at 300 K, Figure 1(b) normalized errors at 1500 K.

Figure 1 shows the results of the performance test for assumed uncertainties of 10^{-4} and 10^{-5} in travel time when extrapolated up to 25 GPa. From the 10^{-4} uncertainty in travel time, we conclude that

(1) ρ can be determined to better than 10^{-3} up to 25 GPa and 1500 K; this corresponds to the uncertainty of the X-ray method (e.g., MAO et al., 1991; WANG et al., 1991).

(2) M can be determined well ($\sim 10^{-2}$) up to 25 GPa and 1500 K.

(3) v_p and v_s can be determined to better than 5×10^{-3} up to 25 GPa and 1500 K; this corresponds to the uncertainty of the optical methods over their experimental range (~ 13 GPa; BROWN et al., 1992).

(4) Φ can be determined with 10^{-1} uncertainty up to 15 GPa.

(5) M' can be determined with 10^{-1} uncertainty up to 25 GPa.

(6) Some information is available for M'' up to 25 GPa, for $(\partial M/\partial T)_P$ up to ~ 10 GPa, and Φ' up to 4 GPa.

Note that the present analysis in the extrapolated region was made by means of the expanded Birch-Murnaghan EOS, because the synthetic data were produced by means of that EOS. This means that the ambiguity due to the selection of the empirical EOS was minimized in the present analysis, and thus it may be the most optimistic case. We will consider this matter in the next section.

Distinguishing between Different Empirical EOSs

The selection of an empirical EOS (EOS describing the pressure-volume relation in a simple functional form, such as the Murnaghan EOS and the Birch-Murnaghan

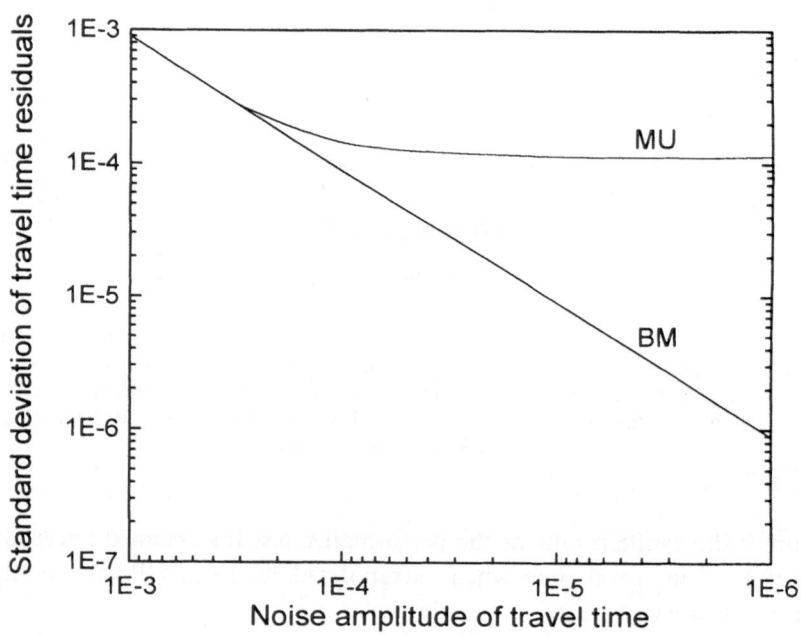

Figure 2

Standard deviations of travel-time residuals for the Birch-Murnaghan EOS (BM) and the Murnaghan EOS (MU) in the range from 0 to 4 GPa and from 300 to 1500 K are shown as a function of the noise amplitude in the travel times. We used these EOSs as examples because of their frequent use. The procedure for the calculation is quite similar to that for the extrapolation problem (see Fig. 1 and Appendix) except the following 4 items: (1) Normally distributed random noise is added to the synthesized travel-time data (amplitude 10^{-6} to 10^{-3}), to α and C_p at 0 GPa (amplitude 10^{-2}), and to ρ_0 and l_0 at 0 GPa and 300 K (amplitude 10^{-4}). (2) K_S, K_T, and μ are calculated up to 4 GPa and 1500 K. (3) The least squares method is used to determine K_{S_0}, ξ_{K_S}, η_{K_S}, K_{T_0}, ξ_{K_T}, η_{K_T}, and μ_0, ξ_μ, η_μ (see Appendix). (4) The travel times are recalculated up to 4 GPa and 1500 K and standard deviations are determined from the differences between travel times with random noise and the recalculated travel-time values. Note that the noisy data in step (1) are a substitute of experimental data. We repeated these calculations 10 times for all data points; the average of the standard deviations is shown.

EOS, etc.) to describe experimental data has been difficult, because the accuracy of experimental data has not been sufficient to differentiate between different EOSs.

The combination of the CT-EOS and accurate experimental data over a sufficient range of temperature and pressure makes it possible to compare different EOSs. In Figure 2 we compare two such equations of state. As in the previous section, noisy (10^{-6} to 10^{-3}) model data were generated with the expanded Birch-Murnaghan EOS. These noisy data were then fit with both the expanded Birch-Murnaghan (BM) and the Murnaghan (MU) equation of state. The travel-time residuals for different noise levels are plotted for both fits in Figure 2. Since the model data were generated by using the BM EOS, it is not surprising that BM EOS gives smaller travel-time residuals than the MU. Note that the purpose of this analysis is to determine the accuracy necessary to distinguish between two empirical EOSs and not to find a superior EOS, which can only be done with real experimental data; thus do not consider the BM EOS superior to the MU EOS. At 10^{-3} noise, the fits to the two EOSs give similar residuals; i.e., either empirical EOS is adequate for the analysis of the model data. At smaller noise levels, the two empirical EOSs yield substantially different residuals, and thus the selection of a specific EOS becomes feasible. For the present case, travel-time accuracy of 10^{-4} or better is required over 4 GPa and from 300 to 1500 K in order to differentiate between the suitability of the BM and the MU EOSs. Note that this unambiguous discussion can only be made because of the completeness of the CT-EOS.

Appendix: Synthesis of Model Travel-time Data and Procedure for Error Analysis

The model travel times used in this work satisfy the following requirements:
(1) The data should be representative of real minerals.
(2) The data should be "normal" at least up to 25 GPa and 1500 K (here "normal" means that neither the pressure derivative nor the temperature derivative of any parameter examined (see Table 2) changes sign).

In order to describe the pressure dependence of elastic moduli, we used the expanded Birch-Murnaghan EOS (YONEDA, 1990), because it was easy to generate "normal" data to 25 GPa. The expanded Birch-Murnaghan EOS in elastically isotropic material is written as

$$P = K_{T_0}\left(\frac{2\varepsilon}{3} + 1\right)^{5/2}\left\{\varepsilon + \left(\frac{\xi_{K_T}}{2}\right)\varepsilon^2 + \left(\frac{\eta_{K_T}}{6}\right)\varepsilon^3\right\} \tag{A1}$$

or

$$K_T = K_{T_0}\left(\frac{2\varepsilon}{3} + 1\right)^{5/2}\left\{\frac{7\varepsilon}{3} + 1 + \left(\frac{\xi_{K_T}}{2}\right)\left(\frac{9\varepsilon}{3} + 2\right)\varepsilon + \left(\frac{\eta_{K_T}}{6}\right)\left(\frac{11\varepsilon}{3} + 3\right)\varepsilon^2\right\} \tag{A2}$$

where

$$\varepsilon = \tfrac{3}{2}(V^{-2/3} - 1) \quad V: \text{scaled volume normalized by } V_0(T).$$

$$\xi_{K_T} = K'_{T_0} - 4$$

$$\eta_{K_T} = K_{T_0} K''_{T_0} + K'^2_{T_0} - 7K'_{T_0} + \frac{143}{9},$$

$$K_S = K_{S_0}\left(\frac{2\varepsilon}{3} + 1\right)^{5/2}\left\{\frac{7\varepsilon}{3} + 1 + \left(\frac{\xi_{K_S}}{2}\right)\left(\frac{9\varepsilon}{3} + 2\right)\varepsilon + \left(\frac{\eta_{K_S}}{6}\right)\left(\frac{11\varepsilon}{3} + 3\right)\varepsilon^2\right\} \quad (A3)$$

where

$$\xi_{K_S} = \frac{K_{T_0}}{K_{S_0}} K'_{S_0} - 4$$

$$\eta_{K_S} = \frac{K^2_{T_0}}{K_{S_0}} K''_{S_0} + \frac{K_{T_0}}{K_{S_0}} K'_{T_0} K'_{S_0} - 7\frac{K_{T_0}}{K_{S_0}} K'_{S_0} + \frac{143}{9},$$

$$\mu = \mu_0\left(\frac{2\varepsilon}{3} + 1\right)^{5/2}\left\{\frac{7\varepsilon}{3} + 1 + \left(\frac{\xi_\mu}{2}\right)\left(\frac{9\varepsilon}{3} + 2\right)\varepsilon + \left(\frac{\eta_\mu}{6}\right)\left(\frac{11\varepsilon}{3} + 3\right)\varepsilon^2\right\} \quad (A4)$$

where

$$\xi_\mu = \frac{K_{T_0}}{\mu_0} \mu'_0 - 4$$

$$\eta_\mu = \frac{K^2_{T_0}}{\mu_0} \mu''_0 + \frac{K_{T_0}}{\mu_0} K'_{T_0}\mu'_0 - 7\frac{K_{T_0}}{\mu_0} \mu'_0 + \frac{143}{9}.$$

Note that the expanded Birch-Murnaghan EOS is the same as the "so-called" finite strain EOS (e.g., DAVIES, 1974; ANDERSON, 1989; JEANLOZ, 1992) in spite of the different expression. Each set of equations in the expanded Birch-Murnaghan EOS must be determined at each temperature, and the scaled volume, V, is normalized to 0 GPa and a given temperature, $V_0(T)$. Note that this describes a P-V-T relation through a P-V relation at constant T. This is an alternative to a P-T relation at constant V utilizing the thermal pressure. By using the functional form of Eqs. (A3) and (A4) for K_S and μ, and the input parameters in Table 1, all the parameters were calculated in steps of 0.1 GPa from 0 to 25 GPa, and every 30 K from 300 to 1500 K. Note that: (1) in order to calculate derivatives of elastic constants and other parameters, such as ρ, α, C_P, and γ, the numerical differentiation was applied on the grid data of 0.1 GPa and 30 K intervals and that (2) K_T was calculated from K_S by means of the thermodynamic identity of Eq. (8). We checked that all of the calculated results were "normal" in the entire range examined.

The present error analysis was conducted as follows:

(1) Synthetic travel-time data were generated every 0.1 GPa from 0 to 4 GPa and every 30 K from 300 to 1500 K (41 × 41 points) by the method described above.

(2) Normally distributed random noise added to the synthesized travel-time data (amplitude: 10^{-4} or 10^{-5}), to α and C_P at 0 GPa (amplitude 10^{-2}), and to ρ and l at 0 GPa and 300 K (amplitude 10^{-4}).

(3) K_S and μ were calculated up to 4 GPa and 1500 K by using Eqs. (A3) and (A4), and the input parameters in Table 1. K_T was calculated from K_S by using Eq. (3).

(4) The least squares method was applied to the above "data" of K_S, μ and K_T, to determine the coefficients (K_{S_0}, ξ_{K_S}, η_{K_S}, μ_0, ξ_μ, η_μ, K_{T_0}, ξ_{K_T}, η_{K_T}) which appear in Eqs. (A1)–(A4).

(5) All parameters were calculated up to 25 GPa by using the nine coefficients of the three parameters (K_S, μ, and K_T) determined at (4).

We repeated this calculation 20 times, using new noise values (step 2 above) each time, and obtained standard deviations for all the parameters at a given pressure and temperature. Note that we used the three parameter (K_S, K_T and μ) extrapolation in the present calculation, although the two parameter (K_S and μ) extrapolation is possible theoretically. The three parameter extrapolation was much simpler in calculating the finite strain, ε, because Eqs. (A1) and (A2) give directly the relation between ε and P through K_T. We found that this much enhanced numerical stability in extrapolated regions. Further, in the three parameter extrapolation, C_P and γ can be calculated either through Eqs. (7) and (8), or Eqs. (7) and (13). The latter was used in the present calculation so as to avoid numerical instabilities due to Eq. (8).

Acknowledgments

This paper and its companion paper (previous paper in this issue) are dedicated to the late Professor Edward Schreiber (see more extensive acknowledgment in the companion paper). Thanks to Professors Orson L. Anderson and Ian Jackson for constructive criticism. We hope to have used their comments wisely and improved the clarity of the paper. This work was supported by the National Science Foundation through grant EAR–8916327 and a matching grant from the University of Colorado. One of the authors (AY) received additional support from Yamada Science Foundation and the Seismological Society of Japan while visiting CIRES at the University of Colorado.

REFERENCES

ANDERSON, D. L., *Theory of the Earth* (Blackwell Scientific Publications, Boston 1989) pp. 91–92.

ANDERSON, O. L., ISAAK, D. L., and ODA, H. (1991), *Thermoelastic Parameters for Six Minerals at High Temperature*, J. Geophys. Res. *96*, 18,037–18,046.

BASSETT, W. A., SHIMIZU, H., and BRODY, E. M., *Pressure dependence of elastic moduli of forsterite by Brillouin scattering in diamond cell*. In *High-pressure Research in Geophysics* (eds. Akimoto, S., and Manghanani, M. H.) (Center for Academic Publications Japan, Tokyo 1982) pp. 115–124.

BROWN, J. M., SLUTSKY, L. J., ABRAMSON, E. (1992), *Applications of Impulsive Stimulated Scattering in Mineral Physics: Elasticity, Equation of State, Thermal Diffusivity, and Structural Relaxation*, Eos Trans. Amer. Geophys. Union, Fall Meeting Supplement, 555.

BROWN, J. M., SLUTSKY, L. J., NELSON, K. A., and CHENG, L. T. (1989), *Single-crystal Elastic Constants for San Carlos Peridot: An Application of Impulsive Stimulated Scattering*, J. Geophys. Res. *94*, 9485–9492.

DANDEKAR, D. P. (1970), *Iterative Procedure to Estimate the Values of Elastic Constants of a Cubic Solid at High Pressure from the Sound Wave Velocity Measurements*, J. Appl. Phys. *41*, 667–672.

DAVIES, G. F. (1974), *Effective Elastic Moduli under Hydrostatic Stress — I. Quasi-harmonic Theory*, J. Phys. Chem. Solids *35*, 1513–1520.

GETTING, I. C., and KENNEDY, G. C. (1970), *Effect of Pressure on the emf of Chromel-alumel and Platinum-platinum 10% Rhodium Thermocouples*, J. Appl. Phys. *41*, 4522–4562.

GETTING, I. C., and SPETZLER, H. (1993), *Gas-charged piston-cylinder apparatus for pressures to 4 GPa*. In *High-pressure Science and Technology* (eds. Schmidt, S. C., Shaner, J. W., Samara, G. A., and Ross, M.) (American Institute of Physics, Woodbury, New York 1993) pp. 1581.

GOTO, T., and ANDERSON, O. L. (1988), *An Apparatus for Measuring Elastic Constants of Single Crystals by a Resonance Technique up to 1825 K*, Rev. Sci. Instrum. *59*, 1405–1408.

ISAAK, D. G., ANDERSON, O. L., GOTO, T., and SUZUKI, I. (1989), *Elasticity of Single-crystal Forsterite Measured to 1700 K*, J. Geophys. Res. *94*, 5895–5906.

JACKSON, I., and NIESLER, H., *The elasticity of periclase to 3 GPa and some geophysical implications*. In *High-pressure Research in Geophysics* (eds. Akimoto, S., and Manghnani, M. H.) (Center for Academic Publications Japan, Tokyo 1982) pp. 93–113.

JACKSON, I., NIESLER, H., and WEIDNER, D. J. (1981), *Explicit Correction of Ultrasonically Determined Elastic Wave Velocities for Transducer-bond Phase Shift*, J. Geophys. Res. *86*, 3736–3748.

JEANLOZ, R., *Differential finite strain equation of state*. In *High-pressure Research: Application to Earth and Planetary Sciences* (eds. Syono, Y., and Manghnani, M. H.) (TERRAPUB, Tokyo/American Geophysical Union, Washington, D.C. 1992) pp. 147–156.

KITTEL, C., *Introduction to Solid State Physics*, 6th edition (John Wiley & Sons, New York 1986) pp. 83–98.

MAO, H. K., HEMLEY, R. J., FEI, Y., SHU, J. F., CHEN, L. C., and BASSETT, W. A. (1991), *Effect of Pressure, Temperature, and Composition on Lattice Parameters and Density of (Fe, Mg)SiO$_3$-Perovskites to 30 GPa*, J. Geophys. Res. *96*, 8069–8079.

NIESLER, H., JACKSON, I., and EDWIN, C. M. (1988), *Calibration and Intercomparison of Minalpha and Manganin Resistance Pressure Gauge to 3 GPa*, High Temp.-High Pressures *20*, 495–508.

SPETZLER, H., CHEN, G., WHITEHEAD, S., and GETTING, I. C. (1993), *A New Ultrasonic Interferometer for the Determination of Equation of State Parameters of Sub-millimeter Single Crystals*, Pure Appl. Geophys., *141*, 341–377.

SPETZLER, H., SAMMIS, C. G., and O'CONNELL, R. J. (1972), *Equation of State of NaCl: Ultrasonic Measurements to 8 kbar and 800°C and Static Lattice Theory*, J. Phys. Chem. Solids *33*, 1727–1750.

SUZUKI, I. (1975), *Thermal Expansion of Periclase and Olivine, and their Anharmonic Properties*, J. Phys. Earth *23*, 145–159.

WANG, Y., WEIDNER, D. J., LIEBERMANN, R. C., LIU, X., KO, J., VAUGHAN, M. T., ZHAO, Y., YEGANEN-HAERI, A., and PACALO, R. E. G. (1991), *Phase Transition and Thermal Expansion of MgSiO$_3$ Perovskite*, Science *251*, 410–413.

WATANABE, H. (1982), *Thermomechanical properties of synthetic high-pressure compounds relevant to the earth's mantle*. In *High-pressure Research in Geophysics* (eds. Akimoto, S., and Manghnani, M. H.) (Center for Academic Publications Japan, Tokyo 1982) pp. 441–464.

WEIDNER, D. J. (1975), *Elasticity of Microcrystals*, Geophys. Res. Lett. *2*, 189–192.

YONEDA, A. (1990), *Pressure Derivatives of Elastic Constants of Single Crystal MgO and MgAl$_2$O$_4$*, J. Phys. Earth *38*, 19–55.

YONEDA, A., SPETZLER, H., and GETTING, I. (1993), *Implication of the complete travel time equation of state for a new pressure scale*. In *High-pressure Science and Technology* (eds. Schmidt, S. C., Shaner, J. W., Samara, G. A., and Ross, M.) (American Institute of Physics, Woodbury, New York 1993) pp. 1609.

(Received April 13, 1993, revised October 1993, accepted November 1993)

PAGEOPH, Vol. 141, No. 2/3/4 (1993)

0033–4553/93/040393–22$1.50 + 0.20/0

The Elastic Properties of Single-crystal Fayalite as Determined by Dynamical Measurement Techniques

Donald G. Isaak,[1] Earl K. Graham,[2] Jay D. Bass[3] and Hong Wang[3]

Abstract—We present new elasticity measurements on single-crystal fayalite and combine our results with other data from resonance, pulse superposition interferometry, and Brillouin scattering to provide a set of recommended values for the adiabatic elastic moduli C_{ij} and their temperature variations. We use a resonance method (RPR) with specimens that were previously investigated by pulse superposition experiments. The nine C_{ij} of fayalite are determined from three new sets of measurements. One set of our new C_{ij} data is over the range 300–500 K. We believe that the relatively large discrepancies found in some C_{ij} are due in large part to specimen inhomogeneities (chemical and microstructural) coupled with differences in the way various techniques sample, rather than only systematic errors associated with experimental procedures or in the preparations of the specimens.
Our recommended C_{ij}'s (GPa) and $(\partial C_{ij}/\partial T)_P$ (GPa/K) are:

ij	11	22	33	44	55	66	12	13	23
C_{ij}	266	168	232	32.3	46.5	57	94	92	92
	(4)	(6)	(7)	(0.5)	(0.3)	(1)	(1)	(8)	(5)
$\left(\dfrac{\partial C_{ij}}{\partial T}\right)_P$	−0.051	−0.041	−0.043	−0.0092	−0.0080	−0.020	−0.023	−0.012	−0.0072
	(0.001)	(0.010)	(0.005)	(0.0010)	(0.0015)	(0.002)	(0.004)	(0.001)	(0.0068)

The resulting values for the isotropic bulk and shear moduli, K_S and μ, and their temperature derivatives: $K_S = 134(4)$ GPa; $\mu = 50.7(0.3)$ GPa; $(\partial K_S/\partial T)_P = -0.024(0.005)$ GPa/K; and $(\partial \mu/\partial T)_P = -0.013(0.001)$ GPa/K. An important conclusion is that K_S increases as the Fe/(Fe + Mg) ratio in olivine is increased.

Key words: Fayalite, elastic moduli, olivine, resonance, pulse-superposition, Brillouin scattering.

[1] Center for Chemistry and Physics of Earth and Planets, Institute of Geophysics and Planetary Physics, UCLA, Los Angeles, CA 90024-1567, U.S.A. Also at Azusa Pacific University.
[2] Department of Geosciences Pennsylvania State University, University Park, PA 16802, U.S.A.
[3] Department of Geology, University of Illinois, Urbana, IL 61806-2999, U.S.A.

1. Introduction

Olivine $(Mg, Fe, Mn, Ca)_2 SiO_4$ is the major mineral constituent in most representations of upper mantle petrology (e.g., HARTE, 1983; BASS and ANDERSON, 1984; WEIDNER, 1985; BINA and WOOD, 1986; and DUFFY and ANDERSON, 1989). Although the forsterite $Mg_2 SiO_4$ end-member is of major compositional interest, an accurate description of the physical properties of the fayalite $Fe_2 SiO_4$ fraction is essential for a complete understanding of the state and dynamics of the lithosphere and outermost mantle. Not only is $Fe/(Mg + Fe)$ in the olivine solid-solution a critical factor in compositional modeling (YAN *et al.*, 1989), but iron content is also relevant to the oxygen fugacity and deformation properties of the upper mantle (MACKWELL and KOHLSTEDT, 1990; MACKWELL, 1992). In order to accurately describe the pertinent material properties of intermediate olivine compositions, it is imperative that the Mg and Fe end-members be characterized as well as possible. While there is an extensive set of accurate data describing the elastic properties of natural and synthetic forsterite (e.g., SUMINO and ANDERSON, 1984; ISAAK *et al.*, 1989), the complementary data for the fayalite end-member is sparse and somewhat discrepant.

SUMINO (1979) reported data on the nine adiabatic elastic moduli C_{ij} of single-crystal fayalite up to 673 K at ambient pressure using the rectangular parallelepiped resonance (RPR) technique. GRAHAM *et al.* (1988) used pulse superposition interferometry to obtain the C_{ij} of fayalite at ambient conditions (298 K), and up to pressures of 1.0 GPa (at ambient T) and over a temperature range of 273–313 K (at ambient P). The final C_{ij} values given by GRAHAM *et al.* (1988) represent the most consistent fit of many redundancy checks (24 data sets with 17 different modes) using three different single-crystal specimens cut from a single synthetic boule. While the results of GRAHAM *et al.* (1988) are generally consistent with those of SUMINO (1979), there are some significant differences relative to the stated uncertainties. These were pointed out and discussed in the GRAHAM *et al.* (1988) paper, but the specific causes of the differences are unclear. In particular, discrepancies in the two data sets were noted in the two compressional moduli, C_{22} and C_{33}, and in two off-diagonal moduli, C_{13} and C_{23}. Thus the isotropic adiabatic bulk modulus K_S at ambient P, T conditions found by SUMINO (1979), 138.1(0.8) GPa (Hashin-Shtrikman average using Sumino's C_{ij} data; Sumino reports $K_S = 137.9(1.4)$ GPa using the Voigt-Reuss-Hill average), is larger than the K_S found by GRAHAM *et al.* (1979), 127.9(0.6) GPa.

The differences in measured values of K_S for fayalite cited above are troubling for three reasons. First, an 8% discrepancy in the measurement for K_S is large relative to the estimated uncertainties in the experimental techniques. Different dynamic techniques were used by SUMINO (1979) (RPR) and GRAHAM *et al.* (1988) (pulse superposition). However, both of these methods are widely recognized as being extremely precise; differences of 8% between measured moduli are considerably larger than those expected from precision considerations (less than 1% for both methods). Second, since K_S for forsterite is very near 129 GPa, the uncertainty in the fayalite results makes it unclear whether increasing iron in the olivine solid solution results in an increase in K_S. Third, the range in measured values of K_S for fayalite (128–138 GPa) is larger than usually found for other minerals when comparing results from different laboratories. We illustrate this third point by referring to the data of three minerals: MgO, Mg_2SiO_4, and grossular garnet. Values of K_S for MgO range from 161.8–163.9 GPa at ambient conditions when results from pulse superposition, resonance, and Brillouin experiments are compared (BOGARDUS, 1965; ANDERSON and ANDREATCH, 1966; CHANG and BARSCH, 1969; SPETZLER, 1970; SUMINO et al., 1976; JACKSON and NIESLER, 1982; SUMINO et al., 1983; ISAAK et al., 1989; BASS, 1989; YONEDA, 1990). In the case of forsterite (Mg_2SiO_4), values for K_S at ambient conditions range from 126.5–129.2 GPa as seen from two pulse superposition and three resonance studies (GRAHAM and BARSCH, 1969; KUMAZAWA and ANDERSON, 1969; SUMINO et al., 1977; SUZUKI et al., 1983; ISAAK et al., 1989). Finally, ISAAK et al. (1992) used the resonance technique and found $K_S = 167.8$ GPa at ambient conditions for a near end-member grossular garnet, compared to 168.4 GPa found by BASS (1989) using the Brillouin method.

In response to the discrepancies noted between the fayalite data of SUMINO (1979) and GRAHAM et al. (1988), further measurements on fayalite were made by WANG et al. (1989) and by ISAAK and ANDERSON (1989). WANG et al. (1989) used Brillouin scattering to measure the C_{ij} of a natural specimen. The preliminary report of ISAAK and ANDERSON (1989) is of particular interest insofar as they used the RPR technique to measure one specimen that was also reported on in detail by GRAHAM et al. (1988). Although ISAAK and ANDERSON (1989) reported only one set of measured C_{ij} values, three complete sets of data for the nine C_{ij} moduli were eventually obtained on two of the specimens described by GRAHAM et al. (1988). One set of our RPR data is over the range 300–500 K.

The purpose of the present paper is threefold. First, we wish to provide a more complete description of the C_{ij} data reported by ISAAK and ANDERSON (1989), thereby expanding the overall data base for the elastic properties of fayalite. Second, we wish to discuss possible reasons for the observed discrepancies in the measured values of some of the C_{ij} for fayalite in light of the precision and accuracy of the different techniques that have been used. A third goal is to provide a recommended set of fayalite C_{ij} values together with their temperature derivatives as determined from all the available dynamic data.

2. New RPR C_{ij} Data for Fayalite

a. Experimental Technique and Specimens

In the RPR method single-crystal adiabatic elastic moduli C_{ij} are determined from the natural resonant frequencies of a specimen together with information on the specimen's edge lengths and density (OHNO, 1976; SUMINO *et al.*, 1976; ISAAK *et al.*, 1989). The specimen must be oriented and cut along the principle crystallographic axes into the shape of a rectangular parallelepiped. In principle, nine modal frequencies are required to uniquely determine the nine C_{ij} of materials such as fayalite with orthorhombic symmetry; every modal frequency contains information about each of the nine C_{ij}. Typically several redundant modal frequencies are observed and the C_{ij} are determined by a least-squares fitting procedure.

ISAAK and ANDERSON (1989) report C_{ij} values for one synthetic specimen used by GRAHAM *et al.* (1988) (referred to as $A1$ in the paper by GRAHAM *et al.*). In fact, two complete sets of RPR measurements were made using this specimen at two different parallelepiped volumes. We first prepared specimen $A1$ into a rectangular parallelepiped with edge lengths 2.627(3), 2.112(4), and 2.982(4) mm corresponding to the respective [100], [010], and [001] crystallographic axes. This prepared specimen is referred to as RPR-$A1a$. We observed 42 resonant modes for RPR-$A1a$ at ambient P, T conditions from which the nine independent C_{ij} are computed. Specimen RPR-$A1a$ was then reduced in size to 2.605(2), 2.044(3), and 1.366(5) mm along the respective [100], [010], and [001] axes. This smaller specimen was labelled RPR-$A1b$, and 47 resonant modes were observed at ambient conditions in order to determine the C_{ij}.

A third set of resonant data consisting of 47 modal frequencies was obtained on a rectangular parallelepiped specimen (identified here as RPR-B) prepared from specimen B described by GRAHAM *et al.* (1988). GRAHAM *et al.* (1988) characterized specimen B, but did not include elasticity data from this specimen in their results because of its small size. SOPKIN (1982) reports values for the C_{11}, C_{55} and C_{66} moduli at ambient P, T for this specimen. GRAHAM *et al.* (1988) determined only the [100] crystal axis of B (see their Table 3) using a Laue back-reflection X-ray camera. We used the Laue X-ray method to complete the orientation of this specimen along the [010] and [001] axes. RPR-B has edge lengths of 2.770(1), 3.850(2), and 2.576(1) mm along the respective [100], [010], and [001] axes.

When calculating the C_{ij} moduli from the resonant frequencies f_i, the specimen edge lengths l_i, and the density ρ, we used $\rho = 4.387(6)$ gm/cm^3 as determined by GRAHAM *et al.* (1988) to represent the density for both RPR-$A1a$ and RPR-$A1b$, and $\rho = 4.377(6)$ gm/cm^3 for RPR-B. Our use of ρ from specimens $A1$ and B of GRAHAM *et al.* (1988) requires the assumption that the specimens are homogeneous since the volumes of $A1$ and B are larger than RPR-$A1a,b$ and RPR-B. Specimens $A1$, RPR-$A1a$, and RPR-$A1b$ have respective volumes of approximately 1000, 17, and 7 mm^3. The volume of RPR-B is about 27 mm^3 and about four times smaller than the original B. Neither of the specimens $A1$ and B described by GRAHAM *et al.* (1988) was characterized for the presence of Fe^{+3}.

b. Results

The resonance spectra for *RPR-A*1*a*, *RPR-A*1*b*, and *RPR-B* were used to generate three separate sets of the nine C_{ij} moduli of fayalite. These are given in Table 1. We also include in Table 1 values for the isotropic bulk quantities, K_S, μ, V_p, and V_s, where V_p and V_s are the compressional and shear wave velocities, respectively. The errors indicated in Table 1 represent the standard deviations determined from the differences between each measured frequency and that frequency calculated from the final set of C_{ij}. Since each modal frequency contains information on all the C_{ij}, the difference in measured and calculated frequencies for a particular mode associates an uncertainty with the C_{ij}. The final standard deviation of a given C_{ij} represents the root-mean-square average of these uncertainties over all the measured frequencies. Thus, the standard deviations reflect how well the measured redundant frequencies correlate with each other.

We include in Table 1 (last column) the C_{ij}, K_S, μ, V_p, and V_s values representing the weighted averages of the three data sets. *RPR-A*1*a* and *RPR-A*1*b* are considered separate in this scheme since they represent significantly different sampling volumes. We find that the values of the C_{ij} for *RPR-A*1*a* and *RPR-A*1*b* are generally within the overlap of these errors. Exceptions are the C_{12}, C_{44} and C_{55} moduli, although the absolute differences between *RPR-A*1*a* and *RPR-A*1*b* for these two moduli are very small. We find a tendency for the C_{ij} to increase slightly as the specimen volume is decreased from about 17 mm³ to less than half of its size

Table 1

Measured elastic properties (new RPR results) of fayalite

	Specimen			
	1 *RPR-A*1*a*	2 *RPR-A*1*b*	3 *RPR-B*	weighted† avg. of 1–3
ρ (g/cm³)	4.387(0.006)	4.387(0.006)	4.377(0.006)	
C_{11} (GPa)	267.7(1.4)	270.3(2.5)	270.2(1.6)	269.0(1.6)
C_{22} (GPa)	171.7(1.0)	173.1(1.5)	170.6(1.2)	171.6(1.4)
C_{33} (GPa)	229.8(1.5)	234.7(2.4)	234.7(1.6)	232.5(2.8)
C_{44} (GPa)	32.13(0.06)	32.73(0.06)	32.28(0.05)	32.4(0.3)
C_{55} (GPa)	46.19(0.11)	46.86(0.10)	46.50(0.10)	46.5(0.3)
C_{66} (GPa)	57.53(0.14)	57.65(0.16)	57.26(0.12)	57.4(0.3)
C_{12} (GPa)	93.1(1.1)	95.6(1.7)	95.7(1.4)	94.4(1.5)
C_{13} (GPa)	95.1(1.5)	98.2(2.5)	98.5(1.7)	96.9(1.9)
C_{23} (GPa)	92.8(1.2)	94.2(1.9)	94.5(1.3)	93.7(0.9)
K_S (GPa)	134.6(0.7)	136.9(1.3)	136.6(0.8)	135.8(1.3)
μ (GPa)	50.8(0.5)	51.3(0.5)	50.8(0.5)	50.9(0.3)
v_p (km/s)	6.79(0.02)	6.84(0.02)	6.83(0.02)	6.82(0.03)
v_s (km/s)	3.40(0.02)	3.42(0.02)	3.41(0.02)	3.41(0.01)

† Uncertainties determined by $\sqrt{\Sigma (x_i - \bar{x})^2/(N-1)}$, where $\bar{x}$ is weighted average value and N is 3.

at a volume of 7 mm^3. We believe that the reasons for the differences between the C_{ij} for *RPR-A*1*a* compared to *RPR-A*1*b* are related primarily to the presence of small inhomogeneities (compositional and microstructural) that could affect the results since the volumes are different by more than a factor of two, and to a lesser extent unavoidable errors in how well the specimens were oriented and prepared as rectangular parallelepipeds. As discussed later, the fact that the loading force on the specimens during RPR experiments is nonzero may contribute to these differences as well.

3. Comparison of Fayalite C_{ij} Data from Dynamical Experiments

Table 2 includes the results for C_{ij}, K_S, μ, V_p, and V_s found by the three different dynamical methods of measurement considered in this study and synthesis: ultrasonic pulse superpositions (GRAHAM *et al.*, 1988); Brillouin scattering (WANG *et al.*, 1989); and resonance (RPR) (SUMINO *et al.*, 1979 and the present new results). These measured values for C_{ij}, K_S, and μ are also illustrated in Figures 1–4. We do not include the three C_{ij} reported by SOPKIN (1982) for Graham's specimen *B* in Table 2 since they are an incomplete data set. However, we note that the Sopkin values of $C_{11} = 266.8$ GPa, $C_{55} = 46.6$ GPa, and $C_{66} = 56.9$ GPa are in near agreement with our present values for specimen *RPR-B*, which was cut from

Table 2

Values of elastic moduli of fayalite: four sets of dynamical measurements and recommended averaged values

	(1)	(2)	(3)	(4)	Recommended
ρ (g/cm^3)	4.400(0.009)	4.387(0.006)	4.38(0.05)	4.387(0.006)	4.393
		4.377(0.006)		4.377(0.006)	
C_{11} (GPa)	267.0(1.9)	265.9(0.4)	260.3(1.6)	269.0(1.6)	266(4)
C_{22} (GPa)	173.6(1.1)	160.3(0.1)	167.6(1.5)	171.6(1.4)	168(6)
C_{33} (GPa)	239.2(1.4)	222.4(0.4)	233.6(1.8)	232.5(2.8)	232(7)
C_{44} (GPa)	32.4(0.1)	31.6(0.2)	32.9(0.4)	32.4(0.3)	32.3(0.5)
C_{55} (GPa)	46.7(0.1)	46.7(0.2)	46.0(1.0)	46.5(0.3)	46.5(0.3)
C_{66} (GPa)	57.3(0.1)	57.2(0.1)	55.1(0.6)	57.4(0.3)	57(1)
C_{12} (GPa)	95.2(1.5)	92.4(1.4)	93.9(2.1)	94.4(1.5)	94(1)
C_{13} (GPa)	98.7(1.6)	80.6(1.2)	92.5(3.0)	96.9(1.9)	92(8)
C_{23} (GPa)	97.9(1.2)	88.4(1.4)	88.5(3.0)	93.7(0.9)	92(5)
K_S† (GPa)	138.1(0.8)	127.9(0.6)	132.2(2.0)	135.8(1.3)	134(4)
μ† (GPa)	51.0(0.5)	50.3(0.6)	50.5(0.7)	50.9(0.3)	50.7(0.3)
v_p† (km/s)	6.84(0.02)	6.67(0.02)	6.75(0.05)	6.82(0.03)	6.77(0.07)
v_s† (km/s)	3.40(0.02)	3.39(0.02)	3.40(0.03)	3.41(0.01)	3.40(0.01)

(1) SUMINO (1979); (2) GRAHAM *et al.* (1988); (3) WANG *et al.* (1989); (4) Present data (Table 1).
† Hashin-Shtrikman averages

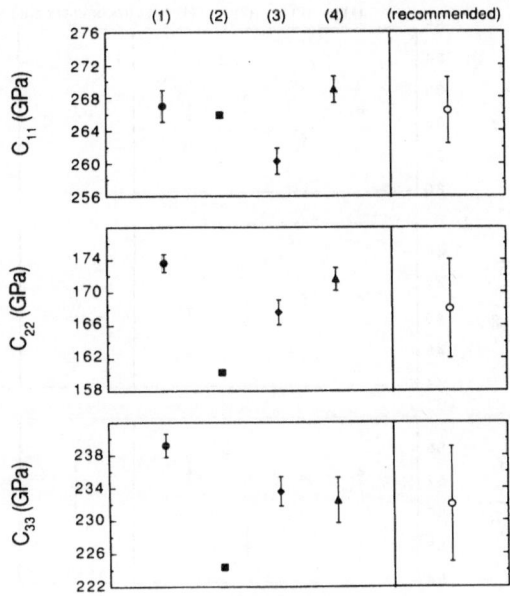

Figure 1

C_{11}, C_{22}, and C_{33} for fayalite at ambient P, T from SUMINO (1979) (1); GRAHAM et al. (1988) (2); WANG et al. (1989) (3); and the present RPR data (4). The value on the far right is the recommended value from the average of (1)–(4).

the original specimen B. The C_{11} (270.2(1.6) GPa) and C_{66} (57.3(0.1) GPa) values from RPR-B are marginally larger than the SOPKIN (1982) values for these moduli.

We have also provided in Table 2 (last column) and Figures 1–4 our recommended values for the C_{ij} of fayalite at ambient P, T based on a simple averaging of the four separate data sets. We do not use a weighted averaging scheme because: (1) each of the experimental methods is characterized by unclear systematic errors; and (2) even for a particular experimental technique the uncertainties can have different meaning when comparing results from different reports. For example, the errors cited by SUMINO (1979) (column 1 of our Table 2) indicate the extent of self-consistency between the 28 vibrational modes that were observed. The standard deviations in columns 1–3 of Table 1 for specimens RPR-A1a, RPR-A1b, and RPR-B should be interpreted in a similar manner. However, the uncertainties of the best set of C_{ij} found from our new RPR data (column 4 in Tables 1 and 2) represent the standard deviation of the average C_{ij} values for RPR-A1a, RPR-A1b, and RPR-B.

We find in most cases good agreement between the C_{ij} of the different sets of data given in Table 2 and Figures 1–4. The agreement in the calculated isotropic shear moduli between the data sets is excellent. However, there are some instances in which the values from one of the four measurements lies significantly outside the

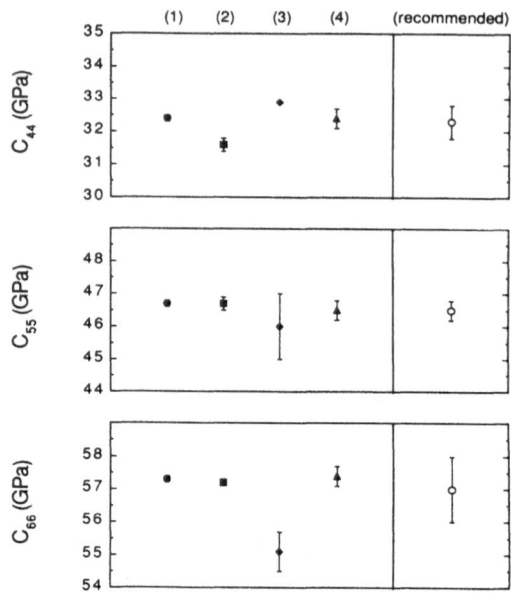

Figure 2

C_{44}, C_{55}, and C_{66} for fayalite at ambient P, T from SUMINO (1979) (1); GRAHAM *et al.* (1988) (2); WANG *et al.* (1989) (3); and the present RPR data (4). The value on the far right is the recommended value from the average of (1)–(4).

error bars of the other three. The relatively large differences between the GRAHAM *et al.* (1988) values for C_{22}, C_{33}, and C_{13} and the values for these moduli reported from the present RPR work are especially perplexing since one of the specimens is common to both techniques. It is primarily the lower values found by GRAHAM *et al.* (1988) for these three moduli that result in their lower K_S value. These discrepancies in C_{22}, C_{33}, C_{13}, and therefore, K_S, are larger than might be expected from the indicated precisions of the experimental techniques. It is worth discussing whether systematic errors in the RPR and pulse superposition techniques could produce these discrepancies.

Large errors in the calculated C_{ij} can result when interpreting RPR frequency spectra if there are few or no redundant modes, or if some modal frequencies are misidentified (i.e., incorrect assignments of measured frequencies are made relative to the characteristic type of resonant modes). However, when many frequencies are used, the effect on the C_{ij}'s of switching two or more modes tends to be averaged out by the other modes. Furthermore, it usually is apparent when mode assignments are reversed since the differences between the measured and calculated frequencies for these modes are anomalously large and have opposite signs. For these reasons and the fact that we used over forty modes in our measurements on each of specimens *RPR-A1a*, *RPR-A1b*, and *RPR-B*, we conclude that it is highly

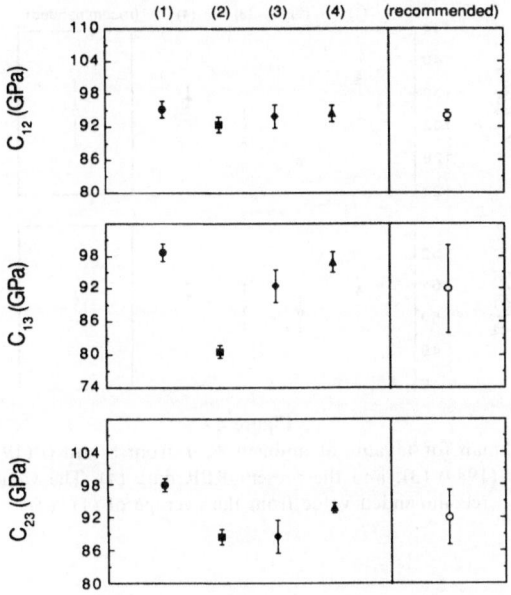

Figure 3

C_{12}, C_{13}, and C_{23} for fayalite at ambient P, T from SUMINO (1979) (1); GRAHAM *et al.* (1988) (2); WANG *et al.* (1989) (3); and the present RPR data (4). The value on the far right is the recommended value from the average of (1)–(4).

improbable that discrepancies in the C_{22}, C_{33}, and C_{13} values are due to modal misidentification in the RPR experiments.

Imperfect crystallographic orientation and specimen parallelepiped shape errors can also affect the C_{ij} values in RPR experiments. It is unlikely these are primary factors for discrepancies as large as those observed for the C_{22}, C_{33}, and C_{13} moduli for the following reasons. The orientations of the specimens were made initially by GRAHAM *et al.* (1988). For the RPR experiments with *RPR-A*1*a*, *RPR-A*1*b*, and *RPR-B* these same initial orientations were used (the orientation of *RPR-B* was carried out as described previously). The accuracy of orientation is typically to within 0.5–1°. In the event that anomalous errors occurred while orienting the crystal axes both the RPR and the pulse superposition data would probably be affected similarly. For a given orientation the RPR data reduction procedure assumes a perfect rectangular parallelepiped specimen. By carefully measuring at several positions each specimen's edge lengths, we were able to establish that all sides are parallel to each other to within 0.1°. The only exception is for *RPR-A*1*b*, in which one pair of sides is parallel to about 0.25°. This precision in preparing the parallelepipeds, and the fact that reasonably consistent values for the C_{ij} are observed among the *RPR-A*1*a*, *RPR-A*1*b*, and *RPR-B* specimens, imply that

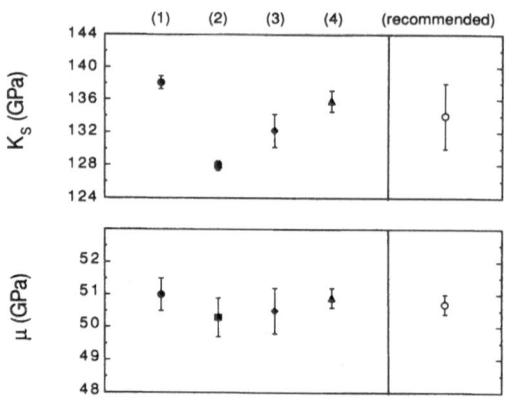

Figure 4

K_S and μ (Hashin-Shtrikman for fayalite at ambient P, T from SUMINO (1979) (1); GRAHAM *et al.* (1988) (2); WANG *et al.* (1989) (3); and the present RPR data (4). The value on the far right is the recommended value from the average of (1) (4).

inaccuracies in RPR specimen geometry make reasonably insignificant contributions to the major discrepancies between RPR and pulse superposition results.

It is important to address one other possibility for large systematic errors in RPR measurements. This source is associated with the nonzero force required to hold the specimen between transducers. The data reduction in the RPR technique assumes zero stress on the specimen. We estimate the resonant frequencies at zero stress by measuring the effect of loading force on the frequencies over a range of holding forces, from around 8 down to 0.5 gram wt., and then extrapolate to zero gram wt. (SUMINO *et al.*, 1976). It is reasonable to expect that any errors, if significant at all, when extrapolating from a nonzero loading force to zero stress will affect the data of small specimens more than large ones since localized strains associated with the holding force at the specimen corners would involve a relatively larger portion of the crystal. The results shown in Table 1 for specimens *RPR-A1b*, *RPR-A1a*, and *RPR-B* are obtained from parallelepiped volumes of approximately 7, 17, and 27 mm^3, respectively. With the possible exception of specimen *RPR-A1b*, which has C_{ij} moduli results that are marginally lower than those for the other two specimens, the three data sets are compatible within the estimated errors (which are invariably less than 1%). It is not entirely clear whether the slightly lower values of specimen *RPR-A1a* result from loading specimen inhomogeneities in either composition or crystal defect structure, or preparation (geometric) errors. In any case, the internal inconsistencies among the three RPR specimens are negligible relative to the discrepant C_{ij} moduli observed between the RPR, ultrasonic, and Brillouin scattering results (in particular, C_{22}, C_{33}, and C_{13} for the ultrasonic data, and C_{11} and C_{66} for the scattering results; see Figures 1–3).

There are also possible systematic errors associated with the pulse superposition measurements that could be large enough to cause the differences in values of the C_{22}, C_{33}, and C_{13} moduli compared to RPR. In the pulse superposition method (MCSKIMIN, 1961) the delay time of a series of rf pulses traveling through a crystal is measured. The time interval T_r between pulses is given by (SCHREIBER et al., 1973)

$$T_r = p\delta - \frac{p\gamma}{360f} + \frac{n}{f},$$

where δ is the true round trip delay time through the crystal, γ is a phase angle associated with the transducer-crystal bond, f is the frequency of the rf pulse, p is an integer representing the number of round trips through the crystal, and n is an integer which relates the phasing of the rf pulses to each other. For the case that $p = 1$ and γ is not significant, we find that

$$T_r = \delta + \frac{n}{f}.$$

Clearly δ is found if $n = 0$. If $n \neq 0$ for a particular mode (wave polarization, propagation direction) then an erroneous C_{ij} (or combination of C_{ij}'s) will result. This is the most likely source of relatively large errors when using the pulse superposition method. However, there are several means to cross-check the data so as to significantly reduce the possibility of n being other than zero. GRAHAM et al. (1988) used both the method of shifting carrier frequencies (MCSKIMIN, 1961) and varied the crystal path length (JACKSON and NIESLER, 1982) to check for the $n = 0$ conditions of all primary modes. Furthermore, the final results of GRAHAM et al. (1988) represent the overall best fit of 24 sets of measurements carried out on three different specimens. Relative to all of the multiple cross checks that were obtained their data are internally self-consistent. Indeed, SOPKIN (1982) demonstrates that when some n conditions are adjusted so as to bring the GRAHAM et al. (1988) results into better coincidence with the SUMINO (1979) RPR data, the internal consistency of the Graham results diminishes significantly. Therefore, in terms of the total assemblage of measured modes considered in the GRAHAM et al. (1988) study, the n conditions chosen satisfy the observations the best. This does not, however, entirely preclude the possibility of an improper selection, insofar as the 24 sets of measurements representing 17 different modes were collectively obtained on three different specimens cut from the original synthetic boule. Compositional and textural differences (e.g., occurrence of microcracks) between the specimens could have caused a systematic bias in the measurement set.

The errors associated with the RPR results in Table 1 are generally a factor of two or more larger than found for the C_{ij} of forsterite (ISAAK et al., 1989). Exceptions are the C_{44}, C_{55}, and C_{66} moduli, which have very small errors for both the forsterite and fayalite data. We find larger errors in the fayalite data in spite of using up to 20 more modal frequencies to constrain the C_{ij}. The degree to which the specimens approach perfect parallelepipeds with sides perpendicular to the crystallographic axes is comparable between the forsterite and fayalite samples. These larger errors in the fayalite results therefore suggest the presence of crystal inhomogeneities, possibly both chemical and structural, in the specimens.

The last set of results we consider on the elasticity of fayalite are those obtained using the Brillouin scattering technique (Table 2, column 3). The basis of this technique involves the interaction of visible light (photons) with thermally gener-ated sound waves (acoustic phonons). Acoustic velocities are measured as a function of direction in a single crystal, for which the C_{ij} are calculated via a least-squares algorithm. Due to the optical nature of the technique, no mechanical contacts with the sample are required. Moreover, a very small volume of the specimen (scattering volume $\approx 3 \times 10^{-6}$ mm^3) is sampled. The formal uncertainties (standard deviations) of the resultant C_{ij}, reflecting only the internal consistency of the data, are generally in the range 0.5–3%, whereas the accuracy of the moduli is approximately two standard deviations (BASS, 1989).

The data of WANG *et al.* (1989) are for a natural specimen that is about 95% fayalite and 5% tephroite (Mn_2SiO_4). We find good agreement in most of the C_{ij} when compared with the recommended values, but there are large differences between the recommended averages in Table 2 and the WANG *et al.* values for C_{11} and C_{66}. We do not expect that frequency dispersion is responsible for the difference in these C_{ij} since any frequency-dependent effects on the C_{ij} would almost certainly be evident in all the moduli, not just C_{11} and C_{66}. Dispersion is unlikely to be noticeable far from the melting point at the relatively high frequencies of all the techniques. Furthermore, no dispersion was resolved between ultrasonic and Brillouin studies on identical or well-characterized samples of enstatite and MgO (WEIDNER *et al.*, 1978; BASS, 1989). Macroscopic imperfections such as voids are also an unlikely source of the discrepancies in the C_{11} and C_{66} moduli. Cracks or voids within the scattering volume are evident by a high intensity of elastically scattered Rayleigh light, and a high background, thus making them easily identified and usually avoidable. More importantly, the Brillouin experiment generally does not average over the sample and the imperfection, thus the results will not be systematically biased in the way that the RPR or pulse superposition techniques are affected.

The discrepancies in C_{11} and C_{66} moduli could possibly derive from one or more of the following effects. First is the effect of Mn on the elasticity of fayalite (SUMINO, 1979). Tephroite (Mn_2SiO_4) has a lower value of C_{11} than does fayalite, so a lower value of this modulus would be expected for the solid solution used by WANG *et al.* (1989). However, C_{66} appears to be equal for fayalite and tephroite (SUMINO, 1979) within the uncertainty of the data, so the lower value found by WANG *et al.* (1989) in this particular modulus is not readily explained as a straightforward chemical effect without invoking unusual variations of elastic properties within the solid solution (as possibly present in garnets, BASS, 1989). Furthermore, the value for the WANG et al. (1989) C_{33} modulus is not lower than the C_{33} from the other fayalite data (see Table 2). This result is also contrary to what is expected if the C_{ij} of the WANG et al. specimen are noticeably influenced by the presence of tephroite and if the C_{ij} behave in a consistent (or approximately linear) way with Fe-Mn substitution on the fayalite-tephroite join, since it appears that C_{33} for tephroite is considerably lower than for fayalite (SUMINO, 1979). A second possibility for the lower values found by WANG et al. (1989) in the C_{11} and C_{66} moduli is that since the respective experimental techniques each have a finite resolution, the results listed in Table 2 may be in error by $\pm 2\sigma$ simply due to random errors. Third, we cannot rule out the possibility of a fine-scale domain structure, twinning, chemical heterogeneity, or other microscopic defects, although these were not readily apparent by X-ray technique.

Although their actual sources are problematical, the discrepancies in C_{ij} moduli that occur among the four data sets that are in excess of the indicated errors in precision would appear to arise from unrecognized systematic errors. In this regard we include experimental procedure errors, orientation and preparation of specimens, and in particular, compositional and microstructural inhomogeneities in the specimens. Evidence for some compositional variation between the specimens used in the GRAHAM et al. (1988) study is noted in their paper. Moreover, discrepancies in their pressure derivative $(\partial C_{ij}/\partial P)_T$ results were attributed to the possible unrecognized occurrence of microcracks. Similar inhomogeneities in the specimen RPR-A1a relative to RPR-A1b may have contributed to the small differences in the C_{ij} moduli observed in Table 1, insofar as the latter was cut and prepared from the former.

It should be noted that each experimental method samples a specimen and its inhomogeneities in a different way. For instance, even if the same specimen is used when measuring the C_{ij}'s by resonance and by pulse superposition techniques, the results (irrespective of procedural experimental error) will not exactly coincide. In particular, the RPR method effectively samples the entire specimen. The presence of an inhomogeneity anywhere within the sample will contribute to C_{ij}, but if the inhomogeneity is small the effect will not be significant, since it will be averaged out. The pulse superposition method effectively samples paths through the center of the specimen. Inhomogeneities that are present in the specimen but not along such a path will be effectively missed. However, inhomogeneities that are small relative to the specimen size could make significant contributions to the results if strategically located along the path of the reflecting plane waves. Even though for fayalite we find somewhat better consistency between two RPR results (present results and those of SUMINO, 1979) on different crystals, than between results from different experiments (ultrasonic and RPR) using the same crystals, it does not necessarily follow that large differences will generally occur when comparing between results obtained from these techniques. In fact, when comparing between data obtained from pulse superposition, resonance, and Brillouin experiments for other materials, i.e., MgO, forsterite, and garnet mentioned previously, we find results that are much closer in agreement than is the case for fayalite. We believe the unique way each technique samples a specimen, coupled with localized microstructural strains and/or chemical inhomogeneities that are often present in iron-rich minerals, is at least partially responsible for the differences in the C_{ij} data of GRAHAM *et al.* (1988) and our present results (Table 2, column 4).

We again emphasize that although the individual measurements (and even internal consistency) that characterize the dynamic experimental methods discussed in this study are highly precise (estimated standard deviations generally less than 1%; see Table 2 and Figures 1–4), unrecognized systematic problems can result in errors in absolute accuracy that are considerably larger (e.g., up to about 3% for Brillouin scattering and 5% for pulse superposition interferometry for an individual C_{ij} modulus). When the prospects of these error sources are taken into consideration, the results presented in the four fayalite C_{ij} moduli data sets in Table 2 are quite compatible. Moreover, because the occurrence and cause of such systematic errors are not generally clear, we have elected to directly average the C_{ij} results from the individual data sets to obtain our final recommended values in Table 2. This minimizes the effect of a possibly inaccurate value without arbitrarily and unjustifiably deleting it.

4. The Temperature Variation of the C_{ij} of Fayalite

Both SUMINO (1979) and GRAHAM *et al.* (1988) report $(\partial C_{ij}/\partial T)_P$ data for fayalite. The measurements of SUMINO (1979) relate to the temperature T range 293–673 K; whereas, GRAHAM *et al.* (1988) used the narrower T span of 273–313 K. Good agreement is found between some of the T derivatives reported by these authors (see Table 12 of GRAHAM *et al.*). GRAHAM *et al.* urge caution when comparing these two sets of $(\partial C_{ij}/\partial T)_P$ values because of the different T ranges over which the measurements were carried out. The data of SUMINO (1979) clearly show nonlinearity in the T dependence of the shear moduli C_{44}, C_{55}, and C_{66}. There is a consistent decrease in the magnitude of the slopes of these moduli as T increases. Thus we expect the mean values for $(\partial C_{ij}/\partial T)_P$ ($ij = 44, 55, 66$) and $(\partial \mu/\partial T)_P$ over 298–473 K (SUMINO, 1979) to be less in magnitude than found by GRAHAM *et al.* (1988) at 298 K. This is indeed the case. However there are some differences between the T derivatives reported by SUMINO (1979) and GRAHAM *et al.* (1988) that do not seem to be due to nonlinearity (e.g., C_{22} and C_{23}). As a result of the differencs in some of the $(\partial C_{ij}/\partial T)_P$, GRAHAM *et al.* (1988) report $(\partial K_S/\partial T)_P = -0.030(0.004)$ GPa/K at 25°C compared to $-0.0205(0.0028)$ GPa/K (mean value over the range 20° to 200°C) found by SUMINO (1979).

We have obtained new RPR data at elevated temperatures using specimen *RPR-B* in order to further investigate the T dependence of the C_{ij} moduli of fayalite. We measured 47 modal frequencies at each T in regular increments of 20 K from 300–500 K. The experimental approach described by GOTO and ANDERSON (1988) was followed when obtaining frequency spectra at elevated temperatures. The highest temperatures were not sufficient to require the use of buffer rods during the experiment; but, as a matter of convenience, we used the same apparatus (with the buffer rods) described by ISAAK (1992). All of our fayalite resonance data were obtained under atmospheric oxygen fugacity conditions. The thermal expansion α data of SUZUKI et al. (1981) are used to determine the edge lengths and density of specimen *RPR-B* when calculating the C_{ij} at each T from the frequency spectrum.

The C_{ij} values obtained at different temperatures for specimen *RPR-B* are listed in Table 3. Although the 200 K range in T used in the present measurements is just over half that used by SUMINO (1979), it is sufficient to investigate the nonlinearity in the T dependence of the moduli reported by Sumino. Figures 5–8 illustrate the T dependence of the elastic moduli of specimen *RPR-B* relative to the measurements of SUMINO (1979) (results for his specimen *TA* are shown). Also shown in Figures 5–8 are the $(\partial C_{ij}/\partial T)_P$ results reported by GRAHAM et al. (1988) from measurements over 273–313 K (short dashed lines) extrapolated to 400 K (long dashed lines). The data of GRAHAM et al. (1988) are shifted somewhat in Figures 5–8 so that the moduli coincide with the 300 K C_{ij} values of specimen *RPR-B* reported here.

Table 3

Temperature dependence of C_{ij}, K_S, and μ for fayalite specimen RPR-B over 300–500 K

T (K)	ρ	C_{11}	C_{22}	C_{33}	C_{44}	C_{55}	C_{66}	C_{12}	C_{13}	C_{23}	K_S‡	μ‡
300	4.377	270.2	170.6	234.7	32.28	46.50	57.26	95.7	98.5	94.5	136.6	50.8
	(0.006)	(1.6)	(1.2)	(1.6)	(0.05)	(0.10)	(0.12)	(1.4)	(1.7)	(1.3)	(0.8)	(0.5)
320	4.375	269.3	170.0	233.8	32.09	46.32	56.86	95.3	98.3	94.4	136.1	50.5
340	4.372	268.3	169.5	233.1	31.91	46.17	56.46	95.1	98.3	94.5	135.9	50.3
360	4.370	266.9	168.6	231.9	31.76	46.05	56.02	94.4	97.7	94.1	135.2	50.0
380	4.368	266.0	168.1	231.1	31.62	45.95	55.64	94.1	97.7	94.2	134.9	49.8
400	4.365	264.9	167.3	230.1	31.52	45.85	55.28	93.6	97.3	94.0	134.3	49.6
	(0.006)	(1.4)	(1.0)	(1.4)	(0.04)	(0.08)	(0.10)	(1.2)	(1.5)	(1.2)	(0.8)	(0.5)
420	4.363	264.0	166.6	229.2	31.43	45.77	54.92	93.1	97.1	93.9	133.9	49.4
440	4.360	263.1	166.1	228.5	31.35	45.69	54.59	92.8	97.0	94.1	133.6	49.2
460	4.357	262.1	165.3	227.5	31.30	45.63	54.26	92.3	96.7	93.9	133.1	49.0
480	4.355	261.0	164.6	226.9	31.24	45.58	53.90	91.7	96.5	93.9	132.6	48.8
500	4.352	259.9	163.8	225.9	31.20	45.52	53.58	91.1	96.1	93.7	132.0	48.6
	(0.006)	(1.4)	(1.0)	(1.4)	(0.04)	(0.09)	(0.10)	(1.2)	(1.5)	(1.2)	(0.8)	(0.5)

C_{ij}, K_S, and μ units are GPa; density ρ units are g/cm³.

‡K_S and μ found by Hashin-Shtrikman averaging method.

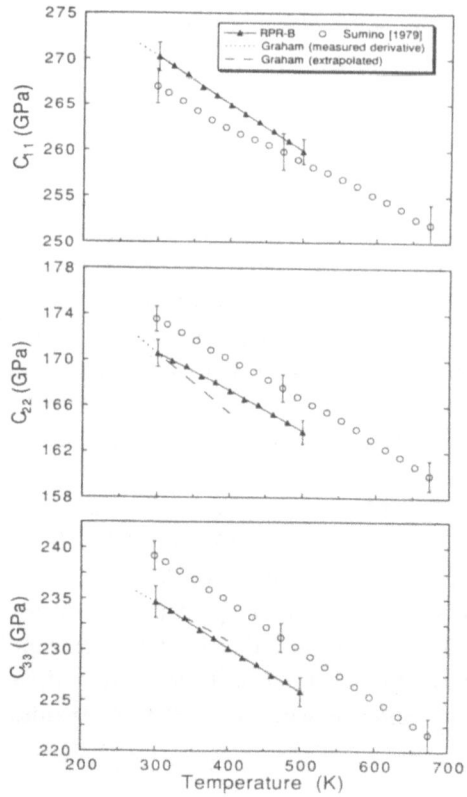

Figure 5

Measured T dependence of C_{11}, C_{22}, and C_{33} of fayalite. The linear derivatives of GRAHAM *et al.* (1988), obtained from direct measurements over the restricted range 273–313 K, have been extrapolated to 400 K for comparison to the results of the present study and those of SUMINO (1979).

The solid lines through the *RPR-B* data in Figures 5–8 represent first- or second-order least squares fits to the data. We find generally very good agreement between the T variation of the elastic moduli of *RPR-B* and that reported by SUMINO (1979). In particular, the nonlinearities found by SUMINO (1979) in C_{44}, C_{55}, C_{66}, and μ are accurately reproduced. Accurate extrapolations of these moduli to high T require consideration of the nonlinear T dependence. Furthermore, the derivatives found by GRAHAM *et al.* (1988) for these C_{44}, C_{55}, C_{66}, and μ moduli at 300 K coincide with results from the resonance experiments. However, the Graham data indicate a stronger temperature dependence for the C_{22} and C_{23} moduli than are found by the RPR results. These differences are the primary reasons why the magnitude of $(\partial K_S / \partial T)_P$ is larger in the pulse superposition than in the RPR experiments.

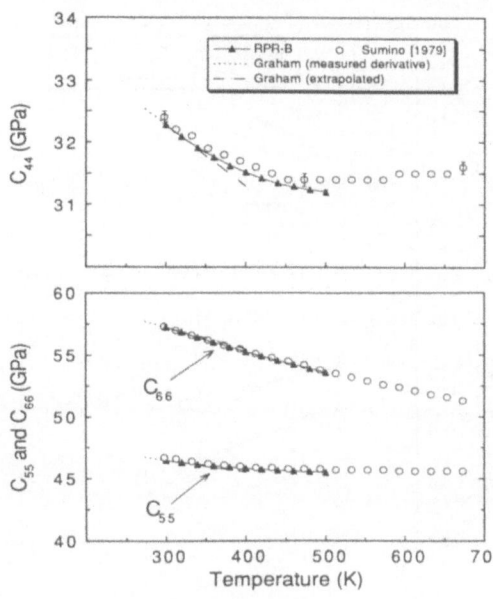

Figure 6

Measured T dependence of C_{44}, C_{55}, and C_{66} of fayalite. The linear derivatives of GRAHAM et al. (1988), obtained from direct measurements over the restricted range 273–313 K, have been extrapolated to 400 K for comparison to the results of the present study and those of SUMINO (1979).

The on-diagonal moduli, C_{11}, C_{22}, and C_{33}, and the three off-diagonal moduli, C_{12}, C_{13}, and C_{23}, are all adequately represented by a linear dependence with T in the range 300–500 K. We suspect that specimen TA used by SUMINO (1979) undergoes some unexplained physical change just before 400 K that affects the elastic properties associated with the [100] direction. There is excellent agreement in C_{11}, C_{12}, and C_{13} temperature derivatives when comparing our data with those of SUMINO (1979) up to about 375 K (see Figures 5–7). Just before 400 K, the magnitudes of the SUMINO (1979) slopes for these three moduli decrease suddenly. At temperatures higher than approximately 550 K the Sumino derivatives for the C_{11}, C_{12}, and C_{13} moduli are similar in value to the Sumino derivatives below 400 K. This behavior may involve the annealing of regions of stress in the range 375–550 K within the crystal used by SUMINO (1979), but of course, this supposition requires verification. In any case, for this reason we do not include the Sumino data for the C_{11}, C_{12}, and C_{13} moduli beyond 373 K when averaging the SUMINO (1979) and GRAHAM et al. (1988) T results with the present data to arrive at a recommended set of temperature derivatives.

Table 4 indicates the values for $(\partial C_{ij}/\partial T)_P$ and $(\partial^2 C_{ij}/\partial T^2)_P$ obtained by appropriate first- or second-order polynomial fits to the RPR-B temperature data. Also provided in Table 4 are our recommended values for each of the T derivative

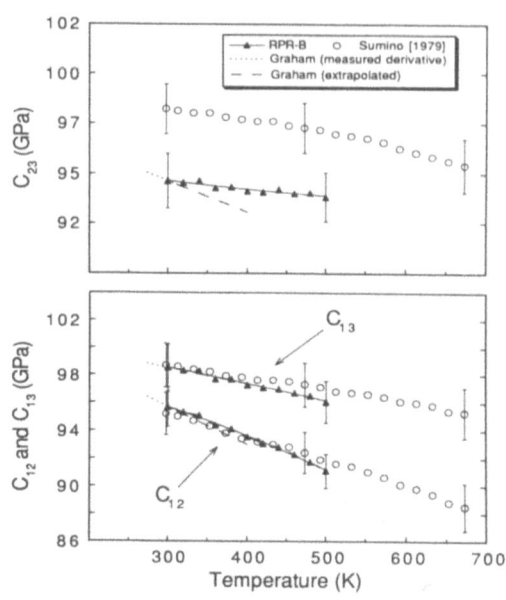

Figure 7

Measured T dependence of C_{12}, C_{13}, and C_{23} of fayalite. The linear derivatives of GRAHAM *et al.* (1988), obtained from direct measurements over the restricted range 273–313 K, have been extrapolated to 400 K for comparison to the results of the present study and those of SUMINO (1979).

quantities. These were derived by averaging (unweighted) the derivatives reported by GRAHAM *et al.* (1988) with those calculated from the data of SUMINO (1979) and the present *RPR-B* data. The SUMINO (1979) derivatives in our Table 4 are computed by fitting the Sumino data as follows: (a) C_{11}, C_{12}, and C_{13} to a linear fit in T from 298–373 K for reasons discussed above; (b) C_{22}, C_{33}, C_{23}, and K_S to a linear fit in T from 298–673 K; and (c) C_{44}, C_{55}, C_{66}, and μ to a second-order fit in T from 298–673 K. The *RPR-B* temperature derivatives given in Table 4 are obtained by linear fits from 300–500 K for all the moduli except C_{44}, C_{55}, C_{66}, and μ. The derivatives for the C_{44}, C_{55}, C_{66}, and μ moduli are found from second-order polynomial fits.

5. Conclusions

We present three sets of data on the nine C_{ij} of single-crystal fayalite at ambient P, T conditions, and one set of data from 300–500 K. We use a resonance method (RPR) and use specimens that were previously measured using pulse superposition interferometry. Our new results are combined with elasticity data on fayalite from pulse superposition, Brillouin scattering, and other resonance experiments to

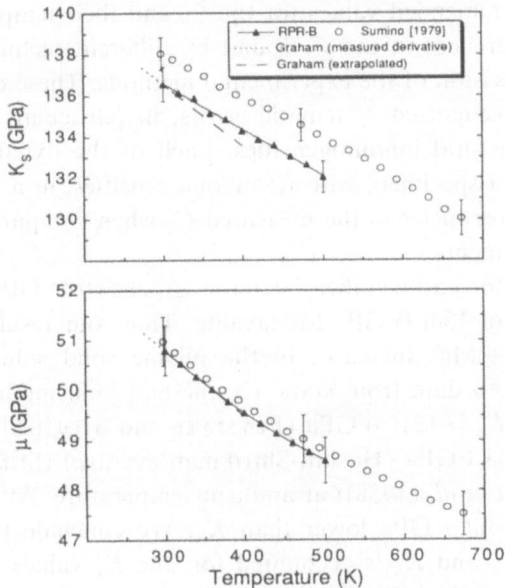

Figure 8

Measured T dependence of K_S and μ of fayalite. The linear derivatives of GRAHAM *et al.* (1988), obtained from direct measurements over the restricted range 273–313 K, have been extrapolated to 400 K for comparison to the results of the present study and those of SUMINO (1979).

Table 4

Temperature derivatives of the elastic moduli of fayalite at 300 K

ij	$\left(\dfrac{\partial C_{ij}}{\partial T}\right)_P$ (GPa/K)				$\left(\dfrac{\partial^2 C_{ij}}{\partial T^2}\right)_P$ (10^{-5} GPa/K^2)		
	Graham	Sumino*	RPR-B	Recommended†	Sumino	RPR-B	Recommended‡
11	−0.052	−0.050**	−0.051	−0.051(0.001)			
22	−0.052	−0.036	−0.034	−0.041(0.010)			
33	−0.038	−0.047	−0.044	−0.043(0.005)			
44	−0.0098	−0.0080	−0.0098	−0.0092(0.0010)	3.3	4.5	3.9(0.8)
55	−0.0096	−0.0066	−0.0079	−0.0080(0.0015)	2.2	3.2	2.7(0.7)
66	−0.018	−0.020	−0.021	−0.020(0.002)	2.1	2.9	2.5(0.6)
12	−0.027	−0.019**	−0.023	−0.023(0.004)			
13	−0.012	−0.011**	−0.012	−0.012(0.001)			
23	−0.015	−0.0028	−0.0037	−0.0072(0.0068)	−2.3		−2.3
K_S	−0.030	−0.021	−0.0224	−0.024(0.005)			
μ	−0.013	−0.0125	−0.0138	−0.013(0.001)	1.9	2.8	2.4(0.6)

* Using data from 298 673 K unless noted otherwise; linear fit used if no $(\partial^2 C_{ij}/\partial T^2)_P$ value provided.

** Linear fit of data for $T \leq 373$ K only (see text).

† Average of GRAHAM *et al.* (1988), SUMINO (1979), and *RPR-B* values.

‡ Average of SUMINO (1979) and *RPR-B* values.

provide a set of recommended values for the C_{ij} and their temperature derivatives. There are some differences in the C_{ij} found by different techniques that are well beyond the usual precision of the experimental methods. These differences are most likely related to unrecognized systematic errors, in particular specimen compositional and microstructural inhomogeneities. Each of the dynamical methods discussed here samples a specimen, with its inhomogeneities, in a different way so as to contribute to discrepancies in the measured C_{ij} when comparing between results from different experiments.

The value of K_S for end-member forsterite is 128.8(0.5) GPa compared to our recommended value of 134(4) GPa for fayalite. Thus, our results indicate that K_S increases as Fe/(Fe + Mg) increases in the olivine solid solution, a point that has been unresolved to date from static (isothermal bulk modulus $K_T = 127$ GPa (YAGI *et al.*, 1975); $K_T = 121(5)$ GPa (PLYMATE and STOUT (1990)) and dynamic experiments ($K_S = 138.1$ GPa (Hashin-Shtrikman average) (SUMINO, 1979); $K_S = 127.9$ GPa (GRAHAM *et al.* (1988)) at ambient temperature. At 300 K the value of K_T for fayalite is about 1 GPa lower than K_S. We conclude that even when the difference between K_T and K_S is accounted for, the K_T values found in the static experiments (YAGI *et al.*, 1975; PLYMATE and STOUT, 1990) are too low. It is generally recognized that dynamic measurement techniques (resonance, pulse superposition, Brillouin) are more accurate than is static compression in determining the value of the bulk modulus. It is possible that tradeoffs between K_T and $(\partial K_T/\partial P)_T$ in fitting the static compression data contribute to the low value for K_T from the static experiments.

We find very good agreement in the values of all the $(\partial C_{ij}/\partial T)_P$ at room P, T conditions when comparing our new data with the resonance data of SUMINO (1979). We also confirm the departure from linear T dependence of the C_{44}, C_{55}, C_{66}, and μ moduli of fayalite observed by SUMINO (1979) where our temperature range (300–500 K) overlaps with that used by SUMINO (393–693 K). There is general agreement in the $(\partial C_{ij}/\partial T)_P$ found by the resonance experiments and those found by pulse superposition interferometry over a much narrower range in T. Two notable exceptions are the values of $(\partial C_{ij}/\partial T)_P$ for C_{22} and C_{23} which show a much stronger T dependence in the pulse superposition data. Thus the calculated value for $(\partial K_S/\partial T)_P$ is -0.030 GPa/K from GRAHAM *et al.* (1988), compared to values nearer -0.022 GPa/K from the resonance data.

Since we are unable to identify the precise cause of these and other differences the recommended set of values is obtained from a simple average of the four different sets of data: (1) resonance (RPR) (SUMINO, 1979); (2) pulse superposition interferometry (GRAHAM *et al.*, 1988); (3) Brillouin scattering (no temperature dependence) (WANG *et al.*, 1989); and (4) our present resonance (RPR) data. We find no reason to put more weight on a set of values from one laboratory than on any other set of values we consider. It may well be that the relative magnitudes of the systematic errors from different techniques do not correlate with the relative precision (reproducibility of random errors) of the various techniques. We have no way at present to identify the size of the systematic errors. Therefore, it is most reasonable and appropriate to take a straight average.

Acknowledgements

We are indebted to Ed Schreiber for his pioneering work in physical acoustics. We thank Orson L. Anderson (UCLA) for the use of laboratory facilities and Peter Dunn of the Smithsonian Institution for providing a sample of natural fayalite. This work was partially supported by NSF grants EAR 91–17280 and EAR 90–18676, and by ONR grant 014–92–5–1354. IGPP contribution no. 3942.

REFERENCES

ANDERSON, O. L., and ANDREATCH, P. (1966), *Pressure Derivatives of Elastic Constants of Single-crystal MgO at 23° and −195.8°C*, J. Am. Ceram. Soc. *49*, 404–409.

BASS, J. D. (1989), *Elasticity of Grossular and Spessartite Garnets by Brillouin Spectroscopy*, J. Geophys. Res. *94*, 7621–7628.

BASS, J. D., and ANDERSON, D. L. (1984), *Composition of the Upper Mantle: Geophysical Tests of Two Petrological Models*, Geophys. Res. Lett. *11*, 237–240.

BINA, C. R., and WOOD, B. J. (1986), *The 400 km Discontinuity*, Nature *11*, 449–451.

BOGARDUS, E. H. (1965), *Third-order Elastic Constants for Ge, MgO, and Fused SiO_2*, J. Appl. Phys. *36*, 2504–2513.

CHANG, Z. P., and BARSCH, G. R. (1969), *Pressure Dependence of the Elastic Constants for Single-crystalline Magnesium Oxide*, J. Geophys. Res. *74*, 3291–3294.

DUFFY, T. S., and ANDERSON, D. L. (1989), *Seismic Velocities in Mantle Minerals and the Mineralogy of the Upper Mantle*, J. Geophys. Res. *94*, 1895–1912.

GOTO, T., and ANDERSON, O. L. (1988), *An Apparatus for Measuring Elastic Constants of Single Crystals by a Resonance Technique up to 1,825 K*, Rev. Sci. Instrum. *59*, 1405–1408.

GRAHAM, E. K., and BARSCH, G. R. (1969), *Elastic Constants of Single-crystal Forsterite as a Function of Temperature and Pressure*, J. Geophys. Res. *74*, 5949–5960.

GRAHAM, E. K., SCHWAB, J. A., SOPKIN, S. M., and TAKEI, H. (1988), *The Pressure and Temperature Dependence of the Elastic Properties of Single-crystal Fayalite Fe_2SiO_4*, Phys. Chem. Minerals *16*, 186–198.

HARTE, B., *The kimberlite sample*. In *Continental Basalts and Mantle Xenoliths* (eds. Hawkesworth, C. J., and Norry, M. J.) (Shiva Publ. Ltd., 1983) pp. 46–91.

ISAAK, D. G. (1992), *High Temperature Elasticity of Iron-bearing Olivines*, J. Geophys. Res. *97*, 1871–1885.

ISAAK, D. G., and ANDERSON, O. L. (1989), *Elasticity of Single-crystal Fayalite* (abstract), EOS Trans. AGU *70*, 1418.

ISAAK, D. G., ANDERSON, O. L., and GOTO, T. (1989), *Measured Elastic Moduli of Single-crystal MgO up to 1800 K*, Phys. Chem. Minerals *16*, 704–713.

ISAAK, D. G., ANDERSON, O. L., and GOTO, T. (1989), *Elasticity of Single-crystal Forsterite Measured to 1700 K*, J. Geophys. Res. *94*, 5895–5906.

ISAAK, D. G., ANDERSON, O. L., and ODA, H. (1992), *High-temperature Thermal Expansion and Elasticity of Calcium-rich Garnets*, Phys. Chem. Minerals *19*, 106–120.

JACKSON, I., and NIESLER, H., *The elasticity of periclase of 3 GPa and some geophysical implications.* In *High Pressure Research in Geophysics* (eds. Akimoto, S., and Manghnani, M. H.) (Center for Academic Publications, Tokyo, Japan 1982) pp. 93–113.

KUMAZAWA, M., and ANDERSON, O. L. (1969), *Elastic Moduli, Pressure Derivatives, and Temperature Derivatives of Single-crystal Olivine and Single-crystal Forsterite,* J. Geophys. Res. *74*, 5961–5972.

MACKWELL, S. J. (1992), *Oxidation Kinetics of Fayalite (Fe₂SiO₄),* Phys. Chem. Minerals *19*, 220–228.

MACKWELL, S. J., and KOHLSTEDT, D. L. (1990), *Diffusion of Hydrogen in Olivine: Implications for Water in the Mantle,* J. Geophys. Res. *95*, 5049–5088.

McSKIMIN, H. J. (1961), *Pulse Superposition Method for Measuring Ultrasonic Wave Velocities in Solids,* J. Acoust. Soc. Am. *33*, 12–16.

OHNO, I. (1976), *Free Vibration of a Rectangular Parallelepiped Crystal and its Application to Determination of Elastic Constants of Orthorhombic Crystals,* J. Phys. Earth *24*, 355–379.

PLYMATE, T. G., and STOUT, J. H. (1990), *Pressure-volume-temperature Behavior of Fayalite Based on Static Compression Measurements at 400°C,* Phys. Chem. Minerals *17*, 212–219.

SCHREIBER, E., ANDERSON, O. L., and SOGA, N., *Elastic Constants and Their Measurement* (McGraw-Hill, New York 1973).

SOPKIN, S. M. (1982), *Ultrasonic Determination of the Elastic Properties of Single-crystal Fayalite Fe₂SiO₄,* M.S. thesis, The Pennsylvania State University.

SPETZLER, H. (1970), *Equation of State of Polycrystalline and Single-crystal MgO up to 8 Kilobars and 800 K,* J. Geophys. Res. *75*, 2073–2087.

SUMINO, Y. (1979), *The Elastic Constants of Mn₂SiO₄, Fe₂SiO₄, and Co₂SiO₄ and the Elastic Properties of Olivine Group Minerals at High Temperature,* J. Phys. Earth *27*, 209–238.

SUMINO, Y., and ANDERSON, O. L., *Elastic constants of minerals.* In *CRC Handbook of the Physical Properties of Rocks* (ed. Carmichael, R. S.) (CRC Press, Inc., Boca Raton, FL 1984) pp. 39–138.

SUMINO, Y., ANDERSON, O. L., and SUZUKI, I. (1983), *Temperature Coefficients of Elastic Constants of Single Crystal MgO Between 80 and 1300 K,* Phys. Chem. Minerals *9*, 38–47.

SUMINO, Y., NISHIZAWA, O., GOTO, T., OHNO, I., and OZIMA, M. (1977), *Temperature Variation of Elastic Constants of Single-crystal Forsterite Between −190 and 400°C,* J. Phys. Earth *25*, 377–392.

SUMINO, Y., OHNO, I., GOTO, T., and KUMAZAWA, M. (1976), *Measurement of Elastic Constants and Internal Friction on Single-crystal MgO by Rectangular Parallelepiped Resonance,* J. Phys. Earth *24*, 263–273.

SUZUKI, I., ANDERSON, O. L., and SUMINO, Y. (1983), *Elastic Properties of a Single-crystal Forsterite Mg₂SiO₄, up to 1,200 K,* Phys. Chem. Minerals *10*, 38–46.

SUZUKI, I., SEYA, K., TAKAI, H., and SUMINO, Y. (1981), *Thermal Expansion of Fayalite,* Phys. Chem. Minerals *1*, 60–63.

WANG, H., BASS, J. D., and ROSSMAN, G. R. (1989), *Elastic Properties of Fe-bearing Pyroxenes and Olivines* (abstract), EOS Trans. AGU *70*, 474.

WEIDNER, D. J. (1985), *A Mineral Physics Test of a Pyrolite Mantle,* Geophys. Res. Lett. *12*, 417–420.

WEIDNER, D. J., WANG, H., and ITO, J. (1978), *Elasticity of Orthoenstatite,* Phys. Earth Planet. Inter. *17*, P7–P13.

YAGI, T., IDA, Y., SATO, Y., and AKIMOTO, S. (1975), *Effect of Hydrostatic Pressure on the Lattice Parameters of Fe₂SiO₄ Olivine up to 70 Kbar,* Phys. Earth Planet. Inter. *10*, 348–354.

YAN, B., GRAHAM, E. K., and FURLONG, K. P. (1989), *Lateral Variations in Upper Mantle Thermal Structure Inferred From Three-dimensional Seismic Inversion Models,* Geophys. Res. Lett. *16*, 449–452.

YONEDA, A. (1990), *Pressure Derivatives of Elastic Constants of Single Crystal MgO and MgAl₂O₄,* J. Phys. Earth *38*, 19–55.

(Received April 13, 1993, revised August 17, 1993, accepted October 6, 1993)

PAGEOPH, Vol. 141, No. 2/3/4 (1993)

0033-4553/93/040415-30$1.50 + 0.20/0

Effects of Cation Disordering in a Natural MgAl$_2$O$_4$ Spinel Observed by Rectangular Parallelepiped Ultrasonic Resonance and Raman Measurements

HYUNCHAE CYNN,[1] ORSON L. ANDERSON[2] and MALCOLM NICOL[3]

Abstract — At moderate temperatures, the elastic properties of natural MgAl$_2$O$_4$ spinel differ in several significant ways from properties of synthetic spinels. Below 1000 K, the ultrasonic resonant frequencies of an ordered natural spinel change significantly after heat treatment; at higher temperatures, both types of spinels have similar resonant responses. The temperature derivatives of the elastic constants of an ordered spinel also differ from those of disordered spinels at moderate temperatures; again, at higher temperatures, both types of spinels have similar behaviors. The Raman spectra also differ below 1000 K for ordered natural and disordered spinels and are similar at higher temperatures and after cooling to ambient temperature. We associate these changes in ultrasonic resonance and Raman spectra of spinel with cation disordering at high temperature which may be quenched by cooling. We deduce estimates of the inversion parameter from the relative intensities of the two A$_{1g}$ Raman modes in very good agreement with estimates made from other measurements. We find that C_{11} and C_{12} decrease by 4 and 8%, respectively, with 20% inversion in spinel; C_{44} is less sensitive to cation order. These results imply that previous measurements of the adiabatic elastic constants of spinels at ambient conditions have been affected by the state of cation disorder of the specimen.

Key words: Spinel, cation disordering, Raman, ultrasonic resonance, inversion, elasticity.

Introduction

The spinel structure is based upon cubic closest packing of oxygen anions. The cubic structure (Fd3m) has only three independent elastic moduli, C_{11}, C_{12}, and C_{44}. The structure accommodates 8 chemical formulas (that is, 8 MgAl$_2$O$_4$'s) in the conventional cubic unit cell. The primitive cell of ordered spinel contains 2 formulas and supports only 5 Raman active vibrational modes. In MgAl$_2$O$_4$ spinel, the 64 tetrahedral and 32 octahedral sites are partially filled by 8 Mg^{2+} and 16 Al^{3+} ions. Thus, only 1 of 8 tetrahedral sites are occupied by Mg^{2+}; and Al^{3+} ions occupy half

[1] Department of Earth and Space Sciences, University of California, Los Angeles, CA 90024-1564, U.S.A. Current Address: Department of Geology, University of California, Davis, CA 95616, U.S.A.

[2] Center for Chemistry and Physics of the Earth and Planets, Institute of Geophysics and Planetary Physics, University of California, Los Angeles, CA 90024-1564, U.S.A.

[3] Department of Chemistry and Biochemistry, University of California, Los Angeles, CA 90024-1569, U.S.A.

of the octahedral sites in the relatively open spinel structure. For a setting that places the origin at an inversion center, the tetrahedral ions are at $(1/8, 1/8, 1/8)$; and the octahedral ions are at $(1/2, 1/2, 1/2)$. The oxygens are at (u, u, u) with u, the oxygen parameter, approximately equal to $1/4$, although the oxygen packing in $MgAl_2O_4$ spinel is not perfect because Mg^{2+} ions are too large for the tetrahedral sites formed by perfect cubic closest packing of the oxygen ions. Thus, oxygen ions shared by Mg^{2+} and Al^{3+} ions are displaced to accommodate the large Mg^{2+} ions. In order to maintain the symmetries of the tetrahedron and octahedron, the oxygen ions must be displaced along a 3-fold axis. The value of u depends on the bond length between a tetrahedral cation and an oxygen and on the unit cell parameter.

$MgAl_2O_4$ spinel is one of the best known examples showing cation disordering among minerals found in the earth. In a normal spinel, Mg^{2+} ions are tetrahedrally coordinated, and Al^{3+} are octahedrally coordinated. In an intermediate spinel, both Mg^{2+} and Al^{3+} are 4- and 6-fold coordinated. Changes in the coordination numbers, or cation disordering, are usually induced by high temperatures.

Several techniques have been used to study properties related to cation order-disorder and structure of $MgAl_2O_4$ and other spinels: X-ray scattering (HÄFNER and LAVES, 1966; NARASIMHAN and SWAMY, 1980; YAMANAKA and TAKEUCHI, 1983; VIÑUELA and AREÁN, 1987); neutron scattering (THOMPSON and GRIMES, 1978; PETERSON *et al.*, 1991); ^{27}Al magic-angle-spinning nuclear magnetic resonance (WOOD *et al.*, 1986; MILLARD *et al.*, 1990); ^{17}O nuclear magnetic resonance (MILLARD *et al.*, 1992); calorimetry (NAVROTSKY and KLEPPA, 1967); Cr^{3+} electron spin resonance (SCHMOCKER and WALDNER, 1976); dilatometry (SUZUKI and KUMAZAWA, 1980); and lattice energy calculations (O'NEILL and NAVROTSKY, 1983). These studies of spinel can be summarized in terms of a transition temperature defined by rather sudden changes in unit cell parameters, the oxygen parameter, infrared spectra, degree of inversion, linear thermal expansion, and estimates of the cation disorder. It appears that the transition temperatures of natural spinels (1000 to 1200 K) are higher than those of synthetic spinels (around 900 K). Differences among these "transition temperatures" might be due to the effects of the degrees of inversion, heating or cooling rates, and compositions of the starting materials on the equilibrium or transport energetics of spinels. Some relevant results of these studies are described in the following paragraphs.

THOMPSON and GRIMES (1978) found sound velocities with time-of-flight inelastic neutron scattering which were higher than ultrasonic measurements by CHANG and BARSCH (1973). The differences between these two sets of measurements may result from differences in the degrees of inversions of the specimen because the sample used for neutron scattering was heated to 673 K in order to obtain high populations of acoustic phonons near the Brillouin zone boundary. The neutron scattering study also showed that the acoustic and optic phonon branches cross at several places in the Brillouin zone which could be the source of the mismatch of the acoustic and thermal Grüneisen parameters (ANDERSON, 1988).

Using Brillouin scattering, YEGANEH-HAERI and WEIDNER (1990) observed that the shear velocity along [001] decreased in a synthetic spinel recovered after heat cycling to 1200 K and related the change of velocity to cation disorder. ASKARPOUR et al. (1991) also reported a change in the temperature dependence of $(C_{11} - C_{12})/2$ at 923 K. However, LIU et al. (1975) observed no cation order-disorder effects up to 423 K in their light-sound scattering experiment.

NAVROTSKY and KLEPPA (1967) reported that an order-disorder transition occurs in natural spinel in the temperature range 973 to 1300 K. An order-disorder transition usually implies a change between a completely ordered distribution of cations into a disordered arrangement. We argue below that this is not what happens in spinel, and we prefer to attach the more general description, cation disordering, to this transition and the associate changes in the properties of spinel. On the basis of lattice energy calculations, O'NEILL and NAVROTSKY (1983) suggested that the enthalpy of spinel depended quadratically on the degree of cation disorder. While their model agrees well with measurements of the degree of cation disorder from neutron data (PETERSON et al., 1991), the model is inconsistent with the abrupt changes of cation disorder of natural spinels estimated by SCHMOCKER and WALDNER (1976).

Elastic properties of spinel at ambient temperature and high pressures have been measured to 1.0 GPa by the pulse-echo method (CHANG and BARSCH, 1973) and to 6.16 GPa by the pulse-echo-overlap method (YONEDA, 1990). Synthetic spinels were used for both studies. Other vibrational modes of natural MgAl$_2$O$_4$ spinel at high pressures and ambient temperatures have been followed to higher pressures in diamond-anvil cells by Raman, infrared, and vibronic spectroscopies (CHOPELAS and HOFMEISTER, 1991). X-ray single crystal diffraction experiments to 4.0 GPa with the same specimen (FINGER et al., 1986) show that the oxygen parameter decreases with pressure. This decrease was rationalized in terms of differences of the compressibilities of the MgO$_4$ and AlO$_6$ units.

The structural similarity of aluminate spinel with the high pressure polymorphs of olivine [β (wadsleyite) with its modified spinel structure and γ (ringwoodite) with a spinel structure] has been the basis for using aluminate spinel as a model for silicate spinels. The analogy may not be so good for elastic properties. Differences between the elastic properties of MgAl$_2$O$_4$ spinel and wadsleyite have been noticed from in situ ultrasonic measurements of wadsleyite at ambient temperature and high pressures (GWANMESIA et al., 1990). Indeed, the structures of the two minerals differ. Although both structures are based on cubic closest packing of oxygen atoms, all of the tetrahedra in the aluminate spinel are isolated from each other, while two tetrahedra sharing an oxygen provide the basic framework of wadsleyite structure. Another interesting suggestion by DUFFY and ANDERSON (1989) is that forsterite, wadsleyite, and ringwoodite have similar elastic properties even though their structures differ. Their systematics indicate that the temperature and pressure derivatives of K_S, the adiabatic bulk moduli, and μ the isotropic shear moduli, of the three olivine polymorphs are almost identical. Gwanmesia *et al.* (1990) determined the pressure derivatives, $(\partial\mu/\partial P)_T$ and $(\partial K_S/\partial P)_T$, of β-Mg$_2SiO_4$; their values are almost the same as values predicted by DUFFY and ANDERSON (1989). However, the values for $(\partial K_S/\partial P)_T$ and $(\partial\mu/\partial P)_T$ of wadsleyite from DUFFY and ANDERSON (1989) are not in the ranges for 400 km discontinuity suggested by GWANMESIA *et al.* (1990).

The various interpretations of the effects of cation disorder on the elastic properties of spinels are not consistent. YEGANEH-HAERI and WEIDNER (1990) interpreted Brillouin results to suggest that cation disordering lowers the shear velocity along [001]. ASKARPOUR *et al.* (1991) suggested that a change in the temperature dependence of $C_s = (C_{11} - C_{12})/2$ at 923 K indicates that the crystal structure of synthetic spinel changes, yet YEGANEH-HAERI and WEIDNER found no anomaly near 900 K. SUZUKI (1982, unpublished) also has studied the temperature dependencies of the adiabatic elastic moduli of synthetic spinel by the RPR (rectangular parallelepiped resonance) technique. He found no anomaly in C_{44} and an apparent anomaly in C_{11} at 920 K.

Raman spectra of MgAl$_2$O$_4$ spinels often contain more bands than group theory predicts for the Fd3m structure. Many interpretations have been proposed for the origin of extra features like the band near 720 cm^{-1}. FRAAS *et al.* (1973) observed a band at 715 cm^{-1} and associated it with an infrared band at 735 cm^{-1} (SLACK *et al.*, 1966) and a vibronic band at 745 cm^{-1} (WOOD *et al.*, 1968). O'HORO *et al.* (1973) assigned the 727-cm^{-1} mode as a combination band, not a fundamental, and ISHII *et al.* (1982) adopted the same assignment for a 725-cm^{-1} band in a nonstoichiometric synthetic spinel. MCMILLAN and HOFMEISTER (1988) compared the relative intensities of the 727-cm^{-1} band and the 772-cm^{-1} A$_{1g}$ fundamental between stoichiometric (O'HORO *et al.*, 1973) and nonstoichiometric spinel (ISHII *et al.*, 1982) and concluded that the 727-cm^{-1} band is not simply associated with either normal-inverse disorder or nonstoichiometry. These comparisons have not been extended to include natural spinels. From analysis of Raman spectra excited at several wavelengths between 450 and 520 nm, CHOPELAS and HOFMEISTER (1991) claimed that the mode of natural spinel at 718 cm^{-1} is intrinsic to the spinel and not intrinsically associated with the 727-cm^{-1} mode observed by FRAAS *et al.* (1983), O'HORO *et al.* (1973) and ISHII *et al.* (1982). CHOPELAS and HOFMEISTER (1991) could not determine the symmetry of this weak, broad 718-cm^{-1} mode of their natural spinel, although O'HORO *et al.* (1973) have assigned it as A$_{1g}$ based upon polarization studies with synthetic spinel.

The purpose of this study is to dertermine how cation disordering affects the optic and acoustic vibrational modes of spinel as detected by *in situ* high temperature measurements of Raman and ultrasonic resonance spectra. Although common wisdom suggests that disordering atoms in a crystal may not greatly affect its elastic properties, inconsistencies among previously reported values for the adiabatic elastic constants of spinel suggested that this system may be an exception to the rule. However, HILBERT and GRAHAM (1989) noticed that the systematic elastic discrepancies among the literature data may reflect differing degrees of specimen disorder involving the distribution of the Al cations between tetrahedral and octahedral structural sites.

The RPR technique has been successfully used to investigate high temperature elastic properties of α-SiO$_2$ (OHNO, 1990), MgO (ISAAK et al., 1989b), Al$_2$O$_3$ (GOTO and ANDERSON, 1989), NaCl (YAMAMOTO et al., 1987), KCl (YAMAMOTO and ANDERSON, 1987), α-Mg$_2$SiO$_4$ (ISAAK et al., 1989a), α-(Mg$_{0.9}$Fe$_{0.1}$)$_2$SiO$_4$ (ISAAK, 1992), α-Fe$_2$SiO$_4$ (ISAAK, 1992), and garnet (ISAAK et al., 1992). Although none of these studies revealed anomalous resonance spectra, the anomalous Raman spectra suggested that cation disorder might produce anomalous or extra resonances for spinel.

In this report, the high temperature elastic constants of spinel estimated from these RPR results are characterized. Although we looked for extra resonances, we found none by comparing RPR spectra of less disordered and highly disordered specimen. The ordered and highly disordered spinels, however, showed considerably different shifts in the resonance frequencies at high temperatures and after quenching to ambient temperature. We also noticed that the intensities of two A$_{1g}$ Raman modes might be used to estimate the extent of cation disordering in MgAl$_2$O$_4$ spinel and compared estimates based on these intensities with estimates based upon NMR and neutron scattering data. The cation disordering phenomena in spinel are characterized in terms of these Raman and ultrasonic data.

Experimental

A natural magnesium aluminate spinel provided by J. Rosenfeld was used for these RPR and Raman measurements. The geological location of the specimen was unknown. It was transparent with a purple tint as a bulk; however, a small chip of 200 μm in size appeared colorless. When illuminated by a blue argon laser, it fluoresced orange-red. A microprobe analysis and high-temperature Raman measurements were made using small chips of the same specimen. Three synthetic magnesium aluminate spinels were included in the Raman measurements to understand better the spectral features associated with cation disordering: Linde (Union Carbide) spinels were studied at ambient and high temperatures; spinel powders provided by WOOD et al. (1986) were studied to compare inversion parameters estimated from NMR data and from Raman intensities; and a spinel which had been characterized by neutron scattering (PETERSON et al., 1991) also was studied.

The electron microprobe analyses were done using a Cameca Camebax equipped with crystal spectrometers and energy dispersive analyzers. The spectrometers were operated at 15 kV and 15 nA. Natural chromites were used as standards for determining Fe and Cr. A natural garnet and a synthetic forsterite were used to dertermine Al and Mg, respectively. The conventional ZAF corrections were applied to the final results. Chemical compositions of the natural spinel are: Al_2O_3 (71.27 wt.%), MgO (27.72 wt.%), FeO (0.27 wt.%), and Cr_2O_3 (0.10 wt.%). The chemical formula of this spinel, assuming 4 oxygens per formula unit, is $(Mg_{.985}Fe_{.005})(Al_{2.004}Cr_{0.002})O_4$.

The optical quality of each specimen was checked using a petrographic or stereo zoom ($45\times$) microscope. The optical quality of the natural spinel was excellent in the sense that there were no inclusions, cracks, or residual stress. The specimens were oriented along a crystallographic direction using a back reflection Laue camera, Polaroid XR-7 Land diffraction cassette, $9 \times 12 \, cm^2$ high-speed (3000 ASA) Polaroid film, and polychromatic X-radiation from a copper X-ray tube operating at 40 kV and 15 mA. A four-fold axis was sought, and the specimen could be oriented well within half a degree. The specimen was polished with alumina powders and diamond paste. Three faces were oriented in shaping a specimen. This procedure for orienting and polishing appeared adequate for the RPR measurements.

After these preliminaries were completed, the edge lengths and density were measured. We used a digital micrometer with round anvils to measure the edge lengths at the center and four corners. For the natural spinel, the measured dimensions were: 2.5614 (0.0021), 2.5194 (0.0022), and 1.8872 (0.0022) mm. Opposing faces were estimated to be flat to less than 5 μm which is slightly greater than the reported instrumental error of the micrometer, $<2 \, \mu m$ at $20°C$. The density of the specimen was estimated by immersion methods and compared with the density calculated from the mass and edge lengths. Both determinations were in excellent agreement; for the natural spinel 3.576 (0.003) g/cm^3 was determined by immersion and 3.572 (0.003) g/cm^3 was calculated.

A heater designed for a Rigaku Denki differential thermal analyzer was used to heat the specimen. The furnace consisted of a 2.54-cm (i.d.) pure alumina tube wrapped with Pt13%Rh wire along a 7 cm length. Both the ac current through and voltage across the heater were monitored throughout the experiments. The temperature of the specimen was probed using one Pt-Pt13%Rh thermocouple placed just above the center of the specimen and less than 1 mm from the specimen. According to D. ISAAK (personal communication), the maximum temperature variation across a 3-mm long specimen in this geometry is less than 5 K; thus, we assume the precision of the temperatures reported here to be ≤ 5 K to 1000 K. Because of the difficulty in reverting the equilibrium state of cation disorder, RPR scans were always made from ambient to steadily increasing higher temperatures. Before each scan at high temperatures, the sample was annealed for 6 to 12 hours.

The RPR technique for estimating elastic and anelastic properties of solids has been described elsewhere (SUMINO *et al.*, 1976). Two improvements in signal processing, developed at Los Alamos National Laboratory and incorporated for this study, greatly increased the sensitivity of the RPR measurements. One improvement involves heterodyne mixing of the signal from a specimen and the attenuated carrier from a separate frequency generator and amplifying the difference signal with a tuned amplifier. Data collection also was enhanced by controlling the drive and measurement electronics from a personal computer through an Iotech IEEE-488 bus system as indicated in Figure 1. No difference was found when resonant frequencies measured by the improved computer-controlled system were compared with frequencies scanned manually.

The improved RPR measurement system used two frequency synthesizers. A Hewlett Packard (model 3325A) synthesizer drove the transducer at nominal voltages below 10 volts which was boosted when desirable by an optional ENI (model 240L) rf power amplifier. An Anritsu (model MG440A) synthesizer provided the intermediate frequency (1 KHz higher than the drive frequency) to the mixer of the modified superheterodyne receiver (MIGLIORI et al., 1991). The output of the mixer was amplified as necessary by a combination of three linear amplifiers ($\times 10$, $\times 100$, and $\times 200$) and filtered with a high frequency cut-off filter in the rms voltmeter (Rhode & Schwarz model 342.1214.02) which also averaged the output signal. The output from the mixer also was monitored on an oscilloscope (Hewlett Packard model 140A). The averaged output of the rms voltmeter was stored in a personal computer and simultaneously displayed on the monitor of the computer.

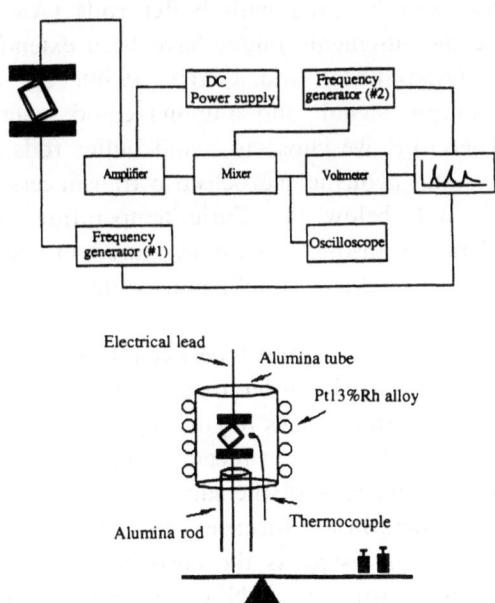

Figure 1
Schematic diagrams of RPR setup to show electron devices and the furnace-sample-transducer assembly coupled with a modified balance.

Each scan was repeated twice to check reproducibility. We averaged the measured resonant frequencies from the two scans to compensate the effects on the C_{ij} due to the temperature fluctuation while scanning.

A modified balance was used to measure the variation of the resonance frequencies of a specimen with the applied force needed to clamp the specimen between the transducers. In previous RPR experiments, forces were generated by weights from one to ten grams; and the measurements were extrapolated to zero applied force to estimate the frequencies of the free vibrations. In this experiment, measurements at ambient temperatures were made with applied weights as small as 0.1 g to reduce the error in the extrapolation. It was difficult to measure precise resonance frequencies with such small weights because the signal responses weakened and the system often started to ring. However, the new signal processing equipment overcame these problems. Measurements were extended from 0.9 to 2.5 MHz at 20 to 60 Hz intervals. Uncertainties associated with the scan rates were minor compared with those due to temperature fluctuations.

Several test runs were performed to select optimal combinations of sample orientation, piezoelectric transducers, input voltages, and clamping forces for these measurements. The transducers were operated in a thickness mode. At ambient conditions, PZT-5a ceramic (shear mode) transducers (Valpy-Fisher 5500, 1.063 mm thick, gold plated) with a parallel resonance frequency at 1 MHz were used. These transducers have relatively low Curie temperatures, 638 K, and for high temperatures must be used with buffer rods (ANDERSON and GOTO, 1989). Although these measurements might have been extended to 1800 K using alumina buffer rods (ANDERSON and GOTO, 1989), the buffer rods significantly attenuate the output signal; and alumina could react with the spinel at high temperatures. Therefore, we chose to avoid buffer rods and to use $LiNbO_3$ (shear mode) single crystal acoustic piezoelectric transducers to extend the measurements to 1068 K, well below the Curie temperature of $LiNbO_3$, 1473 K. A rotated, X-cut $LiNbO_3$ transducer plate was used in the thickness mode to excite the sample; the $5 \times 5 \times 2.805$ (thick) mm^3 transducer plate resonated near 0.7 MHz.

Samples in the furnace were clamped between $LiNbO_3$ transducers with the force applied by a 4-gram load. In order to correct the frequency shifts for this load, "zero-mass" resonance frequencies for the natural spinel specimen, measured at ambient conditions with PZT-5a transducers clamped by a 0.1-gram force, were compared with resonant frequencies of the same specimen measured with $LiNbO_3$ transducers under 4-gram load in the furnace. The differences in resonant frequencies between these conditions, listed as the correction in Table 1, were used to correct measurements made with the $LiNbO_3$ at other temperatures. For most modes, these corrections are less than 0.1%

The elastic constants of the specimen were determined from the observed free resonance frequencies, f_n^{obs}, by an iterative procedure. The nth resonance frequency,

Table 1

Zero mass frequencies (in Hz) for natural spinel at high temperatures estimated by correcting the measured raw frequencies under applied mass by the amount shown for each mode (refer to the text)

	Correction	298 K	348 K	375 K	415 K	459 K
1. B3u-1	1120	977240	974600	972800	970680	967580
2. Au-1	2040	1055600	1053820	1053630	1052220	1049880
3. B2u-1	1200	1121300	1118860	1117360	1115980	1110840
4. Ag-1	220	1150420	1146190	1144080	1139930	1136960
5. B1u-1	200	1172240	1168900	1167040	1164400	1160860
6. Ag-2	960	1266560	1262910	1260360	1257440	1253520
7. B3u-2	1280	1281240	1277640	1276300	1273660	1270300
8. Ag-3	20	1351800	1347770	1345360	1342160	1337980
9. Au-2	1950	1497070	1493430	1492230	1490090	1485870
10. B2g-1	1960	1516000	1512180	1510460	1508820	1505600
11. B1g-1	1700	1522340	1519140	1518160	1513940	1510720
12. B3g-1	1400	1533840	1530200	1528780	1527440	1524100
13. Ag-4	1580	1725120	1721220	1717960	1713750	1708900
14. B2u-2	570	1741580	1736710	1733850	1729690	1724910
15. B1u-2	960	1750420	1745240	1742940	1738280	1733640
16. Ag-5	1800	1773620	1768840	1765080	1762330	1757100
17. B3g-2	20	1809720	1804840	1802080	1797900	1793380
18. B1u-3	700	1820020	1816020	1813400	1809600	1805060
19. B2g-2	480	1834010	1831020	1829280	1826200	1822740
20. B2u-3	1490	1845840	1841370	1838410	1834050	1829470
21. B1g-2	180	1859040	1855020	1853190	1859040	1847520
22. B3u-3	2640	2018020	2014100	2011140	2005980	2003160
23. B1u-4	2420	2074180	2068840	2065540	2060600	2055580
24. B3g-3	320	2084240	2078920	2075980	2069040	2065420
25. B2u-4	2220	2106040	2099520	2097240	2093580	2086980
26. B1u-5	1440	2149860	2145220	2143080	2139340	2133420
27. B1g-3	2110	2160200	2155950	2153930	2147930	2144270
28. B2u-5	1190	2171300	2166350	2163970	2158850	2154010
29. B2g-3	1420	2194880	2191060	2188100	2183740	2179160
30. B3g-4	840	2358600	2353420	2350560	2346480	2343900
31. Au-3	540	2406040	2399510	2397390	2393180	2387580
32. Ag-6	3060	2448500	2440580	2436840	2432120	2425380
33. B3g-5	200	2465140	2458860	2455540	2450180	2442760
34. Au-4	1720	2493680	2488180	2484440	2479820	2473760

f_n, of an elastic rectangular parallelepiped can be represented as a function of the elastic constants, C_{ij}, density, ρ, and three edge lengths, L_k ($k = 1, 2, 3$) as:

$$f_n = f_n(C_{ij}, \rho, L_k) \quad (n = 1, 2, \ldots).$$ (1)

The first step of the procedure was to identify the resonances at ambient temperature by comparing the observed free resonance frequencies with frequencies, f_n^{calc}, calculated iteratively from trial elastic constants and measured density and edge lengths. For a cubic crystal, the difference, Δf_n, between f_n^{obs} and acoustic resonance

Table 1 (*Contd*)

	488 K	577 K	580 K	610 K	661 K	723 K	775 K
1.	965480	959720	959520	957560	954110	949460	946040
2.	1048500	1044960	1044870	1043760	1041420	1038840	1036800
3.	1108320	1102440	1102060	1100600	1097820	1090260	1086600
4.	1133780	1125480	1125140	1123040	1119380	1115040	1110230
5.	1158460	1151500	1151260	1148800	1144660	1139200	1135120
6.	1250820	1243300	1243040	1239960	1235220	1229100	1224540
7.	1267780	1261220	1260940	1258630	1254520	1249420	1245580
8.	1335250	1326980	1326700	1323700	1318120	1312240	1307260
9.	1483350	1475830	1475640	1473690	1469130	1463730	1459590
10.	1503740	1497300	1497200	1495160	1490480	1485350	1481360
11.	1508380	1502020	1501840	1499500	1495540	1491160	1487620
12.	1521040	1514540	1514340	1512460	1508020	1503460	1499800
13.	1705670	1695800	1695520	1692370	1686220	1678900	1673170
14.	1721490	1711450	1711170	1707510	1701210	1693470	1687500
15.	1730280	1720240	1719900	1716240	1710000	1702260	1696320
16.	1753560	1743520	1743060	1739700	1733220	1725780	1719900
17.	1789900	1779740	1779460	1775740	1769380	1761820	1755760
18.	1801940	1793100	1792880	1789640	1783940	1777220	1772060
19.	1820520	1813300	1812960	1809570	1804860	1797960	1792140
20.	1826050	1816490	1815910	1812910	1807960	1802570	1798510
21.	1845300	1838350	1838100	1835380	1831320	1825980	1821780
22.	1999680	1990480	1990140	1986720	1980780	1974120	1968840
23.	2051560	2041100	2040660	2036440	2029300	2021260	2014900
24.	2062240	2050940	2050600	2046520	2039440	2030980	2024260
25.	2082960	2071480	2071020	2067060	2059500	2051220	2044520
26.	2130720	2120020	2119920	2116020	2108880	2101860	2095920
27.	2141270	2131710	2131070	2128130	2121050	2114390	2108930
28.	2150290	2139890	2139490	2135650	2129050	2121070	2115010
29.	2175980	2166480	2166200	2162090	2156480	2149520	2143880
30.	2339760	2330020	2329680	2326560	2320140	2314260	2308860
31.	2383860	2372860	2372460	2368800	2361180	2353680	2347280
32.	2420340	2410960	2409960	2405040	2398620	2389920	2382960
33.	2440120	2426200	2425940	2420380	2414200	2404900	2397100
34.	2471000	2459580	2459180	2455040	2447540	2439500	2432240

frequencies, f_n^{cal}, calculated with a set, C_{ij}^{cal}, of elastic constants may be related to a set of trial elastic constants, C_{ij}^{trial}, by a matrix representation of a Taylor expansion, as follows:

$$\begin{bmatrix} \partial f_1/\partial C_{11} & \partial f_1/\partial C_{12} & \partial f_1/\partial C_{44} \\ \partial f_2/\partial C_{11} & \partial f_2/\partial C_{12} & \partial f_2/\partial C_{44} \\ \partial f_3/\partial C_{11} & \partial f_3/\partial C_{12} & \partial f_3/\partial C_{44} \\ \vdots & \vdots & \vdots \\ \partial f_n/\partial C_{11} & \partial f_n/\partial C_{12} & \partial f_n/\partial C_{44} \end{bmatrix} \begin{bmatrix} \Delta C_{11} \\ \Delta C_{12} \\ \Delta C_{44} \\ \vdots \end{bmatrix} = \begin{bmatrix} \Delta f_1 \\ \Delta f_2 \\ \Delta f_3 \\ \vdots \\ \Delta f_n \end{bmatrix} \qquad (2)$$

or $T\Delta C_{ij} = \Delta f_n$ where $\Delta C_{ij} = C_{ij}^{\text{cal}} - C_{ij}^{\text{trial}}$, T is the matrix of derivatives of $\partial f_n/\partial C_{ij}$, and $\Delta f_n = f_n^{\text{obs}} - f_n^{\text{cal}}$. The derivatives, $\partial f_n/\partial C_{ij}$, were evaluated by a finite difference

Table 1 (Contd)

	820 K	908 K	943 K	999 K
1.	942860	936380	933800	929780
2.	1035090	1031820	1030380	1027980
3.	1083120	1076100	1073280	1068840
4.	1105820	1098740	1095440	1090160
5.	1131280	1123180	1120060	1115080
6.	1220280	1211330	1207820	1202300
7.	1241920	1234660	1231720	1226940
8.	1302580	1292800	1288960	1282780
9.	1455870	1448730	1445490	1440690
10.	1477520	1469780	1466540	1461320
11.	1484500	1478680	1476160	1471700
12.	1496260	1489480	1486540	1481800
13.	1667620	1657000	1652620	1645540
14.	1681770	1669950	1665330	1657550
15.	1690560	1678950	1674360	1666800
16.	1714380	1703220	1698840	1691780
17.	1749940	1737820	1733260	1725700
18.	1767020	1756940	1752980	1746380
19.	1786780	1775460	1771020	1764060
20.	1794490	1787050	1783990	1778710
21.	1817760	1809420	1806180	1800540
22.	1963500	1953485	1949460	1942980
23.	2008480	1996540	1991440	1983400
24.	2017720	2004340	1999360	1991080
25.	2037840	2024940	2019840	2011800
26.	2089890	2076780	2073720	2066280
27.	2103170	2092850	2088590	2081390
28.	2108950	2097190	2092690	2085370
29.	2138120	2126960	2122850	2115740
30.	2303280	2292780	2288760	2281440
31.	2340540	2327940	2323020	2314700
32.	2374980	2362080	2356620	2349780
33.	2389420	2374300	2368060	2358520
34.	2425640	2412500	2407220	2398780

method with an interval of 0.001 for $\Delta C_{ij}/C_{ij}$. The calculation was iterated by a least-squares method, adjusting C_{ij}^{trial} until $\Delta C_{ij} = (T'T)^{-1}T'\Delta f_n$ was zero, where T' is the transpose of the T matrix, for a reasonable identification of the resonances and minimum value of the sum, $\sum_n \Delta f_n^2$.

The shifts of the observed mode frequencies were carefully traced as we changed the temperature of the specimen; and no changes of the mode identifications were detected, even after abrupt temperature (and frequency) jumps at high temperatures or cooling to ambient conditions. The density and edge lengths of the specimen at a given temperature were then calculated from values measured at ambient temperature and the coefficient of linear thermal expansion, α_L, of spinel which we took

Table 1 (*Concluded*)

	298 K(Q)	990 K	1003 K	1062 K	1068 K
1.	986420	936480	934880	932480	933260
2.	1054620	1029400	1028580	1026420	1026060
3.	1131120	1076380	1074660	1072080	1073280
4.	1161680	1097460	1095740		
5.	1183370	1122500	1120840		1119260
6.	1280160	1211680	1209780		1208340
7.	1289080	1233740	1232140	1228900	1230040
8.	1364680	1291220	1289320		1287460
9.	1504470	1447450	1445790	1442370	1442670
10.	1517120	1467240	1465820	1469480	1462220
11.	1528840	1474760	1473520	1475200	1469620
12.	1537060	1486340	1485040	1481080	1481320
13.	1738780	1656380	1654420	1649680	1650460
14.	1755690	1668190	1666230	1661190	1662270
15.	1764900	1677460	1675440	1670340	1671420
16.	1788300	1703440	1701420	1697160	1697940
17.	1824460	1736360	1734520	1719010	1730020
18.	1830920	1754340	1752680	1747520	1748000
19.	1836180	1775140	1773240	1768500	1769280
20.	1858930	1782230	1780930	1775710	1775170
21.	1861680	1802800	1801740	1796160	1795680
22.	2025960	1951000	1949520		1944060
23.	2098780	1995140	1993420		1988020
24.	2122180	2003000	2001220		1995700
25.	2158620	2023780	2022060		2016300
26.	2169720				2050140
27.	2182130	2089830	2088530		2085470
28.	2197870	2093870	2092570		
29.	2202200	2122308	2121200		
30.	2321340	2195080	2193120		
31.	2364180	2287720	2286600		
32.	2411640	2319160	2317980		
33.	2467000	2364620	2363260		
34.	2479700	2368740	2367380		

* 298(Q) represents the frequencies measured at ambient conditions after cooling to room temperature rapidly.

to be those recommended by TOULOUKIAN *et al.* (1977). The elastic constants at high temperatures then were calculated by repeating the iterative computation for the new set of f_n^{obs}'s, ρ and L_k's.

Many measurements suggest that α_L of spinel changes discontinuously above 900 K; however, there is less agreement about this size of the change. SUZUKI and KUMAZAWA (1980) observed by dilatometry a discontinuous change of α_L about 1.0×10^{-6} K^{-1} near 993 K. On a reversal run, they observed a smaller change, about 0.5×10^{-6} K^{-1} at around 920 K. They estimated that the unit cell parameter decreased by about 0.09 pm and the density increased by 0.03% at the discontinuity. YAMANAKA and TAKEUCHI (1983) made similar observations in a single crystal X-ray diffraction study of a synthetic spinel. They reported that the oxygen u parameter decreased abruptly by 8 (S.D. = 2) $\times 10^{-4}$ between 873 and 993 K where the unit cell parameter also changed in a nonlinear manner. However, these changes of synthetic spinels are significantly smaller than changes reported by HÄFNER and LAVES (1966) who found the unit cell parameter of a natural MgAl$_2$O$_4$ increased by 0.4 pm, increasing the density by 0.15%, between 1073 and 1173 K.

Nonlinear or discontinuous changes of α_L affect the calculated elastic constants in principle; however, ANDERSON and GOTO (1989) suggest that the estimated adiabatic elastic constants are not very sensitive to differences in α_L. ISAAK et al. (1989a) drew a similar conclusion in their study of forsterite. Therefore, even at temperatures above 900 K, we decided to use thermal expansion data from TOU-LOUKIAN et al. (1977) in computing elastic constants from RPR frequencies.

The f_n^{obs}'s were not corrected for shifts with driving amplitudes because we estimated that, for the relatively small amplitudes we used, these corrections would insignificantly affect the data reduction. In principle, these corrections would reduce errors in the f_n^{obs}'s extrapolated to zero applied mass by least-squares analysis of the variation of the resonance frequencies with applied force. However, errors contributed by amplitude-dependent frequency shifts were much smaller than differences between specific calculated and the estimated zero-applied-mass frequencies observed in the subsequent calculations of the elastic constant from these frequencies.

Raman spectra at ambient conditions were measured with a conventional double monochromator spectrometer with laser excitation and conventional photon-counting detection. The argon-ion laser (Spectra Physics, model 171) was operated at 515, 488, 477, or 458 nm with output power measured at the sample less than 50 mW. The output from the laser was dispersed by two Brewster angle borosilicate prisms and focused with a $25\times$ microscope objective at a 25 μm spatial filter (Newport, model 900) to eliminate plasma lines and to clean up the beam profile. The prisms and spatial filter were separated by about 1 meter. The output from the spatial filter was recollimated with a $10\times$ microscope objective, passed through an optional quater-wave plate for rotating the polarization of the beam, and refocused into the specimen. Radiation scattered by 90° was collected and focused with a pair of achromatic lenses on the entrance slit of a 0.75-m double monochromator (Spex, model 1400) modified with a computer-controlled stepper motor drive. The slits of the monochromator were set at 150, 200, and 150 μm to provide 2-cm^{-1} spectral resolution near 500 nm; and spectra were collected in 0.01-nm steps at from 5 to 8 sec per step.

For high-temperature Raman spectra, the specimens were mounted in a micro-furnace heater on the translation stage of a Leitz Ortholux-I microscope. The spectra were excited by the 458-nm line of a Spectra-Physics (model 165) argon-ion laser operated to provide 20 to 50 mW at the specimen. These spectra were collected by a Spex Triplemate spectrometer with a charge-coupled detector (CYNN et al., 1992).

Results

A typical resonance spectrum of a spinel is shown in Figure 2. Besides the sharp resonance signals from the specimen, several system responses are evident as broad bands with very low intensity near 1 MHz. With smaller clamping forces, the intensities of the specimen resonances decrease significantly; however, the system responses increase when the amplitude of the signal driving the transducer is increased to compensate for the lower intensities of the resonance signals. The quality of the estimated elastic constants might be jeopardized if these system responses or low resonance intensities covered a sample resonance, especially if the computation for one constant was very sensitive to the missing mode. The procedure of including many modes measured at several clamping forces in the data reduction scheme is designed to reduce possible errors caused by missing modes. Through the processes to find the optimum conditions to measure the resonant frequencies at ambient conditions and the data reduction as previously discussed, the identification of the resonant modes of the sample was confirmed. In other words, no system responses were included in the data reduction. We found that the uncertainties of the C_{ij} become smaller when we included as many modes as possible in the data reduction. We also kept the number of the modes constant at each temperature except the last few runs (Table 1).

Table 1 gives the resonant frequencies of the natural spinel measured with the $LiNbO_3$ transducers at several temperatures as the sample was first heated to 1000 K and then was reheated beyond 1000 K and subsequently quenched to ambient temperature to determine the effects of cation disordering on the measure-

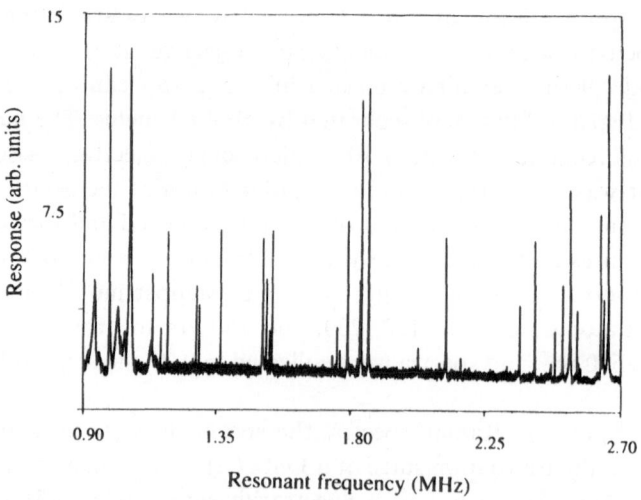

Figure 2
A resonance spectrum of natural spinel collected at ambient conditions.

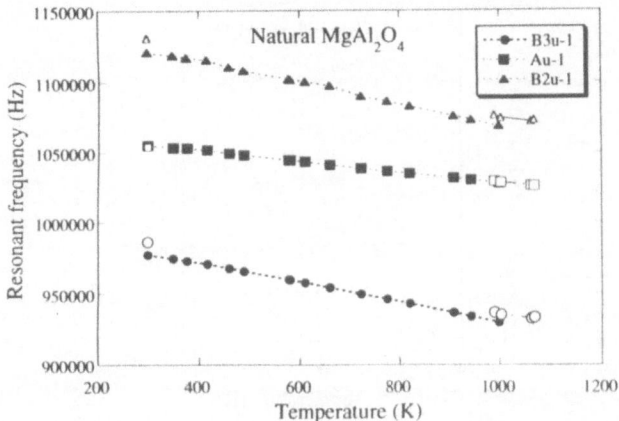

Figure 3

A plot showing the temperature dependencies of the resonant frequencies of natural spinel at high temperatures. Open symbols are used to separate the disordered spinel above 1000 K and after quenching to room temperature.

ments. The response from the $LiNbO_3$ transducers weakened significantly at 1062 K, and frequency measurements were discontinued at 1068 K. The temperature dependencies of the corrected resonance frequencies of this specimen are shown in Figure 3. Note the almost abrupt jump in the frequencies and the change of slope near 1000 K of B3u-1 and B2u-1 modes. The frequencies of most modes, except Au-3, Ag-6, B3g-4, and Au-4, are higher after quenching than before heating (Table 1). An attempt to revert the specimen at around 1000 K failed.

Three independent adiabatic elastic moduli, $K_s = (C_{11}^s + 2C_{12}^s)/3$, $C_s = (C_{11}^s - C_{12}^s)/2$, and C_{44}^s, were computed from these measured frequencies by the iterative inversion method described above. C_{11}^s and C_{12}^s were estimated from the computed values of K_s and C_s; and the isotropic shear moduli, μ_{HSA}, were calculated by the Hashin-Shtrikman averaging scheme (HASHIN and SHTRIKMAN, 1962). These values are listed in Table 2 and shown in Figure 4. The elastic constants of spinel determined for high temperatures are given in Table 2 and Figure 4. The resonant frequencies measured at 990, 1062, and 1068 K were not used to determine elastic constants because some modes were missing; however, these frequencies were included in the data plotted in Figure 3. The errors in the elastic moduli included in Table 2 are the root-mean-square deviations calculated by the least-squares methods.

Figure 5 depicts Raman spectra of the natural spinel before and after heat treatment; that is, before and after cycling the sample from ambient temperature to 1000 K or higher and back to ambient. The upper spectrum was measured at ambient conditions before heating. The lower spectrum, also taken at ambient conditions, was measured after RPR measurements had been made with the

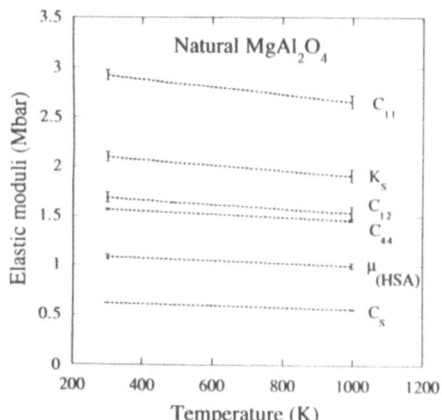

Figure 4
A plot showing the temperature dependencies of the adiabatic elastic constants of natural spinel at ambient pressure. Error bars are shown at ambient and the highest temperatures.

Table 2

Elastic constants of a natural spinel at high temperatures (in 10^2 GPa)[a]

T (K)	C_{11}^s	C_{12}^s	C_{44}^s	C_s	K_s	$\mu(\text{HSA})$[b]
298	2.925	1.689	1.566	0.618	2.101	1.083
	(0.051)	(0.053)	(0.01)	(0.003)	(0.052)	(0.025)
348	2.903	1.673	1.559	0.615	2.083	1.077
375	2.892	1.667	1.556	0.613	2.075	1.075
415	2.881	1.661	1.550	0.610	2.068	1.070
459	2.859	1.646	1.544	0.607	2.050	1.065
488	2.846	1.637	1.536	0.603	2.040	1.060
577	2.820	1.626	1.526	0.597	2.024	1.051
580	2.817	1.624	1.525	0.600	2.022	1.051
610	2.803	1.614	1.521	0.594	2.011	1.047
661	2.787	1.607	1.513	0.590	2.000	1.041
723	2.763	1.594	1.503	0.585	1.983	1.033
775	2.724	1.584	1.496	0.580	1.971	1.027
820	2.723	1.571	1.489	0.576	1.955	1.021
908	2.689	1.553	1.476	0.568	1.932	1.010
943	2.675	1.545	1.470	0.565	1.922	1.006
999	2.659	1.539	1.461	0.560	1.913	0.998
	(0.065)	(0.065)	(0.013)	(0.004)	(0.065)	(0.027)
1003	2.623	1.508	1.469	0.557	1.880	0.999
	(0.033)	(0.033)	(0.007)	(0.002)	(0.033)	(0.024)
298(Q)[c]	2.821	1.554	1.562	0.633	1.976	1.128
	(0.039)	(0.041)	(0.009)	(0.003)	(0.041)	(0.023)

[a] Errors are given in parentheses as root-mean-square deviations.
[b] $\mu(\text{HSA})$ is the isotropic shear modulus calculated from Hashin-Shtrikman averaging scheme.
[c] 298(Q) represents measurements after heat treatment.

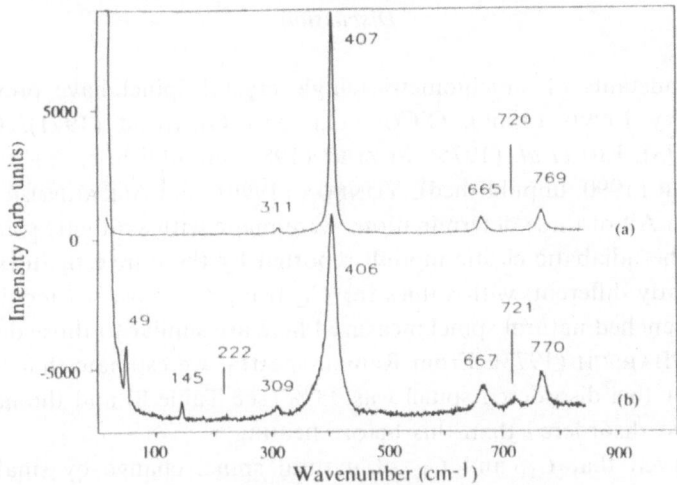

Figure 5

Raman spectra of (a) natural spinel before heating, (b) natural spinel after heating, both taken at ambient conditions.

specimen to 1060 K and the temperature was rapidly quenched to freeze in a highly disordered state by turning off the power to the heater and quickly removing the accessory alumina thermal insulating capsules. The quenching rate, estimated from the two thermocouple readouts, was roughly -0.5 K s^{-1} from 1060 to 600 K and -2 K s^{-1} thereafter to ambient temperature.

The frequencies of the Raman modes of the quenched, disordered spinel are not changed by heat treatment, although the relative intensity of the 720-cm^{-1} band increases, all bands broaden significantly, and the spectrum of the disordered spinel contains extra bands. The 720-cm^{-1} band is not evident in the Raman spectrum of natural spinels but is readily apparent for synthetic and heat-treated natural spinels. The E_g band at 406 cm^{-1} also changes significantly with heat treatment; this band is narrow and symmetric in the spectra of ordered natural spinels and broad and asymmetric for disordered spinels. In disordered spinels (see Fig. 5), the 406-cm^{-1} band also appears to be superimposed on a less intense broader band at nearly the same frequency. A band at 222 cm^{-1} also is evident in the spectra of disordered spinels. These observations are similar to results of three series of *in situ* Raman measurements of spinels, CYNN *et al.* (1992) performed at temperatures to 1424 K. In the first series, spectra of a natural specimen were collected at several, irregularly spaced temperatures from ambient to 1234 K and after quenching the sample to ambient. For the second series with the natural specimen and the series with the synthetic spinel, the maximum temperature was extended to 1434 K; and spectra were collected at several more or less evenly spaced intermediate temperatures during both the heating and cooling phases of the cycle.

Discussion

Elastic constants of stoichiometric single crystal spinel have previously been determined by LEWIS (1966), O'CONNELL and GRAHAM (1971), CHANG and BARSCH (1973), LIU *et al.* (1975), SUZUKI (1982, unpublished), YEGANEH-HAERI and WEIDNER (1990, unpublished), YONEDA (1990), and ASKARPOUR *et al.* (1992, unpublished). All of these determinations were made with synthetic spinels. As seen in Table 3, the adiabatic elastic moduli reported by these investigators are slightly but significantly different, with values for C_{12} being the most scattered. The elastic moduli of quenched natural spinel measured here are similar to those determined by CHANG and BARSCH (1973). From Raman spectra, we estimate that the inversion parameter for this disordered spinel was 23% (see Table 8) and the natural spinel was much less disordered than this before heating.

We observed that C_{11} and C_{12} of natural spinel change by small but easily measurable amounts during heat treatment. By about 20% inversion, C_{11} and C_{12} are 4 and 8% lower, respectively, than in the ordered specimen. C_{44} is less sensitive to cation disorder. The values of adiabatic elastic moduli of the natural spinel before and after heating seem to bound the lowest and upmost values previously determined for synthetic spinels, except for those reported by LEWIS (1966).

We also estimated the elastic constants at ambient conditions for the disordered synthetic spinel examined by SUZUKI (1982, unpublished) whose inversion parameter was 36% as estimated from Raman spectra we collected from the heat-treated sample. We collected two sets of RPR measurements, before and after polishing the sample to a smaller size to examine size effects on estimated C_{ij}, using our newly

Table 3

Elastic constants of spinels at ambient conditions using various techniques (in GPa)

	C_{11}^s	C_{12}^s	C_{44}^s	Techniques
LEWIS	279	153	153	ultrasonic pulse-echo
O'CONNELL and GRAHAM	282.1	155.1	154.2	ultrasonic interferometer
CHANG and BARSCH	282.48	154.91	154.68	ultrasonic pulse-superposition
LIU *et al.*	281.4	154.6	154.3	light-sound scattering
SUZUKI	286.9	160.2	154.7	RPR (before heating)
SUZUKI # 1	282.7	156.4	154.7	RPR (this study)
SUZUKI # 2	281.3	154.5	155.2	RPR (this study)
YONEDA	282.9	155.4	154.8	ultrasonic pulse-echo-overlap
ASKARPOUR *et al.*	290.99	159.79	156.22	Brillouin
Natural spinel*	292.5	168.9	156.6	RPR (this study)
Disordered spinel*	282.1	155.4	156.2	RPR (this study)

* Both were determined from the same natural specimen before and after heat treatment. SUZUKI # 1 and # 2 represent before and after reducing the size of the heat-treated synthetic spinel by SUZUKI (unpublished, 1982).

developed RPR system. We obtained an excellent agreement between the two sets of C_{ij} (SUZUKI #1 and 2 in Table 3) within the experimental errors. The ambient elastic constants of the disordered natural spinel after heat treatment ($i = 23\%$) and the disordered spinel ($i = 36\%$) are indistinguishable, although there is a 13% difference of the inversion parameter between the two samples. The changes of C_{11} and C_{12} through the heating cycle of this synthetic spinel were smaller than we measured for the ordered natural spinel, although the elastic constants of both samples changed in the same direction during heat treatment.

The resonant frequencies of a specimen are proportional to its size. Thus, changes of the resonant frequencies at ambient conditions through the heating cycle might result from a reduction of the unit cell parameter at the transition. However, this trivial effect of crystal or unit cell dimension on frequency cannot explain the fact that the measured resonant frequencies of some higher harmonics change more than lower resonances (Table 1). The trends of the changes of the measured resonance frequencies of a natural spinel at around 1000 K are similar to changes observed between a less (before heating) and more (after cooling) disordered natural spinel at ambient temperature. These observations suggest that all of these effects relate primarily to the degree of cation disorder, although changes of sample sizes may contribute to a smaller degree.

Within the precision of the experiments, we noticed that the derivatives $(\partial f_n/\partial T)_P$ had different values below and above the transition temperature. This suggests that the temperature derivatives of the elastic properties of spinel also will differ across the transition. Although the shear properties as indicated by C_{44} appear to be relatively insensitive to cation disorder, C_s at ambient temperature increases after heat treatment because C_{12} decreases more than C_{11}. After the ordered natural spinel had been heated above the 1000 K transition, C_{11}, C_{12}, and K_S at ambient temperature had decreased by 4, 8, and 6%, respectively.

Several properties of spinel can be computed from these data. The isothermal bulk modulus at constant pressure, K_T, is estimated from the relationship:

$$K_T = K_S/(1 + \alpha\gamma T) \tag{3}$$

where T is the temperature, α is the volume thermal expansion coefficient, and γ is the thermal Grüneisen parameter at ambient pressure (Table 4). A small decrease of K_T is apparent at the transition, and a more pronounced change is evident between the values of K_T at ambient temperature between the ordered natural spinel and the disordered, heat-treated spinel.

The isotropic longitudinal and transverse wave velocities, V_p and V_s, respectively, are calculated using:

$$V_p = [K_S + (4\mu_{HSA}/3)]^{1/2}\rho^{-1/2} \tag{4}$$

and

$$V_s = (\mu_{HSA}/\rho)^{1/2}. \tag{5}$$

Table 4

The isothermal bulk moduli, isotropic velocities, and Grüneisen parameters of a natural spinel

T (K)	α $(10^{-6}/K)$	ρ (g/cm^{-3})	C_p $(J/Kmol)$	K_T $(10^2 GPa)$	V_p (km/s)	V_s (km/s)	γ
298	21.05	3.576	0.815	2.081 (0.052)	9.956 (0.120)	5.502 (0.064)	1.52
348	21.77	3.572	0.896	2.061	9.926	5.492	1.42
375	22.15	3.570	0.933	2.051	9.913	5.487	1.38
415	22.71	3.567	0.980	2.042	9.898	5.477	1.34
459	23.30	3.563	1.022	2.022	9.869	5.468	1.31
488	23.69	3.561	1.046	2.010	9.848	5.457	1.30
577	24.83	3.553	1.103	1.988	9.819	5.439	1.28
580	24.87	3.553	1.105	1.985	9.815	5.438	1.28
610	25.24	3.550	1.120	1.972	9.796	5.431	1.28
661	25.85	3.546	1.144	1.958	9.775	5.417	1.28
723	26.57	3.540	1.168	1.936	9.743	5.402	1.28
775	27.16	3.535	1.188	1.920	9.721	5.390	1.28
820	27.64	3.531	1.203	1.900	9.692	5.378	1.27
908	28.55	3.522	1.231	1.870	9.648	5.355	1.27
943	28.90	3.518	1.241	1.857	9.630	5.346	1.27
999	29.43	3.513	1.253	1.843 (0.063)	9.609 (0.149)	5.330 (0.065)	1.28
1003	29.47	3.512	1.253	1.812 (0.032)	9.563 (0.097)	5.334 (0.065)	1.26
298(Q)	21.05	3.576	0.815	1.958 (0.049)	9.794 (0.102)	5.522 (0.059)	1.43

* Uncertainties are given in parentheses.

The results show that V_p decreases from 9.96 to 9.56 km/s and V_s decreases from 5.50 to 5.33 km/s between 298 and 999 K (Table 4). Over the same range of temperature, the ratio, V_p/V_s, decreases from 1.88 to 1.80; and the seismic parameter, $v = (\partial \ln V_s / \partial \ln V_p)_P$, is about 0.9. Temperature derivatives of the elastic moduli, V_p, and V_s of a natural spinel at ambient pressure are given in Table 5. The values for these properties, estimated from least-squares fits for the natural spinel, are significantly higher than other estimates for synthetic spinels (ASKARPOUR *et al.*, 1991; CHANG and BARSCH, 1973; LIU *et al.*, 1975).

The acoustic Debye temperature, Θ_D, for an isotropic solid is given by:

$$\Theta_D = 3h\rho N V_m/[(4\pi M)^{1/3}k] \tag{6}$$

where h is Planck's constant, k is Boltzmann's constant, N is Avogadro's number, M is the mean atomic weight, and V_m is the mean sound velocity:

$$V_m = [1/3(V_p^{-3} + 2V_s^{-3})]^{-1/3}. \tag{7}$$

Θ_D decreases from 862 to 830 K over the temperature range, 298 to 999 K (Table 6).

Table 5

Temperature derivatives of the elastic moduli (in 10^{-2} GPa/K) and the isotropic velocites (in km/s/K) of a natural spinel at ambient pressure

	Present results	ASKARPOUR et al. (300–1273 K)	CHANG and BARSCH (308–328 K)	LIU et al. (293–423 K)
$\partial C_{11}/\partial T$	−3.79 (0.03)	−3.00	−2.62	−2.58
$\partial C_{12}/\partial T$	−2.12 (0.04)	−1.28	−0.99	−1.07
$\partial C_{44}/\partial T$	−1.50 (0.01)	−1.41	−0.87	−1.01
$\partial K_S/\partial T$	−2.68 (0.03)	−1.80	−1.5	−1.6
$\partial K_T/\partial T$	−3.40 (0.03)	−2.4		
$\partial C_s/\partial T$	−0.84 (0.02)	−0.78		
$\partial \mu/\partial T$	−1.21 (0.01)			
$\partial V_p/\partial T$	−4.94 (0.05)	−3.7		
$\partial V_s/\partial T$	−2.46 (0.01)	−2.33		

* Uncertainties are given in parentheses.

Table 6

Debye temperatures, thermal pressures, and dimensionless parameters for natural spinel

T (K)	Θ_D (K)	αK_T*	ΔP_{TH}	δ_T	δ_S	$(\delta_T - \delta_S)/\gamma$	Γ
298	862	43.81	0.0	7.76	6.06	1.12	5.31
348	860	44.86	0.2217	7.58	5.91	1.18	5.16
375	859	45.44	0.3436	7.48	5.83	1.20	5.08
415	857	46.36	0.5272	7.33	5.71	1.21	4.98
459	855	47.10	0.7328	7.22	5.61	1.23	4.88
488	853	47.60	0.8702	7.14	5.55	1.22	4.82
577	850	49.35	1.3016	6.89	5.33	1.22	4.64
580	850	49.36	1.3164	6.89	5.33	1.22	4.64
610	848	49.77	1.4651	6.83	5.28	1.21	4.58
661	846	50.61	1.7211	6.72	5.18	1.20	4.50
723	843	51.43	2.0374	6.61	5.09	1.19	4.41
775	841	52.13	2.3067	6.52	5.01	1.18	4.34
820	839	52.52	2.5422	6.47	4.96	1.19	4.29
908	834	53.38	3.0082	6.37	4.86	1.19	4.20
943	833	53.68	3.1955	6.33	4.82	1.19	4.16
999	830	54.25	3.4977	6.27	4.76	1.18	4.12
1003	803	53.40	3.5192				
298(Q)	864	41.22					

* αK_T is in 10^{-4} GPa/K.

The thermal Grüneisen parameter at constant pressure, γ, and three other dimensionless parameters were determined to 1.16 Θ_D using the following relations:

$$\gamma = \alpha K_S/(\rho C_P) \tag{8}$$

$$\delta_S = -(\alpha K_S)^{-1}(\partial K_S/\partial T)_P \tag{9}$$

$$\delta_T = -(\alpha K_T)^{-1}(\partial K_T/\partial T)_P \tag{10}$$

$$\Gamma = -(\alpha\mu)^{-1}(\partial\mu/\partial T). \tag{11}$$

By using an equation recommended by ROBIE *et al.* (1978) to estimate the isobaric heat capacity, C_P, at high temperatures, γ was shown to decrease from 1.52 to 1.28 over the temperature range 298 to 999 K (Table 4). Estimates of δ_S, δ_T, and Γ are given in Table 6. Finally, the change of the thermal pressure, $\Delta P_{TH} = P_{TH}(T) - P_{TH}(298) = \int \alpha K_T \, dT$, of spinel varies essentially linearly with T near Θ_D, as for other minerals (ANDERSON *et al.*, 1992).

In many regions of the Raman spectrum of disordered spinel, we find two bands where only one band occurs in the Raman spectrum of an ordered natural spinel. In these cases, group theory for an "averaged" O_h^7 (Fd3m) structure predicts only a single band. Some of these extra bands have simple polarization characteristics; for example, two bands in the spectrum of disordered spinel have the polarization characteristics of an A_{1g} band, although analysis for the Fd3m factor group predicts only one A_{1g} band. Many explanations have been offered for the extra features of the Raman spectra of spinel. There have been several conjectures (WHITE and KERAMIDAS, 1972; STRIEFLER and BARSCH, 1975) that the extra features imply a structural change accompanies cation disordering. In particular, one-to-one ordering of T cations at tetrahedral sites would lower the space group from centric O_h^7 (Fd3m) to acentric T_d^2 (F43m). The irreducible representations for the T_d^2 factor group permit 13 Raman active modes in contrast to the 5 Raman active modes allowed for O_h^7 symmetry.

We interpret the extra features in the spectra of quenched spinel in terms of weak coupling between the disordered cations rather than a change of space group. When Mg and Al ions are randomly distributed at both M and T sites in a disordered spinel, the Raman-active vibrational modes do not have to average the behavior of these ions. Whether average or two-mode behavior is observed as well as the separations between the two modes, depends upon the degree of the coupling between the motions of the various cations. Two modes and two bands occur when ions at symmetrically equivalent sites are weakly coupled; average modes are detected when motions of different ions at similar sites are strongly coupled. In the weak-coupling limit, the Raman spectrum of spinel can be described in terms of approximately local vibrations of MO_4 tetrahedra and MO_6 octahedra. Whether or not a particular vibrational mode of a disordered system is averaged or split depends on the nature of the motions involved and the range of the relevant potentials. Thus, optic vibrations active in the Raman spectrum of spinel may split while acoustic vibrations in the RPR spectra may be averaged so that the number of RPR resonances detected does not change with cation disordering.

The temperature dependencies of the frequencies of the highest energy A_{1g} mode in the Raman spectra are shown in Figure 6 for natural and synthetic spinels. The low and high temperature regions of the frequency-vs-temperature plots have different slopes for both types of spinel; however, the temperature where the break occurs depends upon the origin and history of the specimen. When the ordered natural specimen was heated for the first time, the slope of the frequency-vs-

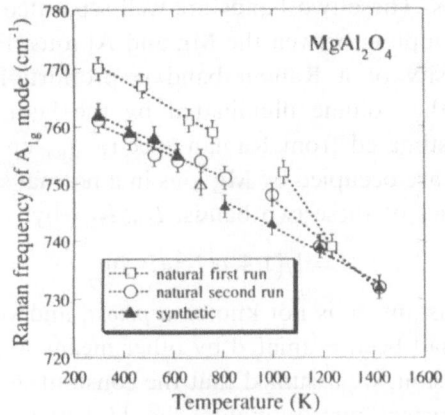

Figure 6
A plot showing Raman frequencies versus temperature of natural and synthetic spinels. Linear fits are given to approximate the break in slopes.

temperature plot started to break at 1000 K, while the break for the synthetic spinel occurs around 650 K. (We believe that the differences shown between the frequencies of the natural and synthetic specimen are due to a systematic error in identifying the olivine bands used for calibration while taking the spectra of the synthetic spinel and the second run for the natural spinel. Although this calibration error shifts all of the spectra for the synthetic spinel by the same amount, these shifts should not change the slopes of the frequency-vs-temperature plot.) The nearly 400 K difference in the onset temperature for the transition indicated by the change of slope is significant. We interpret the change of slope of the frequency-vs-temperature plot for this Raman band as a signal of the onset of cation disordering comparable to the signal provided by the sudden changes of the acoustic frequencies in the RPR study of the natural spinel. We attribute the difference to differences in the initial states of cation disorder of the relevant specimen.

This interpretation is consistent with observations by HÄFNER and LAVES (1966) and SCHMOCKER and WALDNER (1976). For their natural spinel, they found that the cations do not disorder if the specimen was quenched from temperatures below about 1073 K. Transition temperatures for synthetic spinels have been shown to be lower by YAMANAKA and TAKEUCHI (1983) and SUZUKI and KUMAZAWA (1980) as well as our Raman studies (CYNN et al., 1992) and possibly to depend upon the initial degree of inversion of the specimen and heating and cooling rates. The Raman data reported here suggest that disordered natural spinels have low transition temperatures which may depend on the initial degree of inversion and heating rates.

The spectra of the two A_{1g} bands near 720 and 770 cm⁻¹ provide an opportunity to determine whether or not the degree of cation disorder can be evaluated from Raman intensities. These two bands are well separated at ambient conditions because of the weak coupling betwen the Mg and Al ions. For fixed geometry and wavelength, the intensity of a Raman band is proportional to the number of scattering centers in the volume illuminated by the laser excitation. Thus, the inversion parameter estimated from Raman spectra, i_R, or fraction of Al ions in tetrahedral sites which are occupied by Mg ions in a normal spinel should be related to the relative intensities of these two bands, I_{770}/I_{720}, by:

$$i = 1/[1 + b(I_{770}/I_{720})]. \tag{12}$$

Unfortunately, the constant, b, is not known *a priori*, and very few samples whose inversion parameters had been estimated by other means were available to us. For purposes of this discussion, we assumed that the constant, b, was unity and include the data needed to correct our estimates, should better values of the constant become available.

We estimated the relative intensities of the 720 and 770-cm^{-1} bands by using Spectra Calc (© Galactic Industries Corp.) routines to resolve the intensity above 600 cm^{-1} into a smooth backgroud and three mixed Gaussian-Lorentzian peaks. The lower cutoff was chosen to avoid difficulties with background closer to the laser frequency and with bands at lower wavenumbers, although other approaches involving more bands and broader spectral regions were examined. The parameters of the Spectra Calc resolving progam (center position, peak intensity, bandwidth, and Gaussian character) were iterated until a reasonable correlation factor was obtained which was insensitive to further minor adjustments of the parameters. The original and resolved spectra for a quenched natural spinel at ambient temperature are shown in Figure 7, and the parameters for the fits for four specimens are listed

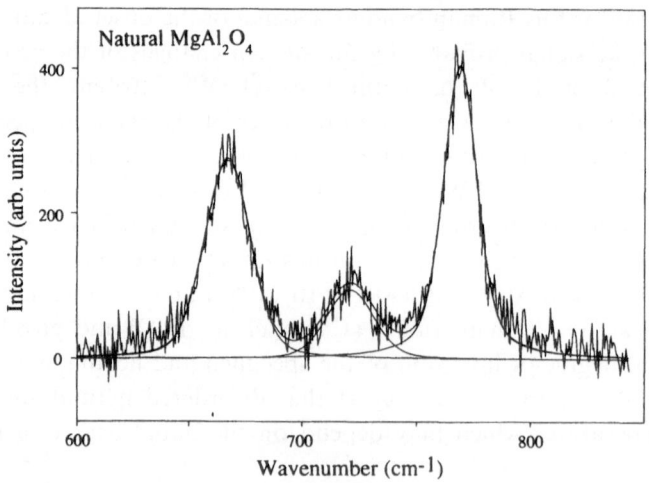

Figure 7
A resolved Raman spectrum of a quenched natural spinel.

in Table 7. Inversions estimated by Equation (12) are compared in Table 8 with values estimated for these samples by other techniques. Our values are similar to values interpolated from *in situ* neutron scattering analysis (PETERSON *et al.*, 1991) at the temperatures from which our samples were quenched, which suggest that the assumption about *b* for these A_{1g} bands is reasonable. Different estimates of the

Table 7

Parameters for curve fits and inversion parameters for spinels[a]

	Center	Height	Width	Type	Area	*i*
natural	667	734	12.6	G_0L_1	14416	
before	717	25	11.8	$G_{0.9}L_{0.1}$	286	
heating	768	733	12.6	G_0L_1	13670	2%
natural	667	274	28.4	$G_{0.8}L_{0.2}$	7872	
after	721	95	28.8	$G_{0.8}L_{0.2}$	2799	
quenching	769	404	20.1	$G_{0.6}L_{0.4}$	9389	23%
WOOD #5	667	651	32	$G_{0.8}L_{0.2}$	23896	
	717	143	26.1	$G_{0.8}L_{0.2}$	4325	
	766	452	19.4	$G_{0.6}L_{0.4}$	11848	26.7%
SUZUKI	671	294	33.0	$G_{0.8}L_{0.2}$	9754	
	725	122	37.1	$G_{0.8}L_{0.2}$	5051	
	770	427	21.1	$G_{0.6}L_{0.4}$	9155	36%

[a] The peak positions are given in cm^{-1} and the inversion parameters (*i*) are calculated with equation (12) by taking the area of the two higher energy A_{1g} modes as measures of their relative intensities. The spectral analysis was performed using a Spectra Calc software.

[b] G represents Gaussian and L represents Lorentzian; and the best mixing parameters were chosen by trial and error.

Table 8

Values of the inversion parameters, i (in percentages) estimated or interpolated for a particular temperature from Raman, NMR, or neutron techniques[a]

	Raman				
Specimen	A_{1g}	E_g	NMR[b]	Neutron[c]	T_{max} (K)
SUZUKI	36			28(2)	1100(?)
Natural	23			27(2)	1060
WOOD #5	27		26(4)	25(2)	983
WOOD #8	30		35(4)	29(2)	1123
WOOD #9	32	17	37(4)		

[a] Wood #5 was analyzed by Raman, NMR, and neutron techniques. The rest of the *i*'s were quoted form NMR and neutron data for comparison at the quenched temperature of a specimen (T_{max}).

[b] From WOOD *et al.* (1986).

[c] From PETERSON *et al.* (1991).

inversion parameter are obtained from the relative intensities of the E_g modes near 400 cm^{-1} which suggests that the corresponding constant for those modes has a different value.

By this method, i_R for the synthetic sample studied to 1100 K by SUZUKI (unpublished, 1982) and reexamined here was estimated to be 36%. This sample was cut from the same boule as the sample for which SUZUKI and KUMAZAWA (1980) found anomalous thermal expansion. i_R was estimated 28% which is greater than the estimate made from neutron analysis at 1100 K (PETERSON *et al.*, 1991). Repeated heat cycling of this specimen during RPR measurements, possibly higher than 1100 K, may have caused this difference in the inverion of the sample. i_R was estimated to be 23% (Fig. 7 and Table 8) for the natural spinel which had been heated to 1060 K for RPR measurements and then cooled to ambient temperature. This value is about 4% lower than the value of the inversion parameter for spinel at 1060 K as interpolated from the neutron analysis (PETERSON *et al.*, 1991) and might easily be due to different quenching rates. These results suggest that estimates of the inversion parameter from Raman spectra are reasonable, although further quantification work should be done to improve the precision of the measurements. The changes of values of the inversion parameter during heat treatment also are consistent with our argument that the changes of elastic properties and Raman spectra of spinel during heat treatment are due to cation disordering.

We did not estimate the uncertainty of the cation disorder because of the assumption of the equality of the scattering factors for the Raman bands. However, we provide approximate values for the uncertainty from the comparison between the different techniques used for the same samples. The i (cation inversion) and the uncertainties for the samples, WOOD #5 and #8 (Table 8) using the NMR technique (WOOD *et al.*, 1986) and for WOOD #5 using the neutron powder diffraction (PETERSON *et al.*, 1991) were compared with i from our Raman measurements. The quoted uncertainties from the NMR study are 4% and the neutron diffraction are 2%. For WOOD #5, our i is 1 or 2% higher than theirs. For WOOD #8, our i is closer to that estimated from the neutron study. We believe that our method is comparable to the above techniques to measure i and so the uncertainties are. To be conservative, our method appears to be better than 5% regarding the uncertainties. MYSEN *et al.* (1982) reported that the area ratio determinations have small, but finite, uncertainties (approximately 5%).

Conclusions

By using two independent techniques on the same samples, we compared effects of thermal cycling and cation disordering on the vibrational dynamics of spinel. We measured the frequencies of more than 30 acoustic resonances of a natural spinel at high temperatures and found, for the first time, significant changes in the resonance frequencies before and after heating to 1000 K and higher. The temperature derivatives of the resonant frequencies also change abruptly near 1000 K; however, neither the extra number of resonances nor other acoustic anomalies were detected either at high temperature or throughout a heating-cooling cycle. We interpret the changes near 1000 K in terms of the onset of cation disordering, which is consistent with previous studies of natural spinels.

From the resonance frequencies, values of the elastic moduli and other thermodynamic properties of natural and synthetic spinels were determined to more than 1000 K with much greater precision than previously available. Although the statistical errors of these measurements are still larger than for other mantle materials which have been studied by RPR methods, the effects of cation disordering in spinel are large enough to identify trends in these properties. By using a natural spinel with less than 2% inversion, we showed that C_{11} and C_{12} decrease significantly with cation disordering and C_{44} is relatively insensitive to disordering. Below 1000 K, the temperature derivatives of the elastic moduli of ordered natural MgAl$_2$O$_4$ spinel differ significantly from those of synthetic spinels; however, the elastic constants of natural spinel disordered at high temperature are similar to the elastic constants of synthetic spinels, even after recovery at ambient temperature. The spread among most previous determinations of the elastic constants of spinel is within the range of the present measurements, which suggests that differences among the measurements were due to different degrees of inversion of the samples.

The transition in natural spinel defined by discontinuous changes of acoustic resonance frequencies near 1000 K also can be detected by changes of the temperature derivative of the Raman band near 770 cm^{-1} and changes in the relative intensities of Raman bands near 720 and 770 cm^{-1}. Similar changes also occur at lower temperatures in disordered spinels and also are associated with cation disordering. Thus, we showed that the degree of inversion of spinel can be estimated from the intensities of two Raman-active totally symmetric stretching vibrations at 720 and 770 cm^{-1}. The values we obtained from Raman data are consistent with those from ^{27}Al MAS-NMR and neutron scattering data. This new method for estimating the degree of inversion from Raman intensities is simpler than methods used previously, is nondestructive, and can readily be adapted for *in situ* high temperature and high pressure studies.

Acknowledgments

We thank Drs. J. L. Rosenfeld, A. L. Montana, J. T. Wasson, H. St. C. O'Neill, and B. J. Wood for providing valuable specimens and Drs. W. Dollase, H. Oda, D. Isaak, J. Sarrao, T. Cooney, S. H. Sharma, V. Askarpour, and C. S. Yoo for helpful discussions, and Mr. B. Jones for the electron probe microanalysis. This work was supported by National Science Foundation DMR90–22076. CHC also thanks Dr. A. Hofmeister for support while preparing this manuscript.

REFERENCES

ANDERSON, O. L., and GOTO, T. (1989), *Measurement of Elastic Constant of Mantle-related Minerals at Temperatures up to 1800 K*, Phys. Earth Plan. Int. *55*, 241–253.

ANDERSON, O. L., ISAAK, D. G., and ODA, H. (1992), *High Temperature Elastic Constant Data on Minerals Relevant to Geophysics*, Rev. Geophys. *30*, 57–90.

ASKARPOUR, V., MANGHNANI, M. H., FASSBENDER, S., and YONEDA, A. (1991), *Single-crystal Elastic Properties of Spinel MgAl₂O₄ up to 1273 K by Brillouin Spectroscopy* (abstract), EOS Trans. AGU *72*, 435.

CHANG, Z. P., and BARSCH, G. R. (1973), *Pressure Dependence of Single-crystal Elastic Constants and Anharmonic Properties of Spinel*, JGR *78*, 2418–2433.

CHOPELAS, A., and HOFMEISTER, A. M. (1991), *Vibrational Spectroscopy of Aluminate Spinels at 1 Atm and of MgAl₂O₄ to over 200 kbar*, Phys. Chem. Minerals *18*, 279–293.

CYNN, H., SHARMA, S., COONEY, T., and NICOL, M. (1992), *High-temperature Raman Investigation of Order-disorder Behavior in the MgAl₂O₄ Spinel*, Phys. Rev. B *45*, 500–502.

DUFFY, T. S., and ANDERSON, D. L. (1989), *Seismic Velocities in Mantle Minerals and the Mineralogy of the Upper Mantle*, J. Geophys. Res. *94*, 1895–1912.

FINGER, L. W., HAZEN, R., and HOFMEISTER, A. M. (1986), *High-pressure Crystal Chemistry of Spinel (MgAl₂O₄) and Magnetite (Fe₃O₄): Comparisons with Silicate Spinels*, Phys. Chem. Minerals *13*, 215–220.

FRAAS, L. M., MOORE, J. E., and SALZBERG, J. B. (1973), *Raman Characterization Studies of Synthetic and Natural MgAl₂O₄ Crystals*, J. Chem. Phys. *58*, 3585.

GOTO, T., and ANDERSON, O. L. (1989), *Elastic Constants of Corundum to 1825 K*, J. Geophys. Res. *94*, 7588–7602.

GWANMESIA, G. D., RIGDEN, S., JACKSON, I., and LIEBERMANN, R. C. (1990), *Pressure Dependence of Elastic Wave Velocity for β-Mg₂SiO₄ and the Composition of the Earth's Mantle*, Science *250*, 794–797.

HÄFNER, S., and LAVES, F. (1966), *Ordnung/Unordnung und Ultrarotabsorption, III. Die Systeme MgAl₂O₄-Al₂O₃ und MgAl₂O₄-LiAl₅O₈*, Z. Krist. *115*, 321–330.

HASHIN, Z., and SHTRIKMAN, S. (1962), *A Variational Approach to the Theory of the Elastic Behavior of Polycrystals*, J. Mech. Phys. Solids *10*, 343–352.

HILBERT, E. G., and GRAHAM, E. K. (1989), *Elastic Properties of Stoichiometric MgAl₂O₄ Spinel* (abstract), EOS Trans. AGU *70*, 1368.

ISAAK, D. G. (1992), *High-temperature Elasticity of Iron-bearing Olivines*, J. Geophys. Res. *97*, 1871–1885.

ISAAK, D. G., ANDERSON, O. L., GOTO, T., and SUZUKI, I. (1989a), *Elasticity of Single-crystal Forsterite Measured to 1700 K*, J. Geophys. Res. *94*, 5895–5906.

ISAAK, D. G., ANDERSON, O. L., and GOTO, T. (1989b), *Measured Elastic Moduli of Single-crystal MgO up to 1800 K*, Phys. Chem. Minerals *16*, 704–713.

ISAAK, D. G., ANDERSON, O. L., and ODA, H. (1992), *High Temperature Thermal Expansion and Elasticity of Calcium-rich Garnets*, Phys. Chem. Minerals *19*, 106–120.

ISHII, M., MIRASHI, J., and YAMANAKA, T. (1982), *Structure and Lattice Vibrations of Mg-Al Spinel Solid Solution*, Phys. Chem. Minerals *8*, 64–68.

LEWIS, M. F. (1966), *Elastic Constants of Magnesium Aluminate Spinel*, J. Acous. Soc. Am. *40*, 728–729.

LIU, H. P., SCHOCK, R. N., and ANDERSON, D. L. (1975), *Temperature Dependence of Single-crystal Spinel (MgAl₂O₄) Elastic Constants from 293 to 423 K Measured by Light-sound Scattering in the Raman-Nath Region*, Geophys. J. R. Astr. Soc. *42*, 217–250.

MCMILLAN, P. R., and HOFMEISTER, A. M., *Infrared and Raman spectroscopy in spectroscopic methods in mineralogy and geology*. In *Reviews in Mineralogy* (ed. Hawthorne, F. C.) (Mineralogical Society of America, Washington D.C. 1988) pp. 99–159.

MIGLIORI, A., STEKEL, A., SARRAO, J. L., VISSCHER, W. M., BELL, T., and LEI, M. (1991), *Techniques and processes for the measurement of the resonances of small single crystal*. In Proc. 28th Annual Technical Meeting of the Society of Engineering Sciences, Nov. 6–8, 1991, Gainesville, FL.

MILLARD, R. L., PETERSON, R. C., and HUNTER, B. K. (1990), *Temperature Dependence of Cation Disorder in MgAl$_2$O$_4$ Spinel Using Aluminum-27 MAS NMR* (abstract), EOS Trans. AGU *71*, 653.

MILLARD, R. L., PETERSON, R. C., and HUNTER, B. K. (1992), *Temperature Dependence of Cation Disorder in MgAl$_2$O$_4$ Spinel Using ^{27}Al and ^{17}O Magic-angle Spinning NMR*, Am. Min. *77*, 44–52.

MYSEN, B. O., FINGER, L. W., VIRGO, D., and SEIFERT, F. A. (1982), *Curve-fitting of Raman Spectra of Silicate Glasses*, Am. Min. *67*, 686–695.

NARASIMHAN, C. S., and SWAMY, C. S. (1980), *Studies on the Solid State Properties of the Solid Solution System MgAl$_{2-x}$Fe$_x$O$_4$*, Phys. Stat. Sol. (a) *59*, 817–826.

NAVROTSKY, A., and KLEPPA, O. J. (1967), *The Thermodynamics of Cation Distributions in Simple Spinels*, J. Inor. Nuc. Chem. *29*, 2701–2714.

O'CONNELL, R. J., and GRAHAM, E. K. (1971), *Equation of State of Stoichiometric Spinel to 10 Kbar and 800°C* (abstract), EOS Trans. AGU *71*, 359.

OHNO, I. (1990), *Rectangular Parallelepiped Resonance Method for Piezoelectric Crystals and Elastic Constants of Alpha-Quartz*, Phys. Chem. Minerals *17*, 371–378.

O'HORO, M. P., FRISILLO, A. L., and WHITE, W. B. (1973), *Lattice Vibrations of MgAl$_2$O$_4$ Spinel*, J. Phys. Chem. Solids *34*, 23.

O'NEILL, H. St. C., and NAVROTSKY, A. (1983), *Simple Spinels: Crystallographic Parameters, Cation Radii, Lattice Energies, and Cation Distribution*, Am. Min. *68*, 81–194.

PETERSON, R. C., LAGER, G. A., and HITTERMAN, R. L. (1991), *A Time-of-flight Neutron Powder Diffraction Study of MgAl$_2$O$_4$ at Temperatures up to 1273 K*, Am. Min. *76*, 1455–1458.

ROBIE, R. A., HEMINGWAY, B. S., and FISHER, J. R. (1978), *Thermodynamic Properties of Minerals and Related Substances at 298.15 K and 1 Bar (10^5 Pascals) Pressure and at Higher Temperatures*, Geol. Surv. Bull. 1452.

SCHMOCKER, U., and WALDNER, F. (1976), *The Inversion Parameter with Respect to the Space Group of MgAl$_2$O$_4$ Spinels*, J. Phys. C *9*, L235–L237.

SLACK, G. A., HAM, F. S., and CHRENKO, R. M. (1966), *Optical Absorption of Tetrahedral Fe^{2+} (3d^6) in Cubic ZnS, CdTe, and MgAl$_2$O$_4$*, Phys. Rev. *152*, 376–402.

STRIEFLER, M. E., and BARSCH, G. R., *Lattice dynamics of MgAl$_2$O$_4$ in relation to the space group of spinel*. In Proc. *Lattice Dynamics* (ed. Balkanski, M.) (Flammarion Pub. Co. Paris 1978) pp. 75–76.

SUMINO, Y., OHNO, I., GOTO, T., and KUMAZAWA, M. (1976), *Measurement of Elastic Constants and Internal Frictions on Single-crystal MgO by Rectangular Parallelepiped Resonance*, J. Phys. Earth *24*, 263–273.

SUZUKI, I., and KUMAZAWA, M. (1980), *Anomalous Thermal Expansion in Spinel MgAl$_2$O$_4$*, Phys. Chem. Minerals *5*, 279–284.

THOMPSON, P., and GRIMES, N. W. (1978), *Observation of Low Energy Phonons in Spinel*, Solid State Comm. *25*, 609–611.

TOULOUKIAN, Y. S., KIRBY, R. K., TAYLOR, R. E., and LEE, T. Y. R., *Thermal Expansion, Nonmetallic Solids: Thermophysical Properties of Matter*, *13* (Plenum, New York–Washington, 1977).

VIÑUELA, J. S. D., and AREÁN, C. O. (1987), *Distribution of Copper Ions among Octahedral and Tetrahedral Sites in Cu$_x$Mg$_{1-x}$Al$_2$O$_4$ Spinels*, Phys. Stat. Sol. (a) *101*, 57–61.

WHITE, W. B., and KERAMIDAS, V. G. (1972), *Application of Infrared and Raman Spectroscopy to the Characterization of Order-disorder in High Temperature Oxides*, National Bureau of Standards Special Publication *364*, 113–126.

WOOD, D. L., IMBUSCH, G. F., MACFARLANE, R. M., KISLIUK, P., and LARKIN, D. M. (1968), *Optical Spectrum of Cr^{3+} Ions in Spinels*, J. Chem. Phys. *48*, 5255–5263.

WOOD, B. J., KIRKPATRICK, Y., and MONTEZ, B. (1986), *Cation Order-disorder Phenomena in MgAl$_2$O$_4$*, Am. Min. *71*, 999–1006.

YAMAMOTO, S., and ANDERSON, O. L. (1987), *Elasticity and Anharmonicity of Potassium Chloride at High Temperature*, Phys. Chem. Minerals *14*, 332–340.

YAMAMOTO, S, OHNO, I., and ANDERSON, O. L. (1987), *High Temperature Elasticity of Sodium Chloride*, J. Phys. Chem. Solids *48*, 143–151.

YAMANAKA, T., and TAKEUCHI, Y. (1983), *Order-disorder Transition in* $MgAl_2O_4$ *Spinel at High Temperatures up to* $1700^\circ C$, Z. Krist. *165*, 65–78.

YAMANAKA, T., and ISHII, M. (1986), *Raman Scattering and Lattice Vibrations of* Ni_2SiO_4 *Spinel at High Temperature*, Phys. Chem. Minerals *13*, 156–160.

YEGANEH-HAERI, A., and WEIDNER, D. J. (1990), *Single-crystal Elastic Properties of* $MgAl_2O_4$ *Spinel up to 1200 K* (abstract), EOS Trans. AGU *71*, 620.

YONEDA, A. (1990), *Pressure Derivatives of Elastic Constants of Single Crystal MgO and* $MgAl_2O_4$, J. Phys. Earth *38*, 19–55.

(Received April 12, 1993, revised August 2, 1993, accepted October 6, 1993)

PAGEOPH, Vol. 141, No. 2/3/4 (1993)

0033–4553/93/040445–22$1.50 + 0.20/0

A High-pressure, High-temperature Apparatus for Studies of Seismic Wave Dispersion and Attenuation

Ian Jackson[1] and M. S. Paterson[1]

Abstract — An apparatus is described which provides for the investigation of viscoelasticity/anelasticity in geologic and related materials under conditions of high pressure and temperature. Cylindrical specimens are tested in torsion—a geometry particularly well suited to shear mode observations at the low strain amplitudes of the linear regime. Forced oscillation experiments allow the measurement of dispersion and attenuation at the low frequencies of teleseismic wave propagation. The conduct of complementary forced oscillation and creep tests allows recoverable anelastic strains to be distinguished from those of permanent viscous deformation. It has been demonstrated that robust measurements can be made at strain amplitudes below 10^{-5} and frequencies of 1 mHz–1 Hz, under P-T conditions to 300 MPa and 1200°C. The prospects for further development of this facility are outlined.

Key words: Viscoelasticity, anelasticity, dispersion, attenuation, shear modulus, creep, forced oscillation methods.

Introduction

Acoustic waves propagate without dispersion, that is, with a frequency independent phase speed $v = \omega/k$ in *perfect (i.e., defect-free) crystalline material*, provided only that the wavelength $\lambda = 2\pi/k$ is long compared to the interatomic spacing—a condition satisfied for frequencies $v = \omega/2\pi$ less than about 10^{12} Hz (1 THz). Support for this contention derives both from lattice dynamics theory and from experimentally measured phonon dispersion curves (e.g., ASHCROFT and MERMIN, 1967, pp. 440–441). Thus one would expect to measure the same elastic wave speed whether at the MHz–GHz frequencies of ultrasonic wave propagation and Brillouin scattering, or at the much lower frequencies of teleseismic wave propagation (< 1 Hz). Another important property of such defect-free (and relatively coarse-grained) crystalline material is very high strength of order one tenth of the shear modulus, or about 10 GPa for oxides and silicates (e.g., POIRIER, 1985, pp. 38–39).

Real materials however, generally deform at stresses lower by a factor of about 100 as a result of the motion of defects, especially dislocations. The mobility of

---

[1] Research School of Earth Sciences, Australian National University, Canberra ACT 0200, Australia.

crystal defects, including also point and planar defects and grain boundaries, leads inevitably to frequency dependent 'elastic' moduli and wave speeds. At very high frequencies, the period of the alternating stress will be too short to allow the migration of the defect and the realization of the associated strain, and the unrelaxed stiffness modulus (equal to the ratio stress/strain for the perfect crystal) will result. At sufficiently low frequencies, the strain resulting from the application of stress will include an additional contribution from defect migration, yielding a lower, relaxed stiffness modulus. Recognition that the (complex) moduli governing 'elastic' wave propagation in the real materials of the earth's interior must be frequency dependent, even under subsolidus conditions as a result of the thermally activated migration of crystal defects (GOETZE, 1977), has provided the motivation for the development of apparatus which provides for the study of dispersion and attenuation within the seismic frequency band.

The discussion above and seismological observations (reviewed by KARATO and SPETZLER, 1990) suggest that departures from ideal elastic behaviour should be most pronounced in shear. Torsional forced oscillation and creep tests are particularly well suited to the experimental investigation of shear mode viscoelasticity at frequencies within and below the seismic band (JACKSON, 1986). The mechanical advantage achieved by off-axis location of strain sensors makes it practicable to perform such measurements at strain amplitudes below the threshold of $10^{-6}–10^{-5}$ for linear behaviour (BRENNAN and STACEY, 1977). This approach has been employed in high-temperature measurements at atmospheric pressure (BERCKHEMER et al., 1982; GUEGUEN et al., 1989; GETTING et al., 1990, 1991), and has also been adopted for the high-pressure, high-temperature apparatus described here.

The goal of this effort has been to develop a versatile apparatus for studies of shear mode viscoelasticity in geological and related materials at high subsolidus and supersolidus temperatures. The use of pressure to minimise the creation of porosity through thermal microcracking, and to simplify the behaviour of interfaces within the experimental assembly, was also a vital design consideration. Temperatures to 1200°C at 300 MPa confining pressure are currently accessible, allowing exploration of high subsolidus temperatures even in refractory anhydrous ultramafic rocks. Forced oscillation and creep tests are conducted at frequencies below 1 Hz and maximum strain amplitudes in the range 10^{-8} to 10^{-5}. Provision has been made for the future control of pore fluid pressure. Previously we have reported on various stages in the development and application of this facility (JACKSON et al., 1984; JACKSON and PATERSON, 1987; JACKSON et al., 1992; JACKSON, 1993a–d). Here it is our purpose to describe the apparatus itself and the calibration of its performance in more detail.

The Principle

The principle underlying the study of shear mode viscoelasticity or anelasticity (distinguished below) through the conduct of torsional forced oscillation and creep

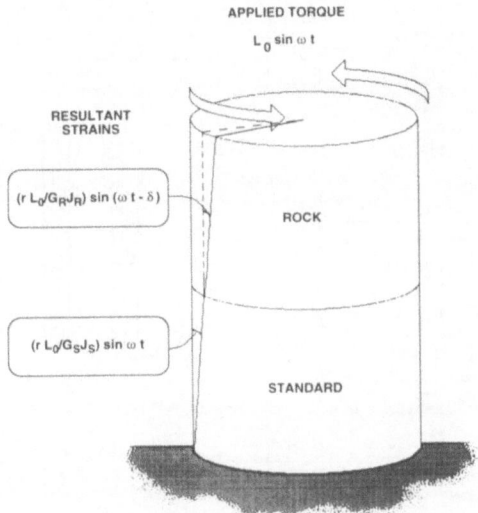

Figure 1

The principle underlying the study of shear mode viscoelasticity/anelasticity through the observation of forced torsional oscillations. At sufficiently low amplitudes of the applied torque $L_0 \sin \omega t$, the relationship between stress and strain and their respective time derivatives is linear. Under these circumstances, the strain ε_R which increases linearly with distance r from the torsional axis, is everywhere related to the local stress through the same shear modulus G_R and loss angle δ. J_R is the polar moment of inertia of the cylindrical rock specimen.

tests is illustrated in Figure 1. An elastic element and the unknown are connected mechanically in series such that, at sufficiently low frequencies, each supports the same steady or oscillating torque. The resulting deformation of the elastic standard provides a measure of the amplitude, and for the forced oscillation tests, the phase of the applied torque. Viscoelastic or anelastic behaviour of the unknown will be manifested as a phase lag between the applied torque which is supported by a radial distribution of shear stress, and the resulting strain—the phase angle being known as the loss angle. The tangent of this angle is variously referred to as the dissipation factor, the internal friction or the attenuation, and is the reciprocal of the quality factor Q. The amplitude of the strain in the unknown relative to that in the standard provides a comparative determination of the shear modulus. In a creep test on the same material, a time-dependent non-elastic strain would be observed—in addition to the elastic strain which appears instantaneously. Monitoring also of the extent of recovery following the removal of the steady applied torque, allows recoverable strictly anelastic strains to be distinguished from the permanent strains of viscous deformation.

Implementation

A cylindrical specimen (150 mm length, 15 mm diameter) is located within the hot zone of an internal furnace (Figure 2). Mechanical coupling to the steel (British

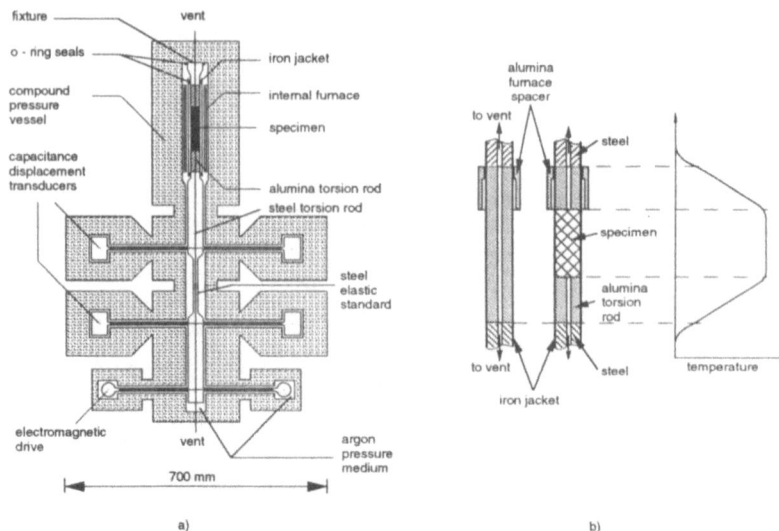

Figure 2
(a) The experimental configuration for the conduct of low amplitude torsional forced oscillation and creep tests under conditions of high pressure (to 300 MPa) and temperature (to 1200 C). (b) Specimen assemblies employed in comparative tests. Vector subtraction of the complex compliances of these two assemblies (Figure 6) allows the elimination of the contribution to the overall compliance from those (common) parts of the assembly outside the furnace hot zone (modified after JACKSON, 1993c).

Standard En 25) torsion rods above and below is achieved through hollow cylinders (15 mm O.D. × 2 mm I.D.) of high-grade alumina ceramic (Morgan Matroc Duramic 998) which span the regions of strong temperature gradient. Both specimen and ceramic spacers are isolated from the argon pressure medium by enclosure within a thin-walled (0.25 mm) iron jacket sealed with an O-ring against each of the steel torsion rods.

The lower torsion rod, including a waisted section of torsional compliance 4.6×10^{-4} rad/N.m which serves as the elastic standard, is drilled transversely in order to accommodate hollow stainless steel lever arms on which the moving elements of the electromagnetic drivers and parallel-plate capacitance displacement transducers are mounted off-axis for the necessary mechanical advantage (Figure 2). The entire assembly except for the specimen itself is drilled axially. The resulting internal space is isolated from the pressure medium by O-ring seals and vented to atmosphere—at the upper end directly, and at the lower end via a flexible and torsionally compliant coil of high-pressure tubing fed through the closure of the pressure vessel. This arrangement ensures that all interfaces within the assembly are subject to a compressive normal stress equal to the confining pressure. In this way the likelihood of significant interfacial compliance, which has plagued anelasticity measurements at atmosphreic pressure, is minimized. Such access from each end to the pore space of the specimen also provides for the future control of pore pressure.

The lowermost part of the lower torsion rod is precision ground to fit snuggly within the bore of the pressure vessel, thereby guaranteeing an axial location. Friction associated with the rolling/sliding of the assembly within the bore of the pressure vessel will modify the amplitude and phase of the resultant torque delivered to the specimen assembly, but does not affect the relative amplitudes and phases of the angular distortions measured at the stations above and below the elastic standard.

Pressure Vessel

A compound pressure vessel has been constructed to house the assembly described above with minimal pressurized volume (Figure 2). That part which houses the internal furnace is machined from a forging of vacuum-melted 5% Cr-Mo-V steel (AISI type H13) hardened and tempered to about 48 Rockwell C (PATERSON, 1970; in future construction a lower hardness near 42 Rockwell C would be sought for greater toughness). This part of the compound pressure vessel is coated with a thin layer of epoxy resin in order to protect it from the water which circulates within spiral grooves machined within a cylindrical mild steel cooling jacket. The remaining parts of the pressure vessel were fabricated from a similar AISI H13 steel, hardened and tempered to 42 Rockwell C.

Seals between the different parts of the compound pressure vessel and at the closures and the electrical feedthroughs are effected mostly with NBR rubber O-rings (ASTM designation D1418, Ludowici & Son, Ltd.) presqueezed to $\sim 85\%$ of their cross-sectional diameter, and supported where appropriate with spun-cast phosphor-bronze anti-extrusion rings (PATERSON, 1962). The electrical feedthroughs for furnace power and strain measurement consist of cylindrical En 25 steel plugs supported within and insulated from the wall of the pressure vessel by magnesia-partially stabilized zirconia disks (Nilcra Pty Ltd.) which prove to be remarkably tough. Silicone polymer O-rings used in this application guarantee a particularly high degree of electrical insulation. The electrical feedthroughs to the electromagnetic drivers consist of En 25 steel cones insulated from the pressure vessel by a thin layer (~ 0.5 mm) of UV-cured dental cement (Kulzer Estilux Posterior C).

Argon from a normal storage cylinder, purified as described by PATERSON (1970), is delivered to the vessel at pressures up to 300 MPa by a two-stage air-driven gas booster (Haskell model AGD-152H)—Figure 3. The second stage has been locally modified to reach higher pressures through reduction of the diameter of the bore of the gas barrel and matching piston, and by the construction of a more robust valve block. The gas seal against the moving piston comprises an O-ring mounted on a deformable flange (of molybdenum disulphide impregnated nylon, Polypenco Nylatron GS) which is closely fitted to the piston, and is

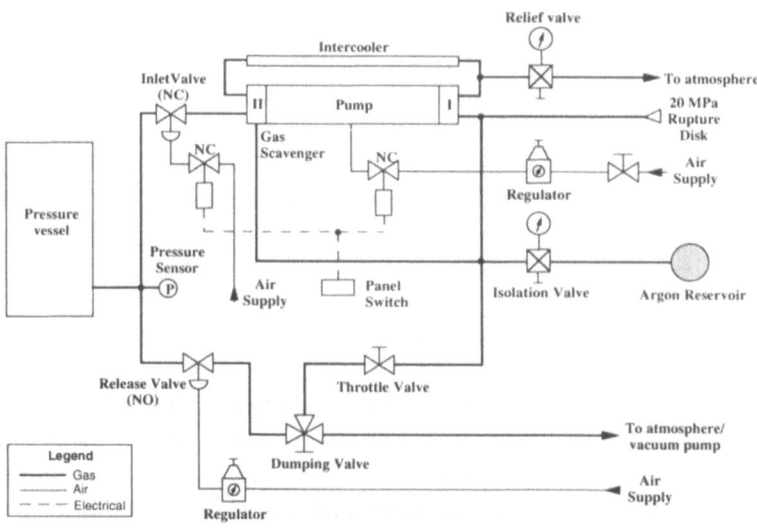

Figure 3

Attenuation apparatus: arrangements for pressurization with argon. Normally closed and normally open air- and solenoid-operated valves are labelled 'NC' and 'NO', respectively.

supported within the gas barrel by hardened Be-Cu anti-extrusion rings and a close-fitting hardened steel collar. The gas delivered by the first stage of the pump passes through an intercooler before reaching the inlet valve for the second stage. Pressure is measured and servocontrolled via a Hardwood manganin resistance gauge, calibrated against a Heise Bourdon-tube pressure gauge.

Generation of Torque

A pair of electromagnetic drivers housed within the pressure vessel provides for the generation of the torque. Each driver consists of a stationary AlNiCo permanent magnet mounted within a soft iron pole piece and a moving coil of enamelled copper wire wound on a pure aluminium former supported upon a bracket which locates within the outer end of the hollow stainless steel lever arm.

The appropriate variation of current with time for a creep or forced oscillation test of prescribed period is computer synthesized, low-pass filtered (1 Hz) to suppress the quantization error signal arising from the 12-bit digital/analogue conversion, and amplified before application to the drivers (Figure 4). The sense of current flow is usually arranged so that the two drivers act cooperatively to produce the applied torque. However, adjustment of series resistors for null response with the drivers switched to act in opposition, provides for equalization of their performance.

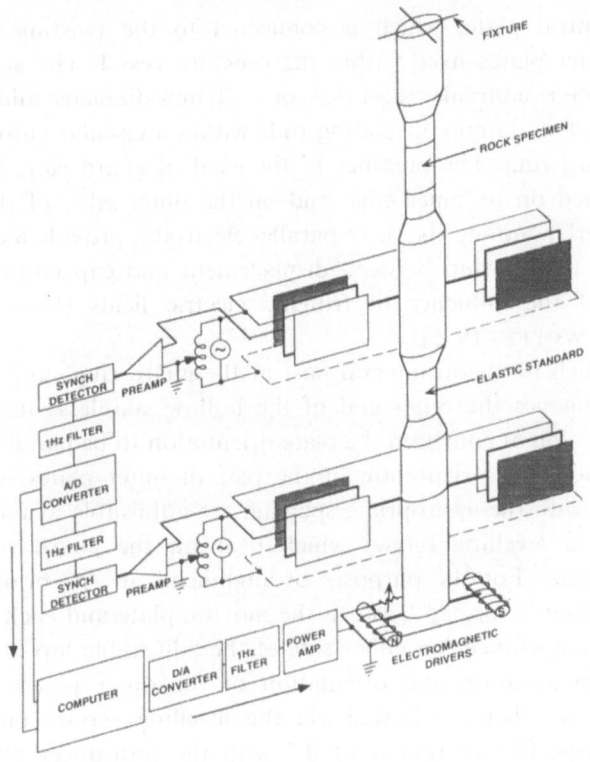

Figure 4

Attenuation apparatus: arrangements for computer control of the applied torque and data acquisition.

These arrangements provide for the generation of a steady or oscillating torque (periods of 1–1000 s) of amplitude adjustable from ~0.001 to ~0.1 Nm, corresponding to maximum strain amplitudes at the periphery of the cylindrical specimen within the range 10^{-8}–10^{-5} for typical specimen compliances. Larger torques could be generated if the AlNiCo magnets were replaced by magnetic material such as rare-earth cobalt alloys of higher remanent magnetization.

Measurement of the Angular Distortions of the Elastic Element and of the Specimen Assembly

The twist of the assembly induced by the electromagnetically applied torque is measured at two stations, respectively above and below the elastic standard, by pairs of parallel plate capacitance transducers. The design and operation of these transducers closely follows that of BRENNAN and STACEY (1977, see also BRENNAN, 1981). Each consists of three closely spaced (0.5 mm) parallel plates arranged so that the central plate, which is connected to the twisting assembly, moves between the outer plates fixed within the pressure vessel. The active part of each plate or electrode is a circular steel disk of ~ 50 mm diameter and 4 mm thickness, supported by radial alumina insulating rods within a circular cut-out in a coplanar rectangular guard ring. The presence of the earthed guard ring, and of re-entrant grooves machined on its inner edge and on the outer edge of the electrode, and appropriate overlap among the three parallel electrodes provide a close approach to linearity of the relationship between displacement and capacitative impedance by minimization of the influence of fringing electric fields (STACEY et al., 1969; GLADWIN and WOLFE, 1975).

A spigot which forms an integral part of the guard ring supporting the central moving plate engages the outer end of the hollow stainless steel lever arm; lugs machined on the spigot constrain the plate orientation to be parallel to the torsional axis. The position and orientation of the pair of outer plates, which are rigidly bolted together with the appropriate spacing, are adjustable against spring loading through a set of levelling screws which bear on the guard ring at azimuthal separations of 120°. For the purposes of alignment, an AC bridge is constructed from the capacitances formed between the moving plate and each of the two outer plates, and the inductances on either side of the adjustable tap of a 6-decade ratio transformer. The position and orientation of the outer plates within the open pressure vessel are then readjusted via the levelling screws until the bridge is balanced at a transformer ratio r of 0.5 with the transducer plates parallel and equally spaced as viewed from the outer end. The transformer ratio r is that fraction of its impedance between the earthed tap and a chosen end.

For measurement of the angular distortion of the assembly, the two capacitance transducers at each station are connected diagonally in parallel (Figure 4) in order to discriminate against any flexural mode displacement (BRENNAN, 1981). For each of these transducers ($i = 1, 2$), the reactive impedances between the central moving plate and each of the fixed outer plates are proportional to the plate separations, respectively d_i and $D - d_i$, so that the bridge is balanced for a transformer ratio $r_i = d_i/D$, D being the range of allowed displacement of the central plate. It follows that the transformer ratio $r_\parallel$ for bridge balance for the parallel combintion of transducers is given by

$$r_\parallel = r_1 r_2 (2 - r_1 - r_2)/[r_1(1 - r_1) + r_2(1 - r_2)]. \tag{1}$$

Displacements corresponding to imbalances of δr_1 and δr_2, result in a bridge imbalance for the parallel combination of transducers given by

$$\delta r_\parallel = (\partial r_\parallel/\partial r_1)\, \delta r_1 + (\partial r_\parallel/\partial r_2)\, \delta r_2$$

which can be written as

$$\delta r_\parallel = (\partial r_\parallel/\partial r_1 + \partial r_\parallel/\partial r_2)[(\delta r_1 + \delta r_2)/2] + (\partial r_\parallel/\partial r_1 - \partial r_\parallel/\partial r_2)[(\delta r_1 - \delta r_2)/2], \tag{2}$$

that is, in the form

$$\delta r_{\parallel} = A_T \, \delta r_T + A_F \, \delta r_F. \tag{3}$$

$\delta r_T = (\delta r_1 + \delta r_2)/2$ and $\delta r_F = (\delta r_1 - \delta r_2)/2$ are the torsional and flexural mode components of the bridge imbalances, and $A_T = \partial r_{\parallel}/\partial r_1 + \partial r_{\parallel}/\partial r_2$ and $A_F = \partial r_{\parallel}/\partial r_1 - \partial r_{\parallel}/\partial r_2$ are the torsional and flexural mode sensitivities of the parallel combination of transducers, respectively. Evaluation of the partial derivatives $\partial r_{\parallel}/\partial r_1$ and $\partial r_{\parallel}/\partial r_2$, through equation (1), yields

$$A_T = 1 + (r_1 - r_2)^2 [(1 - r_1)(1 - r_2) + r_1 r_2]/[r_1(1 - r_1) + r_2(1 - r_2)]^2$$

and

$$A_F = (r_2^2 - r_1^2)(1 - r_1 - r_2)(2 - r_1 - r_2)/[r_1(1 - r_1) + r_2(1 - r_2)]^2. \tag{4}$$

It is evident from equations (4) that perfect discrimination against flexure ($A_F = 0$) and the ideal torsional mode sensitivity ($A_T = 1$) apply only for $r_1 = r_2$. In practice, the alignment of the transducer plates is compromised by mainly pressure-induced distortion of the specimen assembly and of the pressure vessel. However, such distortion is sufficiently reproducible to allow the transducer plate spacings to be offset from the ideal configuration so as to anticipate the changes induced by pressurization. It is thus feasible to operate consistently under conditions for which neither $|A_F|$ nor $|A_T - 1|$ is greater than 0.02. The value of A_T calculated from the values of r_1 and r_2, which are routinely measured under high P-T conditions, allows the appropriate correction to be made for the torsional mode sensitivity during calibration of the displacement records (described below).

The AC ratio transformer bridges, designed and built by the Physics Department of the University of Queensland, are operated at 10 kHz with noise rejection provided by synchronous detection. The out-of-balance signal thus detected is integrated (0.2 ms time constant), and low-pass filtered (1 Hz) to suppress the signal arising from resonant oscillations of the specimen assembly (fundamental mode near 12 Hz) and electromagnetic interference from the switching of current in the furnace windings. The low-pass filters for the two measurement stations are closely matched by requiring that there is no difference (as observed with a sensitive null detector) between their outputs for a common 10 V p-p 1 Hz input signal. This procedure ensures that the systematic error introduced into the amplitude ratio and relative phase (radians) of the two displacement time series is no greater than 10^{-5}. The 0.2 ms integrators are not comparably well matched; prior measurement (with a common input to the two preamplifiers of Figure 4) of the phase angle between the two detected displacement-time sinusoids yields a small correction (currently $2.8 \times 10^{-4}/T_o$ rad, T_o being the oscillation period in s) which is routinely applied to the phase angle measured between the two displacement-time sinusoids.

The filtered signals undergo 12-bit analogue/digital conversion and digital data acquisition at an appropriate sampling frequency. Calibration of the displacement records thus obtained is achieved by measuring the change of out-of-balance signal δV resulting from a switching of the transformer ratio through a known increment $\delta r_{\parallel}$ in the absence of an applied torque. The same signal δV would be generated by a torsional mode displacement of magnitude $D\,\delta r_{\parallel}/A_T$, yielding a voltage/displacement calibration factor f given by

$$f = A_T \cdot \delta V/(D\,\delta r_{\parallel}). \qquad (5)$$

The Internal Furnace

The successful operation of a vertical-axis furnace within the gas pressure medium of high density and heat capacity requires that the influence of convection be controlled (PATERSON, 1970). In order to accommodate the unusually large specimens employed in this apparatus within a hot zone of uniform temperature, a four-zone furnace has been developed. Each winding is of molybdenum wire of diameter either 0.5 or 0.8 mm, depending upon the desired electrical resistance. The wire is wound onto the spiral-grooved exterior of an alumina ceramic tube (Morgan Matroc Duramic 998), embedded within a thick-walled cylindrical alumina-silica insulating sleeve (Zircar type ASH) which is radially compressed within a thin-walled stainless steel can. Excessive convective heat loss from the upper end of the furnace is prevented by the use of alumina and stainless steel inserts which fit snuggly within the annular region between the jacketed specimen assembly and the bore of the furnace.

The partitioning of power among the four windings necessary for the attainment of uniform hot-zone temperatures is achieved as follows. The furnace windings are connected in parallel to the output (50 Hz, 0–60 V RMS) of an autotransformer powered by the secondary of a 2 kVA 4:1 step-down transformer. A $Pt-Pt_{87}Rh_{13}$ thermocouple normally located in contact with a high-grade alumina spacer at the upper end of the specimens provides for the measurement and control of temperature, through use of a current-adjusting PID controller (Leeds and Northrup Electromax V). The current flow within each winding is determined by the firing of a dedicated triac with a duty cycle that is determined by the fraction of the 0–5 V control voltage selected locally by a potentiometer which is preset as described below. Phase-angle firing of the triacs is preferred over the more commonly employed burst firing strategy because the electromagnetic interference between the switching of furnace current and the measurement of the torsional mode displacements by capacitance micrometry consists of harmonics rather than subharmonics of the 50 Hz mains frequency. The harmonics are readily attenuated by low-pass filtering whereas the subharmonics, having periods comparable with those of the forced oscillations, are not.

For the purposes of arranging appropriate power partitioning among the four furnace windings, the specimen is replaced by a hollow cylinder of high-grade alumina of the same external dimensions (Figure 2(b)), allowing the axial thermocouple to traverse the entire length of the furnace while establishing the desired temperature profile. With the autotransformer set to deliver the maximum 60 V RMS from the secondary of the step-down transformer, and the variable 0–5 V control signal replaced in manual mode by a fixed DC level of 5 V, the settings of the four potentiometers, and hence the partitioning of power among the respective windings, are adjusted to realise a uniform temperature some 20–50°C higher than the desired operating temperature. This procedure guarantees sufficient reserve power for effective control about the set point. Uniformity within ± 10°C over the hot zone of 150 mm length is readily achieved and is sustainable for periods of at least several hours without variation of the potentiometer settings; however, drift in furnace performance on longer time scales, attributed to the gradual ageing of furnace components, results in somewhat larger uncertainties, typically ± 20°C in average temperatures above 1100°C, It is anticipated that the improved furnace design should allow future access to stable temperatures of at least 1300°C.

Mechanical Testing Protocols, Data Acquisition and Analysis

Two complementary types of mechanical test, forced oscillation and creep, are currently performed. The forced torsional oscillation test involves measurement of the dynamic torsional compliance of the specimen assembly from which the complex shear modulus of the specimen itself is ultimately derived as described below. The amplitude and phase of this latter quantity are the conventional shear modulus and associated loss angle which control the phase speed and attenuation for the corresponding travelling shear wave. The torsional creep test involves measurement of the response of the specimen assembly to the sudden application of a steady torque. The time-dependent twist per unit torque is the creep function from which the dynamic torsional compliance, measured more directly in the forced oscillation test, may be extracted by Fourier transformation, provided that the behaviour is linear (NOWICK and BERRY, 1972; JACKSON, 1993b). The complementarity between the creep and forced oscillation tests allows the recoverability, as well as the linearity, of the mechanical behaviour to be examined. Thus, a creep test, in which the response is monitored both during the period of application of the steady stress and following its removal, provides an opportunity to distinguish the truly anelastic (recoverable) and viscous (permanent) contributions to the total non-elastic deformation. These two contributions are not separable in the torsional oscillation test in which the specimen is usually exposed to a symmetrical history of positive and negative torque application.

Forced Oscillation Test

For a forced torsional oscillation test, the computer is programmed to synthe-size a sinusoidal current-time waveform which is low-pass filtered, amplified and then converted to torque by the electromagnetic drivers. The resulting distortion of the assembly is measured by capacitance micrometry at the stations above and below the elastic standard, as described above. The bridge out-of-balance signals are digitally sampled (usually at 128 samples per cycle of oscillation) for an integral number of consecutive cycles (usually 16 or 32), following a delay sufficient to ensure steady-state conditions (usually the larger of 2 minutes and 12 cycles of oscillation, a criterion which is probably unduly conservative at long periods). A typical record is reproduced in Figure 5(a). The amplitudes and relative phase of the two displacement versus time records are extracted from the amplitude and phase of the discrete Fourier transform at the driving period, giving a very significant gain in effective signal-to-noise ratio (see Figure 5(b)). The fact that each record comprises an integral number n of cycles of the oscillation guarantees an unambiguous association of integer frequency $v = n$ with the forcing frequency and no spread of energy into neighbouring frequencies. A correction is routinely applied to the relative phase to allow for the interlaced sampling of the two channels of information and for a small time delay attributable to mismatched time constants of the integrators through which the respective synchronously detected bridge imbalance signals are passed.

Processing of the two calibrated digital displacement versus time records yields the amplitudes and the relative phase of the angular distortions in the specimen assembly and in the elastic standard which is of known compliance. From these is calculated the complex torsional compliance for the compound specimen assembly which is given by

$$S^*(\omega) = \phi^*(t)/L^*(t) = |S^*(\omega)| \exp\{-i \tan^{-1}[1/Q_{\mathrm{app}}(\omega)]\}$$

$$\approx |S^*(\omega)| \exp\{-i/Q_{\mathrm{app}}(\omega)\} \tag{6}$$

where $\phi^*(t)$ and $1/Q_{\mathrm{app}} \ll 1$ are respectively the angular distortion and the ap-parent internal friction of the compound assembly, and $L^*(t)$ is the applied torque.

The procedure for the ultimate extraction of the shear modulus G and internal friction Q^{-1} for the *specimen itself* from comparative forced oscillation measure-ments on the two assemblies of Figure 2(b) is illustrated in Figure 6. The vector difference between the complex torsional compliances of the two assemblies de-scribes the behaviour of the jacketed specimen *relative* to that of jacketed alumina at the same hot-zone temperature. An absolute determination of the complex shear modulus for the specimen can therefore be made only if the corresponding quantity is known for each of the iron jacket and the polycrystallline alumina. For temperatures up to and including 1000°C, the departures of the iron-jacketed

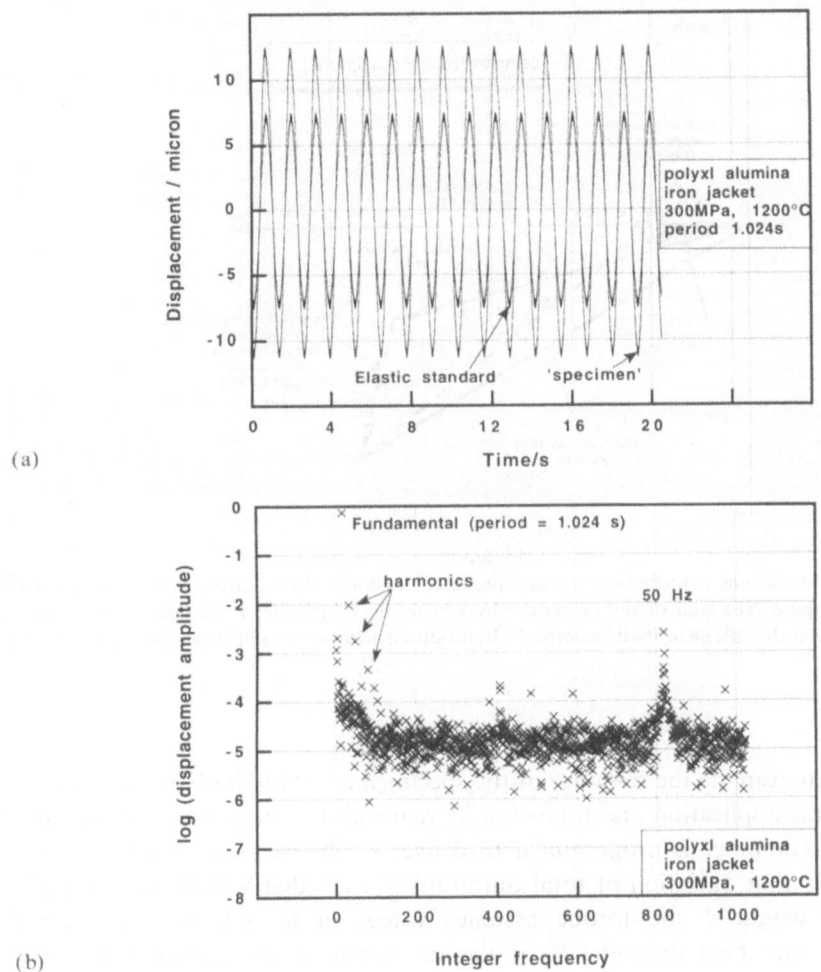

(a)

(b)

Figure 5

(a) The displacements associated with the distortion of the elastic standard and the alumina-iron reference assembly (labelled 'specimen') for a forced sinusoidal oscillation experiment at 1.024 s period. The displacement versus time record is sampled at 128 samples per oscillation period for exactly 16 consecutive periods. (b) The amplitude of the discrete Fourier transform of the displacement versus time record for the specimen assembly of Figure 5(a), showing a signal-to-noise ratio approaching 10^4 (~ 80 dB) at the fundamental forcing frequency. Harmonics resulting from non-linearity in the conversion of electric current into torque in the electromagnetic drivers are also evident, as is interference at the mains frequency 50 Hz and its harmonics consequent upon the switching of large furnace currents (Reproduced, with permission, from the *Annual Review of Earth and Planetary Sciences*, vol. 21, © 1993 by Annual Reviews Inc.).

alumina reference assembly from perfectly elastic behaviour are small and can be neglected (JACKSON *et al.*, 1992). This is definitely not the case at higher temperatures. Progress in the documentation of the viscoelastic behaviour of iron and alumina at temperatures to 1200 C has been described by JACKSON (1993c).

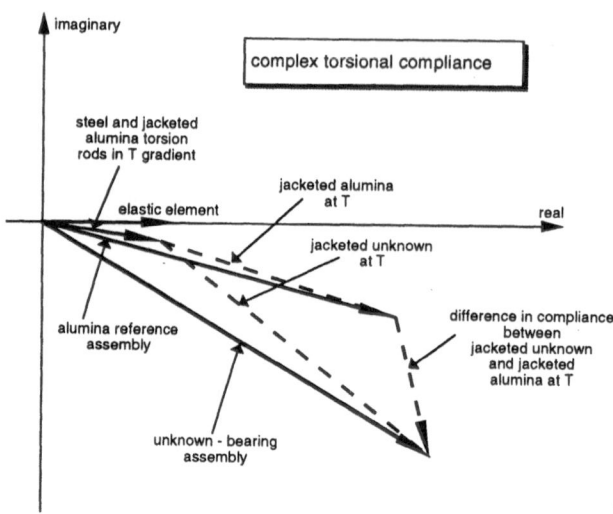

Figure 6
Illustration of the various contributions to the measured torsional compliances of the two specimen assemblies of Figure 2(b), and of the procedure by which an absolute determination of the complex shear modulus for the specimen itself is derived (Reproduced with permission from JACKSON, 1993c).

Creep Test

In order to examine the response of the specimen assembly both during a period of steady stress application and following its removal, the creep test is designed as follows. The computer is programmed to deliver to the electromagnetic drivers a current versus time variation of total duration $5T$ such that within five successive segments of length T the torque assumes values of 0, $+L$, 0, $-L$, and 0, respectively. Any time dependence of the distortion of the specimen assembly recorded during the second and fourth segments (as in Figure 7) provides clear evidence of non-elastic behaviour. The extent to which the non-elastic distortion is recovered and therefore truly anelastic is revealed by analysis of the third and fifth segments of the record. Any permanent (non-recoverable) distortions induced during the periods of application of the positive and negative torques should be equal in magnitude and opposite in sign, so that a close approach to the original null is expected on completion of the test, provided that sufficient time is allowed during the zero torque sgements for full anelastic recovery.

Since the specimen assembly and the elastic standard are subject to the same torque, and the standard is of known torsional compliance, it is straightforward to convert the measured displacement associated with the distortion of the specimen assembly into an instantaneous torsional compliance, i.e., angular distortion per unit torque. For the second segment of a creep record such as that of Figure 7, for which there is no prior history of stress application, the instantaneous torsional

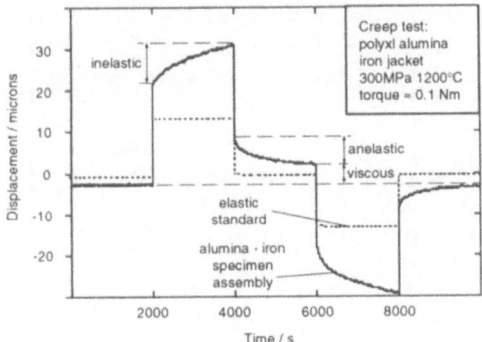

Figure 7

Displacement versus time records representing the distortion of both the alumina-iron reference assembly (Figure 2(b)) and the elastic standard, resulting from a creep test in which the torque assumes values of 0, $+L$, 0, $-L$, and 0, during successive 2000 s time intervals. The peak strain at the periphery of the alumina ceramic within the furnace hot zone is less than 5×10^{-6} (modified after JACKSON, 1993a).

compliance is formally interpretable as the creep function. The well-known Andrade transient creep formula (e.g., POIRIER, 1985)

$$J(\xi) = J_u + \beta\xi^m + \xi/\eta \tag{7}$$

generally provides a superior fit to the data than does the alternative Burgers model, with the same level of parametric economy. In equation (7), J_u is the elastic strain which appears instantaneously, and the term $\beta\xi^m$ with $m \sim 1/3$ describes the initially rapid evolution of inelastic strain rate with increasing time; at longer times, Newtonian viscous deformation with viscosity η dominates. The Andrade model is used in a hybrid numerical/analytical procedure for the Fourier transformation of the creep function to obtain the dynamic compliance. Consistency between this quantity and that more directly measured in the forced oscillation tests is expected (e.g., Figure 8) if the underlying relationship between stress and strain and their time derivatives is linear. Under these circumstances, the remainder of the record is the superposition of the responses to the successive application of a series of Heaviside step function torques, and is processed accordingly (JACKSON, 1993b).

Calibration of the Performance of the Apparatus

A 'Null' Experiment and Torques Imposed by the Gas Pressure Medium

Initial tests of the performance of the apparatus involved the conduct at room temperature and variable pressure of a 'null' experiment in which the specimen was replaced by a cylinder of the same dimensions fabricated from the same steel as is used for the elastic standard. Under these circumstances, the two displacement

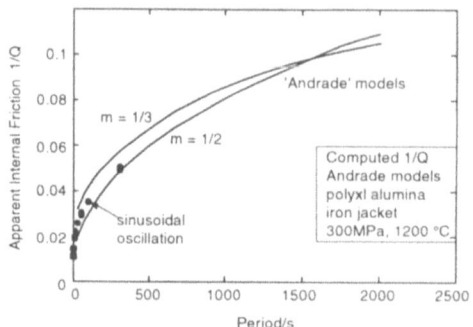

Figure 8

Comparison of the apparent internal friction Q^{-1} for the alumina-iron reference assembly measured in forced oscillation experiments with that computed for the second segment of the creep record of Figure 7. Andrade models with two alternative values of the exponent m have been used in fitting the creep data. Consistency between the computed and observed values is evidence of linearity of the stress-strain relationship (Reproduced, with permission, from the *Annual Review of Earth and Planetary Sciences*, vol. 21, (c) 1993 by Annual Reviews Inc.).

versus time sinusoids should be in phase with an amplitude ratio accurately calculable from the known geometry of the assembly. These two conditions were found to be closely satisfied for pressures above a threshold around 50 MPa, for oscillation periods longer than about 10 s (Figure 9). These findings provided compelling evidence that there is no significant compliance associated with the interfaces within the assembly, at least for normal stresses in excess of 50 MPa (JACKSON and PATERSON, 1987). For oscillation periods less than 10 s, a departure from this ideal was noted which varies systematically with pressure and oscillation period (Figure 9). The possibility was recognised that the relative amplitudes and phase of the two displacement versus time sinusoids might be perturbed significantly by torques associated with fluid flow between the closely spaced plates of the capacitance displacement transducers. It was shown that the force $F(t)$ exerted by a fluid medium of density ρ and viscosity μ upon a circular plate of radius R moving with prescribed displacement $d(t) = d_0 \exp(i\omega t)$ normal to a stationary parallel plate about an equilibrium spacing h is approximately

$$F(t) = [(1+i)\pi\rho\omega^2 R^4/8A\delta] \, d(t)$$

with

$$A = b - 2\tanh(b/2), \quad b = (1+i)h/\delta \quad \text{and} \quad \delta = (2\mu/\rho\omega)^{1/2}. \tag{8}$$

Note that an error was accidentally introduced into the last of these expressions when it was transferred from the appendix to the text of JACKSON and PATERSON (1987). The central moving plate of each displacement transducer experiences a force given by equation (8) as a consequence of its interaction with the stationary

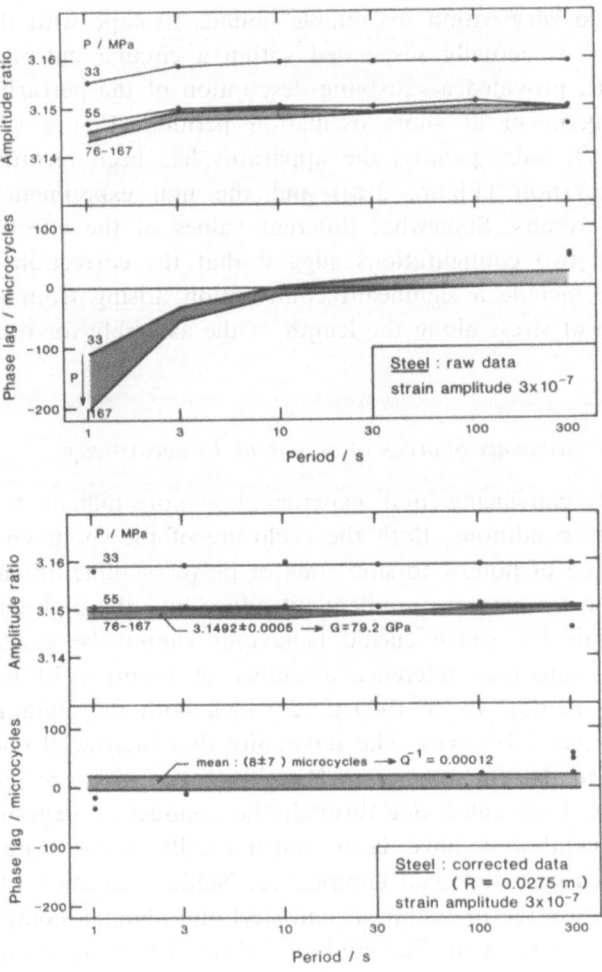

Figure 9

Results of a 'null' experiment in which the rock specimen was replaced by a dimensionally equivalent cylinder of the same low Q^{-1} steel used throughout the remainder of the assembly. The ratio of displacements measured at the two stations and their relative phase are shown both before ('raw') and after ('corrected') correction as explained in the text for the interaction between the transducer plates and the gas pressure medium. The corrected quantities closely approach the values expected for the amplitude ratio (3.14) and phase angle (zero) in the absence of extraneous influences on the apparent anelasticity. The phase lag plotted here is the phase difference between the two measured displacement versus time sinusoids, neither of which provides directly the phase of the applied torque; the connection between phase lag and Q^{-1} is therefore not immediate (Reproduced with permission from JACKSON and PATERSON, 1987).

plate on either side. Since the plate spacings are routinely measured and the pressure dependence of the density and viscosity of argon are well-known, it is thus possible to calculate the resultant torque operative at each measurement station, and accordingly, the perturbation to be expected to the imposed harmonically time-varying displacement. Corrections calculated in this way, allowing the effective plate radius R to vary within reasonable bounds to cope with the fact that the circular electrode is actually suspended within a circular cut-out in a coplanar rectangular plate, provided a satisfying description of the perturbations observed in the null experiment at short oscillation periods (Figure 9; JACKSON and PATERSON, 1987). Subsequently, the apparatus has been reconfigured for high temperature operation (Figure 2(a)) and the null experiment repeated with broadly similar results. Somewhat different values of the effective plate radius inferred for the two configurations suggest that the correction derived in this way might also include a significant contribution arising from phase delays in the propagation of stress along the length of the assembly for oscillation periods less than 10 s.

Other Possible Extraneous Sources of Apparent Viscoelasticity

A thoroughly convincing 'null' experiment is more difficult to arrange under high temperature conditions. Both the enclosure of the specimen within an iron jacket and the use of hollow torsion rods of polycrystalline alumina to span the regions of strong temperature gradient introduce into the high-temperature environment materials for which elastic behaviour cannot be guaranteed. Indeed, tests on the alumina/iron reference assembly of Figure 2(b) reveal significant viscoelastic deformation above 1000°C to which both the alumina and the iron contribute (JACKSON, 1993a,c). The possibility that interfacial compliance might also contribute to the apparent viscoelasticity of the specimen assembly at high temperatures has been ruled out through the conduct of experiments in which the interfacial conditions have been systematically varied without discernible effect on the measured torsional compliance. Neither variation of the quality of the surface finish on the precision ground steel and alumina components, nor the insertion at the interfaces of thin mild steel disks, has a significant impact upon the complex torsional compliance of the alumina/iron reference assembly. Similarly, variation of the radial clearance between the jacketed specimen and the close-fitting annular insert in the upper part of the furnace, produced no noticeable change in the apparent internal friction measured in forced oscillation experiments on the reference assembly at 1200°C.

It therefore appears probable that the apparent viscoelasticity of the reference assembly at temperatures above 1000°C is simply attributable to the viscoelasticity of the iron jacket and of the polycrystalline alumina. Improved constraints on the behaviour of these materials from ongoing experiments on soft iron specimens and a range of alumina (both single-crystal and polycrystalline) will provide for the future derivation of more robust absolute values for the shear modulus and internal friction for the specimen itself from the comparative complex torsional compliances of the two experimental assemblies of Figure 2(b).

Applications

This apparatus was first employed in room temperature studies on a fine-grained granitic rock and of a suite of limestones and marbles (JACKSON *et al.*, 1984; JACKSON and PATERSON, 1987). For these materials, the shear modulus was generally found to increase markedly with increasing pressure below 100 MPa, presumably as a consequence of crack closure. At higher pressures the modulus is less pressure sensitive and usually approaches that calculated for a non-porous aggregate from ultrasonically derived single-crystal elastic moduli for the relevant minerals. The internal friction measured on oven-dried specimens decreases with increasing pressure below 100 MPa; values near 0.001 are the norm at higher pressures. Since comparably low values of Q^{-1} are obtained on rocks tested at ambient pressure only after the most thorough degassing (TITTMANN *et al.*, 1980), it is concluded that the loading of grain-to-grain contacts in the high-pressure experiments must suppress viscoelastic deformation otherwise associated with the presence of adsorbed moisture. For the suite of four calcite rocks of widely varying grain size, the internal friction is insensitive to the variation of both oscillation period and grain size, suggestive of an intracrystalline relaxation mechanism with a wide distribution of relaxation times.

More recently, the realm of high temperature and pressure has been explored in studies of a medium-grained olivine-rich rock (Åheim dunite, JACKSON *et al.*, 1992) and of iron (JACKSON, 1993c,d). In each of these materials, pronounced dispersion of the shear modulus is evident at temperatures above 1000°C for oscillation periods between 1 and 1000 s, along with the concomitant internal friction which typically varies with oscillation period T_o as T_o^{α} with α of order 1/6 to 1/4. The variation with temperature of the shear modulus is complicated by the presence of porosity in the prefired specimens of the ultramafic rock, and by the bcc → fcc transition in iron. However, within the stability field for the bcc phase of iron, the temperature dependence of the shear modulus is much greater at frequencies below 1 Hz than at ultrasonic frequencies. Further experiments are now underway on low-porosity specimens of Ånheim dunite prefired in order to effect dehydration and subsequently annealed at 300 MPa and 1200°C, and on iron specimens spanning a range of carbon content. The viscoelastic relaxation observed in the ultramafic rock is probably of intracrystalline origin, but further work on related materials with systematically manipulated microstructure is necessary in order to establish a mechanistic under-standing of the observed relaxation.

Studies of the viscoelasticity of phase-transforming zirconia-based materials (G. J. FISCHER, personal communication) and of thermally cracked rocks (C. LU, personal communication) are also underway in this laboratory.

Future Developments

The present 300 MPa, 1200°C capability of the apparatus provides access to the entire subsolidus temperature range for even the most refractory ultramafic rocks.

However, the effect of partial melting on the seismic wave dispersion, attenuation and rheology of ultramafic rocks is also of considerable geophysical interest. Accordingly, priority will be given to extension of the temperature capability to at least 1300°C, through measures designed to reduce the convective heat loss from the upper end of the furnace without causing excessive friction between the twisting assembly and the furnace.

A start has also been made towards the implementation of pore pressure control which will facilitate studies at teleseismic frequencies of the inevitably frequency-dependent effective moduli of fluid-saturated rocks. For such studies it would be highly desirable to extend the range of experimentally accessible oscillation periods (current $T_o > 1$ s) to shorter periods. This would allow more comprehensive exploration of the mechanical relaxation expected to result from the progressive increase in the spatial scale of stress-induced fluid flow with increasing period. In addition it would allow a closer approach from below to the frequencies typically employed in the seismological exploration industry (> 50 Hz). The lower limit to the accessible range of oscillation periods is imposed by the strongly frequency-dependent interaction between the twisting assembly and the argon pressure medium, which varies as ω^2 or T_o^{-2} (equation (8)). Quantitative correction for this perturbation to the distortion of the assembly becomes marginal with the present configuration at about 1 s period. This restriction might be eased by the use of more widely spaced transducer plates and/or judicious perforation of the central moving plates of the displacement transducers to allow a short circuit for the otherwise essentially radial flow of the pressure medium without significantly compromising the linearity of the transducer performance.

Other possible future developments include the provision for the generation of larger torque amplitudes, needed to explore the transition to non-linear stress-strain behaviour at strains above $\sim 10^{-5}$, and reconfiguration of drivers and detectors to allow extensional mode measurements. Measurements deriving from extensional and torsional mode observations could then be combined to expose any departures from elastic behaviour in material subject to hydrostatic stress. Such relaxation of the bulk modulus and concomitant dissipation might be expected in material undergoing a first-order polymorphic phase transformation, and in fluid-saturated rocks under certain favourable circumstances (e.g., JACKSON, 1991).

Acknowledgements

Construction of this apparatus has been heavily dependent upon the skill and dedication of members of the RSES mechanical workshop, especially R. M. Waterford. Others who have contributed very substantially during its commissioning and subsequently to its successful operation are H. Niesler, G. R. Horwood, A. Forster, J. Lanc and Z. Guziak. The consistent encouragement of A. L. Hales is gratefully acknowledged.

REFERENCES

ASHCROFT, N. W., and MERMIN, N. D., *Solid State Physics* (Holt, Rinehart and Winston, New York 1976) 826 pp.

BERCKHEMER, H., KAMPFMANN, W., AULBACH, E., and SCHMELING, H. (1982), *Shear Modulus and Q of Forsterite and Dunite near Partial Melting from Forced Oscillation Experiments*, Phys. Earth Planet. Inter. *29*, 30–41.

BRENNAN, B. J., *Linear viscoelastic behaviour in rocks*. In *Anelasticity in the Earth* (eds. Stacey, F. D., Paterson, M. S., and Nicolas, A.) (Geodynamics Series *4*, 1981) pp. 13–22.

BRENNAN, B. J., and STACEY, F. D. (1977), *Frequency Dependence of Elasticity of Rock — Test of Seismic Velocity Dispersion*, Nature *268*, 220–222.

GETTING, I. C., Paffenholz, J., and SPETZLER, H. A., *Measuring attenuation in geological materials at seismic frequencies and amplitudes*. In *The Brittle-ductile Transition in Rocks*, Geophys. Monogr. Ser. 56 (eds. Duba, A. G., Durham, W. B., Handin, J. W., and Wang, H. F.) (AGU Washington 1990) pp. 239–243.

GETTING, I. C., SPETZLER, H. A., KARATO, S., and HANSON, D. R. (1991), *Shear Attenuation in Olivine*, EOS Trans. Am. Geophys. Union *72*, 451.

GLADWIN, M. T., and WOLFE, J. (1975), *Linearity of Capacitance Displacement Transducers*, Rev. Sci. Instrum. *46*, 1099–1100.

GOETZE, C., *A brief summary of our present day understanding of the effect of volatiles and partial melt on the mechanical properties of the upper mantle*. In *High-pressure Research: Applications in Geophysics* (eds. Manghnani, M. H., and Akimoto, S.) (Academic Press, New York 1977) pp. 3–23.

GUEGUEN, Y., DAROT, M., MAZOT, P., and WOIRGARD, J. (1989), Q^{-1} *of Forsterite Single Crystals*, Phys. Earth Planet. Inter. *55*, 254–258.

JACKSON, I., *The laboratory study of seismic wave attenuation*. In *Mineral and Rock Deformation — Laboratory Studies*, Geophys. Monogr. Ser. 36 (eds. Hobbs, B. E., and Heard, H. C.) (Am. Geophys. Union, Washington 1986) pp. 11–23.

JACKSON, I., *The petrophysical basis for the interpretation of seismological models for the continental lithosphere*. In *The Australian Lithosphere* (ed. Drummond, B. J.) (Geol. Soc. Aust. Spec. Publ. 17, 1991) pp. 81–114.

JACKSON, I. (1993a), *Progress in the Experimental Study of Seismic Wave Attenuation*, Ann. Rev. Earth Planet. Sci. *21*, 375–406.

JACKSON, I. (1993b), *Dynamic Compliance from Torsional Creep and Forced Oscillation Tests: An Experimental Demonstration of Linear Viscoelasticity*, Geophys. Res. Lett. *20*, 2115–2118.

JACKSON, I., *The high temperature shear mode inelasticity of iron — An exploratory foray into the fcc field*. In *Defects and Processes in the Solid State: Geoscience Examples* (eds. Boland, J. N., and FitzGerald, J. D.) (Elsevier 1993c) pp. 214–228.

JACKSON, I. (1993d), *Viscoelastic Relaxation in Iron and the Shear Modulus of the Inner Core*, Proc. AIRAPT/APS Conf. High. Press. Sci. Technol., Colorado Springs, June 28–July 2, 1993 (in press).

JACKSON, I., and PATERSON, M. S. (1987), *Shear Modulus and Internal Friction of Calcite Rocks at Seismic Frequencies: Pressure, Frequency and Grainsize Dependence*, Phys. Earth Planet. Inter. *45*, 349–367.

JACKSON, I., PATERSON, M. S., and FITZGERALD, J. D. (1992), *Seismic Wave Dispersion and Attenuation in Åheim Dunite: An Experimental Study*, Geophys. J. Int. *108*, 517–534.

JACKSON, I., PATERSON, M. S., NIESLER, H., and WATERFORD, R. M. (1984), *Rock Anelasticity Measurement at High Pressure, Low Strain Amplitude and Seismic Frequency*, Geophys. Res. Lett. *11*, 1235–1238.

KARATO, S., and SPETZLER, H. A. (1990), *Defect Microdynamics in Minerals and Solid State Mechanisms of Seismic Wave Attenuation and Velocity Dispersion in the Mantle*, Rev. Geophys. *28*, 399–421.

NOWICK, A. S., and BERRY, B. S., *Anelastic Relaxation in Crystalline Solids* (Academic Press, New York 1972) 677 pp.

PATERSON, M. S. (1962), *O-ring Piston Seals for High Pressure*, J. Sci. Instrum. *39*, 173–175.

PATERSON, M. S. (1970), *A High-pressure, High-temperature Apparatus for Rock Deformation*, Int. J. Rock. Mech. Min. Sci. *7*, 517–526.

POIRIER, J.-P., *Creep of Crystals* (Cambridge University Press 1985) 260 pp.

STACEY, F. D., RYNN, J. M. W., LITTLE, E. C., and CROSKELL, C. (1969), *Displacement and Tilt Transducers of 140 dB Range*, J. Phys. E *2*, 945–949.

TITTMANN, B. R., CLARK, V. A., RICHARDSON, J. M., and SPENCER, T. W. (1989), *Possible Mechanism for Seismic Attenuation in Rocks Containing Small Amounts of Volatiles*, J. Geophys. Res. *85*, 5199–5208.

(Received June 1, 1993, revised September 29, 1993, accepted October 6, 1993)

PAGEOPH, Vol. 141, No. 2/3/4 (1993)

0033-4553/93/040467-18$1.50 + 0.20/0
© 1993 Birkhäuser Verlag, Basel

Hot Pressing of Polycrystals of High-pressure Phases of Mantle Minerals in Multi-anvil Apparatus

GABRIEL D. GWANMESIA,[1] BAOSHENG LI,[2] and ROBERT C. LIEBERMANN[2]

Abstract — In the 1960s, E. Schreiber and his colleagues pioneered the use of hot-pressed polycrys-talline aggregates for studies of the pressure and temperature dependence of the elastic wave velocities in minerals. We have extended this work to the high-pressure polymorphs of mantle minerals by developing techniques to fabricate large polycrystalline specimens in a 2000-ton uniaxial split-sphere apparatus. A new cell assembly has been developed to extend this capability to pressures of 20 GPa and temperatures of 1700°C. Key elements in the new experimental design include: a telescopic LaCrO$_3$ for $T > 1200$°C; Toshiba Tungaloy grade F tungsten carbide anvils; and the use of homogeneous glasses or seeded powder mixtures as starting material to enhance reactivity and maximize densities. Cell tempera-tures are linearly related to electrical power to 1700°C and uniform throughout the 3 mm specimens. Pressure calibrations at 25°C and 1700°C are identical to 15 GPa. Cylindrical specimens of the beta and spinel phases of Mg$_2$SiO$_4$, stishovite (SiO$_2$-rutile), and majorite-pyrope garnets have been synthesized within their stability fields in runs of 1–4 hr duration and recovered at ambient conditions by simultaneously decompressing and cooling along a computer-controlled P-T path designed to preserve the high-pressure phase and to relax intergranular stress in the polycrystalline aggregate. These specimens are single-phased, fine-grained (<5 micron), free of microcracks and preferred orientation, and have bulk densities greater than 99% of X-ray density. The successful fabrication of these high-quality polycrystalline specimens has made possible experiments to determine the pressure depen-dence of acoustic velocities in the ultrasonics laboratory of S. M. Rigden and I. Jackson at the Australian National University.

Key words: Polycrystals, hot-pressing, multi-anvil apparatus, acoustic velocities.

Introduction

Following the publication of Birch's landmark papers (BIRCH, 1960, 1961) on the compressional velocities versus pressure for a wide variety of natural rocks [and the follow-up work for shear velocities a few years later by SIMMONS (1964)], it

[1] Center for High Pressure Research* and Department of Physics and Astronomy, Delaware State University, Dover, DE 11901, U.S.A.
[2] Center for High Pressure Research* and Department of Earth and Space Sciences, University at Stony Brook, Stony Brook, NY 11794, U.S.A.
* CHiPR: NSF Science and Technology Center for High Pressure Research.

became evident that one could not hope to obtain reliable values of the pressure and temperature derivatives for velocity from experiments on natural rock specimens, except perhaps at pressures >1 GPa [for example, the later work by WANG (1974) and CHRISTENSEN (1974)]. Consequently, the focus of new work shifted to single crystals (natural and synthetic) as evidenced by the papers of ANDERSON and ANDREATCH (1966) on MgO, of SCHREIBER (1967) on $MgAl_2O_4$, and of KUMAZAWA and ANDERSON (1969) and GRAHAM and BARSCH (1969) on olivine. However, because of the lack of availability of natural single crystals or the difficulty of growing suitable synthetic crystals, there was also considerable attention to the use of high-quality ceramics for measurement of velocities as a function of pressure and temperature. The Mineral Physics group at Lamont Geological Observatory was the principal advocate of this approach and the training and background of Ed Schreiber (who obtained his Ph.D. at the New York State College of Ceramic Engineering at Alfred University) was especially valuable in this effort. The Lamont group, however, did not sinter or hot-press their own specimens; they chose instead to capitalize on existing ceramic specimens such as magnesia and alumina or the ferrite materials such as hematite or nickel ferrite (e.g., ANDERSON and SCHREIBER, 1965; SCHREIBER and ANDERSON, 1966; LIEBERMANN and SCHREIBER, 1968).

High-pressure hot-pressing in the 1970s was primarily limited to pressures less than 1 GPa. For phases stable only at higher pressures, it was left to geophysicists to develop the necessary technology themselves, and major efforts were launched in the laboratories of Akimoto in Tokyo, Japan and Ringwood in Canberra, Australia to hot-press polycrystalline specimens at pressures up to 10 GPa (e.g., MIZUTANI *et al.*, 1970; LIEBERMANN *et al.*, 1975). However, most of these high-pressure polycrystals had some bulk porosity (up to 5%) and also considerable crack porosity as evident in the characteristic velocity-pressure behavior of many of those earlier specimens (e.g., LIEBERMANN, 1972). Consequently, the primary result was velocities at 7 to 10 kbar that were taken as representative of the zero-porosity, elastically-isotropic polycrystalline aggregate.

For a few of the Canberra specimens (e.g., Fe_2GeO_4-spinel), the behavior of the velocity versus pressure data gave some hope of obtaining useful pressure derivatives if more careful ultrasonic work was performed (LIEBERMANN, 1975). This promise was realized by the work of RIGDEN *et al.* (1988) on Mg_2GeO_4-spinel and later by RIGDEN and JACKSON (1990) on a large suite of germanate and silicate spinels synthesized in Ringwood's laboratory.

The promising results from the work on Mg_2GeO_4-spinel and the establishment of the High Pressure Laboratory at Stony Brook gave the impetus to a research program to fabricate a new class of polycrystals at pressures up to 20 GPa. Specimens of both the beta and spinel phases of Mg_2SiO_4 have been synthesized at temperatures below 1200°C (GWANMESIA *et al.*, 1990a; GWANMESIA and LIEBERMANN, 1992), and their velocities measured as functions of pressure to 3 GPa at room temperature (GWANMESIA et al., 1990b; RIGDEN et al., 1991) using ultrasonic interferometry techniques developed in Jackson's laboratory in Canberra (NIESLER and JACKSON, 1989; see also RIGDEN et al., 1992a).

The purpose of the present paper is to report recent technological progress in this hot-pressing program which extends the temperatures to 1700°C at pressures of 20 GPa. Such pressure-temperature conditions are necessary for synthesizing specimens of the majorite ($MgSiO_3$) garnet phase which is only stable above 1600°C (GASPARIK, 1992).

Experimental Techniques

High-pressure Apparatus

Figure 1 is a schematic diagram of the 2000-ton uniaxial split-sphere apparatus (USSA–2000) used in all the experiments to hot-press polycrystalline specimens [see additional details in LIEBERMANN and WANG (1992) and references therein]. Pressure in the USSA–2000 is generated by a two-stage anvil system driven by a 2000-ton uniaxial ram. The first stage is a tool steel sphere split into six parts, glued permanently into upper and lower guideblocks (Fig. 1b) and enclosing a cubic cavity (60 mm on edge) which contains the second stage anvil assembly (Fig. 1c). The second stage (Fig. 2) is assembled outside the press and is electrically insulated from the first stage by phenolic sheets; it consists of eight tungsten carbide cubes separated by gaskets and spacers. Each cube has corners truncated into a triangular face, thus creating an octahedral cavity in which the sample assembly is compressed. The cell assembly is an octahedron made of semi-sintered MgO. The maximum pressure attainable in the USSA–2000 is a function of the truncation edge length (TEL = a in Fig. 1c) of the tungsten carbide anvils and the edge length of the octahedral pressure medium.

Cell Assembly

We have utilized two types of cell assemblies for these experiments (Figs. 3a and 3b). The edge of the MgO octahedral pressure medium for both types of sample assemblies is 14 mm and the triangular truncations on the corner of the tungsten carbide anvils are 7.5 mm. These assemblies differ mainly in the type of resistance furnace used to heat the sample and the grade of tungsten carbide cubes used to generate pressure.

For pressure range up to 18 GPa and temperatures to 1200°C, the second stage consists of eight Kennametal grade KZ313 cubes, and heating is provided by a graphite resistance furnace (Fig. 3a). This cell assembly has been utilized for fabricating polycrystalline aggregates of the beta and spinel phases of Mg_2SiO_4

UNIAXIAL SPLIT SPHERE APPARATUS
USSA-2000

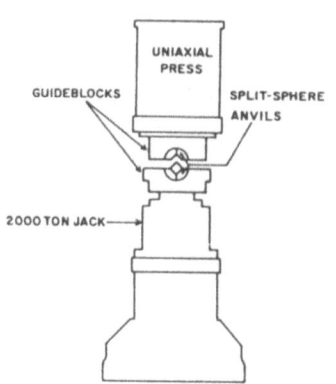

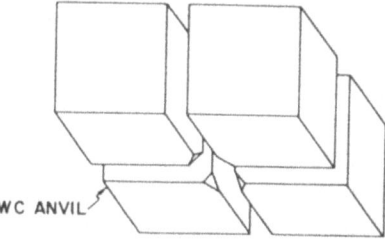

WC ANVIL

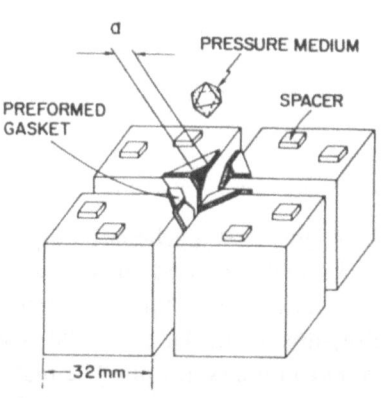

USSA - 2000

UPPER GUIDEBLOCK

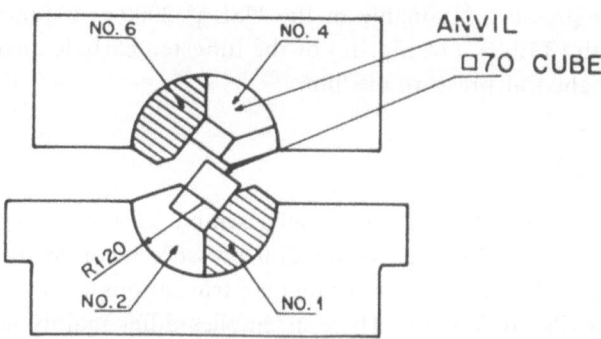

LOWER GUIDEBLOCK

Figure 1

Schematic diagrams for 2000-ton uniaxial split-sphere apparatus (USSA 2000). (a) Hydraulic press with guideblocks and split-sphere anvils in place. (b) Details of the first-stage spherical steel anvils in the guideblocks. (c) Second-stage with the assembly of eight tungsten carbide cubic anvils of truncation edge length (TEL = a) which compress an octahedral pressure medium.

Figure 2

Photo of second-stage assembly of eight tungsten carbide cubes enclosing octahedral pressure medium. Pyrophyllite (black), paper (grey) and teflon (white) strip gaskets provide lateral support for the cubes and improve reproducibility of pressure generation. Balsa wood spacers are for alignment purposes during preparation of the experiment and are passive during the high-pressure run.

(Gwanmesia et al., 1990a; Gwanmesia and Liebermann, 1992) and SiO_2-stishovite (Li et al., 1992).

To attain temperatures above 1200°C at pressures of 20 GPa, we have utilized Toshiba Tungaloy grade F cubes. The heating is provided by a lanthanum chromite resistance sleeve which is capable of generating stable temperatures up to 2900°C. We adopt the notation 14/7.5 TL to identify the new cell assembly (Fig. 3b), where the first number represents the edge length of the ceramic octahedron, the second number is the truncation edge length (TEL) on the corner of the tungsten carbide cubes, and the letters T and L represent the grade (Toshiba) of the second-stage anvils and the furnace material (lanthanum chromite), respectively. Using this notation, the original assembly (Fig. 3a) shall henceforth be referred to as the 14/7.5 KG sample assembly, in which the letters K and G represent Kennametal and graphite, respectively.

In the 14/7.5 TL assembly, the sample is placed inside a platinum capsule which is electrically insulated from the furnace by an MgO sleeve or an NaCl sleeve. Electrical contact is achieved internally within the lanthanum chromite sleeve by two small rings of lanthanum chromite on each end of the furnace. Contact between these rings and the carbide anvils is accomplished via molybdenum alloy (TZM) plugs which fit into the lanthanum chromite rings; the anvils are thermally insulated by sleeves of zirconia outside the heater and at both ends of the cell assembly. This modification of the original assembly was made to accommodate the large resistance and phase changes which occur in lanthanum chromite during pressurization and heating.

Temperature Calibration

Figure 4 shows the cell assembly used to calibrate the sample temperature as a function of the power supplied to the $LaCrO_3$ furnace using procedures similar to those reported by GWANMESIA and LIEBERMANN (1992). In brief, temperature within the cell is monitored by two axial thermocouple wires (W3%Re and W25%Re), in electrical contact through the platinum capsule containing the sample located at the center of the cell assembly; provided that the W3%Re/Pt and Pt/W25%Re junctions are exposed to the same T and P, the net thermo-emf is that of a single W3%Re/W25%Re junction. The two wires emerge at the cold end of the sample chamber through two single-hole alumina ceramic sleeves which are ce-

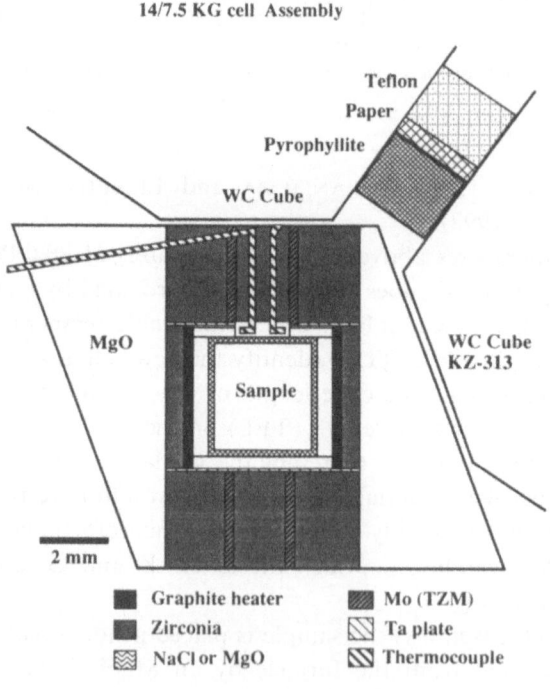

Figure 3(a)

14/7.5 TL Cell Assembly with LaCrO₃ Furnace

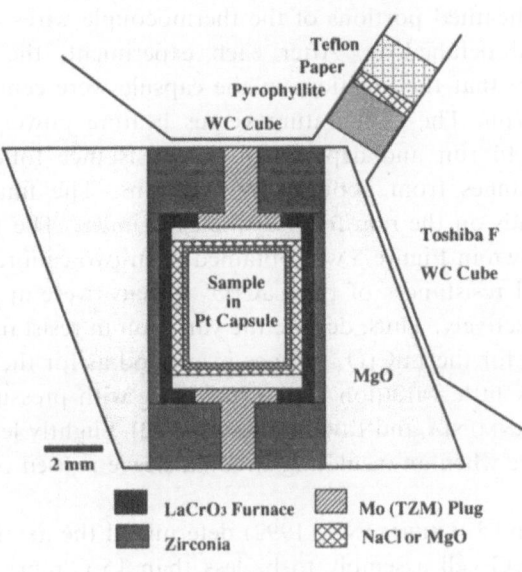

Figure 3(b)

Figure 3

Cross section of the 14 mm octahedral MgO cell assembly for hot-pressing polycrystalline aggregates. (a) For the pressure range up to 18 GPa and temperatures to 1200°C. Carbide cubes are Kennametal grade KZ313 and heating is provided by a graphite resistance furnace; thus, it is termed the 14/7.5 KG assembly. (b) For the pressure range up to 20 GPa at temperatures to 1800°C. Carbide cubes are Toshiba Tungaloy Grade F and heating is provided by a lanthanum chromite resistance sleeve; thus, it is termed the 14/7.5 TL assembly.

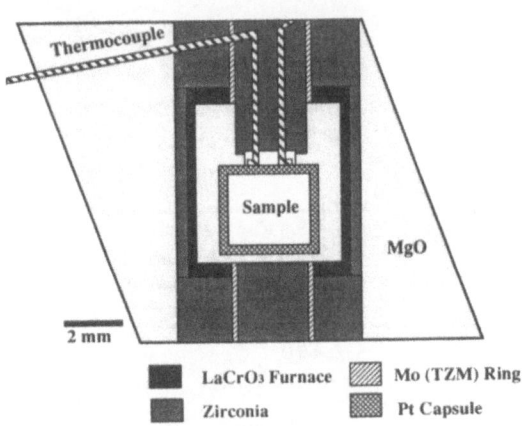

Figure 4

Cell assembly for temperature calibration using the lanthanum chromite furnace and an axial thermocouple located at the hot spot.

mented inside grooves machined in the TZM ring, zirconia end-plug, and MgO octahedron. The sheathed portions of the thermocouple wires exit the assembly at two apices of the octahedron. After each experiment, the thermocouple was examined to ensure that the junctions at the capsule were centrally located within the MgO octahedron. The temperature versus heating power for $LaCrO_3$ varies slightly from run to run and depends on the resistance following compression, which generally ranges from about 11 to 15 ohms. The final resistance of the furnace also depends on the ram force of the experiment. The temperature-heating power relation shown in Figure 5 was obtained from two calibration experiments in which the terminal resistances of the $LaCrO_3$ furnace were approximately 8 ohms and 13 ohms, respectively. Thus, despite the variation in resistance from run to run, the reproducibility for the $LaCrO_3$ furnace is as good as for the graphite furnace in which there is very little variation of the resistance with pressure and temperature (see Fig. 5 of GWANMESIA and LIEBERMANN, 1992). Slightly less power is required at high temperature when an insulating zirconia sleeve is used on the outside of the furnace (Fig. 5).

GWANMESIA and LIEBERMANN (1992) determined the axial temperature gradient in the 14/7.5 KG cell assembly to be less than 15 C/mm. They attributed the uniform temperature distribution within the sample assembly to the box-like nature of the furnace and to the thin tantalum contact plates which provided additional heating at the top and bottom of the furnace to smooth the temperature distribu-

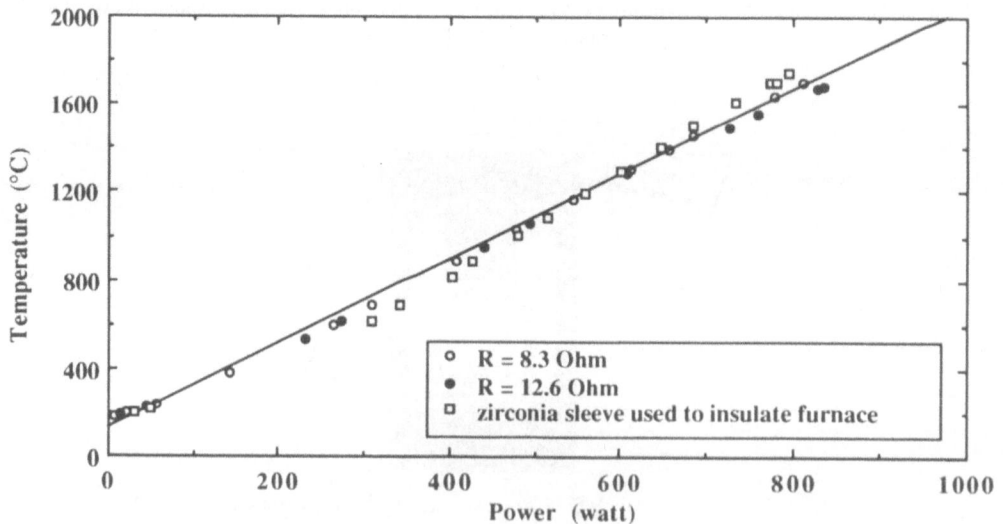

Figure 5

Temperature calibration of the 14/7.5 TL cell assembly: Two different experimental runs demonstrating reproducibility of the power-temperature behavior (open and closed circles); with zirconia sleeve used to insulate the furnace (open squares).

tion within the sample. We have not experimentally determined the temperature gradient in the 14/7.5 TL cell assembly, but it is evident from the uniform average grain size distribution for the specimens hot-pressed in this assembly that the temperature gradient is comparable to that within the KG assembly.

Pressure Calibration

The pressure for 14/7.5 TL cell assembly was calibrated as a function of the ram force using the techniques described by GWANMESIA (1987). The pressure calibrations at room temperature were based on observing changes in electrical resistance associated with phase transformations in various standard materials: Bi (I-II, 2.55 GPa; Bi III-V, 7.7 GPa; LLOYD, 1971), ZnTe (9.5, 12.5 GPa; KUSABA et al., 1993) and ZnS (15.5 GPa; BLOCK, 1978). For the bismuth calibrations, the wires were imbedded in AgCl inside a small circular recess at the center of an MgO insert. In the case of ZnS or ZnTe, the powder was packed inside the recess at the center of the MgO insert. Electrical contact between the sample and the anvils was achieved using platinum strips. All the room temperature phase transitions are reproducible within 2% of the ram force. The 14/7.5 TL assembly was calibrated at high temperatures (1700°C) by reversing the equilibrium phase boundaries between coesite (Cs) to stishovite (St) [9.9 GPa from YAGI and AKIMOTO, 1976; or 10.6 GPa from ZHANG et al., 1993] and between the olivine and beta phases of Mg_2SiO_4 (15.2 GPa; KATSURA and ITO, 1989). Uncertainty in both the room temperature and high-temperature pressure calibrations is estimated to be better than ± 0.5 GPa.

The results of the pressure calibration at room temperature are presented in Figure 6 for the 14/7.5 TL cell assembly (with filled circles). The second curve (with open circles) was obtained by GWANMESIA and LIEBERMANN (1992) using an identical cell assembly, but with Kennametal cubes used for generating pressure. A comparison of the two curves illustrates the effect of the yield strength of the tungsten carbide on the pressure-generating capability for the 14/7.5 cell assembly. For ram loads above 300 tons, the Toshiba cubes clearly generate substantially higher cell pressures than the Kennametal cubes (about 1.5 GPa at 600 tons). GETTING et al. (1993) have performed uniaxial compressive stress-strain measurements on a suite of cylindrical specimens for various commercial grades of tungsten carbide cermets and some maraging steels of interest for use in the high-pressure apparatus. Their results show that the yield strength of Toshiba Tungaloy grade F tungsten carbide is approximately twice that of Kennametal grade KZ313, while the percentage axial strain at failure for the Toshiba carbide is only half that for the Kennametal. LIEBERMANN and WANG (1992) have discussed how the cell pressure within the sample assembly is affected by plastic deformation on the triangular truncations on the anvils for various grades of tungsten carbide cubes. From an examination of the cubes after the pressure calibration experiments at room

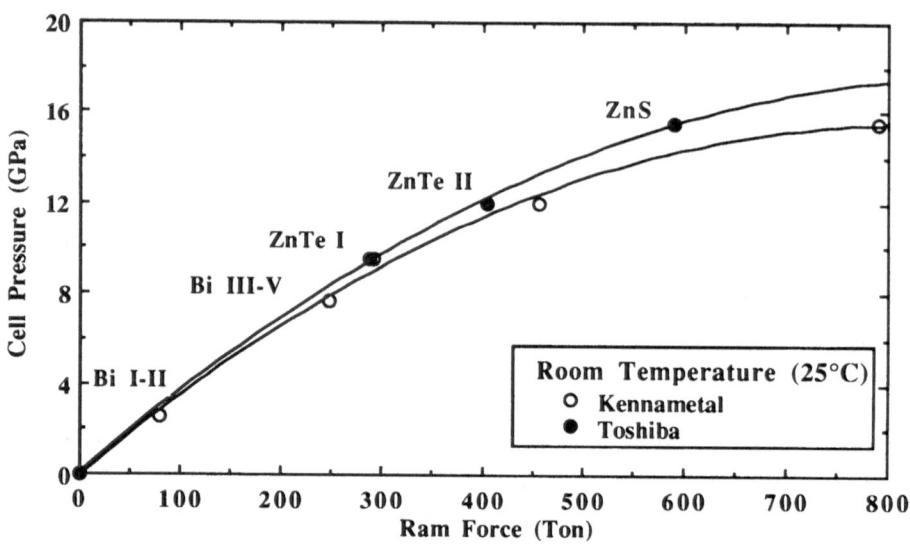

Figure 6

Pressure calibration of 14/7/5 cell assemblies at room temperature using both Kennametal (open circles) and Toshiba (closed circles) tungsten carbide cubes.

temperature, it is clear that above about 8 GPa (300 tons) the Kennametal cubes suffer greater plastic deformation (as a result of their lower yield strength) than Toshiba cubes. By contrast, the Toshiba cubes exhibit no significant deformation until the ram loads exceed 600 tons (15 GPa). However, to avoid failure of the cubes due to stress concentration at the truncations, care must be taken to increase pressure on the sample assembly slowly when using the Toshiba carbide which exhibits a relatively lower axial strain at failure.

The results of the pressure calibrations at high temperatures (1700°C) for the 14/7.5 TL cell assembly are shown by the open squares in Figure 7, in which we also plot the room temperature data (open circles). Using the data of YAGI and AKIMOTO (1976) for the coesite-stishovite phase transformation, we observe that for ram loads of 400 tons, there is a slight decrease (0.4–0.5 GPa) of cell pressure at high temperature; at 625 tons, the room and high temperature calibrations coincide. However, if we use the data of ZHANG *et al.* (1993) for the coesite-stishovite phase boundary, we conclude that both high pressure points coincide with the room temperature pressure calibration curve.

For the 14/7.5 KG cell assembly (Fig. 3a), GWANMESIA *et al.* (1990a) noted significant enhancement of the cell pressure at high temperature, which they attributed to the large thermal pressure generated by the NaCl sleeve immediately surrounding the sample. For the high-temperature pressure calibration experiments in the 14/7.5 TL cell assembly an MgO sleeve was used around the sample; NaCl

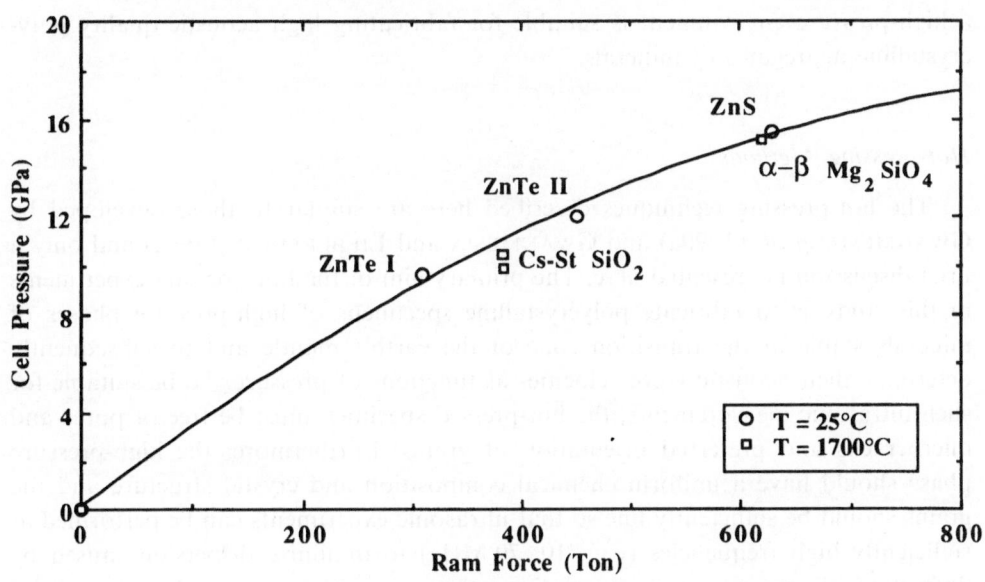

Figure 7
Pressure calibration of the 14/7.5 TL cell assembly at both room temperature and 1700°C.

could not be used at these relatively higher temperatures because of chemical reaction with the lanthanum chromite furnace. Since the thermal expansivity of MgO is less than that for NaCl, the TL cell assembly does not experience similar pressure enhancement as the KG assembly. However, carbide deformation is less significant in the TL assembly and may partially compensate the loss in cell pressure resulting from factors such as flow of the MgO pressure medium during heating and cell volume reduction due to phase change and sample compaction.

GWANMESIA and LIEBERMANN (1992) estimated the differential stress in the 14/7.5 KG cell assembly to be < 10 MPa at high temperatures (> 1000°C). In the study of the ortho/clinoenstatite transition, KANZAKI (1991) placed NaCl adjacent to the sample and by using the grain size versus stress relationship for NaCl developed by GUILLOPE and POIRIER (1979), estimated the differential stress in a split-sphere multi-anvil high-pressure apparatus to be of the order of < 10 MPa at 1000–1400°C. WEIDNER et al. (1992) have also examined the deviatoric stress in a sample of NaCl mixed with Au in a DIA-type cubic-anvil high-pressure apparatus (SAM-85); they observed that the deviatoric stress decreases markedly from a maximum of 0.3 GPa at 1.5 GPa to almost zero upon heating in the first 300°C of heating. Thus, the geometrical configuration of multi-anvil devices pressurization system provides a hydrostatic stress environment for the sample, while the use of NaCl cell should significantly reduce any deviatoric stress within the sample. Such a high-pressure environment is suitable for fabricating high acoustic quality poly-crystalline aggregates of minerals.

Hot-pressing Methods

The hot-pressing techniques described here are similar to those developed by GWANMESIA *et al.* (1990a) and GWANMESIA and LIEBERMANN (1992) and only a brief discussion is presented here. The primary aim of the hot-pressing experiments in this study is to fabricate polycrystalline specimens of high-pressure phases of minerals stable in the transition zone of the earth's mantle and to subsequently determine their acoustic wave velocities as functions of pressure. To be suitable for such ultrasonic measurements, the hot-pressed specimen must be free of pores and microcracks and preferred orientation of grains. Furthermore, the high-pressure phase should have a uniform chemical composition and crystal structure and the grains should be sufficiently fine so that ultrasonic experiments can be performed at sufficiently high frequencies (e.g., 10–70 MHz) to minimize dispersion caused by diffraction effects from grain boundaries and energy reflection from the side walls of the specimen. The elastic properties of such a specimen would then be those for a fully dense, elastically isotropic material.

The nature of the starting material is often a key factor in these hot-pressing experiments. In our previous work on the high-pressure phases of Mg_2SiO_4, we utilized fine-grained crystalline powder of olivine to synthesize directly the beta

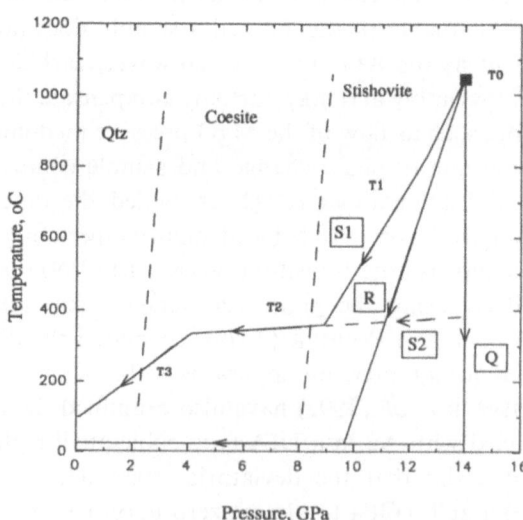

Figure 8
P-T paths with arrows for hot-pressing and recovering high-quality polycrystals of SiO_2-stishovite (from LI, 1993).

phase polycrystals, but needed previously-synthesized seeds of spinel (olivine:spinel ratio of 10:1) to enhance chemical reactivity and to ensure complete transformation to the spinel phase (GWANMESIA et al., 1990a; GWANMESIA and LIEBERMANN, 1992).

In a series of experiments to hot-press polycrystalline aggregates of specimens within the enstatite-pyrope ($En_{100 - x}Py_x$; $x = $ mole% of pyrope) solid solution series having the garnet structure, we have used three different types of starting material (e.g., RIGDEN et al., 1992b). For the end-member majorite garnet (Mg-

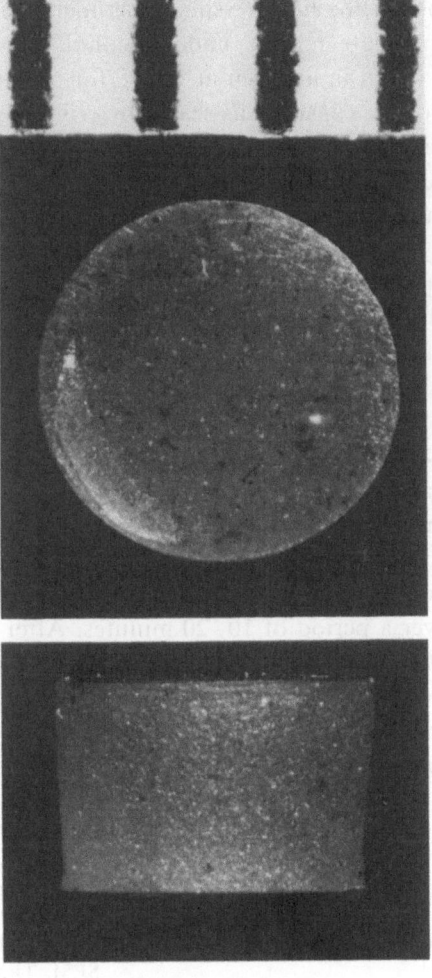

Figure 9

Photos of hot-pressed polycrystalline aggregate of a specimen from the majorite-pyrope solid solution series on recovery at room conditions. (a) View across diameter of specimen; (b) view along axis of specimen. See millimeter scale at top of photos.

SiO_3), the starting material was $MgSiO_3$-orthopyroxene which had been crystallized at 1650°C and 1 atm from glass of the same composition; the resultant product for such high-pressure experiments was clinopyroxene even at pressures within the majorite stability field (GASPARIK, 1992). Under the same *P-T* conditions, reaction to the high-pressure phase was complete within 1 hr when the starting material consisted of a 10:1 mixture of $MgSiO_3$-orthopyroxene and seeds of majorite garnet which had been synthesized previously by J. Zhang in the 10/4 cell assembly of the USSA−2000 apparatus (GASPARIK, 1989). For all of the garnet compositions, we have also achieved successful results with starting materials consisting of glass quenched from high-temperature melts of molar oxide mixtures of the appropriate composition. Before use in the hot-pressing experiments, each glass starting material was ground in an agate mortar under alcohol to a fine-grained ($2-5\ \mu$m) homogeneous powder, dried in an oven at 170°C for 5 hours and then loaded into a Pt capsule. The end of the capsule was sealed by crimping followed by cold-pressing. Reaction rates using these glass materials were extremely rapid.

For certain hot-pressing experiments, the choice of starting material is dictated by the percentage change in volume that occurs due to transformation from the low-pressure to the high-pressure phase. For example, there is about 60% reduction in volume when amorphous silica transforms to SiO_2-stishovite, yielding specimens of the high-pressure phase which are generally too small for ultrasonic experiments. LI *et al.* (1992) (see also LI, 1993) have minimized this volume reduction by using dense fused silica rods of external diameter comparable to the internal diameter of the Pt capsule. The capsule was sealed in air before placing it symmetrically at the center of the MgO octahedron previously dried at 1100°C for 1 hour (see Figure 3), and then hot-pressed along the *P-T-t* paths shown in Figure 8 (from LI, 1993).

Hot-pressing experiments using 14/7.5 TL cell assembly in the USSA−2000 apparatus are conducted by first applying pressure to the sample slowly (over 5−10 hrs) up to the target pressure, after which the sample temperature is increased to the desired value over a period of 10−20 minutes. After maintaining the sample under the experimental conditions for a suitable time interval (1−5 hrs), the pressure is decreased very slowly over periods extending to 36 hours. The sample temperature is also reduced slowly and simultaneously along suitable computer-controlled pressure-temperature (*P-T*) paths. Cooling rates are typically about

Table 1

Sample quality	Method of verification
single-phased	X-ray, TEM
fine-grained	SEM, TEM
homogeneous	SEM, TEM
free of pores/cracks	Optical, TEM, Density
elastically isotropic	TEM, Ultrasonics

$50^\circ C/hr$ for $T > 700^\circ C$, and $70^\circ C/hr$ for $T < 700^\circ C$. For the highest pressure runs ($P > 15$ GPa) requiring long depressurization time, specimens are generally held at $\sim 750^\circ C$ for several hours until the cell pressure is sufficiently low (< 5 GPa), after which the sample pressure and its temperature are again decreased simultaneously for recovery. The objective in utilizing this complex P-T-t path is to relax intergranular stresses in the polycrystalline aggregates. No thermocouple is used in hot-pressing experiments so as to maximize the sample volume and also minimize stresses on the sample that are likely to be introduced by the ceramic tubing used to contain the thermocouple wires (compare Figs. 3b and 4).

Hot-pressed specimens fabricated at high pressures and temperatures recovered at ambient conditions using the procedures described above (see example in Fig. 9) are then analyzed using the experimental techniques listed in Table 1. Portions of the test specimens are used to verify the crystal structure of the run product and to confirm that the product was single-phase by X-ray diffraction using a nondestructive method or after crushing to powder form (see Fig. 10). Cracked specimens are prepared and examined by SEM to estimate the average grain size and determine the grain size distribution within the sample. Polished sections of the test specimens

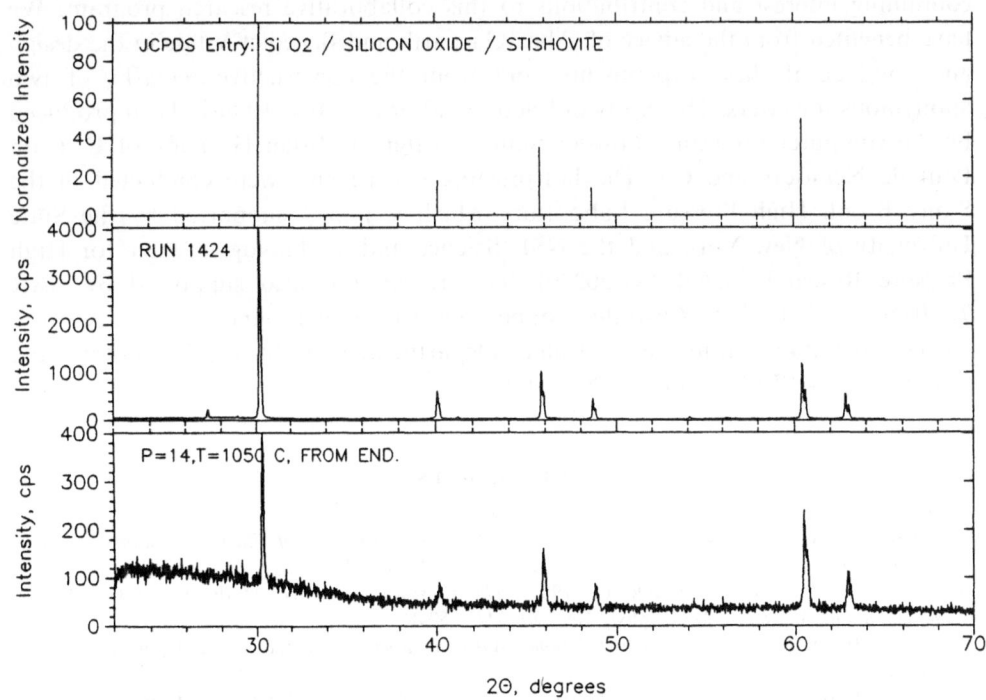

Figure 10

Powder X-ray diffraction patterns for stishovite: (a) standard JCPDS file; (b) from specimen #1424 after crushing to a fine powder; (c) from the end of a cylindrical specimen by nondestructive technique. From Li (1993).

are ion-thinned and examined for microcracks and possible preferred orientation using transmission electron microscopy (TEM). The sample density is determined using Archimedes' method and compared with theoretical X-ray values. For the best specimens (density >99%) the ends are ground and polished flat and parallel using diamond paste to prepare them for measurements of their elastic wave velocities at ambient conditions. Subsequently, the velocities in these polycrystals are determined as a function of pressure by ultrasonic interferometric techniques in a piston cylinder apparatus with a liquid pressure medium (see NIESLER and JACKSON, 1989; and RIGDEN *et al.*, 1992a).

Acknowledgements

We dedicate this paper to the memory of Professor Edward Schreiber from whom one of us (RCL) learned to be an experimentalist; in a very real sense, the other authors of this paper may be considered as Ed's academic grandchildren in their study of the elasticity of polycrystals using ultrasonic techniques. We thank Sally Rigden and Ian Jackson of the Australian National University for their continuing interest and contributions to this collaborative research program. We have benefited from the advice of Tibor Gasparik and Benedict Vitale in the design and conduct of these experiments, and from the constructive remarks of two anonymous reviewers. The X-ray diffraction patterns in Fig. 10 have been produced by the computer program "Powder Suite" written by Brian H. Toby of E. I. du Pont de Nemours and Co. The hot-pressing experiments were conducted in the Stony Brook High Pressure Laboratory which is jointly supported by the State University of New York and the NSF Science and Technology Center for High Pressure Research (EAR 89–20239). This research is also supported by EAR 91–04563 and the U.S.-Australia Cooperative Research Program (NSF INT 89–13363 and funding from the Australian Department of Industry, Technology and Commerce). MPI Contribution No. 109.

REFERENCES

ANDERSON, O. L., and ANDREATCH, P. (1966), *Pressure Derivatives of Elastic Constants of Single Crystal MgO at 23 and −195 C*, J. Am. Ceram. Soc. *49*, 404–408.

ANDERSON, O. L., and SCHREIBER, E. (1965), *The Pressure Derivatives of the Sound Velocities of Polycrystalline Magnesia*, J. Geophys. Res. *70*, 5241–5248.

BIRCH, F. (1960), *The Velocity of Compressional Waves in Rocks of 10 Kilobars*, J. Geophys. Res. *65*, 1083–1102.

BIRCH, F. (1961), *The Velocity of Compressional Waves in Rocks to 10 Kilobars*, J. Geophys. Res. *66*, 2199–2224.

BLOCK, S. (1978), *Round-robin Study of the Phase Transformation in ZnS*, Acta Cryst. *A34*, Suppl 316.

CHRISTENSEN, N. I. (1974), *Compressional Wave Velocities in Possible Mantle Rocks to Pressures of 30 Kilobars*, J. Geophys. Res. *79*, 407–412.

GASPARIK, T. (1989), *Transformation of Enstatite-diopside-jadeite Pyroxene to Garnet*, Contri. Mineral Petrol. *102*, 389–405.

GASPARIK, T. (1992), *Melting Experiments on the Enstatite-pyrope Join at 80–152 kbar*, J. Geophys. Res. *97*, 15181–15188.

GETTING, I. C., CHEN, G., and BROWN, J. A. (1993), *The Strength and Rheology of Commercial Grade Tungsten Carbide Cermets Used in High-pressure Apparatus*, Pure and Appl. Geophys, this issue.

GRAHAM, E. K. Jr., and BARSCH, G. R. (1969), *Elastic Constants of Single-crystal Forsterite as a Function of Temperature and Pressure*, J. Geophys. Res. *74*, 5949–5960.

GUILLOPE, M., and POIRIER, J. P. (1979), *Dynamic Recrystallization during Creep of Single-crystalline Halite: An Experimental Study*, J. Geophys. Res. *84*, 5557–5567.

GWAMESIA, G. D. (1987), *Pressure Calibrations in a Girdle-anvil and a DIA-type High-pressure Apparati at Room Temperature (25°) and High Temperature (1000°)*, M. S. Thesis, State University of New York at Stony Brook.

GWANMESIA, G. D. (1991), *High-pressure Elasticity for the Beta and Spinel Polymorphs of Mg_2SiO_4 and Composition of the Transition Zone of the Earth's Mantle*, Ph.D. Thesis, State University of New York at Stony Brook.

GWANMESIA, G. D., LIEBERMANN, R. C., and GUYOT, F. (199a), *Hot-pressing and Characterization of Polycrystals of Beta-Mg_2SiO_4 for Acoustic Velocity Measurements*, Geophys. Res. Lett. *17*, 1331–1334.

GWANMESIA, G. D., RIGDEN, S. M., JACKSON, I., and LIEBERMANN, R. C. (1990b), *Pressure Dependence of Elastic Wave Velocity for Beta-Mg_2SiO_4 and the Composition of the Earth's Mantle*, Science *250*, 794–797.

GWANMESIA, G. D., and LIEBERMANN, R. C., *Polycrystals of high-pressure phases of mantle minerals: Hot-pressing and characterization of physical properties*. In *High Pressure Research: Application to Earth and Planetary Sciences* (eds. Syono, Y. and Manghnani, M.) (Terra Scientific Publishing Co., and American Geophysical Union, Tokyo and Washington, D.C. 1992) pp. 117–135.

KANZAKI, M. (1991), *Ortho-clinoenstatite Transition*, Phys. Chem. Minerals *17*, 726–730.

KATSURA, T., and ITO, E. (1989), *The System Mg_2SiO_4-Fe_2SiO_4 at High Pressures and Temperatures: Precise Determination of Stabilities of Olivine, Modified Spinel, and Spinel*, J. Geophys. Res. *94*, 15663–15670.

KUMAZAWA, M., and ANDERSON, O. L. (1969), *Elastic Moduli, Pressure Derivatives, and Temperature Derivatives of Single-crystal Olivine and Single-crystal Forsterite*, J. Geophys. Res. *74*, 5961–5972.

KUSABA, K., GALOISY, L., WANG, Y., VAUGHAN, M. T., and WEIDNER, D. J. (1993), *Determination of Phase Transition Pressure of ZnTe under Quasihydrostatic Conditions*, Pure and Appl. Geophys, this issue.

LI, B., RIGDEN, S. M., and LIEBERMANN, R. C. (1992), *Pressure Derivatives of the Elastic Wave Velocities in Polycrystalline Stishovite* (abstract), EOS Trans. AGU *73*.

LI, B. (1993), *Polycrystalline Stishovite: Hot-pressing and Elastic Properties*, M.S. Thesis, State University of New York at Stony Brook.

LIEBERMANN, R. C. (1972), *Compressional Velocities of Polycrystalline Olivine, Spinel and Rutile Minerals*, Earth Planet. Sci. Inter. *17*, 263–268.

LIEBERMANN, R. C. (1975), *Elasticity of Olivine( ), Beta( ), and Spinel( ) Polymorphs of Germanates and Silicates*, Geophys. J. Roy. Astr. Soc. *42*, 899–929.

LIEBERMANN, R. C., and SCHREIBER, E. (1968), *Elastic Constants of Polcrystalline Hematite as a Function of Pressure*, J. Geophys. Res. *73*, 6585–6590.

LIEBERMANN, R. C., and WANG, Y., *Characterization of sample environment in a uniaxial split-sphere apparatus*. In *High Pressure Research: Application to Earth and Planetary Sciences* (eds. Syono, Y., and Manghnani, M.) (Terra Scientific Publishing Co., and American Geophysical Union, Tokyo and Washington, D.C. 1992) pp. 19–31.

LIEBERMANN, R. C., RINGWOOD, A. E., MAYSON, D. J., and MAJOR, A., *Hot-pressing of polycrystalline aggregate at very high pressure for ultrasonic measurements*. In *Proceedings of 4th Conference on High Pressure* (ed. Osugi) (Physico-Chemical Society of Japan, Tokyo 1975) pp. 495–502.

LLOYD, E. C., *Accurate Characterization of the High Pressure Environment* (NBS Spec. Publ. No. 326, Washington D.C. 1971) pp. 1–3.

MIZUTANI, H., HAMANO, Y., IDA, Y., and AKIMOTO, S. I. (1970), *Compressional Wave Velocities of Fayalite, Fe$_2$SiO$_4$ Spinel, and Coesite*, J. Geophys. Res. *75*, 2741-2747.

NIESLER, H., and JACKSON, I. (1989), *Pressure Derivatives of the Elastic Wave Velocities from Ultrasonic Interferometry Measurements on Jacketed Polycrystals*, J. Acoust. Soc. Am. *86*, 1573-1585.

RIGDEN, S. M., and JACKSON, I. (1991), *Elasticity of Germanate and Silicate Spinels at High Pressure*, J. Geophys. Res. *96*, 9999-10006.

RIGDEN, S. M., JACKSON, I., NIESLER, H., RINGWOOD, A. E., and LIEBERMANN, R. C. (1988), *Pressure Dependence of Elastic Wave Velocities of Mg$_2$GeO$_4$*, Geophys. Res. Lett. *15*, 605-608.

RIGDEN, S. M., GWANMESIA, G. D., FITZGERALD, J., JACKSON, I., and LIEBERMANN, R. C. (1991), *High Pressure Elasticity of Mg$_2$SiO$_4$-spinel: Implications for the 520 km Seismic Discontinuity and the Transition Zone of the Earth's Mantle*, Nature *354*, 143-145.

RIGDEN, S. M., GWANMESIA, G. D., JACKSON, I., and LIEBERMANN, R. C., *Progress in high-pressure ultrasonic interferometry, the pressure dependence of elasticity of Mg$_2$SiO$_4$ polymorphs and constraints on the composition of the transition zone of the earth's mantle*. In *High Pressure Research: Application to Earth and Planetary Sciences* (eds. Syono, Y., and Manghnani, M.) (Terra Scientific Publishing Co., and American Geophysical Union, Tokyo and Washington, D.C. 1992) pp. 167-182.

RIGDEN, S. M., GWANMESIA, G. D., and LIEBERMANN, R. C. (1992b), *Pressure Dependence for Acoustic Wave Velocities for Majorite-pyrope Garnet*. EOS, Trans. Amer. Geophys. Un. *73*, 517.

SCHREIBER, E. (1967), *Elastic Moduli of Single-crystal Spinel at 25°C and to 2 kbar*, J. Appl. Phys. *38*, 2508-2511.

SCHREIBER, E., and ANDERSON, O. L. (1966), *Pressure Derivatives of the Sound Velocities of Polycrystalline Alumina*, J. Am. Ceram. Soc. *49*, 184-190.

SIMMONS, G. (1964), *Velocity of Shear Waves in Rocks to 10 kilobars, 1*, J. Geophys. Res. *69*, 1123-1130.

WANG, C. Y. (1974), *Pressure Coefficient of Compressional Wave Velocity for a Bronzitite*, J. Geophys. Res. *79*, 771-772.

WEIDNER, D. J., VAUGHAN, M. T., KO, J., and WANG, Y., *Characterization of stress, pressure and temperature in SAM 85, a DIA-type pressure apparatus*. In *High Pressure Research: Application to Earth and Planetary Sciences* (eds. Syono, Y., and Manghnani, M.) (Terra Scientific Publishing Co., and American Geophysical Union, Tokyo and Washington, D.C. 1992) pp. 13-17.

YAGI, T., and AKIMOTO, S. (1976), *Direct Determination of Coesite-stishovite Transition by in situ X-ray Measurements*, Tectonophys. *35*, 259-270.

ZHANG, J., LIEBERMANN, R. C., GASPARIK, T., HERZBERG, C. T., and FEI, Y. (1993), *Melting and Subsolidus Relations of SiO$_2$ at 9-14 GPa*, J. Geophys. Res., *98*, 19,785-19,793.

(Received August 1993, revised/accepted October 1993)

Diamond-anvil Cell Experiments

PAGEOPH, Vol. 141, No. 2/3/4 (1993)

0033–4553/93/040487–09$1.50 + 0.20/0
(c) 1993 Birkhäuser Verlag, Basel

Hydrothermal Studies in a New Diamond Anvil Cell up to 10 GPa and from −190°C to 1200°C

W. A. Bassett,[1] A. H. Shen,[1] M. Bucknum,[1] and I-Ming Chou[2]

Abstract — The new hydrothermal diamond anvil cell (HDAC) has been designed for optical microscopy and X-ray diffraction at pressures up to 10 GPa and temperatures between −190°C and 1200°C. Laser light reflected from the top and bottom anvil faces and the top and bottom solid sample faces produce interference fringes that provide a very sensitive means of monitoring the volume of sample chamber and for observing volume and refractive index changes in solid samples due to transitions and reactions. Synchrotron radiation has been used to make X-ray diffraction patterns of samples under hydrothermal conditions. Individual heaters and individual thermocouples provide temperature control with an accuracy of ±0.5°C. Liquid nitrogen directly introduced into the HDAC has been used to reduce the sample temperature to −190°C. The $\alpha-\beta$ phase boundary of quartz has been used to calculate the transition pressures from measured transition temperatures. With this method we have redetermined 5 isochores of H_2O up to 850°C and 1.2 GPa at which the solution rate of the quartz became so rapid that the quartz dissolved completely before the $\alpha-\beta$ transition could be observed. When silica solutions were cooled, opal spherules and rods formed.

Key words: Pressure, temperature, hydrothermal, H_2O, equation of state, diamond anvil cell.

Instrument

Until recently, very few studies of hydrothermal reactions or of equations of state of fluids had been made in the diamond anvil cell (VAN VALKENBURG *et al.*, 1987; SHEN *et al.*, 1992a,b, 1993). The hydrothermal diamond anvil cell (HDAC) described in this paper is designed to make such studies reasonably easy and accurate.

The HDAC consists of two platens with diamond anvils and heaters mounted at their centers (Figure 1). The two platens are drawn together by tightening nuts on the threaded ends of three posts, thus applying pressure to a sample held between the diamond anvils. Holes through the centers of the platens allow visual and X-ray access to the sample along the compression axis. Molybdenum wires wrapped

[1] Department of Geological Sciences, Snee Hall, Cornell University, Ithaca, New York 14853, U.S.A.

[2] 959 National Center, U.S. Geological Survey, Reston, Virginia 22092, U.S.A.

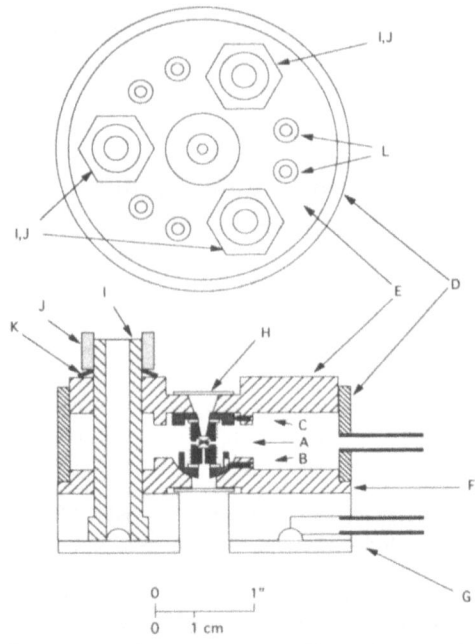

Figure 1
Plan and elevation of the new diamond anvil cell. The diameter of the cell is 3 inches. The height is 2.25 inches. The parts are as follows: *A* sample, diamond anvils, heaters, ceramic heat barriers, *B* ball joint for orienting the lower anvil, *C* sliding disk for positioning the upper anvil, *D* cylinder enclosing inert gas chamber, *E* upper platen, *F* lower platen, *G* base with cooling chamber, *H* upper and lower windows (glass or mica), *I* three posts, *J* nuts on threaded parts of posts for applying force, *K* bellville springs, *L* electric feed-throughs. The base is constructed of brass, the platens, posts, and cylinder are constructed of stainless steel.

around the tungsten carbide seats which support the diamond anvils serve as heaters (Figure 2). These can heat the anvils and the sample very uniformly and very constantly to temperatures in excess of 1200 C. Electrical leads for the heaters and the thermocouples are fed through the platens. The volume containing the heaters, the anvils, and the sample can be enclosed and completely surrounded with a gas to prevent oxidation. We have found that a constant flow of a mixture of Ar with 1% H_2 is very satisfactory for this purpose.

The new HDAC can also be used for low temperature work. By introducing liquid nitrogen into the brass chamber that forms the base of the HDAC we have been able to lower the sample temperature to -132 C and by introducing liquid nitrogen directly into the chamber surrounding the anvils, we have achieved temperatures as low as -190 C. The sample temperature can be controlled by reheating the sample using the resistance heaters and temperature controller to any temperature between -190 C and room temperature.

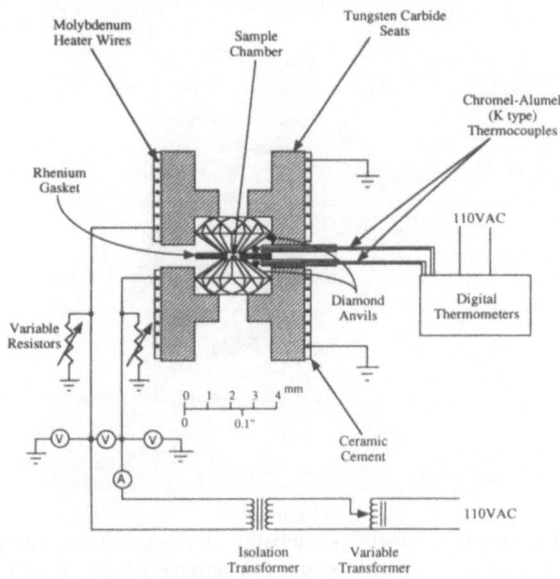

Figure 2
Schematic diagram showing the method of heating and controlling the temperature of the sample.

During all runs, light from a laser is introduced into the top of the diamond cell to produce interference fringes in the sample chamber as well as the solid samples. Interference fringes produced by laser light reflected from top and bottom anvil faces provide a means of monitoring the volume of the sample chamber. Fringes produced by laser light reflected from the top and bottom of the solid sample provide a very sensitive means of observing volume and refractive index changes in samples due to transitions and reactions.

The HDAC has been designed to incorporate several principles important for application to fluid studies and especially for studies of hydrothermal systems: (1) The new cell is designed so that a minimum of dimensional change occurs as the sample is heated. (2) All electric leads are securely fed through the platens. (3) The lower anvil mount consists of a round-bottomed cradle that fits into a round socket for orientational alignment; the upper anvil mount consists of a disk for translational alignment. (4) Molybdenum wire (0.010″ in diameter) wound around the tungsten carbide seats delivers heat mainly by direct thermal conduction to the seats and to the diamond anvils. (5) The thermocouples are cemented in thermal contact with the upper and lower diamond anvils and are shielded from direct line-of-sight to the heaters and cold parts of the HDAC. (6) The power to the upper and lower heaters is controlled individually and the temperatures are measured by individual thermocouples, allowing us to control the temperatures of the two anvils to within less than 0.1 C of each other over the temperature range from $-100°C$ to $380°C$ and within 0.5 C outside of that range. (7) The heating wires, diamond anvils, and

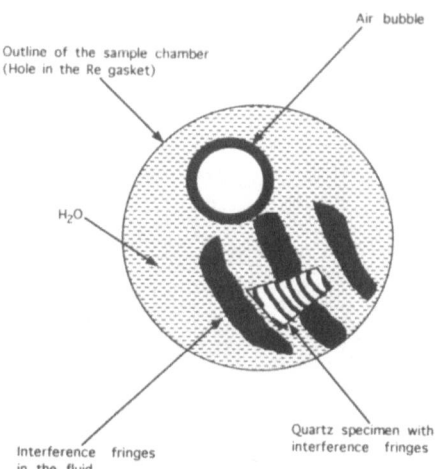

Figure 3
Typical appearance of a sample of quartz in a mixture of water and air. The portion of the sample illuminated with laser light shows coarse fringes due to interference between light reflected from the two anvil faces. The fine fringes in the quartz sample are caused by interference between light reflected from the top and bottom of the quartz platelet.

tungsten carbide seats are enclosed in a chamber into which a mixture of 1% H and 99% Ar is introduced to prevent oxidation. (8) There are two windows that offer visual access to the sample through the diamond anvils and there are two windows that offer visual access to the sides of the heater assemblage. (9) The sample chamber, which is formed by laser drilling a 500 μm hole in a Re foil 125 μm thick is typically filled with a crystal, distilled-deionized water, and an air bubble (Figure 3). (10) The variation of the distance between the anvil faces, determined by laser interferometry, is less than 0.5% and the lateral dimensions undergo negligible change and so the volume of the sample chamber remains nearly constant. The crystal in the sample chamber typically occupies less than 1.5% of the total volume; therefore, the effect of volume change of the crystal at high temperatures and high pressures is negligible.

Precision and Accuracy

The thermocouples are not subject to any pressure-induced emf error because they are external and therefore not pressurized. We believe that temperature can be measured with a precision of ± 0.1°C over the range of temperature from -110°C to 380°C. The accuracy of our measurements is ± 0.5°C over the same range. At temperatures from -190°C to -100°C and above 380°C, the precision is ± 0.5°C and the accuracy is ± 1.5°C.

The accuracy of the pressure determination depends on the accuracies of the temperature measurements and the equation of state of H_2O. The accuracy for our homogenization temperature measurements is believed to be $\pm 0.5°C$, which corresponds to a maximum density uncertainty of ± 0.004 (g/cm³).

Equation of State of H_2O

A single crystal of quartz, polished on top and bottom, was placed in distilled, deionized water along with a bubble (Figure 3). The sample was then heated until the bubble disappeared (homogenization). This is shown as T_{hh} in Figure 4. As the sample was further heated, both the temperature and pressure of the homogeneous supercritical fluid increased. This was continued until sudden motion of laser-produced interference fringes in the crystal was observed indicating the $\alpha - \beta$ transition, T_{tr}. Once this easily recognized phenomenon was observed, the heating was halted and the sample was cooled along an isochore IC_c until a vapor bubble reappeared. The sample was then slowly reheated until the vapor bubble disappeared. This was repeated three times, and the temperature at the moment of bubble disappearance T_{hc} was measured. The isochore density was determined from the homogenization

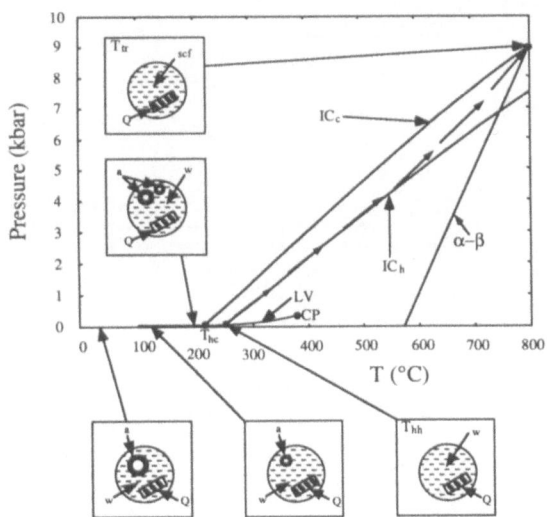

Figure 4

A typical pressure-temperature path during the observation of a phase transition of a sample immersed in H_2O. The labels mean the following: Q quartz sample, W water, scf supercritical fluid, a air bubbles, LV liquid-vapor curve, CP critical point, IC_h isochore at start of heating, IC_c isochore followed during cooling, T_{hh} temperature of homogenization measured in the heating cycle, T_{hc} temperature of homogenization after liquid-vapor separation in a cooling cycle, T_{tr} temperature of transition. The path drifts from one isochore to another during heating, probably due to compression of the gasket. No change in sample volume and therefore density was observed during cooling.

temperature along the liquid-vapor curve and the transition pressure was determined from the temperature of fringe movement in the quartz along that isochore. Our results (SHEN *et al.*, 1992b) show that the equation of state of H_2O of HAAR *et. al.* (1984) is in good agreement with the $\alpha - \beta$ quartz boundary of MIRWALD and MASSONNE (1980) for three different isochores (Figure 5). We have observed that at higher pressures and temperatures the quartz platelets do dissolve in H_2O. However, the effect of dissolving quartz on the transition pressure is minor because when the quartz dissolves, the volume of the solid is reduced and the volume of the fluid is increased. The effect that this has on pressure is secondary, due only to the difference between the partial molar volume change during dissolving.

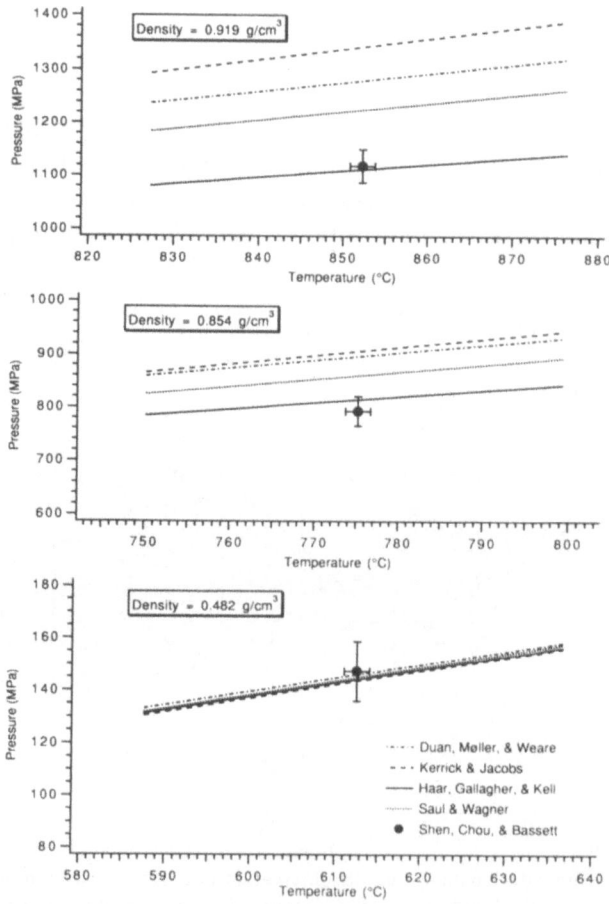

Figure 5

Measurements of the equation of state of H_2O at three densities using the α β quartz transition for calibration. Our data, shown by the solid dots with horizontal and vertical uncertainty bars, are derived from our $\alpha - \beta$ quartz transition temperature measurements combined with the transition T-P relationship given by MIRWALD and MASSONNE (1980). Our data agree well with those of HAAR *et al.* (1984).

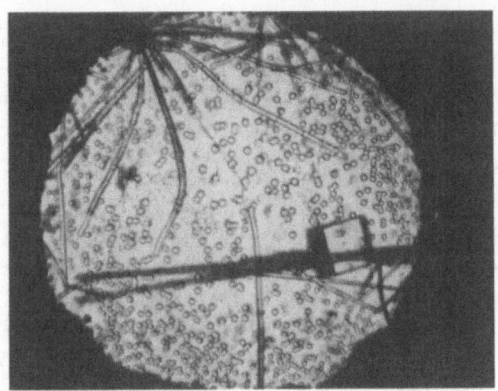

Figure 6
Spherules and rods of opal that formed when an aqueous solution of silica was cooled rapidly from 720 C to 400 C.

Isochores of greater and greater density were studied this way until temperatures required were so high that complete solution of the quartz crystal took place before the $\alpha - \beta$ transition in the quartz could be observed. If the temperature was then slowly lowered before the quartz crystal completely dissolved, the crystal was observed to grow again. If, however, the temperature was lowered rapidly, the silica was observed to precipitate as spherules and rods of opal (Figure 6). Some of the rods showed weak birefringence between crossed polars indicating some degree of crystallinity, either quartz or tridymite. The exact identification will have to await future studies.

Other Applications

We are applying the new diamond anvil cell to an investigation of the morphologies of the various solid phases of H_2O with the remarkable observation that some of the ice phases crystallize with exceptionally well developed crystal forms. The tetragonal-to-cubic transition in $BaTiO_3$ is being mapped with the objective of using it as an internal pressure calibrant, much as we have used the $\alpha - \beta$ transition in quartz.

We have conducted *in situ* X-ray diffraction experiments on the Cornell High Energy Synchrotron Source (CHESS). Using a white beam and the energy dispersive method, we have followed the dehydration of montmorillonite under pressure at temperatures reaching 775 C. Using a monochromatic beam and the Debye-Scherrer method, we have made some preliminary measurements of the lattice parameters of calcite as a function of pressure and temperature. With these two experiments we have shown that it is possible to obtain good diffraction data on a polycrystalline sample that is immersed in water at high temperature and pressure.

We have modified the new design of diamond anvil cell for low temperature work. By introducing liquid nitrogen directly into the brass chamber that forms the base of the diamond cell we have been able to lower the sample temperature to $-132°C$ and by introducing liquid nitrogen into the upper chamber surrounding the anvils, we have been able to reduce the temperature to $-190°C$. These modifications enable us to load carbon dioxide in solid form. It also allows us to measure the temperature of the ice-water transition in H_2O as a function of pressure. This measurement is necessary for studying isochores that have densities in the range 0.98 g/cm^3 to 1.12 g/cm^3. Densities above 1.12 g/cm^3 can also be measured by means of the ice-liquid transition in H_2O but do not require low temperatures. The capability of controlling and measuring temperatures below $25°C$ is necessary for determinations of the solute concentrations in aqueous solutions through freezing point depression measurements. Therefore, this development is essential to our future studies of the equations of state of aqueous solutions.

Conclusions

The new hydrothermal diamond anvil cell has proved to be an excellent method for observing and measuring the pressure-temperature conditions of hydrothermal reactions, phase transitions, and for tracing isochores of fluids at pressures up to 10 GPa and temperatures from $-190°C$ to $1200°C$. Phase boundaries can be mapped with high precision by using the equation of state of H_2O or some other fluids. The technique is capable of very high precision measurements of temperature, pressure, and density. When calibrated with reliable calibrants, the method is capable of very accurate measurements.

Acknowledgments

The project was supported by grant EAR9117637 to Bassett and Chou from the National Science Foundation. We wish to acknowledge the help of Tren Haselton, Terry Wu, and Amy Sun. Critical reviews from Drs. T. Haselton and G. L. Nord, Jr. at U.S. Geological Survey, Reston, Virginia are also acknowledged.

REFERENCES

DUAN, Z.-H., MØLLER, N., and WEARE, J. H. (1992), *An Equation of State for the CH_4-CO_2-H_2O System: I. Pure Systems from 0° to 1000°C and 0 to 8000 bar*, Geochim. Cosmochim. Acta *56*, 2605–2617.

HAAR, L., GALLAGHER, J. S., and KELL, G. S., *NBS/NRC Steam Tables: Thermodynamic and Transport Properties and Computer Programs for Vapor and Liquid States of Water in SI Units* (Hemisphere Publ. Corp., Washington, D.C. 1984).

KERRICK, D. M., and JACOBS, G. K. (1981), *A Modified Redlich-Kwong Equation for H_2O, CO_2, and H_2O-CO_2 Mixtures at Elevated Pressures and Temperatures*, Am. J. Sci. *281*, 735–767.

MIRWALD, P. W., and MASSONNE, H.-J. (1980), *The Low-high Quartz and Quartz-coesite Transition to 40 kbar between 600 C and 1600 C and some Reconnaissance Data on the Effect of $NaAlO_2$ Component on the Low Quartz-coesite Transition*, J. Geophys. Research *85*, 6983–6990.

SAUL, A., and WAGNER, W. (1989), *A Fundamental Equation for Water Covering the Range from the Melting Line to 1273 K at Pressures up to 25000 MPa*, J. Phys. Chem. Ref. Data *18*, 1537–1565.

SHEN, A. H., BASSETT, W. A., and CHOU, I-MING, *Hydrothermal studies in a diamond anvil cell: Pressure determination using the equation of state of H_2O*. In *High-pressure Research: Application to Earth and Planetary Sciences* (eds. Syono, Y., and Manghnani, M. H.) (Terra Sci. Publ. Co./AGU, Tokyo/Washington, D.C. 1992a) pp. 61–68.

SHEN, A. H., CHOU, I-MING, and BASSETT, W. A. (1992b), *Experimental Determination of the Equation of State of H_2O Using α–β Quartz Transition in a Diamond Anvil Cell*, 29th International Geological Congress, Kyoto, Japan, Aug. 23–Sep. 5, 1992, Abstract Volume, 207 (Abstract).

SHEN, A. H., BASSETT, W. A., CHOU, I-MING (1993), *The α–β Quartz Transition at Simultaneous High Temperatures and High Pressures in a Diamond Anvil Cell by Laser Interferometry*, American Mineralogist *78*, 694–698.

VAN VALKENBURG, A., BELL, P. M., and MAO, H. K., *High-pressure mineral solubility experiments in the diamond-window cell*. In *Hydrothermal Experimental Techniques* (eds. Ulmer, G. C., and Barnes, H. L.) (J. Wiley and Sons, New York 1987) pp. 458–468.

(Received March 13, 1993, revised June 25, 1993, accepted October 6, 1993)

PAGEOPH, Vol. 141, No. 2/3/4 (1993)

0033–4553/93/040497–11$1.50 + 0.20/0

Thermal Analysis in the Laser-heated Diamond Anvil Cell

JEFFREY S. SWEENEY[1] and DION L. HEINZ[1,2]

Abstract — A new technique actively controls thermal radiation and monitors sample properties during laser-heating in a diamond anvil cell. The technique can be described as a qualitative application of thermal analysis. Discontinuities in temperature, laser power, visible thermal radiation, or in their derivatives as functions of time can be associated with the enthalpy of phase transitions (such as melting) or with changes in material properties (such as emissivity).

The technique is illustrated with melting experiments on iron-magnesium-silicate perovskite. Temperature corrections associated with these experiments are discussed and the results are briefly reviewed.

Key words: High pressure, high temperature, thermal analysis, diamond anvil cell, laser heating, radiometry.

Introduction

Laser-heating in a diamond anvil cell provides the temperatures (3000 K) and pressures (100 GPa) necessary for studying minerals at conditions equivalent to the earth's lower mantle (MING and BASSETT, 1974; JEANLOZ and HEINZ, 1984). Recent advances allow greater stability and control of temperatures with this apparatus (HEINZ et al., 1991). Despite these advances, exploratory experiments on the melting of iron-magnesium-silicate perovskite indicated that laser-heated $(Fe_{.14}Mg_{.86})SiO_3$ samples were unstable near their melting temperatures. The instability was attributed to Soret diffusion of the iron toward cooler regions of the samples. Soret diffusion considerably altered the emissivity and optical absorption of the samples over time intervals of less than a minute. We required a technique that could measure melting temperatures rapidly: i.e., prior to the onset of Soret diffusion.

Further, we sought alternative criteria, in lieu of textural changes, to identify phase transitions at high temperatures and pressures. In this regard, we were

[1] Department of Geophysical Sciences, University of Chicago, 5734 S. Ellis Avenue, Chicago, IL 60637, U.S.A.

[2] James Franck Institute, University of Chicago, Chicago, IL, U.S.A.

encouraged by previous results from resistively heated iron filaments at ambient pressure (HEINZ *et al.*, 1991). During those experiments, thermal radiation was continuously monitored as electric current was increased through the filaments. Just prior to melting, a decrease in the emissivity of the heated filaments was detected across the γ-to-δ phase transition of iron. Thus, an alternative criterion for determining phase transitions was demonstrated by concurrently monitoring a physical property of a sample (emissivity) and the conditions of an experiment (time, temperature). Such solid-solid phase transitions offer little textural evidence in opaque materials and would be particularly difficult to detect by texture alone.

Accordingly, a new technique was devised to identify melting (and other phase transitions that have thermal signals) during experiments in a laser-heated diamond anvil cell. The technique can be described as a qualitative application of thermal analysis. The laser-heating system and the sample are monitored over time, thermal signals are observed, and the signals are correlated with changes to the properties of the sample. The technique has been used to remeasure the melting temperature of iron-magnesium-silicate perovskite as a function of pressure (SWEENEY and HEINZ, 1993).

Experimental Method

The major components of the heating apparatus were described in detail elsewhere (HEINZ *et al.*, 1991) and are reviewed in Figure 1. Using a feedback loop, this apparatus stabilizes sample temperatures by monitoring visible thermal radiation emitted by laser-heated samples.

The experiments are controlled by an electrical signal (the driving function, 0 V to 5 V) injected into the stabilization system. The driving function increases or decreases the intensity of thermal radiation (monitored by detector D2) in proportion to the applied voltage. Thus, it controls the temperature of the laser-heated sample. The intensity of thermal radiation emitted by a blackbody (emissivity of one) or a greybody (emissivity less than or equal to one and independent of wavelength) is proportional to the fourth power of temperature from the Stephan-Boltzmann equation ($I = \varepsilon\sigma T^4$, where I is intensity, ε is emissivity, σ is the Stephan-Boltzmann constant and T is temperature). Therefore, a linear driving function yields a sample temperature that varies with the fourth root of time and a driving function that varies as the fourth power of time yields a linear temperature change. Note that this relationship is approximate since detector D2 (S20 response) only monitors a portion of the thermal radiation spectrum, and in general the samples are not ideal greybodies.

The driving function, a timing signal from detector controller DC1, the heating-laser power from photodiode D3, and the status of the stabilization system from servo controller SC1 can be monitored from zero to several hundred kilohertz.

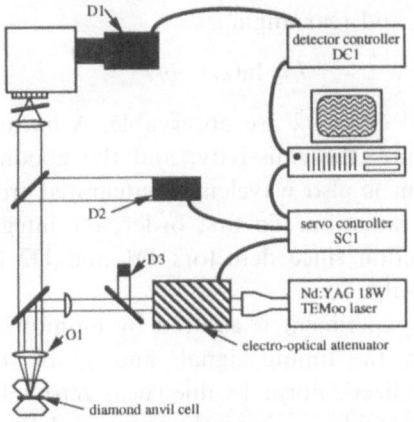

Figure 1

The laser-heating apparatus is constructed around a research-grade microscope with attachments that provide optical access through multiple ports (Nikon Optiphot with quadhead and two episcopic-fluorescence modules). This apparatus directs near-infrared light from the heating laser onto the sample. It gathers thermal radiation from the sample in the visible region of the spectrum and then directs part of the thermal radiation to detector D1 which measures the sample's temperature. The rest of the thermal radiation is monitored by detector D2 which stabilizes the heating laser.

Typically, these signals are monitored at 1024 Hz. Detector D1 (1025 diode optical multichannel analyzer) collects spectra of the thermal radiation at a maximun frequency of 30 Hz (33 ms for each spectrum), and detector controller DC1 is able to store up to 128 sequential spectra (4.27 s of data at 30 Hz). A TTL gating signal from DC1 synchronizes data collection and triggers the driving function.

After an experiment, the spectra of thermal radiation are analyzed using the Planck blackbody radiation function (HEINZ and JEANLOZ, 1987a; HEINZ et al., 1991):

$$I_{\lambda, T} = \frac{\varepsilon c_1}{\lambda^5 [\exp(c_2/\lambda T) - 1]}, \tag{1}$$

where λ is wavelength and c_i are constants. The Wien approximation makes equation (1) more tractable:

$$I_{\lambda, T} = \frac{\varepsilon c_1}{\lambda^5 [\exp(c_2/\lambda T)]}. \tag{2}$$

For the spectral window of these experiments (550–750 nm), the error in temperature from this approximation is vanishingly small below 3000 K, less than a half percent at 4000 K, and about three percent at 6000 K (HEINZ and JEANLOZ, 1987a). If the emissivity (ε) has no wavelength dependence (the greybody approximation) then equation (2) can be made linear in its parameters by taking the logarithm of both sides and rearranging:

$$J = \ln(\varepsilon) - \omega T^{-1}, \tag{3}$$

where $J = \ln(I\lambda^5/c_1)$ and $\omega = c_2/\lambda$ are observable. A linear regression of equation (3) yields the temperature, the emissivity, and the goodness of fit (χ^2) for each spectrum. Each spectrum is also wavelength-integrated from 550–750 nm to estimate the total thermal radiation. To first order, the integrated thermal radiation mimics the driving function since detectors D1 and D2 have approximately the same wavelength sensitivity.

The quality of each experiment is assured by monitoring the driving function, the stabilizer controller, the timing signal, and χ^2 of the blackbody fit. Wide excursions from the stabilizer's normal value (near zero volts) are evidence that the controller is operating outside its dynamic range and that the thermal radiation emitted by the sample is not well controlled.

Interpreting Signals

The thermal signals from an experiment depend on a sample's emissivity, heat capacity, thermal diffusivity, the enthalpy of any phase transitions, and the optical absorption of the sample at the laser's wavelength. These material properties can be functions of temperature, pressure, and composition; while conductive and radiative heat loss also depend on the configuration of the sample assembly and the diamond anvil cell.

Consider a model sample with emissivity independent of wavelength but decreasing discontinuously from 1.0 to 0.5 across a univariant phase transition (solid line, Figure 2b). Assuming that the response of detectors D1 and D2 are equal, the integrated thermal radiation (Figure 2a) increases linearly for a linearly increasing driving function. For blackbody or greybody thermal radiation, the temperature (Figure 2c) increases proportional to the fourth root of the intensity (from the Stephan-Boltzmann equation).

Laser power (Figure 2d) must be added to increase the temperature of the sample due to its heat capacity (approximately $3nR$ above the Debye temperature). Assuming constant heat capacity and volume ($c_r = [\partial U/\partial T]_r$), we find that power due to heat capacity is proportional to the time derivative of the temperature. Laser power is also added to offset increased radiative and conductive heat loss at higher temperatures which depend on the emissivity of the sample, the thermal diffusivities of the sample and the diamond anvils, and the configuration of the diamond anvil cell. Finally, laser power increases or decreases to account for any changes in the optical absorption of the sample at the laser's wavelength.

At the phase transition the emissivity decreases and, from equation (1), the temperature increases discontinuously. Added to the laser power is a peak or trough (a delta function for the case of a univariant transition) whose area is proportional

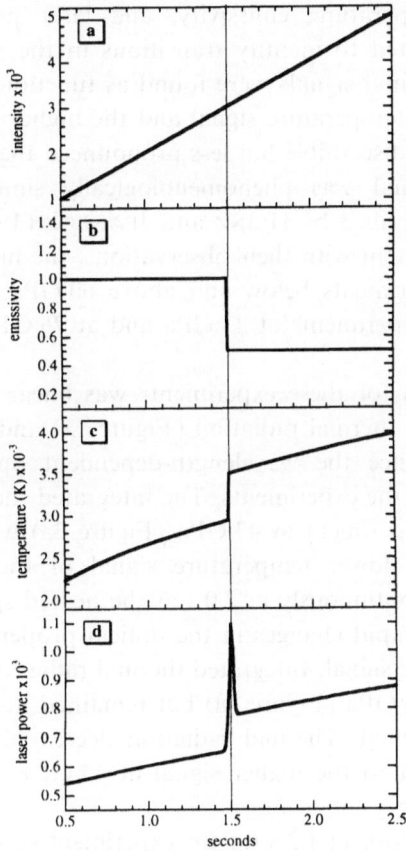

Figure 2
Integrated thermal radiation (intensity (a), emissivity (b), temperature (c), and heating-laser power (d))
for a model experiment with wavelength-independent emissivity. The driving function increases linearly
with time ($V = a + bt$) and is mimicked by the integrated thermal radiation. The emissivity changes
discontinuously across a phase transition at 1.5 s.

to the enthalpy of the transition. This is in contrast to differential thermal analysis
(where temperature remains constant during a phase transition) since we drive the
thermal radiation of a sample and therefore the heating rate is not constant. For
transitions with increasing entropy, such as melting, the enthalpy and the peak in
laser power are positive. Total laser power (Figure 2d) is the sum of these effects.

Results

Numerous melting experiments on iron-magnesium-silicate perovskite were performed with this thermal analysis technique. Discontinuities in the integrated

thermal radiation, temperature, emissivity, and laser power; or in their time derivatives were correlated to identify transitions in the samples (SWEENEY and HEINZ, 1993). Two thermal signals were found as functions of pressure. These are designated as the lower temperature signal and the higher temperature signal. The lower signal was clearly discernible but less pronounced than the higher signal. The higher temperature signal was phenomenologically similar to the melting of $(Fe_{.14}Mg_{.86})SiO_3$ as described by HEINZ and JEANLOZ (1987b) and KNITTLE and JEANLOZ (1989). Consistent with their observations, the higher signal was qualitatively different for experiments below and above 60 GPa (SWEENEY and HEINZ, 1993). Representative experiments at 43 GPa and at 76 GPa are shown in Figures 3 and 4.

The driving function for these experiments was linear with time ($V = a + bt$). However, the integrated thermal radiation (Figures 3a and 4a) only approximated the driving function since the wavelength-dependent optical properties of the samples changed during the experiments. The integrated thermal radiation changed slope at 1.3 s for the experiment at 43 GPa (Figure 3a) with the slope decreasing after 1.3 s. This is the lower temperature signal of the experiments. Thermal radiation increased discontinuously at 2.0 s as the heated spot suddenly brightened at its center indicating rapid changes in the optical properties of the sample. This is the higher temperature signal. Integrated thermal radiation changed slope at 1.7 s for the experiment at 76 GPa (Figure 4a) but remained linear in time. The change in slope is the lower signal. Thermal radiation decreased abruptly at the higher signal (2.6 s) in contrast to the higher signal at 43 GPa where thermal radiation increased discontinuously.

Emissivity changed slope at 1.3 s for the experiment at 43 GPa (Figure 3b), and was not measurable after 2.0 s due to the rapid increase in thermal radiation. Emissivity changed slope at 1.7 s for the experiment at 76 GPa (Figure 4b), and decreased abruptly at 2.6 s, consistent with the lower and higher signals determined from the integrated intensity. Emissivity decreased linearly between 1.3–2.0 s for the experiment at 43 GPa (Figure 3b) and from 1.7–2.6 s for the experiment at 76 GPa (Figure 4b). Such continuous variations in emissivity might result from diffusion (such as Soret diffusion) which could alter the optical properties of a sample over time. In this case, a phase transition is not associated with the change in emissivity. Alternatively, a continuous change in emissivity might result from a phase transition through a multi-phase region (such as incongruent melting). In this case, the enthalpy of the phase transition would contribute to the laser power signal but it would be spread over a time interval (0.7–0.9 s for these examples).

Temperature abruptly changed slope at 1.3 s for the experiment at 43 GPa (Figure 3c) and was not measurable after 2.0 s, as for emissivity. For the experiment at 76 GPa (Figure 4c), the lower signal is not discernible in temperature while the higher signal is clearly present at 2.6 s where temperature abruptly changed slope.

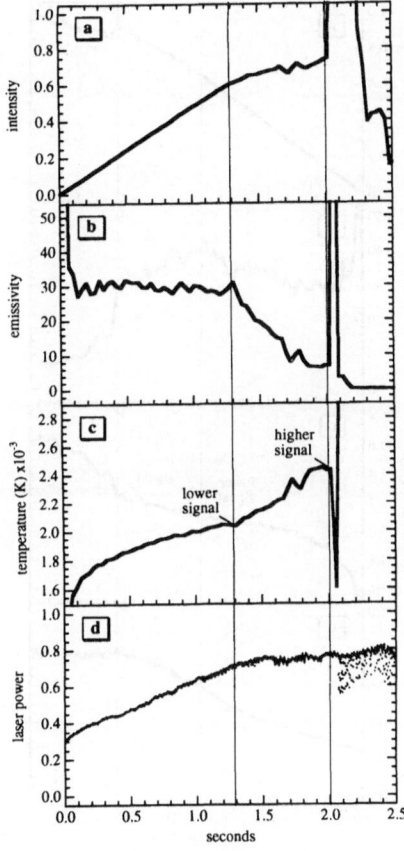

Figure 3

Integrated thermal radiation (intensity (a), emissivity (b), temperature (c), and heating-laser power (d)) for a representative experiment on ($Fe_{.14}Mg_{.86}$)SiO_3 at 43 GPa. For this experiment, the driving function increased linearly for four seconds. Thermal radiation was integrated from 550–750 nm for comparison to the driving function. The integrated thermal radiation approximates the driving function until about 1.3 s (lower signal of SWEENEY and HEINZ, 1993) where the slope changes. Thermal spectra were collected at a rate of 30 Hz and fit to Planck's blackbody function using the greybody and Wien approximations to determine emissivity and temperature. Melting (higher signal of SWEENEY and HEINZ, 1993) occurred at about 2 s.

For the experiment at 43 GPa (Figure 3d), neither signal was discernible in the laser power. The stabilization system lost control of the experiment after 2.1 s as indicated by the scatter in laser power. Laser power abruptly decreases at 2.6 s for the experiment at 76 GPa (Figure 4d). For iron-magnesium-silicate perovskite, integrated thermal radiation, emissivity, and temperature were more diagnostic of transitions than laser power. Laser power is less diagnostic since it is modulated both to stabilize the laser-heating system and to control the experiments. Laser power signals may be masked by the noise of stabilization, particularly for

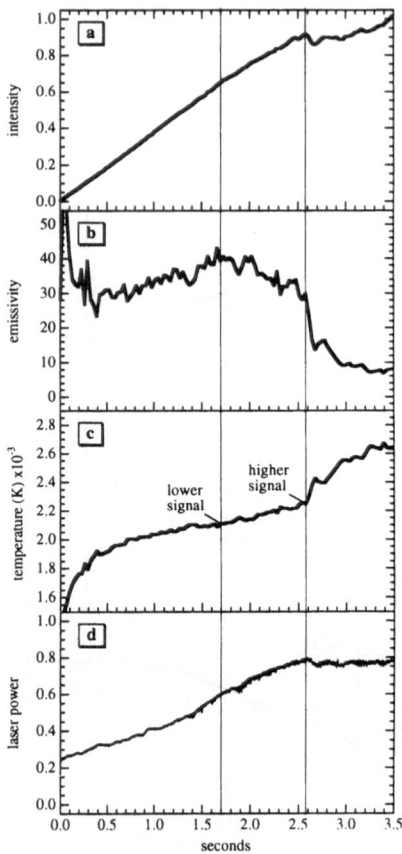

Figure 4

Similar to Figure 3 but at 76 GPa. For the integrated thermal radiation, a change in slope at the lower signal is apparent when a straight edge is placed along the curve. Note the differences at melting (higher signal) for this experiment compared to Figure 3. At 43 GPa, thermal radiation (3a) increased dramatically while emissivity (3b) and temperature (3c) were not measurable due to rapid changes in the optical properties of the sample. At 76 GPa, thermal radiation (a) and laser power (d) decreased abruptly but slightly and emissivity (b) and temperature (c) were measurable on both sides of the transition.

$(Fe_{.14}Mg_{.86})SiO_3$, which couples in a nonlinear fashion with the heating laser (HEINZ and JEANLOZ, 1987b).

For the melting experiments on iron-magnesium-silicate perovskite, thermal radiation was collected from entire laser-heated spots and fit with equation (3) to determine temperatures. Due to thermal gradients, the observed temperatures were less than the peak temperatures of the heated spots (JEANLOZ and HEINZ, 1984; HEINZ and JEANLOZ, 1987a). Ideally, observed temperatures should be corrected to peak temperatures, where phase transitions (such as melting) most likely occur.

Two methods have been reported which consider temperature gradients normal to the optical axis and measure peak temperatures in a laser-heated diamond anvil cell. BOEHLER et al. (1990) apertured the image of a sample so that only the hottest portion was measured. HEINZ and JEANLOZ (1987a) scanned a slit across the image of a sample and applied a tomographic technique to retrieve the radial temperature profile. Both methods are most appropriate for opaque (optically thick) materials where surface temperatures are measured. An extra correction is required for materials which transmit visible light such as $(Fe_{.14}Mg_{.86})SiO_3$ since their temperatures are determined from thermal radiation integrated through the thickness of the sample. Thus, the observed temperatures in transmissive (optically thin) materials depend upon temperature gradients both normal and parallel to the optical axis.

Moreover, for experimental reasons neither method was satisfactory for correcting the melting temperatures of the present study. Chromatic aberration in our system adversely affected temperature measurements when a 100 μm pinhole (4 μm diameter at the sample) was used to aperture the image of the laser-heated spot. The magnitude of the error was about 200–300 K. We attribute the chromatic aberration to the microscope objective O1 (Figure 1, Leitz Wetzlar L 25/0.22 P or UT 40/0.34). This refracting objective compromises chromatic correctness to achieve a long working distance. The effect was discovered when the data were viewed in the normalized coordinates of the Wien approximation, equation (3) (HEINZ et al., 1991). If thermal radiation exhibits significant curvature when viewed in these coordinates, then the sample has a pronounced wavelength-dependent emissivity, or there is chromatic aberration in the system. The first hypothesis was tested by measuring a known blackbody source: Curvature was apparent with the aperture in place. When $(Fe_{.14}Mg_{.86})SiO_3$ samples were melted with the aperture removed, no curvature in the data was observed.

The tomographic technique was discarded in part because of chromatic aberration. The scanning slit is a small aperture, although in only one dimension. Also, with this technique about one minute is required to scan the spots, and the optical properties of iron-magnesium-silicate perovskite near its melting temperature changed within that time due to Soret diffusion. Therefore, uncorrected melting temperatures were reported, Soret diffusion was minimized, and the effects of chromatic aberration were avoided (SWEENEY and HEINZ, 1993).

The results of the experiments suggest two alternative interpretations. The lower signal of Figures 3 and 4 may be the temperature where iron diffuses rapidly. Then, the higher signal would correspond to melting of magnesium-enriched silicate perovskite. Alternatively, the lower signal may be the solidus where $(Fe_{.14}Mg_{.86})SiO_3$ undergoes incongruent melting to an iron-magnesium-enriched solid plus a silica-enriched liquid. Then, the higher signal would be the liquidus. Energy-dispersive X-ray analysis of several quenched samples (beam-line X17C, National Synchrotron Light Source, Brookhaven National Laboratory) provided no evidence of incongruent melting. SWEENEY and HEINZ (1993) discuss these results and the complete set of experiments in more detail.

Applications

Any sample that can be laser-heated in the diamond anvil cell is amenable to thermal analysis by the technique described here. A sample can be laser-heated if it absorbs laser power, similar to $(Fe_{.14}Mg_{.86})SiO_3$. A sample that does not absorb (e.g., stishovite, $MgSiO_3$-perovskite) can be heated by conduction if it is in contact with a material that does absorb. An ideal absorbing material is inert to the sample, and it is stable or its phase transitions are well-known (e.g., tungsten, rhenium, tantalum). A sample can be mixed with an absorbing powder, or it can be loaded on top of an absorbing foil. For a transparent sample with an opaque absorbing material, temperatures are determined by measuring thermal radiation from the surface of the absorber. When the sample undergoes a phase transition, laser power will peak due to the enthalpy of the transition, since it is in thermal contact with the absorber. The peak should be observable provided the laser-power signal is not masked by the noise of stabilization.

An opaque sample that absorbs the heating laser (such as iron) is also amenable to thermal analysis, but it must be insulated from the upper anvil to avoid scorching the diamond. An ideal insulator is transparent to the heating laser and has no wavelength-dependent absorption at the wavelengths where thermal radiation is measured. The insulator should be stable or its phase transitions should be well known, and it should be inert to the sample. Depending on the sample, likely insulators include corundum, periclase, and condensed noble gases such as argon.

A wide range of geophysically important phases can be explored at high temperatures and pressures by this thermal analysis technique. Of interest for the study of the lower mantle are: silicate perovskites, magnesiowüstite, stishovite, iron and iron alloys, as well as assemblages of these phases and their melts. The first application of the technique measured the melting curve of an iron-magnesium-silicate perovskite with a composition thought likely to exist at lower mantle conditions (SWEENEY and HEINZ, 1993).

Acknowledgments

We thank an anonymous reviewer, Tzy-chung Wu, R. Jeanloz, and E. Knittle for frank and helpful comments. This work was funded by NSF grant EAR–9205857 and the Materials Research Laboratory at the University of Chicago.

REFERENCES

BOEHLER, R., VON BARGEN, N., and CHOPELAS, A. (1990), *Melting, Thermal Expansion, and Phase Transitions of Iron at High Pressures*, J. Geophys. Res. *95*, 21731 21736.

HEINZ, D. L., and JEANLOZ, R., *Temperature measurements in the laser-heated diamond cell*. In *High-pressure Research in Mineral Physics* (eds. Manghnani, M. H., and Syono, Y.) (American Geophysical Union, Washington D.C. 1987a) pp. 113–127.

HEINZ, D. L., and JEANLOZ, R. (1987b), *Measurement of the Melting Curve of $Mg_{0.9}Fe_{0.1}SiO_3$ at lower Mantle Conditions and Its Geophysical Implications*, J. Geophys. Res. *92*, 11437–11444.

HEINZ, D. L., SWEENEY, J. S., and MILLER, P. (1991), *A Laser Heating System that Stabilizes and Controls the Temperature: Diamond Anvil Cell Applications*, Rev. Sci. Instrum. *62*, 1568–1575.

KNITTLE, E., and JEANLOZ, R. (1989), *Melting Curve of $(Mg,Fe)SiO_3$ Perovskite to 96 GPa: Evidence for a Structural Transition in Lower Mantle Melts*, Geophys. Res. Lett. *16*, 421–424.

JEANLOZ, R., and HEINZ, D. L. (1984), *Experiments at High Temperature and Pressure: Laser Heating through the Diamond Cell*, J. Phys. (Paris) *45*, 83–92.

MING, L.-C., and BASSETT, W. A. (1974), *Laser Heating in the Diamond Anvil Press up to 2000°C Sustained and 3000°C Pulsed at Pressures up to 260 kilobars*, Rev. Sci. Instrum. *45*, 1115–1118.

SWEENEY, J. S., and HEINZ, D. L. (1993), *Melting of Iron-magnesium-silicate Perovskite*, Geophys. Res. Lett. *20*, 855–858.

(Received March 12, 1993, revised June 21, 1993, accepted October 6, 1993)

PAGEOPH, Vol. 141, No. 2/3/4 (1993)   0033-4553/93/040509-11$1.50 + 0.20/0
(c) 1993 Birkhäuser Verlag, Basel

Deviatoric Stress in a Diamond Anvil Cell Using Synchrotron Radiation with Two Diffraction Geometries

T.-C. Wu[1] and W. A. Bassett[1]

Abstract — Deviatoric stress in a diamond anvil cell with gold as a pressure and stress indicator is measured by two complementary techniques using synchrotron radiation. The first method employs a white X-ray beam using energy dispersive X-ray diffraction. The incident X-ray beam is parallel to the load axis and the diffraction pattern is recorded at a low two-theta angle. Using powder diffraction patterns of polycrystalline gold, we measured the elastic strain of two crystal planes oriented normal to the diffraction vector. Stresses nearly parallel and perpendicular to the load axis can be calculated by stress-strain tensor relationship. The other method uses a monochromatic wiggler X-ray beam. In this case, the diamond cell is oriented so that the incident beam is perpendicular to the load axis. The diffraction pattern is recorded on an image plate area detector. Elastic strains responding to stresses perpendicular and parallel to the load axis can be measured and stresses of the same orientations can be calculated from the strain data. These measurements provide a lower bound of the actual differential stresses in a diamond cell. With these techniques, we can measure stress distribution in a less deviatoric gasketted sample and determine yield strength of mantle materials at high pressures and temperatures.

Key words: Stress, pressure, elasticity, X-ray diffraction, diamond anvil cell, synchrotron radiation.

Introduction

The Diamond Anvil Cell (DAC), when not used with a gasket or a hydrostatic pressure medium, is basically a uniaxial loading device that provides one of the largest nonhydrostatic stresses available (Fig. 1). However, the deviatoric stress distribution across the cell is very unclear, especially since it is material-dependent. Knowing the deviatoric stress is important not only because it causes uncertainties in pressure measurement in the DAC (Fig. 2) (see also MENG *et al.*, 1993), but also produces interesting shear-induced phenomena (Wu *et al.*, 1993) and can be used to estimate the yield strength of materials (KINSLAND and BASSETT, 1976; SUNG *et al.*, 1977; KIMURA *et al.*, 1982; MEADE and JEANLOZ, 1988).

Using pressure gradient measured by the ruby fluorescence method and thickness measured at room temperature, KIMURA *et al.* (1982) and MEADE and

---

[1] Mineral Physics Laboratory, Department of Geological Sciences, Snee Hall, Cornell University, Ithaca, New York 14853, U.S.A.

JEANLOZ (1988) estimated the flow stress hence the yield strength of specimens at high pressure and room temperature. Because one of the principal objectives of our research is to study the rheologic properties of mantle minerals under conditions of the earth's mantle, similar measurements need to be made at high temperatures as well as high pressures. However, it is difficult to accurately determine the thickness of a sample at high temperature in a DAC. SUNG et al. (1977) used pressure

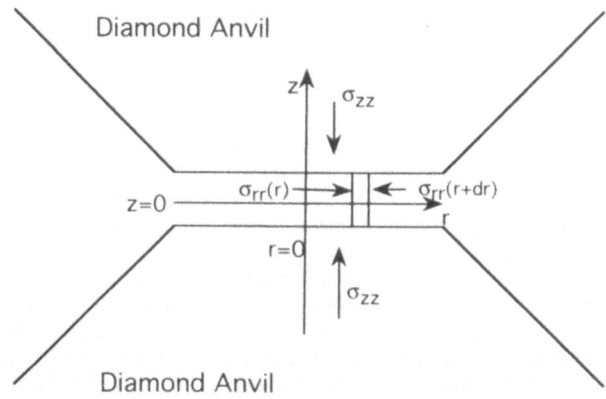

Figure 1

Cross section of a Diamond Anvil Cell (DAC) sample chamber. A cylindrical coordinate system is used, although the azimuth, θ, is not shown in the diagram. σ_{zz} and σ_{rr} are axial stress and radial stress, respectively. Ideally, a concentric symmetry in the sample chamber can be assumed, so that σ_{zz} and σ_{rr} can be treated as functions only of radius.

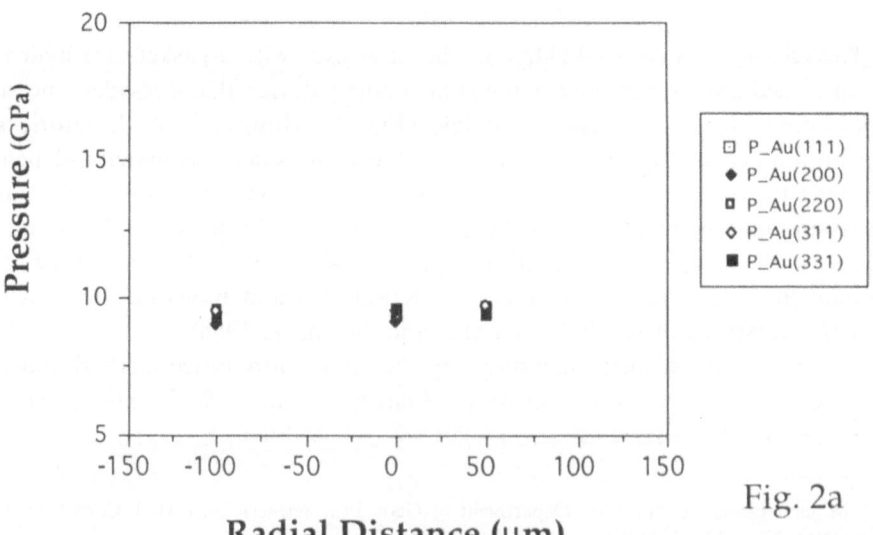

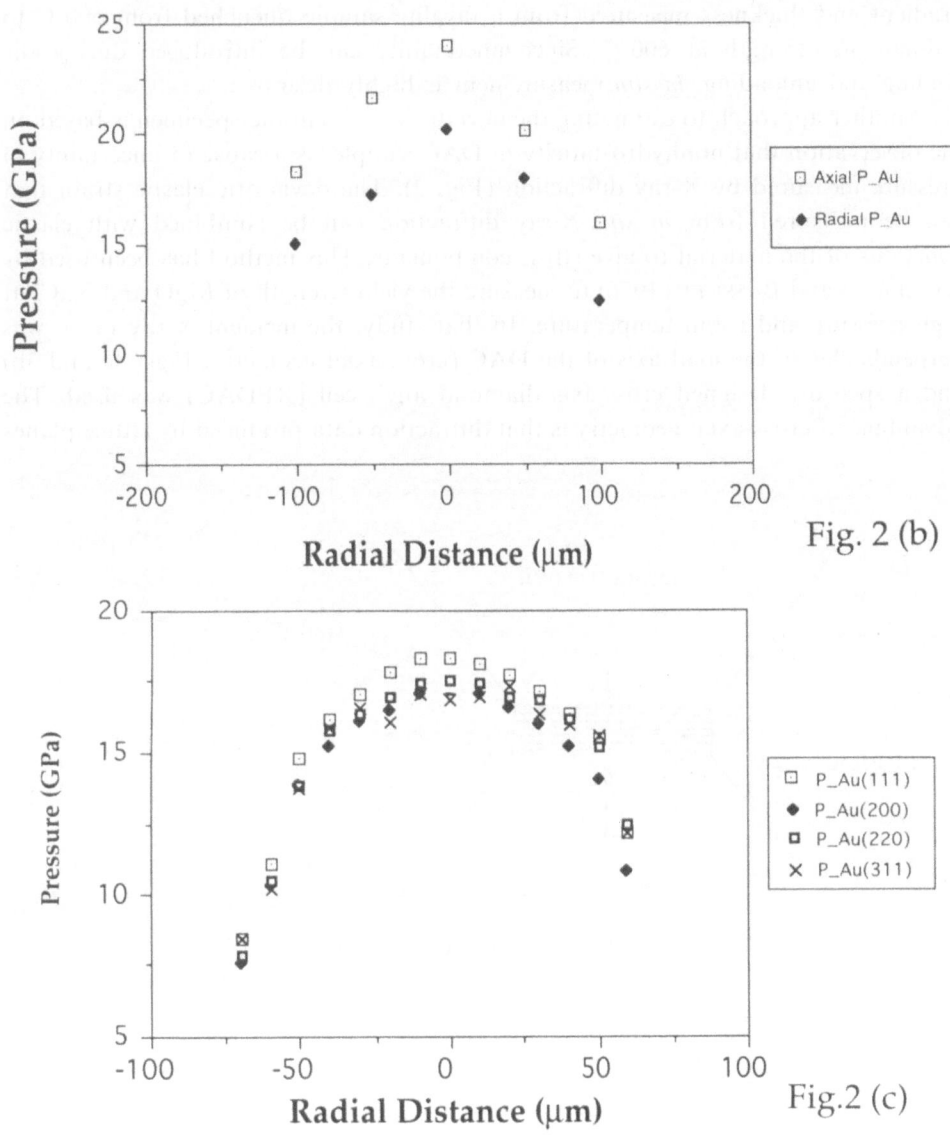

Fig. 2 (b)

Fig.2 (c)

Figure 2
Pressure distribution in fayalite samples in DACs measured by diffraction data of gold in various environments: a) quasihydrostatic, in pressure medium N_2; b) and c) nonhydrostatic, without gasket and pressure medium. Pressure measurements based on the volume changes of an internal calibrant such as gold are notoriously unreliable when the calibrant is subjected to nonhydrostatic stress, as compared to measurements under hydrostatic pressure. The source of these uncertainties can be observed in two situations. First, as shown in Figure 2b, the elastic strain of lattice plane (111) of gold measured at two orientations in cross-axial geometry (see Figure 3) is very different, therefore, the pressures calculated from the cube of linear strains of the different orientations show a large discrepancy. The other uncertainty, as shown in Figure 2c, is due to the anisotropic properties of gold. Pressures are calculated from the cube of linear strains measured for differential lattice planes. These discrepancies of elastic strain can be used as the information source of the state of nonhydrostatic stress.

gradient and thickness measured from a fayalite sample quenched from 600°C to estimate its strength at 600°C. Since uncertainty can be introduced during the cooling and unloading, *in situ* measurement is highly desirable.

Another approach to estimating the deviatoric stress in the specimen is based on the observation that nonhydrostaticity in DAC samples is a cause of uncertainty in pressure measured by X-ray diffraction (Fig. 2). The deviatoric elastic strain that can be measured from *in situ* X-ray diffraction can be combined with elastic constants of the material to give stress components. This method has been used by KINSLAND and BASSETT (1976) to measure the yield strength of MgO and NaCl at high pressure and room temperature. In that study, the incident X-ray beam was perpendicular to the load axis of the DAC (cross-axial geometry, Figs. 3a and 3b) and a specially designed cross-axis diamond anvil cell (XPDAC) was used. The advantage of cross-axial geometry is that diffraction data produced by lattice planes

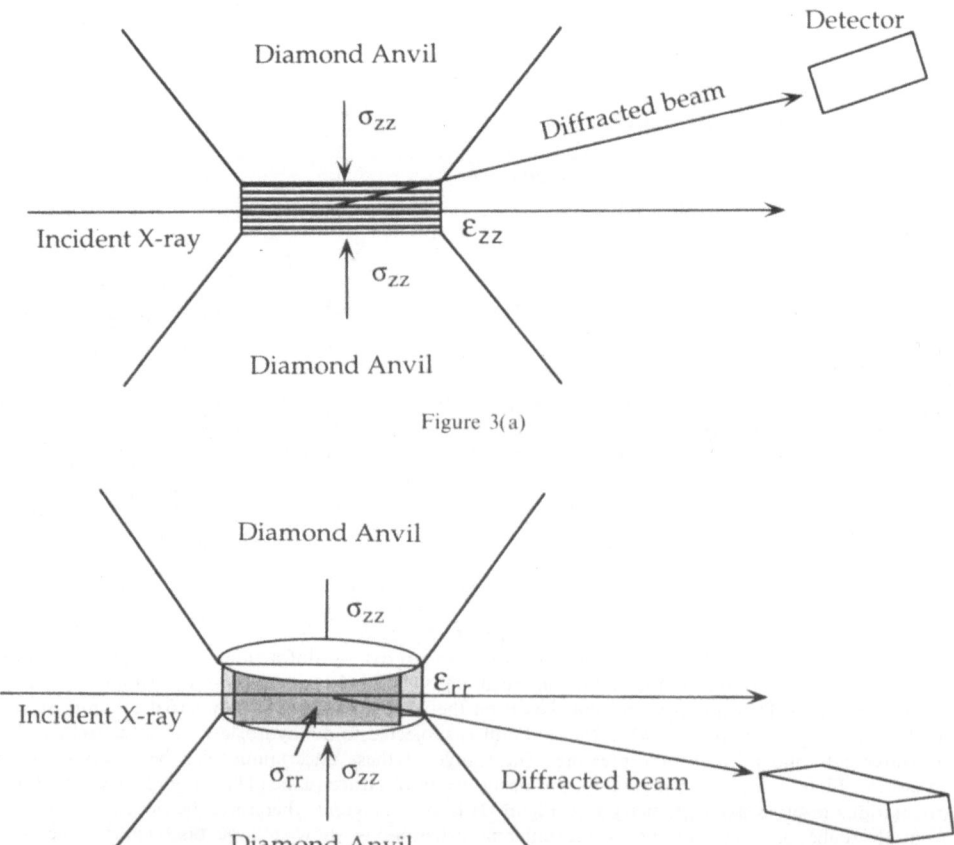

Figure 3(a)

Figure 3(b)

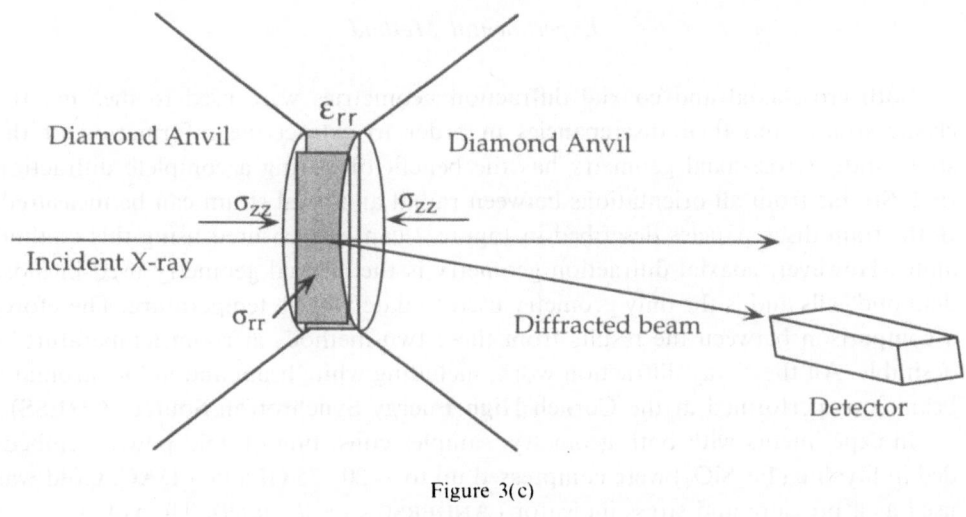

Figure 3(c)

Figure 3

a) and b) Cross-axial X-ray diffraction geometry, c) coaxial X-ray diffraction geometry in DAC experiments and their relationship with the stress and strain orientations. ε_{zz} and ε_{rr} are elastic strains of lattice planes perpendicular to the loading and radial axes, respectively. Note that in the most frequently used coaxial diffraction geometry for the powder sample, only strain close to ε_{rr} is measured. On the other hand, the cross-axial set-up shown in Figures 3a and 3b can be combined if an area detector or a photographic film is used, and both ε_{rr} and ε_{zz} can be measured.

having normals perpendicular and parallel to the loading axis can be collected simultaneously, therefore, strain of both orientations can be measured. Since a wide opening on the side of the DAC is required for X-ray clearance in cross-axial geometry (KINSLAND and BASSETT, 1976), this design places severe constraints on any heating method for high-temperature studies.

WEIDNER et al. (1992) has demonstrated the use of energy dispersive X-ray diffraction (EDXRD) to measure the stress in a large volume pressure device. This approach can also be applied in the DAC. A coaxial geometry is usually used for a DAC with EDXRD set-up. With this geometry, a high-temperature DAC can be used without further modification. However, when combining the EDXRD method and coaxial geometry, only those crystalline planes with their plane normals perpendicular to the load axis can be measured (Fig. 3c). Since most material is elastically anisotropic, different lattice planes respond differently to deviatoric stress (Fig. 2c). Therefore, strain measured from two lattice planes can be used to derive stress components. This paper describes some preliminary results of a comparative study of stress measurement in a DAC using these two geometries at room temperature.

Experimental Method

Both cross-axial and coaxial diffraction geometries were used to measure the elastic strains and their discrepancies in order to extract the information of the stress state. Cross-axial geometry has the benefit of having a complete diffraction ring. Strains from all orientations between radial and axial strain can be measured. Both strain discrepancies described in Figure 2 can be measured using this method alone. However, coaxial diffraction geometry is the normal geometry used in most diamond cells and is the only geometry used to date at high temperature. Therefore, a comparison between the results from these two methods at room temperature is desirable. All the X-ray diffraction work, including white beam and monochromatic beam, was performed at the Cornell High Energy Synchrotron Source (CHESS).

In experiments with both geometry, samples consisting of gold powder embedded in fayalite (Fe_2SiO_4) were compressed up to ~ 20–25 GPa in a DAC. Gold was used as a pressure and stress indicator (ANDERSON et al., 1989). The volume ratio between fayalite and gold was approximately 10:1.

Experiments with Coaxial Diffraction Geometry

In experiments with coaxial geometry, the EDXRD method was used. A white incident synchrotron radiation beam at CHESS B-1 beam-line was collimated to a size of 15 μm. The diffraction beam was collected at $2\theta = 10^{\circ}$ with a Ge solid-state detector. With this small angle, the lattice planes corresponding to the collected diffraction signal are nearly perpendicular to the radial stress, thus their strain can be treated as radial strain ε_{rr} (Fig. 3c). A piston type of DAC was used with a pair of 1/8 carat diamonds. With the small beam size, several diffraction patterns were collected across the sample of 250 μm, thus the strain values were measured and the stress distribution was calculated for the sample.

Experiments with Cross-axial Diffraction Geometry

In the cross-axial diffraction geometry, for collecting the complete diffraction ring, a monochromatic synchrotron radiation beam is used with an image plate area detector (Fig. 4). These experiments were done at CHESS A-2 wiggler beam line. The monochromatic beam from a Si(111) double-crystal monochromator was used with a beam energy of 17 KeV (~ 0.73 Å). The incident X-rays were collimated by a pinhole 40 μm in diameter. The image plate area detector system was used to collect diffraction data because of its exceptional precision compared with photographic film. The XPDAC designed for this geometry by KINSLAND and BASSETT (1976) was used with diamond anvil faces ~ 450 μm in diameter. High-pressure diffraction patterns of different parts of the sample were taken by moving the DAC relative to the incident X-ray beam. An external standard of NaCl placed downstream of the sample was used for camera length calibration. A fragment of

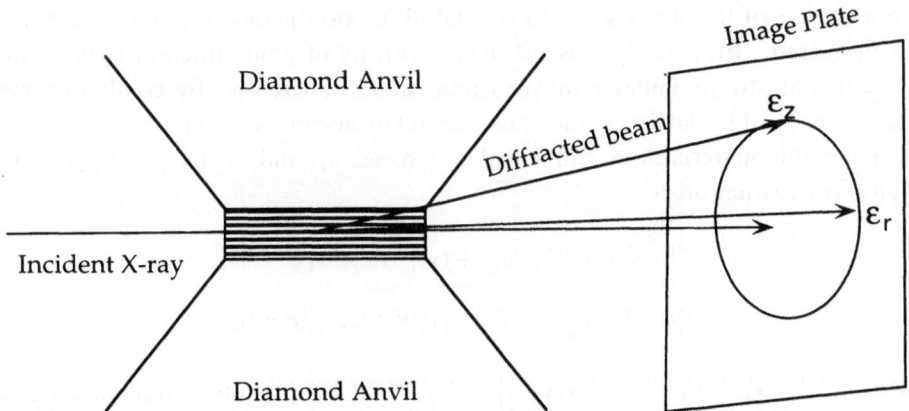

Figure 4
Schematic diagram showing experimental setup in cross-axial geometry with image plate detector.

polycrystalline gold was used for calibration. The variation in camera length determined by these two patterns was within 0.01 mm, even though the diamond cell was removed and replaced between patterns. The gold pattern was also used as a standard for determining line width for making corrections for line broadening due to pressure gradient.

The Image Plate System

Photons of the transmitted and diffracted X-rays in cross-axial geometry were recorded on a phosphor storage plate which was then read by a laser image scanner. The scanner produces an image of 2500×2048 pixels with a pixel size of $100 \times 100 \ \mu m$. The dynamic range that can be detected in our setup is ~ 2000, which is much greater than for photographic film. The exposure time needed for the image plate is $\sim 20\%$ of the time needed for photographic film with similar integrated intensity. The dimensional stability of image plate also eliminates problems that exist in photographic film (KINSLAND and BASSETT, 1976). The only data correction necessary is for the vertical tilting due to the plate holder. The diffraction pattern was measured and intensity for each individual pixel was read out on the display program IPVIEW.

Data Analysis

From the pressure distribution that we measured in samples in the DAC, we know that the assumption of an axially symmetric sample space is valid. Therefore cylindrical coordinates for the sample chamber can be used. The variables σ_{rr}, $\sigma_{\theta\theta}$ and σ_{zz} are radial stress, tangential stress and axial stress, respectively. Due to axial symmetry, stress σ_{rr} and $\sigma_{\theta\theta}$ are azimuth independent at the same radial distance, r; and σ_{rr} is equal to $\sigma_{\theta\theta}$ at the center of the chamber ($r = 0$). We also assume that $\sigma_{rr} \sim \sigma_{\theta\theta}$ when $r > 0$, so that Figure 1 is a two-dimensional representation of the stress state. In coaxial diffraction geometry, only radial strain ε_{rr} is measured. However, because of the anisotropy of gold, different lattice planes have different strains under nonhydrostatic stress. Therefore, by combining measured strain of two lattice planes and the relationship between stress and strain tensor in cubic material, one can calculate stresses σ_{rr} and $\sigma_{::}$ by solving a pair of simultaneous equations:

$$\varepsilon_{rr}^{(h_1k_1l_1)} = \bar{s}_{12}^{(h_1k_1l_1)}\sigma_{::} + (s_{11}^{(h_1k_1l_1)} + \bar{s}_{13}^{(h_1k_1l_1)})\sigma_{rr}$$

$$\varepsilon_{rr}^{(h_2k_2l_2)} = \bar{s}_{12}^{(h_2k_2l_2)}\sigma_{::} + (s_{11}^{(h_2k_2l_2)} + \bar{s}_{13}^{(h_2k_2l_2)})\sigma_{rr}$$

where $s_{11}^{(h_1k_1l_1)}$, $\bar{s}_{12}^{(h_1k_1l_1)}$, $\bar{s}_{13}^{(h_1k_1l_1)}$ and $s_{11}^{(h_2k_2l_2)}$, $\bar{s}_{12}^{(h_2k_2l_2)}$, $\bar{s}_{13}^{(h_2k_2l_2)}$ are compliance tensors for two lattice planes referred to the sample chamber coordinate system. Note that $\bar{s}_{12}^{(h_1k_1l_1)}$, $\bar{s}_{13}^{(h_1k_1l_1)}$ and $\bar{s}_{12}^{(h_2k_2l_2)}$, $\bar{s}_{13}^{(h_2k_2l_2)}$ are averaged for all possible crystal orientations in a polycrystalline sample with lattice plane normals $(h_1k_1l_1)$ and $(h_2k_2l_2)$ perpendicular to the loading axis z and nearly perpendicular to the incident X-ray beam (i.e., nearly parallel to the diffraction vector). Due to the averaging procedure for a polycrystalline sample, $\bar{s}_{12}$ and $\bar{s}_{13}$ turn out to be numerically the same. Elastic constants of gold were taken from HIKI and GRANTO (1966) and were linearly extrapolated to high pressure.

The diffraction data from cross-axial geometry experiments recorded on the image plate was analyzed on a DEC station 5000 computer. The diameter of diffraction ring was directly read out from the digital image. In this geometry, since X-rays are penetrating the sample from one side of the anvil face to the other side, a diffraction line obtained at high pressure is therefore a composition of diffraction from material under a range of pressures. The position of the center of high pressure lines was determined by subtracting half the natural line width of gold obtained at ambient pressure from the outer (lower d-spacing) edge of the composite diffraction line. The vertical tilting of the plate calculated from the shifting center of the diffraction rings is $\sim 0.5-1^\circ$, and it is corrected for each individual plate. Values for ε_{rr} and $\varepsilon_{::}$ of gold were measured at two diameters of the diffraction ring. Using the strain data and the compliance tensor for a single lattice plane, one can solve $\sigma_{::}$ and σ_{rr} from a pair of equations:

$$\varepsilon_{::}^{(hkl)} = s_{11}^{(hkl)}\sigma_{::} + (\bar{s}_{12}^{(hkl)} + \bar{s}_{13}^{(hkl)})\sigma_{rr}$$

$$\varepsilon_{rr}^{(hkl)} = \bar{s}_{12}^{(hkl)}\sigma_{::} + (s_{12}^{(hkl)} + \bar{s}_{13}^{(hkl)})\sigma_{rr}$$

where $s_{11}^{(hkl)}$, $\bar{s}_{12}^{(hkl)}$ and $\bar{s}_{13}^{(hkl)}$ are compliance tensors of the lattice plane (hkl) referred to the sample chamber coordinate system. The averaging procedure is the same as described above.

Results and Discussion

Distribution of deviatoric stress $(\sigma_{zz} - \sigma_{rr})$ measured from gold embedded in fayalite by two experimental geometries at comparable pressures is shown in Figure 5. The pressure distribution for these measurements is shown in Figure 2. Reflection (111) was used for cross-axial geometry because of its intensity and freedom from interference from other peaks. It also has the smallest 2θ diffraction angle among the lattice planes of gold, so that the stresses calculated from the elastic strain are the closest approximation to σ_{rr} and σ_{zz}. In coaxial geometry, elastic strain of gold (111), (311) pair was used. Both geometries gave a deviatoric stress of ~ 0.7 GPa at the center of the fayalite sample. This value gradually decreased toward the edge. Data taken at the center of the sample at different pressures using coaxial geometry indicated that deviatoric stress increases from ~ 0.1 GPa at a pressure of ~ 1 GPa to ~ 0.7 GPa at a pressure of ~ 10 GPa. The deviatoric stress in gold seems to increase with the pressure up to ~ 10 GPa and to stop increasing above ~ 10 GPa. Therefore, this small stress may represent yield strength of gold and only a minimum value for that of fayalite, and this is in agreement with Weidner (personal communication) and the yield strength of fayalite measured by SUNG et al. (1977).

Comparing Figures 5a and 5b, we also noticed that the deviatoric stress drops more rapidly along the radius in the coaxial geometry run (Fig. 5b). This could be due to the smaller anvil face used in this run. There are other differences between the two methods that should also be considered. First, the 2θ angle used in the two geometries is different. It is fixed at $10°$ for all lattice planes in coaxial geometry while in cross-axial geometry it is $17.83°$ for the (111) plane of gold. Since stresses σ_{zz} and σ_{rr} are approximately equal to the principal stresses close to the center, they are simply the stresses we measured, divided by $\cos \theta$. Calculated differences between the two geometries, caused by the different 2θ values, are less than one percent. The difference is even smaller if differential stresses $(\sigma_{zz} - \sigma_{rr})$ are compared. Secondly, the natural half line width used to determine the high-pressure line position in the cross-axial geometry was based on the gold diffraction pattern collected at ambient pressure and may underestimate the real half line width at high pressure since high stress may cause line broadening. This results in an overestimate of the elastic strain and therefore of the stress because too small a correction is made. Since this effect is similar at all points along a diffraction ring, it is expected to be negligible in differential stress $(\sigma_{zz} - \sigma_{rr})$.

In the strict sense, the stress components σ_{zz} and σ_{rr} measured in the sample can represent the principal (maximum and minimum) stresses only when the strains are measured at the center of the sample. Away from the center, the stress ellipsoid may be at a different orientation and therefore the principal stresses may be at different orientations from the σ_{zz} and σ_{rr} which are always perpendicular and parallel to the

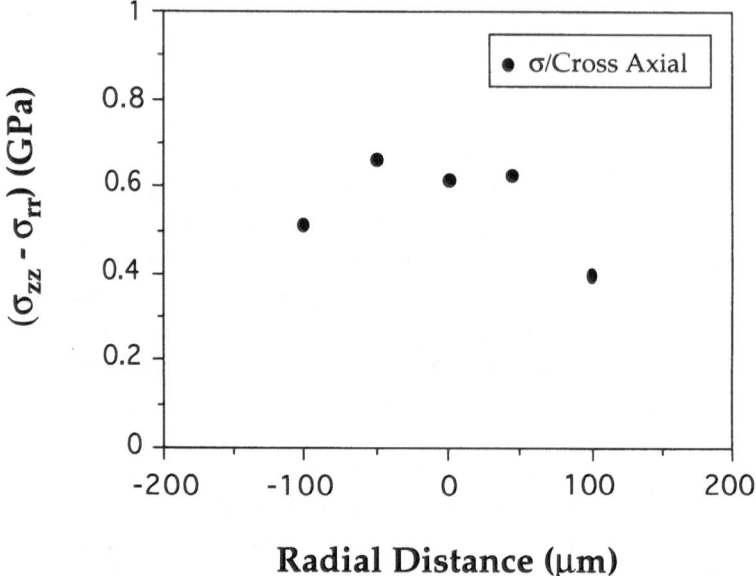

Figure 5(a)

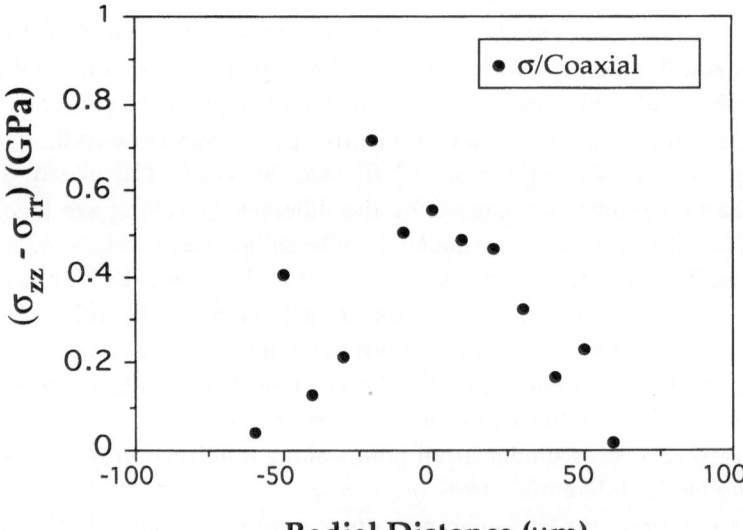

Figure 5(b)

Figure 5

Distribution of deviatoric stress $(\sigma_{zz} - \sigma_{rr})$ measured from gold embedded in fayalite by a) cross-axial geometry and b) coaxial geometry corresponding to pressure distributions shown in Figures 2b and 2c.

loading axis in DAC, respectively (MEADE and JEANLOZ, 1988). For this reason, the stress measured for the part of the sample away from the center can only be treated as a lower bound of the deviatoric stress at those positions. From our measurements, we think that at least within a quarter of the anvil radius from the center, the deviatoric stress is comparable to and possibly even larger than that at the center. This has a profound effect on shear-sensitive processes at high pressure (e.g. WU et al., 1993).

Acknowledgments

We thank Don Weidner of SUNY Stony Brook for providing computer programs and for his encouragement. Critical reviews from D. Heinz and Y. Meng are appreciated. We also thank staff at CHESS for their valuable help. This research is supported by NSF grant EAR-9206004.

REFERENCES

ANDERSON, O. L., ISAAK, D. G., and YAMAMOTO, S. (1989), *Anharmonicity and Equation State for Gold*, J. Appl. Phys. *65*, 1534.

HIKI, Y., and GRANATO, A. V. (1967), *Anharmonicity in Noble Metals; Higher Order Elastic Constants*, Phys. Rev. *144*, 411.

KIMURA, H., AST, D. G., and BASSETT, W. A. (1982), *Deformation of Amorphous $Fe_{40}Ni_{40}P_{14}B_6$ under Compression to 250 kbar*, J. Appl. Phys. *53*, 3523–3528.

KINSLAND, G. L., and BASSETT, W. A. (1976), *Strength of MgO and NaCl Polycrystals to Confining Pressure of 250 kbar at 25°C*, J. Appl. Phys. *48*, 978–985.

MEADE, C., and JEANLOZ, R. (1988), *Yield Strength of MgO to 40 GPa*, J. Geophys. Res. *93*, 3261–3269.

MENG, Y., WEIDNER, D. J., and FEI, Y., *Deviatoric Stress in a Quasi-hydrostatic Diamond Anvil Cell: Effect on the Volume Based Pressure Calibration*, Geophys. Res. Lett. *20*, 1147–1150.

SUNG, C.-M., GOETZE, C., and MAO, H.-K. (1977), *Pressure Distribution in the Diamond Anvil Press and the Shear Strength of Fayalite*, Rev. Sci. Inst. *48*, 1386–1391.

WEIDNER, D., VAUGHAN, M. T., KO, J., WANG, Y., LIU, X., YEGAANEH-HAERI, A., PACALO, R. E., and ZHAO, Y., *Characterization of stress, pressure and temperature in SAM 85, a DIA type high pressure apparatus*. In *High-pressure Research: Application to Earth and Planetary Science* (eds. Syono, Y., and Manghnani, M. H.) (TERRAPUB, Tokyo/AGU, Washington, D.C, 1992) pp. 13–17.

WU, T.-C., BASSETT, W. A., BURNLEY, P. C., and WEATHERS, M. S. (1993), *Shear-promoted Phase Transition in Mg_2SiO_4 and Fe_2SiO_4 and the Mechanism of Deep Earthquakes*, J. Geophys. Res. *98*, 19767–19776.

(Received March 10, 1993, revised July 23/November 1, 1993, accepted November 15, 1993)

Rheological Investigations

PAGEOPH, Vol. 141, No. 2/3/4 (1993)

0033-4553/93/040523-21$1.50 + 0.20/0

Improvements to Griggs-type Apparatus for Mechanical Testing at High Pressures and Temperatures

TRACY N. TINGLE,[1,3] HARRY W. GREEN, II,[1,4] THOMAS E. YOUNG[1] and
TED A. KOCZYNSKI[2]

Abstract —New and improved techniques and apparatus for testing the mechanical properties of materials at high pressures and temperatures are described. These include an improved Griggs-type deformation apparatus designed to operate to 5 GPa and associated servo-controlled hydraulic drive and electronics, the design of hydrostatic (molten alkali halide mixtures) pressure assemblies to measure flow stresses as low as a few MPa, the characterization of temperature gradients and friction in such assemblies, measurement of the melting curve of an alkali halide mixture used as a confining pressure medium, and the measurement of acoustic emissions.

Key words: Pressure, temperature, stress, measurement, acoustic emissions, deformation, techniques.

Introduction

In recent years, the field of high-pressure experimental rock deformation has experienced several advances due to improvements in techniques and equipment. Some of the important advances include (i) development of hydrostatic pressure assemblies utilizing molten alkali halide confining media (GREEN and BORCH, 1989, 1990; GLEASON and TULLIS, 1992, 1993), (ii) measurement of low flow stresses (a few MPa) in such assemblies (BORCH and GREEN, 1989; MEAGHER *et al.*, 1992), (iii) monitoring of acoustic emissions (GREEN *et al.*, 1992; TINGLE *et al.*, 1993), and (iv) quasi-controlled deformation of materials at very high pressures in multianvil apparatus (GREEN *et al.*, 1990, 1992; GREEN and WALKER, 1992; BUSSOD *et al.*, 1992; WEIDNER *et al.*, 1992) and in diamond anvil cells (MEADE and JEANLOZ, 1990, 1991; WU *et al.*, 1993).

[1] Department of Geology, University of California, Davis, CA 95616, U.S.A.

[2] Rock Mechanics, Lamont-Doherty Earth Observatory, Palisades, NY 10984, U.S.A.

[3] Now at Department of Geological and Environmental Sciences, Stanford University, Stanford, CA 94305, U.S.A.

[4] Now at Department of Earth Sciences and Institute of Geophysics and Planetary Physics, University of California, Riverside, CA 92521-0412, U.S.A.

This paper describes new techniques and apparatus for axial deformation at strain rates of 10^{-2} to 10^{-7} s^{-1} in a Griggs-type apparatus, improvements in the design of such apparatus, and the measurement of acoustic emissions.

Experimental Apparatus

5 GPa Deformation Apparatus

Figure 1 depicts the new Griggs-type deformation apparatus. The load frame was designed to sustain confining pressures to 5 GPa, the practical upper limit for deformation imposed by the room-temperature compressive strength of the tungsten carbide (WC) deformation pistons. The maximum pressure at which specimens

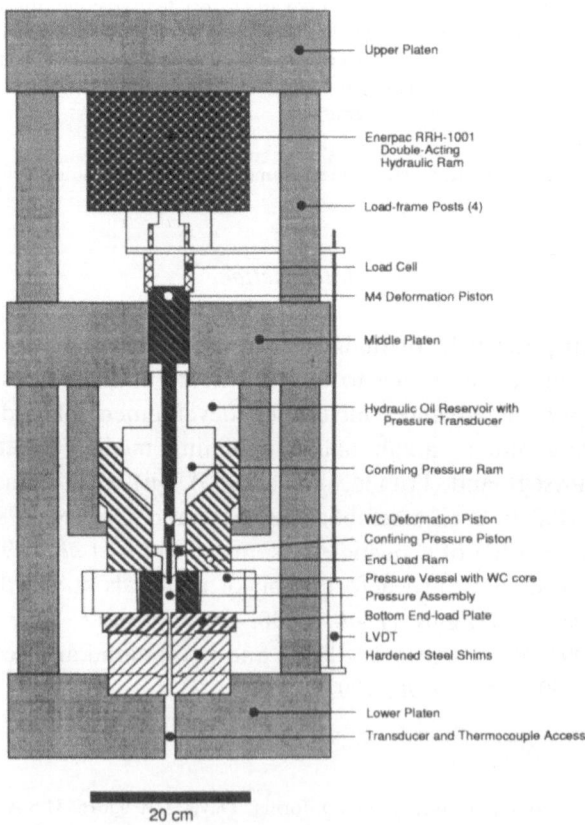

20 cm

Figure 1

Schematic drawing of the 5 GPa load frame, confining pressure ram, hydraulic drive and standard pressure vessel. Note the axial hole in the lower platen and base plate used to introduce the acoustic transducer and additional thermocouples.

have been deformed in the new apparatus is 2800 MPa (TNT and HWG, unpublished results). Details of the new apparatus will be published after the end-load modifications for attainment of maximum pressure are completed (HWG and I. C. Getting, manuscript in preparation).

Data Collection and Analysis

Measurement of all experimental variables (pressure, load or stress, temperature, displacement, and acoustic emissions) is controlled by computer. Confining pressure is calculated from the oil pressure of the confining pressure hydraulic ram (Figure 1), as measured by a pressure transducer (Precise Sensors Model 6550). The axial force applied to the specimen is measured by a load cell between the deformation ram and pistons (Houston Scientific Model 3500; Figure 1). Temperature is measured by two type-B thermocouples adjacent to the specimen (Figure 2) and controlled by a Eurotherm Model 818P temperature controller interfaced (via an RS232 serial communications port) to the computer. This allows the temperature and power output of the controller to be monitored and stored on disk (all temperature controller functions are handled by the computer via the interface). Furnace power (1000–1600 W) is supplied through a 3 kVA AC power transformer (115:8) by a silicon-controlled rectifier (SCR) driven by the temperature controller. Displacement is measured by a linear voltage displacement transducer (LVDT; Schaevitz Model DC-E500).

Output voltages of the various transducers and thermocouples are measured by Omega OM3 series signal conditioners interfaced to analog-digital data acquisition and control boards internal to the computer (IBM-PS2 with IBM Data Acquisition and Control System). The resolution imposed by the 12-bit analog-digital converters corresponds to 1 MPa for the measured confining pressure and piston stress, 1 K for temperature, and 0.3 μm for displacement. Data acquisition is performed by FORTRAN programs written in our laboratory, and data are stored on disk for subsequent analysis. The data are subsequently transferred to Macintosh computers and stress-strain curves are calculated (described below) utilizing Kaleidagraph®, a software graphics and analysis package.

Pressure Assemblies

The high-pressure sample assemblies are designed to accommodate a diversity of experimental conditions (Figure 2). Changes can be made to the size of the specimen, the position of the specimen within the furnace, the position of the two thermocouples, the thicknesses of the furnace tube and ceramic sheaths, the length and composition of the metal sleeve (used to decrease the specimen temperature gradient), and the compositions of the alkali halide mixtures providing the confining pressure. Central holes in the lower platen and base plates provide access to the pressure assembly, so that a piezoelectric transducer may be placed in close proximity to the base of the specimen to record acoustic emissions, or additional thermocouples may be positioned inside the volume normally occupied by the specimen (Figures 1, 2).

Figure 2a shows the standard configuration for moderate temperature (up to 1450 K) experiments. The graphite furnace tube is sandwiched between two cylin-

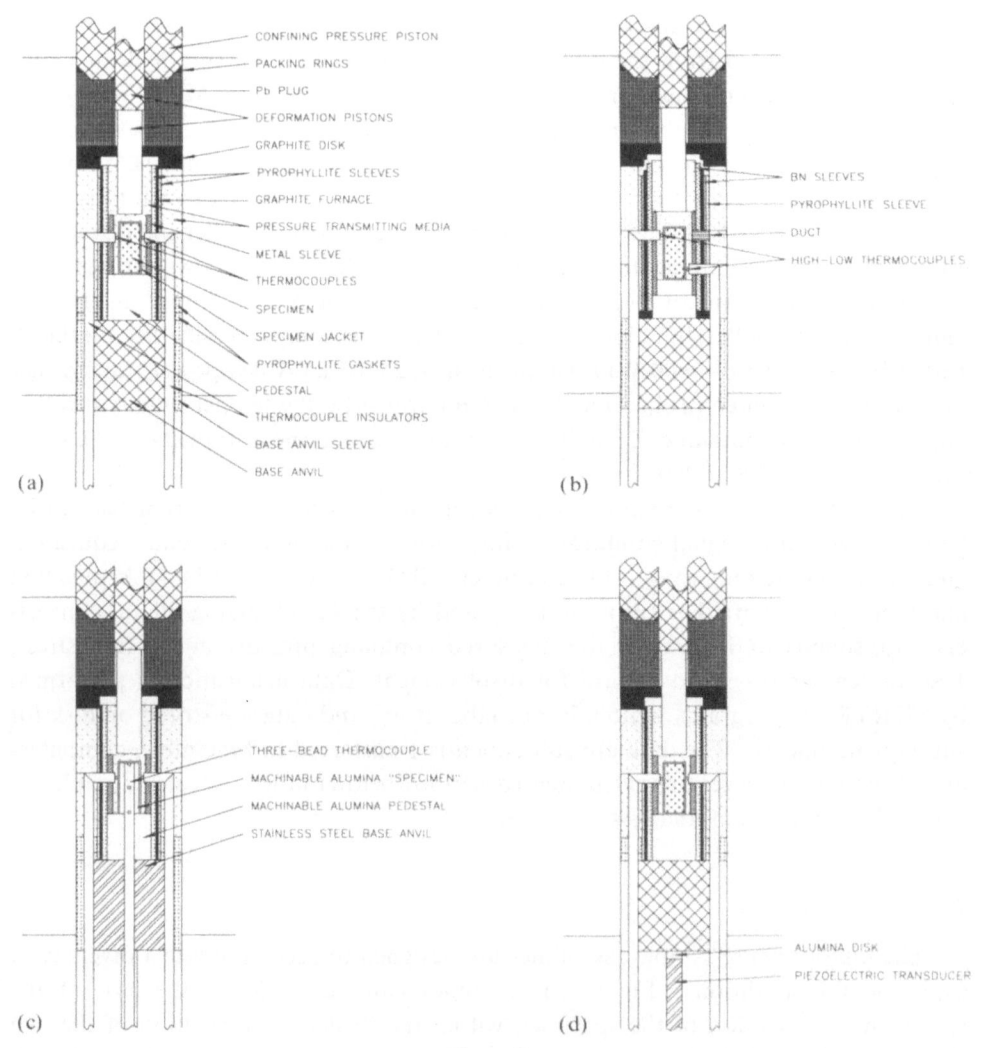

Figure 2
Schematic drawings of pressure assemblies used in the high-pressure deformation experiments. (a) Standard high-pressure assembly. (b) Assembly modified for temperatures above 1450 K. (c) Assembly for measuring temperature distribution. (d) Assembly used for acoustic emission experiments. Components and materials are labeled and indicated. For scale, the diameter of the assemblies is 1.75 cm.

ders of "soft-fired" pyrophyllite (Alsimag Technical Ceramics Machinable Grade A Lava, a natural material composed mostly of pyrophyllite, fired at 1223 K for 1 hr). The porous soft-fired pyrophyllite sleeves are able to compact during pressurization of the assembly, and they protect the furnace tube from damage due to traction imposed on the furnace assembly by the compacting outer salt sleeves. Pyrophyllite reacts to an aluminosilicate phase (mullite, andalusite, or kyanite depending on pressure) + quartz above 700 K (KERRICK, 1968 and references therein). At low pressures, the fired pyrophyllite probably melts incongruently at 1450–1500 K (DE VRIES, 1964); at 2520 MPa, HASKELL and DEVRIES (1964) report the kyanite + quartz eutectic to be 1450 K. Such melting can damage the graphite furnace, necessitating the use of a more refractory material (such as boron nitride or alumina) for long-term experiments at temperatures above 1500 K.

A metal sleeve fabricated from Inconel 600, Ni, or Mo is placed inside the inner ceramic sleeve to provide a more uniform temperature distribution near the specimen (Figure 2). Inconel and Ni have 1-bar melting temperatures of 1627–1686 and 1728 K, respectively (CRC Handbook of Materials Science; CRC Handbook of Chemistry and Physics). However, at moderate pressures (<2 GPa) and temperatures above 1450 K, both metals dissolve extensively in the molten salts, which leads to alloying and embrittlement of the Pt specimen jacket and metal precipitation in the cooler regions of the molten volume. Mo is used above 1450 K because of its high melting temperature (2883 K at 1 bar, CRC Handbook of Chemistry and Physics) and resistance to reaction and dissolution in the molten salts. However, its cost and the difficulty of machining limit the desirability of Mo for use in all experiments. GLEASON and TULLIS (1993) employed a graphite sleeve as a thermal conductor in their 2.54 cm high-pressure molten salt assembly.

Two type-B ($Pt_{94}Rh_6$-$Pt_{70}Rh_{30}$) thermocouples in ceramic thermocouple sheaths are placed in direct contact with the Pt specimen jacket. The thermocouples are inserted through holes in the ceramic, graphite, and metal sleeves. Where the thermocouples penetrate the sleeves (Figure 2), sintered high-purity alumina (McDanel's >99.5% Al_2O_3) is used as the sheath material to minimize chemical contamination of the thermocouple bead. Outside the furnace in the lower-temperature region, mullite is used because it compacts and collapses around the thermocouple wires and prevents extrusion of the molten salt confining pressure medium. An unfired pyrophyllite gasket is placed at the top of the "hard-fired" (1423 K for 1 hr) pyrophyllite base-anvil sleeve to prevent the sheaths from being extruded. The thermocouple wires are glued (Zircar[R] alumina cement) into the sheaths, and the sheath is glued into the base anvil sleeve to further prevent extrusion and "blowouts." The tips of the thermocouples also are coated with Zircar[R].

The Pt-encapsulated specimen is cast into the metal sleeve with the desired alkali halide confining-pressure medium using a micro-oxy-acetylene torch and special jigs for aligning the specimen. The cast specimen and metal sleeve rest on a sintered alumina pedestal, which in turn rests on a WC (Federal grade FC-3) base anvil

(Figure 2). The base anvil serves as an electrical lead for the graphite furnace and is sleeved in hard-fired pyrophyllite to insulate the anvil from the pressure vessel, which serves as the other lead. A sintered alumina deformation piston is situated above the specimen. This piston is pushed by a movable WC piston that enters the high-pressure volume through an axial hole in the confining pressure piston (Figure 1). The WC piston enters through an Al-bronze 600 packing ring and a Pb plug at the top of the assembly. The alumina piston moves through the Pb plug and graphite disk, which are intended to distribute the force from the confining pressure piston evenly to the various components of the pressure assembly and the inner salt (which may be solid or molten around the specimen, depending upon the requirements of the experiment). The graphite disk also serves as the electrical contact between the furnace tube and the pressure vessel.

The salt outside the furnace is KCl or NaCl, hydrostatically cold-pressed to 200 MPa in an evacuated rubber bladder placed in a large-bore (5 cm) pressure vessel filled with hydraulic fluid, and machined to final dimensions. The composition of the inner salt is NaCl, the 1-bar NaCl-KCl eutectic, or the lowest-melting 1-bar eutectic in the K-Na-Ca-Ba-Cl system (referred to hereafter as E4; LEVIN *et al.*, 1969). The inner salt is cast around the specimen in the metal sleeve as described above. The same salt mixture is cast around the alumina piston above the specimen, or a separate piece is isostatically cold-pressed to final dimensions in a special hardened steel jig. For most experiments, the composition of the inner salt is chosen to produce melting around the specimen and extensive, but not complete, melting above it. Thus, for relatively low temperature experiments (800–1400 K), the E4 mixture is used, whereas for temperatures above 1400 K, NaCl or the KCl-NaCl eutectic mixture is used. For high-stress experiments (>600 MPa), a solid salt must be used to avoid axial splitting of the alumina piston. The outer salt (NaCl or KCl) is chosen so that little, if any, melt is produced. KCl is generally used below 1500 MPa, and NaCl is used above 1500 MPa; the KCl I–II phase transition at ~1800 MPa (CLARK, 1959) may cause some anomalous compaction problems that lead to furnace damage.

Figure 2b shows some modifications that have been made to increase high-temperature furnace life and to improve knowledge of the temperature profile during deformation. The material of the sleeves adjacent to the graphite furnace is boron nitride (Union Carbide HBR-grade BN), which has a much higher melting temperature than pyrophyllite. Additionally, BN alters the chemical environment inside the furnace such that the Mo sleeve does not react or dissolve in the molten salt. A pyrophyllite sleeve is used outside of the outer BN sleeve to provide mechanical protection to the brittle and relatively weak BN during pressurization. The BN, in turn, protects the graphite furnace from pyrophyllite melting at high temperatures. To further protect the sleeves and furnace from damage during pressurization, a graphite ring is placed at the base of the sleeves for "cushioning."

For most experiments, the thermocouples are placed in a high-low arrangement to deduce the specimen temperature profile (the true temperature distribution is a complicated function of the exact placement and length of all the components and their thermal diffusivities, the amount of power being applied to the furnace, and numerous other variables). The temperature profile can change with time during an experiment due to movement of the deformation piston and, at very high temperatures, due to convection of the molten salt and degradation of the furnace with time at temperature. Experiments performed with the purpose of determining the temperature distribution are described in a later section.

A ceramic duct (sintered alumina four-hole thermocouple tube) is placed opposite the upper thermocouple to provide a conduit for equalization of pressure between the inner and outer salt volumes during advancement and retraction of the deformation piston. It is intended to avoid over- or underpressurizing the inner volume, which can lead to fracture of the furnace and ceramic sleeves.

Servo-controlled Hydraulic Drive

A servo-controlled hydraulic drive, similar to that developed by SCHOLZ and KOCZYNSKI (1979), was installed on the new apparatus prior to performing the acoustic emissions tests (described below) in order to eliminate mechanical noise inherent in a gear train driven by a stepping-motor. However, there are other advantages to such a system, including ease of operation and precise control of the deformation piston based on either the measured displacement or load.

The deformation pistons are advanced and retracted by a double-acting hydraulic cylinder (Enerpac RRH-1001) controlled by an electro-hydraulic servo-controller monitoring either the LVDT (displacement feedback) or the load cell (load feedback) to close the servo loop. The double-acting hydraulic ram is driven by a Vickers (Model TK20, 20 MPa, 19 L min^{-1}) hydraulic pump and accumulator. The servo-controller also is interfaced to the computer, thus providing a fail-safe (in addition to those built into the servo) in the event of power or furnace failure and the capability to automate and program the deformation cycles. Additional details about the servo-controlled hydraulic drive system can be found in SCHOLZ and KOCZYNSKI (1979). Preliminary experiments suggest that it will be feasible to conduct creep (constant stress) tests with the hydraulic drive, although systematic exploration has not yet been undertaken.

Characterization of Friction on the Deformation Pistons

The strength of the specimen is measured external to the high-pressure volume (Figure 1). Hence, the measured stresses necessarily include forces acting on the deformation pistons other than those imposed by the mechanical strength of the specimen. In this and subsequent discussions, these additional forces are referred to as "friction;" in fact, some of them are viscous.

Forces acting on the deformation pistons include (i) friction on the deformation piston packing ring, (ii) friction between the WC deformation piston and Pb plug, (iii) friction between the alumina deformation pistons and the Pb plug, (iv) friction between the alumina piston and the graphite disk used to make the upper furnace contact, (v) friction between the alumina piston and solid salt in the upper, colder, part of the furnace, (vi) viscous drag between the alumina piston and molten salt in the high-temperature region, (vii) the viscosity of the molten salt between the sample and the alumina piston, and (viii) the ductile strength of the lid of the Pt capsule. For experiments performed at constant piston-displacement rate or constant strain rate, as is the usual practice in our laboratory, these forces are either independent of or vary smoothly as a function of piston displacement (or time). The contribution of these additional forces to the measured stress is determined for each deformation cycle as described below.

At the start of an experiment, the WC and alumina deformation pistons are in direct contact (no Pb between them), and the alumina piston is 1–2 mm from the specimen. The abrupt rise in force when the alumina piston encounters the specimen establishes the onset of the deformation cycle (see Figures 9, 10), referred to later as the "hit-point." The confining pressure increases as the deformation piston advances. The contribution of the increased confining pressure to the measured stress is accounted for by subtracting the measured confining pressure from the piston stress, and the result is referred to as the "differential piston stress." The confining pressure measured during a deformation cycle also is subject to uncertainties due to friction between the confining pressure piston and packing ring and the finite strength of the confining pressure media. The steady advance of the deformation piston causes slow, steady retraction of the confining pressure piston. Hence the experiments correspond to the "hot piston-out" technique, and uncertainties in the measured pressure are typically less than 5% (e.g., JOHANNES *et al.*, 1971; MIRWALD *et al.*, 1975; HOLLAND, 1980). The differential piston stress is observed to be approximately constant or to increase slightly with piston displacement before the specimen is contacted. The differential piston stress measured immediately prior to the hit-point is assumed to represent the total friction on the deformation pistons throughout the cycle. Reduction of data showing attainment of steady-state flow at high temperatures confirms the general validity of this procedure (see Figure 10). In some cases, the change of total friction with displacement (time) is not trivial (with respect to the specimen strength), and a linear or quadratic fit to the pre-hit force record is extrapolated through the cycle and subtracted from the differential piston stress to obtain the stress on the specimen (see Figure 9).

If friction is not constant or varying slowly before specimen contact, significant uncertainties may be introduced into the stress measurements, and the data are discarded because it generally signifies internal problems. For example, when the alumina deformation piston is in contact with the specimen, and Pb is present between the WC and alumina pistons, the differential piston stress may increase dramatically during piston advance. Moreover, the differential piston stress measured prior to the hit-point does not include the friction which must develop when that alumina piston begins to move and deformation of the sample begins. There is no way to deconvolve the friction on the alumina piston from the measured stress, and the data provide only an upper bound on the specimen strength. It is this effect that has introduced large systematic overestimates of strengths in previous studies with Griggs apparatus in which solid confining media were used (GREEN, 1992; GLEASON and TULLIS, 1992, 1993).

A number of modifications to the pressure assemblies have decreased the total friction. These improvements include using Al-bronze 600 (instead of mild or stainless steel) packing rings, replacing the upper salt wafer of the original molten salt assembly described by GREEN and BORCH (1989, 1990) with graphite, and utilizing lower-melting temperature alkali halide mixtures, thereby decreasing the amount of solid salt in contact with the alumina deformation piston. The absolute level of friction is of some concern because it does limit the maximum pressure at which deformation experiments can be performed.

Figure 3 compares the differential piston stress as a function of piston displacement measured at 500, 1000 and 1500 MPa confining pressure for Al-bronze 600 versus mild steel packing rings. The bore of the pressure vessel was filled with Pb and contained an axial 4 mm diameter alumina piston as in the standard pressure assemblies. As expected, the differential piston stress (total friction) increases with increasing confining pressure. However, the total friction is ~100 MPa lower in the experiments utilizing the Al-bronze packing rings. The total friction in the experi-

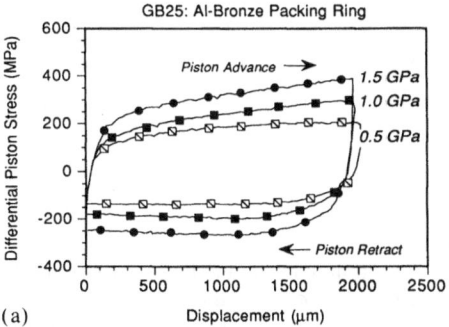
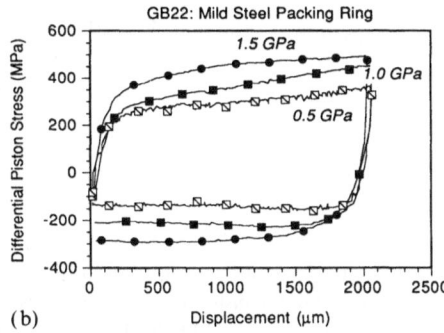

Figure 3

Differential piston stress-displacement data for experiments utilizing (a) an Al-bronze packing ring on the WC deformation piston (GB25) and (b) a mild steel packing ring (GB22). The data for each piston advance and retraction cycle are indicated: 0.5 GPa (square with diagonal), 1.0 GPa (solid square), and 1.5 GPa (solid circles); symbols were plotted every 10th data point.

ments above (Figure 3) is comparable to that measured in actual deformation experiments performed at high temperatures with molten and solid salt present. Although there are significant differences between these experiments and the deformation experiments, the data suggest that the friction exerted by the packing rings and Pb on the WC piston is the dominant source of friction in these high-pressure assemblies. The Al-bronze packing rings have been used successfully to a confining pressure of 2800 MPa.

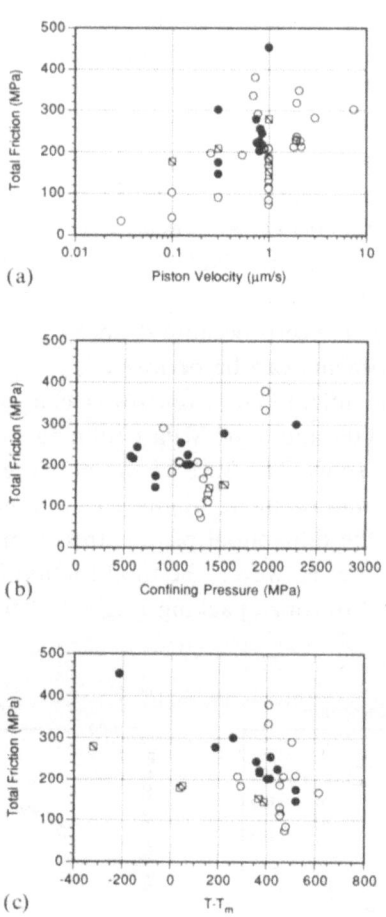

Figure 4

Total friction (differential piston stress at the hit-point) as a function of (a) piston displacement rate, (b) confining pressure for piston displacement rates in the range 0.8–1.2 μm s^{-1} and (c) specimen temperature relative to the melting temperature of the inner alkali halide mixture $(T - T_m)$ for piston displacement rates in the range 0.8–1.2 μm s^{-1}. Symbols represent pressure assemblies (see Figure 2) utilizing the following: (1) upper NaCl disk, pyrophyllite inner furnace sleeve, and E4 inner salt (open circles), (2) upper graphite disk, pyrophyllite inner furnace sleeve, and E4 inner salt (solid circles), (3) upper graphite disk, BN inner furnace sleeve, and E4 inner salt (open circle and dot), (4) graphite disk, pyrophyllite inner furnace sleeve, and NaCl-KCl eutectic inner salt (square with diagonal).

The use of a graphite disk in place of the NaCl disk utilized in previous assemblies (GREEN and BORCH, 1989, 1990) reduced the differential piston stress (friction) by approximately 40–60 MPa, with a greater reduction noted at higher confining pressure (Figure 4). The total friction depends on the confining pressure, piston displacement rate, temperature, and details of the specimen assembly (Figure 4). Temperature is important because it determines the volume fraction of molten salt present; a higher volume of molten salt reduces the total friction. It has been our experience that the best results in multiple strain cycle experiments are obtained when the sample temperature is > 200 K above the melting temperature of the alkali halide confining-pressure medium inside the graphite furnace. This decreases the contact area between solid salt and the alumina piston and helps ensure that the WC and alumina pistons remain in contact throughout the experiment. If the friction on the alumina piston is too high, Pb intrudes between the two pistons during retraction of the WC piston between deformation cycles. If one desires to perform multiple strain cycles (as are needed to precisely measure the stress exponent or activation energy), the presence of Pb between the two pistons introduces significant uncertainties to the stress measurement, as discussed above. Similarly, the composition of the furnace sleeves is important because if affects the thermal properties of the assemblies. Note that experiments using BN inner furnace sleeves exhibit lower friction than experiments using pyrophyllite sleeves. This probably is due to the much higher thermal diffusivity of BN (relative to pyrophyllite), and hence, greater volume of molten salt present during the experiment.

Measurement of Temperature Gradients

The mechanical behavior of many materials is a sensitive function of temperature. This is particularly true for partially molten materials, such as peridotite, in which the volume fraction of partial melt, and hence the mode of deformation and its effective flow stress, may change dramatically as a function of temperature.

The temperature distribution of the pressure assembly depicted in Figure 2c was characterized as follows: A single four-hole alumina thermocouple tube was used to sheath three type-B thermocouples whose beads were spaced approximately 2–3 mm apart. The stiffer $Pt_{70}Rh_{30}$ wire was used as the common reference wire for the three-thermocouple device. The thermocouple beads were covered with Zircar[R] alumina cement to inhibit contamination and extrusion of the molten NaCl pressure medium (although see discussion below). The sintered alumina pedestal was replaced by a crushable alumina cylinder with a hole to accommodate the thermocouples and sheath. A Pt capsule with the same dimensions as those used normally in our experiments was fitted around a hollow cylinder of alumina, and the three thermocouples were placed inside. Opposing thermocouples that touched the side of the Pt can were also present, for a total of five thermocouples inside the high-pressure, high-temperature volume. Thermoelectric potential was measured by

Omega OM3-ITC-B modules (linearized type-B thermocouple inputs) calibrated by a millivolt potentiometer source.

Figure 5 shows the measured temperatures for experiment GB41 relative to the controlling thermocouple. The temperature was ramped at 5 K min $^{-1}$ to 1600 K. The perturbation in temperature recorded by the left (opposing), middle and bottom thermocouples at ∼1350 K is perhaps due to melting and the onset of convection in the molten NaCl pressure medium inside the furnace. In experiment GB39 (not shown), melting of the salt (and the associated volume expansion on melting) caused a rupture in the graphite furnace, and the temperature difference increased dramatically to a maximum of −60 to −80 K relative to the control thermocouple, demonstrating the importance of maintaining the integrity of the graphite tube furnace for minimizing the temperature gradient in all high-pressure assemblies. In two experiments, one or more of the central thermocouples failed at 1450–1550 K due to extrusion of molten NaCl into the holes in the sheath for the thermocouple wires. The extrusion stretched and ultimately broke the Pt wires (as discussed above, the sintered alumina thermocouple tubes generally do not collapse around the thermocouple wires and alumina cement is generally not effective at preventing such "blow-outs"). This problem is unique to this particular assembly and is not a general problem for deformation or annealing experiments.

Figure 6 is a compilation of the data from three experiments (GB39, GB41 and GB43); the dashed curve is a second-order polynomial regression of those data. The "hot-spot" is located just slightly below the midpoint in the furnace. The maximum temperature difference in the region occupied by the sample is relatively small (±15 K at 1200 and ±20 K at 1500 K), corresponding to axial temperature

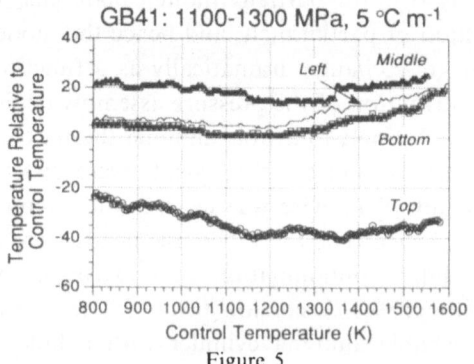

Figure 5

Thermal gradient measurement (experiment GB41) heated at a rate of 5 K min $^{-1}$ at a confining pressure of 1100–1300 MPa (confining pressure increased during heating due to thermal expansion of the pressure assembly components). Upper (open circles), middle (open triangles), bottom (open squares) and left (solid line) thermocouple temperature are plotted relative to the control thermocouple (right thermocouple). The perturbation in the left and middle thermocouples at 1350 K may be due to the onset of melting of the NaCl confining pressure medium.

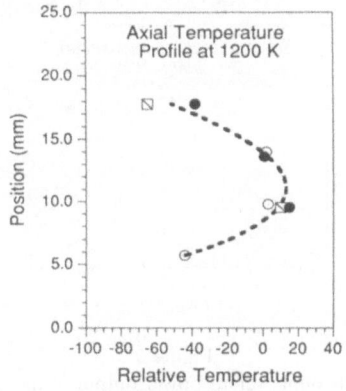

Figure 6

Axial temperatures, measured at 1200 K for experiments GB39 (squares with diagonal), GB41 (solid circles), and GB43 (open circles), plotted as a function of position in the graphite furnace. Thermocouple positions were varied by changing the height of the lower alumina base anvil, and hence the position of the Pt capsule; the position of the left and right control thermocouples remained fixed at a height of 13.6 mm. The maximum temperature difference at 1200 K over the length of an 8 mm specimen is estimated to be ± 15 K; the base of the specimen would be at a height of 7.62 mm. The furnace hot-spot is located slightly below the centerline of the furnace, probably because the assembly is not vertically symmetric and the thermal diffusivities of the various components are different.

gradients of ± 3–5 K mm^{-1}. Lengthening the furnace will further reduce the axial temperature gradient, and this modification is in progress.

Thermal Analysis of Melting of the Alkali Halide Confining Pressure Medium

A number of alkali halide mixtures are utilized as confining pressure media. In particular, the E4 mixture is used frequently, because it possesses a very low melting temperature (700 K) at 1 bar, and it does not extensively dissolve the metal sleeves. Phase relations of these two-, three- and four-phase mixtures are well-known at 1 bar, but most have not been determined at higher pressures.

When the alkali halide confining-pressure medium begins to melt, it is expected that the furnace power will increase to provide the latent heat of fusion. KANZAKI (1990) used this heating curve method to detect melting of Au in a multianvil apparatus. By monitoring the temperature and furnace power closely while heating at a constant rate, it is possible to detect "plateaus" in the temperature-power curves due to the latent heat effects.

Figure 7 shows a well-developed thermal plateau at 935 K that is interpreted to represent first melting of the E4 salt. The melting curve of the E4 salt determined in this way from several experiments is shown in Figure 8. The data were fitted to the Simon equation ($P - P_0 = A[(T/T_0)^c - 1]$; CLARK, 1959), where $P_0 = 0$ and $T_0 = 700$ K. The best-fit parameters were determined to be A = 1636.7 and

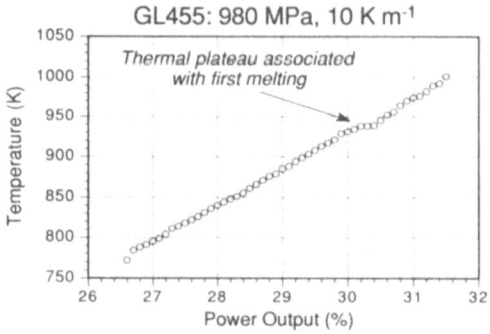

Figure 7
Temperature of the control thermocouple versus power output of the temperature controller to the SCR and the graphite furnace for experiment GL455 (980 MPa confining pressure and 10 K min⁻¹ heating rate). The thermal plateau associated with the first melting of the E4 inner salt is indicated by an arrow. Increased power consumption at the melting point is due to the latent heat of fusion.

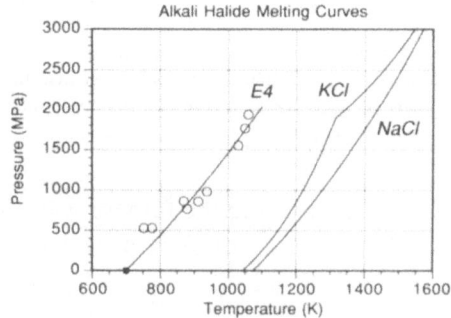

Figure 8
Data obtained by the heating curve method (Figure 7) for the melting of the E4 alkali halide mixture. The black square is the 1 bar eutectic (700 K) in the Na-K-Ca-Ba-Cl system (Figure 3320, LEVIN *et al.*, 1969). The melting curve calculated by the Simon equation (described in the text) is shown as a solid line. The melting curves for KCl and NaCl (CLARK, 1959) are shown for comparison.

$c = 1.788$, where P and T are in MPa and K, respectively. The observed scatter in these data may represent actual deviations in the melting behavior due to changes in the liquidus phase relations, or errors associated with salt preparation, determining the melting temperature from the thermal plateaus, or calibration of the confining pressure.

Measurement of Very Low Flow Stresses

It is possible to measure flow stresses as low as 3–5 MPa at high pressures and temperatures in the molten-salt pressure assemblies described here. This capability was developed as part of our ongoing investigations into the rheology of partially molten peridotite and polycrystalline Ni (MEAGHER *et al.*, 1992). The procedures for making such measurements are described below.

The "zeroth-hit" procedure of GREEN and BORCH (1990) is used in low-stress experiments to (i) ensure that no Pb is present between the alumina and WC deformation pistons, (ii) flatten the lid of the Pt capsule, if present, and (ii) to locate the hit-point precisely. Commonly, the zeroth-hit is performed at 1200–1300 K at a strain rate of 10^{-4} s^{-1} (piston displacement rate of $1-2 \mu$m s^{-1}). After loading the specimen elastically to a specimen stress of 200–400 MPa, the piston is retracted 1–2 mm, and the piston stress is monitored closely to determine if Pb has intruded between the alumina and WC pistons. If Pb does not intrude between the two pistons, the piston stress will decrease rapidly and remain constant at a differential piston stress comparable in magnitude (but negative) to that noted during piston advance. If Pb does intrude, the piston stress will increase 30–80 MPa within a few seconds after the specimen is unloaded, and the procedure above is repeated at a slightly higher temperature ($+50$ K) or to a slightly higher stress. After retracting both pistons intact, the temperature is increased to the desired value, and the low-stress deformation cycle begins.

Figure 9 shows the differential piston stress-displacement record for deformation of a partially molten synthetic pyrolite polycrystal ($\sim 5\%$ partial melt) at 1090 MPa confining pressure, 1500 K, at a strain rate of 1.4×10^{-5} s^{-1} (experiment GB40). In this particular experiment, the hit-point was determined from the previous deformation cycle to be $\sim 8600 \mu$m.

The raw data were smoothed by 100-point averaging to reveal the actual hit-point (Figure 9b). The noise in the stress measurements (1σ standard deviation is ~ 2 MPa) approaches the stresses being measured (3–4 MPa). The raw data were collected at a rate of 600 samples s^{-1} and 1000 samples were averaged to yield each individual data point in Figure 9a. There are several factors that contribute to the noise including analog-digital conversion, radio-frequency signals generated by SCR, and fluctuations or "dithering" of the servo-hydraulic drive. The differential piston stress prior to deformation increased slightly (0.018 MPa μm^{-1}) as the piston was advanced. In this particular experiment, the smoothed data were corrected for this increase in stress by second-order polynomial least-squares fitting of the stress-displacement data up to the hit-point and extrapolating that fit to the final displacement (Figure 9c). The deformation is assumed to be a homogeneous, constant-volume, uniaxial shortening (an excellent approximation in most cases), and the consequent increase in the cross-sectional area of the specimen is used for calculation of the specimen stress (Figure 9c). The resulting stress-strain curve for this experiment is shown in Figure 9d.

After the specimen is removed from the pressure assembly, the Pt capsule is removed, and the final dimensions of the specimen are carefully measured and compared for any discrepancies to those calculated by the stress-strain data

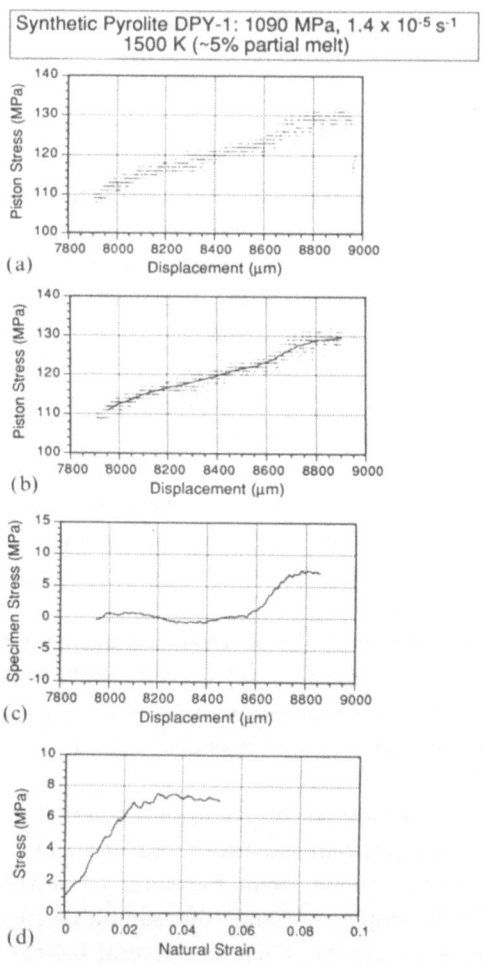

Figure 9
Stress-displacement record for deformation of a partially molten synthetic pyrolite polycrystal at 1090 MPa, 1500 K (~5% partial melt), at a strain rate of $1.4 \times 10^{-5} s^{-1}$ (experiment GB40). The raw data are shown in (a) and smoothed data (100 point averaging) are shown in (b). (c) The smoothed data were corrected for friction by fitting the differential piston stress prior to the hit-point (8621 μm) to a second-order polynomial, and the differential piston stress was converted to specimen stress as described in the text. (d) The resultant stress-strain curve.

reduction. In cases where discrepancies are noted, it is possible to recalculate the stress-strain data using the measured final dimensions of the specimen. In general, such discrepancies introduce only minor errors to the measured flow stresses (< 10%).

For specimens with strengths greater than about 20–50 MPa, data reduction is more straightforward. Figure 10a shows the stress-displacement record for defor-

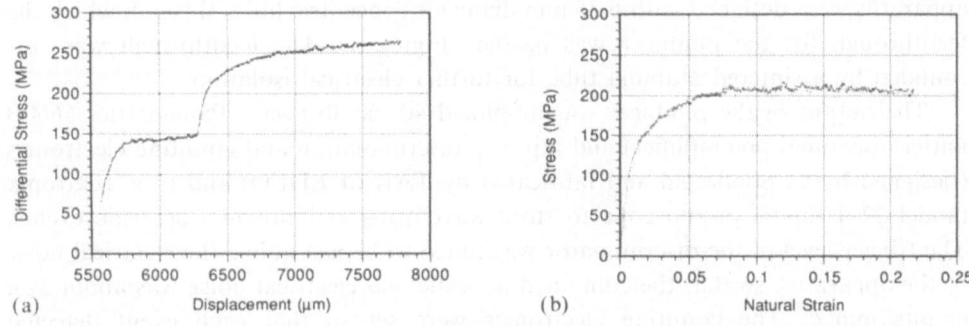

Figure 10

Stress-displacement record for deformation of Mg_2GeO_4 olivine polycrystal at 822 MPa, 1400 K at a strain rate of $4.5 \times 10^{-5}\ s^{-1}$ (experiment GB65). The raw data are shown in (a). The resultant stress-strain curve in (b) was calculated as described in the text.

mation of a Mg_2GeO_4 olivine polycrystal (see Figure 1, TINGLE *et al.*, 1993) at 822 MPa confining pressure, 1400 K at a strain rate of $4.5 \times 10^{-5}\ s^{-1}$ (experiment GB69). A zeroth hit was performed at the same conditions to 7% natural strain. The subsequent hit thus exhibited a sharp hit-point and a steep elastic slope (Figure 10a) as expected. The specimen flowed at a stress of 207 ± 4 MPa (the error is the 1σ standard deviation of the measured steady-state flow stress) to 22% natural strain. The stress-strain curve in Figure 10b was calculated using the measured initial dimensions of the specimen without smoothing or correcting the raw data for any time-dependence of friction. The steady-state flow stress calculated using the final specimen dimensions was 208 MPa, in excellent agreement with that obtained using the initial dimensions.

Measurement of Acoustic Emissions at High Pressure and Temperature

To measure acoustic emissions inside the pressure assembly, a miniature piezo-electric transducer ("pinducer") was spring-loaded against the bottom of the WC base anvil of the pressure assembly (Figure 2d). The pinducers are 2 mm diameter (Valpey-Fisher model VP1093) with a frequency response (-3 db) of 1.2 MHz and are high-pass filtered to have sensitivity in the range 0.2–5 MHz. The WC base anvil serves as an electrical lead for the graphite furnace, so the pinducer was insulated electrically by bonding a thin disk of alumina (~ 1 mm thick, polished flat and parallel) to its tip with silver-impregnated epoxy (probably any bonding agent will suffice). The pinducer was inserted through a 3 mm diameter hole in the WC insert in the base plate of the apparatus (Figure 1) and was electrically insulated from the base plate by wrapping the pinducer with a single layer of Scotch Transparent" tape (a number of other methods, paints, coatings, and teflon tape proved unreliable). Below the base plate, the shims and lower platen of the apparatus were designed with 6.35 mm diameter concentric holes through which the feedthrough for the pinducer was passed (Figure 1). The feedthrough was surrounded by a sintered alumina tube for further electrical isolation.

The output of the pinducer was amplified 40–60 db (via a Panametrics 5660B battery-operated preamplifier) and input to discrimination and counting electronics (designed by C. Sondergal and fabricated by TAK at LDEO) and to a Tektronix model 2211 digital oscilloscope to store waveforms of individual acoustic events. The trigger level of the discriminator was tuned to be just above the electrical noise of the apparatus, so that the combined acoustic and electrical noise was about 3–4 counts min^{-1}. The counting electronics were set so that each event detected incremented the output voltage by ~10 mV (1024 events equaled 10 Vdc). This voltage was input through an OM3 signal conditioner to the computer and stored on disk.

The measurement of acoustic emissions (AE) generated at high pressure and temperature was attempted as a part of ongoing investigations of the high-pressure faulting behavior of Mg_2GeO_4 olivine and its implications for the mechanism of deep earthquakes (e.g., GREEN and BURNLEY, 1989, 1990; BURNLEY *et al.*, 1991). Those results have been reported by GREEN *et al.* (1992) and TINGLE *et al.* (1993) and are briefly discussed here to illustrate the technique.

Acoustic events were monitored continuously during the experiments. During pressurization to 600 MPa, AE events typically were abundant, due most likely to compaction of the components of the pressure assemblies. Above 600 MPa during pressurization, AE was still observed but was relatively less frequent. After achieving a confining pressure approximately 150–200 MPa below the intended pressure of the experiment, the temperature was ramped to 900–1200 K at 5–10 K min^{-1}; AE events were observed at a low rate as during pressurization above 600 MPa, probably due to differential thermal expansion and continued compaction of the pressure assembly. After reaching the desired temperature and pressure, AE reached a stable level that was typically 3–4 counts min $^{-1}$. During piston advance, the AE remained at background levels. In some experiments, the piston was advancing through solid NaCl; the absence of any AE above background indicates that the NaCl flowed ductilely around the advancing piston. In others, the alumina piston was placed in contact with the specimen, and the WC piston advanced through Pb. No change in AE rate was observed during loading of the specimens until the stress drop associated with the high-pressure faulting was observed. During the stress drop, a burst of AE occurs which has been interpreted to be due to accelerations within the thin superplastic layer generated during nucleation and growth of the fault (TINGLE *et al.*, 1993). Lack of AE before the onset of failure demonstrates that the high-pressure faulting is not a brittle phenomenon. Some AE during the stress drop could be due to microfracturing associated with the dynamic mechanical instability, although microstructural evidence to support this has not been observed. After rupture, some specimens were allowed to slide along the fault and it was noted that the AE diminished to levels only slightly above those recorded prior to failure, indicating that microfracture and comminution of grains in the faults as occurs during brittle sliding does not occur in specimens that have failed by the high-pressure faulting mechanism.

The digital oscilloscope was used to record individual acoustic events (see Figure 4, TINGLE *et al.*, 1993). Unfortunately, it was not possible to collect and analyze a large number of events with a single digital oscilloscope, because the storage capacity of the oscilloscope was limited and the data had to be transferred to another computer before a new waveform could be acquired. In the future, high-speed signal digitizers will be used to characterize the waveforms of the AE associated with the high-pressure faulting mechanism. The waveforms observed to date suggest that there may be two distinct populations of AE; the first population is impulsive and exhibits exponential decay, as is typical of brittle AE. The second population is not impulsive; the frequency of these events is strikingly monotonic, but the rise and decay envelope is asymmetric.

Summary

The techniques and equipment described here for experimental rock deformation at high pressures and temperatures are being used to address several important geophysical problems, among them the mechanism of deep-focus earthquakes, the rheology of partially molten rock and segregation of partial melt, the effects of nonhydrostatic stress on the kinetics of phase transformations, and the rheology of the upper mantle, transition zone, and the lower mantle as elucidated by the study of analogs to deep-mantle phases. The validity of such measurements has been established, and such measurements provide important constraints for geophysical models that attempt to describe dynamic processes in the Earth's interior. There is every reason to expect that techniques and apparatus for high-pressure rock deformation will continue to improve.

Acknowledgements

Various aspects of these developments have been supported by NSF Grants INT83–03077, EAR85–11809, EAR89–05059, EAR89–15938, OCE90–12941, EAR92–19369, and DOE grant GO3–88ER45360. We owe a special debt to J. Abril for design, fabrication, and maintenance of many components of the high-pressure apparatus and equipment used in this laboratory. We would like to thank M. Van de Water for TIG-welding the Pt capsules. P. Waterstraat has been an invaluable source of knowledge and help in automating and computer-interfacing the various electronic components of the high-pressure apparatus. The following individuals have contributed ideas or designs that have led to our current apparatus and procedures: L. Anderson, R. S. Borch, P. M. Burnley, I. C. Getting, and P. Vaughan. M. Kanzaki and R. W. Luth shared information about interfacing the temperature controller to the computer. The philosophy underlying the development of this apparatus has been influenced by discussions over many years between HWG and Ed Schreiber, who was a good friend and stimulating colleague. He will be sorely missed.

REFERENCES

BORCH, R. S., and GREEN, H. W. (1989), *Deformation of Peridotite at High Pressure in a New Molten Salt Cell: Comparison of Traditional and Homologous Temperature Treatments*, Phys. Earth Planet. Int. *55*, 269–276.

BURNLEY, P. C., GREEN, H. W., II, and PRIOR, D. J. (1991), *Faulting Associated with the Olivine to Spinel Transformation in Mg_2GeO_4 and Its Implications for Deep-focus Earthquakes*, J. Geophys. Res. *96*, 425–443.

BUSSOD, G. Y., KATSURA, T., and RUBIE, D. C. (1992), *A New Method to Experimentally Determine the Rheologic Properties of Transition Zone Minerals at High P-T*, EOS, Trans. Amer. Geophys. Union *73*, 556.

CLARK, S. P. (1959), *Effect of Pressure on the Melting Points of Eight Alkali Halides*, J. Chem. Phys. *31*, 1526–1531.

CRC Handbook of Chemistry and Physics (R. C. Weast and M. J. Astle, eds.) 63rd edition (CRC Press, Boca Raton, Florida 1984).

CRC Handbook of Materials Science (C. T. Lynch, ed.) vol. II, *Metals, Composites, and Refractory Materials* (CRC Press, Boca Raton, Florida 1980).

DEVRIES, R. C. (1964), *The System Al_2SiO_5 at High Temperatures and Pressures*, J. Am. Ceram. Soc. *47*, 230–237.

GLEASON, G. C., and TULLIS, J. (1993), *Improving Flow Laws and Piezometers for Quartz and Feldspar Aggregates*, Geophys. Res. Letters *20*, 2111–2114.

GLEASON, G. C., and TULLIS, J. (1992), *Use of Molten Salt Cell for Experimental Deformation of Quartz and Feldspar: Preliminary Results*, EOS, Trans. Am. Geophys. Union *73*, 556.

GREEN, H. W. (1992), *Accurate Stress Measurement at Pressures to 5 GPa*, EOS, Trans. Am. Geophys. Union *73*, 556.

GREEN, H. W., and BORCH, R. S. (1989), *A New Molten Salt Cell for Precision Stress Measurement at High Pressure*, Eur. J. Mineral *1*, 213–219.

GREEN, H. W., and BORCH, R. S., *High pressure and temperature deformation in a liquid confining medium*. In *The Brittle-ductile Transition in Rocks, The Heard Volume*, Geophys. Monog. Ser. *56* (eds. Duba, A., Durham, W. B., Handin, J., Logan, J., and Wang, H.) (American Geophysical Union, Washington, D.C. 1990) pp. 195–200.

GREEN, H. W., and BURNLEY, P. C. (1989), *A New, Self-organizing, Mechanism for Deep-focus Earthquakes*, Nature *341*, 733–737.

GREEN, H. W., and BURNLEY, P. C. (1990), *The failure mechanism for deep-focus earthquakes*. In *Deformation Mechanisms, Rheology and Tectonics* (eds. Knipe, R. J., and Rutter, E. H.), Geol. Soc. London Spec. Pub. *54*, 133–141.

GREEN, H. W., and WALKER, D., *The multianvil as a deformation apparatus*. In *Proc. of the 29th Int. Geol. Congress* (August 1992).

GREEN, H. W., II, YOUNG, T. W., WALKER, D., and SCHOLZ, C. H. (1990), *Anticrack-associated Faulting at Very High Pressure in National Olivine*, Nature *348*, 720–722.

GREEN, H. W., II, SCHOLZ, C. H., TINGLE, T. N., YOUNG, T. E., and KOCZYNSKI, T. A. (1992), *Acoustic Emissions Produced by Anticrack Faulting During the Olivine → Spinel Transformation*, Geophys. Res. Letters *19*, 789–792.

HASKELL, R. W., and DeVRIES, R. C. (1964), *Estimate of Free Energy of Formation of Kyanite*, J. Am. Ceram. Soc. *47*, 202–203.

HOLLAND, T. J. B. (1980), *The Reaction Albite = Jadeite + Quartz Determined Experimentally in the Range 600–1200°C*, Am. Mineral. *65*, 129–134.

JOHANNES, W., BELL, P. M., BOETTCHER, A. L., CHIPMAN, D. W., HAYS, J. F., MAO, H. K., NEWTON, R. C., and SEIFERT, F. (1971), *An Interlaboratory Comparison of Piston Cylinder Calibration Using the Albite-breakdown Reaction*, Contrib. Mineral. Petrol. *32*, 24–38.

KANZAKI, M. (1990), *Thermal-analysis in a Multianvil High-P Apparatus*, EOS, Trans. Am. Geophys. Union *71*, 1697.

KERRICK, D. M. (1968), *Experiments on the Upper Stability Limit of Pyrophyllite at 1.8 kb and 3.9 kb Water Pressure*, Am. J. Sci. *266*, 204–214.

LEVIN, E. M., ROBBINS, C. R., and McMURDIE, H. F., *Phase Diagrams for Ceramists 1969 Supplement* (American Ceramic Society, Columbus, Ohio 1969).

MEADE, C., and JEANLOZ, R. (1990), *Deep-focus Earthquakes and Recycling of Water into the Earth's Mantle*, Science *252*, 68–72.

MEADE, C., and JEANLOZ, R. (1991), *The Strength of Mantle Silicates at High Pressures and Room Temperature: Implications for the Viscosity of the Mantle*, Nature *348*, 533–535.

MEAGHER, S., BORCH, R. S., GROZA, J., MUKHERJEE, A. K., and GREEN, H. W. (1992), *Activation Parameters for High-temperature Creep in Polycrystalline Nickel at Ambient and High Pressures*, Acta Metal. Material. *40*, 159–166.

MIRWALD, P. W., GETTING, I. C., and KENNEDY, G. C. (1975), *Low-friction Cell for Piston-cylinder High-pressure Apparatus*, J. Geophys. Res. *80*, 1519–1525.

SCHOLTZ, C. J., and KOCZYNSKI, T. A. (1979), *Dilatancy Anisotropy and the Response of Rock to Large Cyclic Loads*, J. Geophys. Res. *84*, 5525–5534.

TINGLE, T. N., GREEN, H. W., II, SCHOLZ, C. H., and KOCZYNSKI, T. A. (1993), *The Rheology of Faults Triggered by the Olivine-spinel Transformation in Mg_2GeO_4 and Its Implications for the Mechanism of Deep-focus Earthquakes*, J. Struct. Geol. *15*, 1249–1256.

WEIDNER, D. J., WANG, Y., and VAUGHAN, M. T. (1992), *X-ray Strain Measurement and Strength of Materials at High Pressure and Temperature*, EOS, Trans. Am. Geophys. Union *73*, 556.

WU, T.-C., BASSETT, W. A., BURNLEY, P. C., and WEATHERS, M. S. (1993), *Shear-promoted Phase Transitions in Fe_2SiO_4 and Mg_2SiO_4 and the Mechanism of Deep Earthquakes*, J. Geophys. Res. *98*, 19767–19776.

(Received April 2, 1993, revised October 5, 1993, accepted October 19, 1993)

PAGEOPH. Vol. 141. No. 2/3/4 (1993)

0033-4553/93/040545-33$1.50 + 0.20/0
© 1993 Birkhäuser Verlag, Basel

The Strength and Rheology of Commercial Tungsten Carbide Cermets used in High-pressure Apparatus

IVAN C. GETTING,[1] GANGLIN CHEN,[1] and JENNIFER A. BROWN[2]

Abstract — Uniaxial compressive stress-strain curves have been measured on a suite of 26 commercial grades of tungsten carbide cermets and three maraging steels of interest for use in high-pressure apparatus. Tests were conducted on cylindrical specimens with a length to diameter ratio of two. Load was applied to the specimens by tungsten carbide anvils padded by extrudable lead disks. Interference fit binding rings of maraging steel were pressed on to the ends of the specimens to inhibit premature corner fractures. Bonded resistance strain gages were used to measure both axial and tangential strains. Deformation was exremely uniform in the central, gauged portion of the specimens. Tests were conducted at a constant engineering strain rate of $1 \times 10^{-5}\,\text{s}^{-1}$. The composition of the specimens was principally WC/Co with minor amounts of other carbides in some cases. The Co weight fraction ranged from 2 to 15%. Observed compressive strengths ranged from about 4 to just above 8 GPa. Axial strain amplitude at failure varied from $\sim 1.5\%$ to $\sim 9\%$. Representative stress-strain curves and a ranking of the grades in terms of yield strength and strain at failure are presented. A power law strain hardening relation and the Ramberg-Osgood stress-strain equation were fit to the data. Fits were very good for both functions to axial strain amplitudes of about 2%. The failure of these established functions is accompanied by an abrupt change in the trend of volumetric strain consistent with the onset of substantial microcrack volume.

Key words: Tungsten carbide. strength, rheology, high pressure design.

Introduction

Tungsten carbide cermets have been used as structural components in large volume high-pressure apparatus for many decades. The development of these devices has been largely empirical. While supporting steel structures are often amenable to design analysis, the carbide components themselves have most often not been carefully analyzed due to lack of sufficient information about their constitutive relations. Even the choice of grades has been significantly hampered by lack of rheological data. In an effort to rectify this situation, a relatively extensive

[1] Cooperative Institute for Research in Environmental Science (CIRES), Campus Box 216, University of Colorado, Boulder, CO 80309 · 0216, U.S.A.
[2] CIRES and the Department of Geological Sciences, University of Colorado, Boulder, CO 80309, U.S.A.

set of measurements of the uniaxial compressive behavior of tungsten carbide cermets has been made. Grades from a number of manufacturers have been selected to reflect the interests of the high-pressure research community.

Some data on mechanical properties are available from carbide vendors. Parameters such as ultimate compressive strength and transverse rupture strength are in principle useful for grade comparison. They give information on the strength and toughness of grades in question. Compressive strength is taken as the maximum force supported by a specimen divided by the original cross-section area (engineering stress). Transverse rupture strength is the calculated tensile stress at failure on the tensile face of a rectangular prism subjected to bending produced by three point loading. The calculation makes the somewhat erroneous assumption of linear elastic conditions. Both of these quantities are difficult to measure accurately for brittle carbides. Some of these data are measured in accordance with published standards such as those of the International Organization for Standardization (ISO) or the American Society for Testing and Materials (ASTM), e.g., ISO 4506 Hardmetals-Compression test for compressive strength and ISO 3327/0 for transverse rupture strength. Many are not. Comparison of these data from different manufacturers is risky. Without stress-strain curves, it is impossible to determine if a reported strength represents the true ultimate value or a lower stress associated with brittle failure. It is also impossible to know what strain is required to achieve the listed strength. Values of transverse rupture strength can depend on sample dimensions, surface finish, and amount of plastic strain. Values from slightly different tests may have significant systematic differences. Generally, transverse rupture strength is used as a relative measure rather than an absolute value. In short, it is not possible to make a detailed comparison of carbide grades from generally provided parameters.

There is a modest body of literature on the deformation and strength of tungsten carbide cermets. Relatively little of it is directed to the issue at hand in this study. Samples are often described in generic terms without reference to a specific manufacturer and grade. Experimental compositions are often tested in pursuit of information about specific issues. Much of the focus is on deformation mechanism rather than comparative strengths. The study of HAYGARTH and KENNEDY (1967) reports apparent strength for a number of commercial grades. The focus of this work, however, was on the effect of length to diameter ratio on the axial stress supported by cylindrical specimens. Tests fairly similar to those reported here were performed by HARA and IKEDA (1972) on a suite of unidentified specimens ranging from 2.4 to 10.1 weight percent Co. They determined stress-strain curves on cylindrical specimens. Additionally, changes in microstructure, density, hardness, coercive force, Poisson's ratio, and Co phase structure were monitored.

For reference we list a number of generally relevant papers grouped by major subject area. The first several references listed are of general interest. EXNER (1979) discusses the crystal structures and phase relations pertinent to cemented carbides.

A useful review of the parameters affecting the mechanical properties of WC/Co alloys is presented by EXNER and GURLAND (1970). Included are discussions of the effects of composition, structure, porosity, grain size, mean free path in the cobalt phase, thermal stress, residual stress, temperature, sample volume, and test conditions. Traditional testing methods are reviewed by KERPER et al. (1958) and VANDEPUT and MASTRANTONIS (1988). The development and early testing of what has become the ISO 4506 standard specimen is reported by JOHANSSON et al. (1970). They discuss issues of stress concentrations, sample shape, uniformity of load, composition, grain size, and cyclic fatigue. Results are presented for both room temperature and high temperature tests. Fracture and failure are discussed by VEKINIS and LUYCKX (1987), FISCHMEISTER et al. (1988), ROWCLIFFE et al. (1988), JOHANNESSON and WARREN (1988), GODSE and GURLAND (1988), SURESH (1988), HAN and MECHOLSKY (1990, 1991), and SPIEGLER and FISCH-MEISTER (1992). Plastic deformation mechanisms are considered by PELEPELIN (1965, 1967), DOI et al. (1969), SARIN and JOHANNESSON (1975), VASEL et al. (1985), and JAYARAM et al. (1986). Residual stresses and magnetic coercivity are discussed by HANABUSA et al. (1983), KRAWITZ et al. (1986), VEKINIS and LUYCKX (1987), NABARRO and VEKINIS (1988), and KRAWITZ et al. (1989). Mechanical properties at elevated temperatures are investigated by LAUGIER (1988) and SCHMID et al. (1988).

The selection of grades for this study is not meant to represent the spectrum of carbide grades in production; nor are all manufacturers of note represented. This study was motivated by an interest in the mechanical properties of carbides currently used by the high-pressure community. An earnest effort was made to test all of the grades requested by the various funding institutions and interested parties. A suite of fine grain grades with systematically varying Co fraction from 6 to 15 weight percent was also included, the Kennametal KF series. All of the test specimens used were graciously donated by the various manufacturers. Their generosity is gratefully acknowledged. No request was denied.

Results are presented in terms of true stress and true strain. Dilatancy was observed in all but the highest strength grades. Systematics in the relation between yield strength and failure strain are presented. It is anticipated that these relations and the stress-strain curves themselves will be useful in guiding subjective choices of carbide grades.

Several constitutive relations are fit to the data. These include a power law strain hardening relation and the Ramberg-Osgood nonlinear stress-strain equation. The behavior of Poisson's ratio is detailed as a check of conventional assumptions about transverse strain. It is intended that this study provide the rheological information necessary to facilitate realistic numerical analysis of highly stressed structural components constructed from cemented tungsten carbides. The applicability of these data to structural analysis and design is discussed.

Experimental Procedure

Samples

Test specimens were acquired from seven carbide manufacturers from four countries. Twenty-six grades were tested. All were requested as standard commercial grades; no experimental compositions were tested. The choice of grades was guided by current and historical use in multi-anvil, piston-cylinder, and belt type high-pressure apparatus. Relatively low Co fraction grades, typically near 6 weight percent Co, are used in multi-anvil apparatus. Piston-cylinder apparatus often employs grades with Co weight fractions as high as 15% for dies. The majority of the grades were fine grained with grain size around one micrometer. Some coarser grain grades were included. Table 1 presents the grades tested along with the manufacturer's characterizations where available. Co weight fraction ranged from 2 to 15%, grain size from ~ 0.4 to $\sim 8 \mu$m. The grades are listed alphabetically by manufacturer and by increasing Co fraction within each manufacturer. Composition information may be approximate.

Additionally, three grades of 18% Ni, Co based maraging steel from Teledyne Allvac/Vasco were tested. The grades were VascoMax C–350, C–300, and C–250. Specimens were machined from hot rolled, annealed round bar approximately 1 inch in diameter. They were heat treated by aging for 14.4 ks (4 hours) at 758 K (480°C) in air. These data are provided to assist in the analysis of maraging steel support structures for carbide components. The cylindrical carbide specimens were supplied by the manufacturers in finished ground condition. Dimensions and their associated tolerances are given in Table 2.

Test Method

Uniaxial compression tests were performed on cylindrical specimens deformed at a constant engineering strain rate of 1×10^{-5} s^{-1}. This rate is much lower than is used in traditional tensile tests. Approximately one to five hours were required for each test. For a brief discussion of uniaxial tests and some of the concepts referred to later in this paper see, for instance LUBLINER (1990, chapter 2). The uniaxial test fixture is shown in Figure 1. Radially preloaded carbide anvils were used to load the specimens. Both Kennametal KF306 and Sandvik 6UF were used in the anvils with equal success. Axial force was generated by a hydraulic press driven by a positive displacement, single stroke pump. The pump rate was manually controlled to maintain the strain rate constant to about 5%. Axial force was determined from the hydraulic pressure and the ram area with a small, pressure dependent correction of $\sim 1\%$ for ram seal friction. True stress was determined from the axial force and the cross-section area of the deformed specimen calculated from the observed tangential (transverse) strain. Axial force was determined to an absolute accuracy of 0.3%,

Table 1

Grade specifications from the manufacturers

Manufacturer	Grade	Cobalt Fraction wt.%	Other Fraction wt.%	Other Chemistry	Grain Size μm	Density Mg/m³	Hardness HRa	Ultimate Strength GPa	Transverse Rupture Strength GPa
Federal Carbide	FC3M	6.0	—	—	0.4–1	14.90	92.5	6.06	2.26
Federal Carbide	FC10M	10.0	—	—	0.4–1	14.50	91.0	5.34	2.77
Federal Carbide	FC12M	15.0	—	—	0.4–1	14.00	89.5	4.48	3.08
Fujilloy	TJ07	2.0	5.0	TiC + TaC	1.7	14.50	92.5		1.77
Fujilloy	TD05	2.0	—		1.5	15.37	92.5		2.45
Fujilloy	TN05	5.0	2.0	TaC + CrC	1.0	14.93	93.0		2.75
Hertel	K05	5.0	—	—	<1	14.95			2.80
Hertel	KF1	6.0	1.0	TaC	<1	14.90			2.80
Kennametal	K96	5.5	*	TaC	1–6	14.90	92.1	5.28	2.14
Kennametal	K68	5.7	*	TaC	1–4	14.90	92.7	5.90	1.97
Kennametal	KF306	6.0	*	VC	<1	14.90	93.0	5.69	2.24
Kennametal	K313	6.0	*	Cr_3C_2	1+	14.90	93.0		3.10
Kennametal	KF308	8.0	*	VC	<1	14.70	92.5	5.34	2.56
Kennametal	KF310	10.0	*	VC	<1	14.50	92.0	5.00	2.82
Kennametal	KF312	12.0	*	VC	<1	14.30	91.3	4.72	2.97
Kennametal	K3109	12.2	*	TaC	1–8	14.20	88.0	4.38	2.97
Kennametal	KF315	15.0	*	VC	<1	13.90	90.4	4.11	3.42
Sandvik	3UF	3.0	—	—	0.4	15.10	>95	7.85	2.41
Sandvik	H6F	6.0	—	—	0.8	14.90	93.2	7.05	3.75
Sandvik	6UF	6.0	—	—	0.4	14.75	94.3	8.40	3.30
Sandvik	8UF	8.0	—	—	0.4	14.50	93.9	7.70	3.60
Sandvik	H10F	10.0	—	—	0.8	14.50	92.1	6.25	4.10
Sumitomo	F0	5.0	1.0	*	<0.7	14.80	94.5		1.60
Sumitomo	F1	8.5	1.5	*	<0.7	14.50	93.0		2.00
Toshiba	G1F	4.0	—	—	1.1	15.10	92.0	6.08	2.60
Toshiba	F	6.0	2.5	$TaC + Cr_3C_2$	0.5	14.90	93.5	6.86	2.94

* Other carbides are present in proprietary amounts.

Table 2

Test specimen dimensions

Dimension	Units	Value	Tolerance
Diameter	mm (″)	9.525 (0.3750)	0.005 (0.0002)
Length	mm (″)	19.05 (0.750)	0.03 (0.001)
Ends flat and parallel, TIR*	μm (″)	0 (0.0000)	5 (0.0002)
Ends normal to axis, TIR*	μm (″)	0 (0.0000)	5 (0.0002)
Corner radius	μm (″)	70 (0.003)	50 (0.002)
Surface finish, rms	μm (μ″)	0.3 (10)	0.2 (5)

* TIR indicates Total Indicated Runout.

UNIAXIAL COMPRESSION TEST FIXTURE

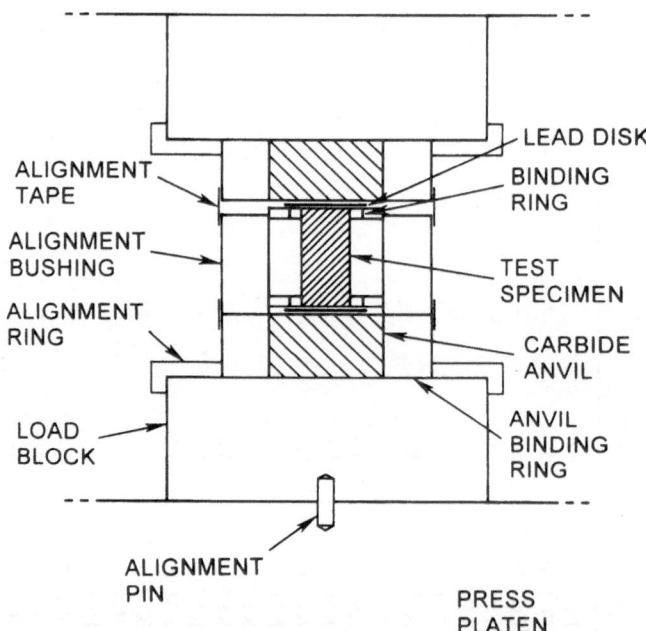

Figure 1

Fixtures for uniaxial compression testing of tungsten carbide cermets. The test specimen is loaded axially by a pair of carbide anvils and has no radial support over most of the length. Premature fracture at the ends of the specimen is inhibited by the use of small press fit binding rings around the specimen which are in turn surrounded by Teflon alignment rings. Extrudable lead pads between the specimen and the anvils redistribute the axial stress concentrating it gently toward the center line. Careful alignment is necessary to achieve uniform deformation.

cross-section area to 0.5%. The nominal true stress was thereby determined to 0.6%.

The cylindrical specimen and anvil geometry used in these tests would normally create a large axial compressive stress concentration at the corners of the specimen due to the indentation of the anvils. This would cause premature brittle failure initiated at the overloaded corners. These failures were inhibited by HAYGARTH and KENNEDY (1967, see their Figures 1 and 2) by inserting the ends of the test specimen into interference fit binding rings of high strength steel. Similar binding rings were employed in these tests. The binding rings were made from 18% Ni maraging steel, VascoMax C–250. They had an O.D./I.D. ratio of 1.67 and an interference of (0.5 ± 0.2) %. These binding rings were in turn surrounded by Teflon alignment rings to center the specimen in the test fixture.

While binding rings help inhibit corner failures, they do not relieve the axial stress concentrations created by the anvils. A novel technique was used in these tests to redistribute the axial stress on the face of the specimen. A 150 μm thick sheet of annealed lead was inserted between each specimen face and its anvil. Early in the loading history, much of the lead is extruded. Extrusion is terminated by the close approach of the corner of the specimen face to the indented shape of the anvil. A "pillow" of lead is trapped between the face of the specimen and the anvil. Inspection of the face of intact specimens after deformation to 5% axial strain reveals a slight residual indentation in the face, ~ 50 μm total indicated runout. This implies a gentle concentration of axial stress toward the centerline of the specimen and a consequent relief of the edge concentration. This procedure significantly reduces the probability of premature failures and has resulted in highly reproducible tests. All but the most brittle grades tested achieve zero or near zero slope of the stress-strain curve (tangent Young's modulus) before failure. This implies that their intrinsic strength has been sampled. Several grades exhibited stable post-peak deformation. One reference test was made on Sandvik 6UF, a very high strength, relatively brittle grade, without the lead pad. The failure stress was reduced 21% from 8.0 to 6.3 GPa.

The anvils had a similar, but much less pronounced residual indentation. The anvils typically survived many runs. No correlation between specimen behavior and anvil history was observed. The useful life of the anvils was typically terminated by their being impaled with fragments generated by catastrophic specimen failures or by brittle fracture initiated at the edge of the loaded circle.

Strains were measured with bonded resistance strain gages. Each of the observed strains, axial and tangential, was measured with two gages on opposite sides of the specimen. The gages occupied the central 17% of the length of the specimens. Constantan foil strain gages with 25 μm thick polyimide foil backs were used, Measurements Group EA–06–125TA–120. Gage factors were listed as accurate to 0.5% by the manufacturer. The gages were bonded to the carbide with Measurements Group AE–10 two part epoxy. The specimen surface was prepared for gage bonding by solvent cleaning, slight abrasion under a mild etchant, and chemical neutralization as per manufacturer recommendations. The bond was cured at a clamping pressure of 200 kPa (~ 2 atm) gage at 323 K (50°C) for 86 ks (24 hours). The resultant bond thickness was approximately 15 μm. The bonding procedure was generally very successful with consistent behavior of the strain gages to at least 5% strain. Strains were determined from the off balance output voltage of a Wheatstone bridge associated with each gage. Corrections were made for nonlinearity in the bridges and for the transverse sensitivity of the strain gages. Engineering strains were converted to true strains. The accuracy of individual strain determinations is affected by the uncertainty in the gage factor, the transverse sensitivity, the resistance determination, and the mechanical behavior of the gage bond. Approximately 100 data points were recorded for each 1% in axial strain. We estimate the absolute uncertainty of individual strain measurements to be 1% of the strain for strains in the 1–5% strain amplitude range.

Elastic finite element modeling of our specimen and binding ring assembly indicates departure of less than 2% from the ideal uniaxial stress state in the central portion of the specimens where strains are measured. Residual dilation of intact specimens subjected to 5% axial strain was uniform over the gauged region to within several percent of the maximum transverse strain. These samples exhibited near zero tangent modulus at 5% axial strain. They were therefore extremely sensitive to any nonuniformity in stress distribution. From these observations, we conclude that the stress state was uniaxial compression to within ~2% in the central portion of our specimens. We estimate the absolute uncertainty of the average axial true stress to be 2%.

It is difficult to achieve uniform strains in uniaxial compression tests. Very slight misalignment of the press components, or even slight asymmetry in the tool stack, can cause a significant difference in the strains measured on opposite sides of the specimen. The press must also exhibit substantial lateral stiffness to produce uniform deformation on the low modulus portion of the stress-strain curve. In our setup, the tool stack was centered in the press to within 0.05% of the press span. All components of the tool stack (anvils, specimens, and load blocks) were collinear to within 100 μm. The axial strains on opposite sides of the specimen were typically equal to within 2–4% of the strain at yield. We estimate the absolute uncertainty of the average axial and tangential strains to be 2% of the strain. A few tests showed strain differences on opposite sides as large as 10% at yield, but still reproduced the average stress-strain curves of duplicate tests very well.

Results

Stress-strain Curves

Stress-strain curves are presented in terms of true stress,

$$\sigma_z = \frac{F_z}{a},$$

where F_z is the axial force and a is the instantaneous cross-section area of the deformed specimen, and true strain

$$\varepsilon_{\text{true}} = \int_{l_0}^{l_1} \frac{dl}{l} = \ln\left(\frac{l_1}{l_0}\right) = \ln(\varepsilon_{\text{engr}} + 1),$$

where l is a dimension of the specimen, l_0 is its original value, l_1 is the deformed value, and $\varepsilon_{\text{engr}}$ is the traditional engineering strain

$$\varepsilon_{\text{engr}} = \frac{l_1 - l_0}{l_0}.$$

The sign convention is used wherein tensile stress and extensile strain are positive, hence axial stresses and strains in these compression tests are negative. Tangential strains, which are extensile, are positive.

The stress-strain curves were highly reproducible. At a given stress, duplicate tests reproduced averaged strains to 1–2% of the strain for strain amplitudes up to several percent. Yield and ultimate strengths and their associated strains reproduced to these same levels. This test method was able to reproduce even the sensitive strain at failure to within $\pm 5\%$ of the strain for most of the cases.

Representative stress-strain curves for six grades are shown in Figure 2. In this somewhat unconventional treatment, we plot three stress-strain curves for each grade: axial (longitudinal), tangential (transverse), and volumetric. Axial strain, ε_z, and tangential strain, ε_θ, are measured directly. Volumetric strain, ε_v, is calculated from them according to the exact expression

$$\varepsilon_v = \varepsilon_z + 2\varepsilon_\theta + 2\varepsilon_z\varepsilon_\theta + \varepsilon_\theta^2 + \varepsilon_z\varepsilon_\theta^2.$$

These three curves are labeled in Figure 2(d). Individual data points are plotted with about one hundred points for each 1% in axial strain. No smoothing is applied.

In general, the axial stress-strain curves are highly curved. The quasilinear portion at low stress gives way to an extended region of gradually decreasing slope. The most brittle, high strength grades have stress-strain curves which are monotonic increasing to failure near 2% compressive strain amplitude, Figures 2(a) and 2(b). At failure, the curves have substantial slopes. In these cases, the intrinsic ultimate strength has not been sampled. Failure by brittle fracture occurs before the stress amplitude reaches its maximum potential value. This phenomenon is highly sensitive to grade and test configuration. Larger strains are observed in lower strength grades. In some of these cases the axial stress-strain curve is also monotonic increasing to failure. In others the stress becomes essentially constant by failure, Figure 2(c). In yet other cases, a maximum in stress amplitude is observed followed by post-peak deformation, Figures 2(d), 2(e), and 2(f). Most of the grades which failed at axial strain amplitudes less than 3% exhibited monotonic increasing behavior. Most which failed beyond 3% exhibited post-peak deformation.

Representative Stress-Strain Curves
To Failure (diamond) or 5% Axial Strain

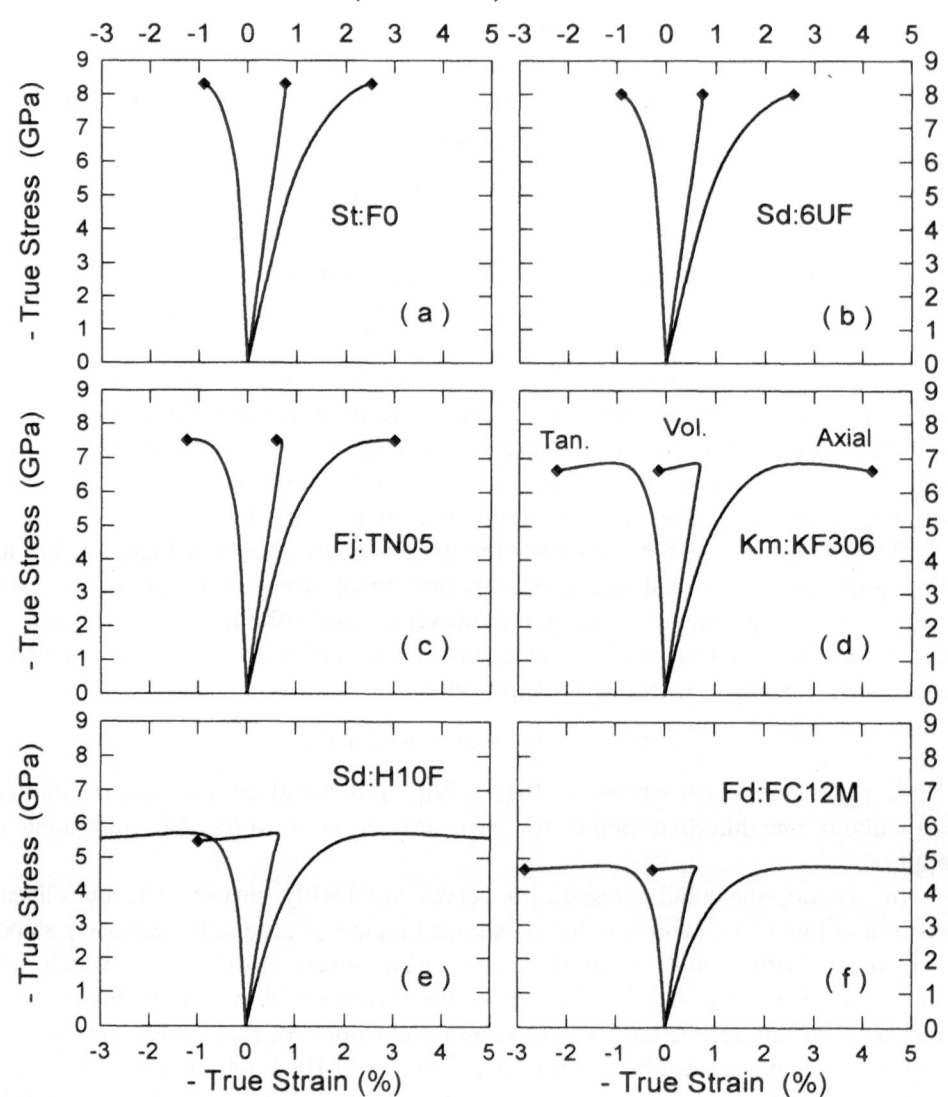

Figure 2
Representative compressive stress-strain curves for six grades. Axial, tangential, and volumetric curves are shown for each grade as identified in (d). The two strongest grades, (a) and (b), failed before reaching their ultimate potential stress as demonstrated by the positive slope of the axial stress-strain curve at failure. The other four grades achieved their maximum potential stress and all exhibit some degree of dilatancy as evidenced by the abrupt corner in their volumetric stress-strain curves and the subsequent increase in volume. Grades are designated on the plots. Manufacturer abbreviations are as follows: St: Sumitomo, Sd: Sandvik, Fj: Fujilloy, Km: Kennametal, Fd: Federal Carbide.

In general, we observe larger axial strain amplitudes than have most previous workers. Our observed ultimate strengths are also about 10% higher on the average than the values supplied by the manufacturers and listed in Table 1. This may be due in part to steady improvements in carbides over time and in part to the quality of our test configuration. The only grades for which we observed ultimate strengths lower than manufacturers' values were the Sandvik grades with Co weight fractions of six percent or greater. Sandvik claims conformance to ISO 4506 in its tests. This standard calls for constant stress rate which implies very high strain rates as the stress-strain curves flatten out. This may lead to the support of higher stresses for a short period of time before failure.

The tangential curves themselves reveal relatively little new information on visual inspection. They reflect the same general behavior as do the axial curves. The volumetric stress-strain curves, on the other hand, exhibit striking behavior. The volumetric stress-strain curves are nearly linear to within a few percent of the ultimate stress. The volume of the specimen decreases over this portion of the curve. In the strongest grades, the volumetric curves are highly linear all the way to failure, Figures 2(a) and 2(b). In lower strength grades the volume of the specimen starts to increase dramatically just below the ultimate stress. The specimens become dilatant. The volumetric curves exhibit an abrupt change in direction as the volume begins to increase. With further axial strain, the volume continues to increase to final failure. Some grades fail at volumes larger than their original volume, Figures 2(d), 2(e), and 2(f). This general behavior, referred to as dilatancy, is well-known in rock mechanics. See PATERSON (1978, p. 114) for a brief discussion with background references. It is generally not common to think of structural materials as exhibiting dilatancy. At the strain levels of conventional structures, it is likely that they do not. Similar tests on maraging steel in this study also reveal dilatant behavior in uniaxial compression beyond axial strain amplitudes of ~2%.

This dramatic increase in volume is presumed to be the result of the generation of microcrack volume. The smoothness and reproducibility of the data implies that this volume is uniformly distributed over a large number of small cracks. If only a small number of large fractures participated in the process, one would not expect the limited region sampled by the strain gages to show the high degree of reproducibility that we observe. This suggests that the failure mechanism for these tests is probably the formation, growth, and coalescence of a large number of microfractures. While some of the tests were terminated prematurely, all of the grades tested exhibited cataclysmic failure. One can inquire as to the relative contribution in the entire deformation process of plastic mechanisms, which maintain the continuum and preserve volume, and microcrack mechanisms, which partition the continuum and create volume. Previous investigations, cited in the introduction grouped roughly by mechanism, reveal plastic deformation in both the Co matrix and in the WC grains as well as fractures within the WC and at the grain boundaries between the WC and Co. Unfortunately, microstructural examination of deformed specimens is beyond the scope of this study. We observe about 0.5%

residual increase in volume on unloading intact specimens from 5% axial strain amplitude. This is in very close agreement with HARA and IKEDA (1972) who observed a residual decrease in density of approximately 0.5% for WC/Co samples taken to similar strains. Microstructural studies by them revealed substantial microcracking.

Appendix A gives the axial stress-strain curves only for all the grades tested. They are grouped by manufacturer to facilitate comparison of the grades offered by each individual vendor. Parameterization of the curves follows in a later section of this paper. Complete, three component curves for all 26 grades plotted on the same scale for direct comparison are available from The National Center for High Pressure Research, State University of New York at Stony Brook, Stony Brook, NY 11794, USA.

While it is anticipated that the stress-strain curves themselves will be of use in grade selection, the systematics in the behavior of the various grades is somewhat inaccessible in this form. We have therefore tried to characterize some measure of strength relative to some measure of ductility. Figure 3 shows the 0.2% compressive

Yield Strength vs. Failure Strain

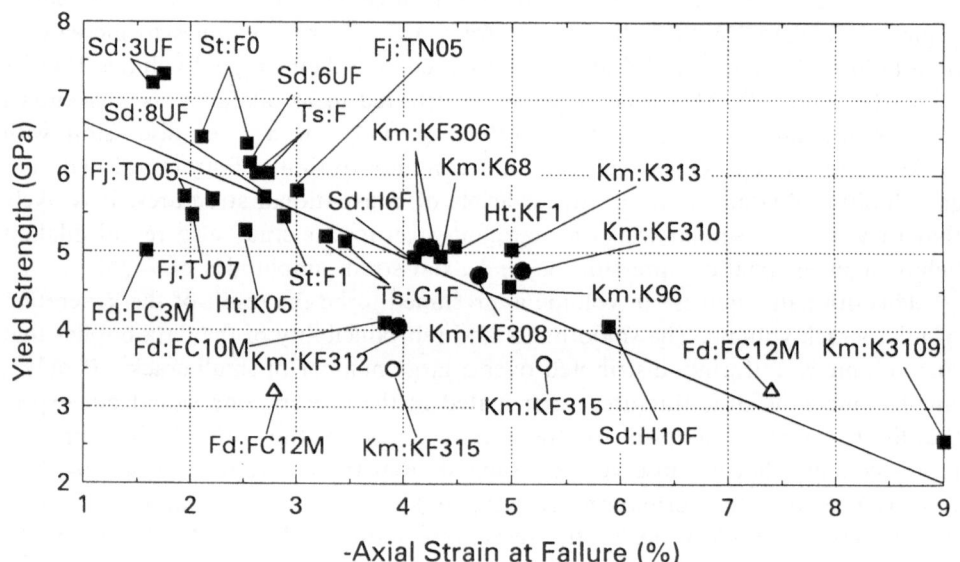

Figure 3

Yield strength as a function of axial strain at failure for all grades. This plot presents the strength-ductility systematics as represented by uniaxial compression tests. Typical scatter is indicated by the six duplicate tests. Better performance would generally be expected at a given strength level from grades with larger strain amplitude at failure. Manufacturer abbreviations used in the grade identifiers are as follows: Fd: Federal Carbide, Fj: Fujilloy, Ht: Hertel, Km: Kennametal, Sd: Sandvik, St: Sumitomo, Ts: Toshiba. The Kennametal KF series, in which only the Co fraction varies, is plotted as circles. Extreme scatter was observed in two 15% Co, submicron grades: Federal Carbide FC12M and Kennametal KF315. These two grades are plotted with open symbols.

offset yield strength plotted against the axial strain at failure. The 0.2% offset strength is the stress at which the observed strain amplitude exceeds that predicted from a linear extrapolation of the initial slope of the stress-strain curve by a strain amplitude of 0.2%. This is a common definition of yield and can be accurately identified even if the material does not exhibit a distinct break in slope at yield. Strain at failure was determined directly from the strain gages for all specimens which failed at less than 5% axial strain amplitude. Beyond 5% the strain gages become unreliable. Mean axial strain amplitudes larger than 5% were determined from the readings of a pair of linear voltage displacement transducers calibrated against the strain gages on each run. Calibration was done on that portion of the stress-strain curve in which both the strain gage and the displacement transducer outputs were essentially linear with stress. This eliminates effects of lead pad extrusion at low stress and localized strain at high stress. A straight line was fit in Figure 3 for the grades which failed consistently at axial strain amplitudes less than 5%. The trend is obvious, higher strength grades fail at lower strains. The scatter, however, contains information of great interest for grade selection. At a given strength level, there are often several grades to choose from. To the extent that uniaxial compression tests reflect the relative behavior in a real application, better performance would be expected at a given strength level from grades with larger strain amplitudes at failure as this implies greater ductility. Detailed strength and strain data for all grades are presented later in Table 3.

Figure 3 should be applied with a measure of caution. While we believe the general patterns to be fairly represented, there is a limit to the experimental resolution. Six pairs of duplicate tests were made spanning the 5% failure strain range. They can be identified in the figure by the double tie lines from the grade identifiers to the plotted points. Strain at failure typically repeats to within a few percent of the strain. From the scatter in these pairs of points, one can make a subjective estimate of the resolution offered by this method. Many of the grades are represented by only one point. There is no demonstration in one measurement of the reproducibility of these grades. Carbide structural components often enjoy the fracture inhibition of polyaxial compressive stress fields in high-pressure applications. This effect is demonstrated by JAYARAM et al. (1986) who produced axial strains of 10% without failure in WC/10 wt.% Co at ~1.5 GPa confining pressure. This is approximately twice the strain at failure which would be anticipated in our tests for such a grade. It is possible that the failure mechanisms under polyaxial compression differ from those in our uniaxial tests. Axial strain at failure in uniaxial tests may fail to some extent to reflect the relative behavior of the grades in other circumstances. We feel, however, that these observations provide a good starting point for grade selection.

Two grades were much less reproducible in their strain at failure. They were both 15% Co, submicron grades: Federal Carbide FC12M and Kennametal KF315. These grades have relatively low yield strength and might therefore have been expected to behave more reproducibly, not less. Three tests were run on each grade with very similar results. One test for each went to 5% axial strain amplitude without failure and was terminated. In the other tests FC12M failed at axial strain amplitudes of 2.8 and 7.4% while KF315 failed at 3.9 and 5.3%. These grades are plotted with unfilled symbols in Figure 3. While we pondered the possibility of two relatively distinct failure mechanisms in these grades, no discernible difference was observed in the post-failure fragments and the large variation in failure strain is not understood at this time.

Modeling of Stress-strain Data

A number of characterizing parameters were extracted from each test. Table 3 presents various quantities associated with the stress-strain curve for each grade. Included are initial values of Young's modulus and Poisson's ratio, yield strength and strain at yield, C'' strength and strain at C'' (associated with dilatancy), ultimate strength and strain at ultimate stress, failure strength and strain at failure. These quantities are defined in detail below. Table 4 shows parameters for two rheological models fit to the stress-strain data. Included are two fit parameters for a power law strain hardening relation, and three fit parameters for the Ramberg-Osgood equation. Also provided are the axial strains to which these two relations have been fit, $\varepsilon_{z\,\text{fit}}$. These quantities are also discussed in detail below. All stresses and strains are true stresses and true strains. Compressive stresses and strains are negative. Strengths, however, are taken as the amplitude of the appropriate compressive stresses and are expressed as positive values. The associated compressive strains are left with their original negative signs.

The rheology of uniform, isotropic, linear engineering materials is typically described by two elastic constants such as Young's Modulus and Poisson's ratio. These same parameters are useful in describing the highly nonlinear response of the carbides but are determined locally along the stress-strain curves from the original, discrete data points. Five-point running differences were used to achieve some smoothing without loss of detail. Their values, E_i and v_i, were assigned to the mid-point of the difference intervals. The value of Young's modulus was calculated from

$$E_i = \frac{\Delta\sigma_z}{\Delta\varepsilon_z} = \frac{\sigma_{z(i+2)} - \sigma_{z(i-2)}}{\varepsilon_{z(i+2)} - \varepsilon_{z(i-2)}} \cong \frac{\partial\sigma_z}{\partial\varepsilon_z}.$$

Poisson's ratio was similarly determined from,

$$v_i = -\frac{\Delta\varepsilon_\theta}{\Delta\varepsilon_z} = -\frac{\varepsilon_{\theta(i+2)} - \varepsilon_{\theta(i-2)}}{\varepsilon_{z(i+2)} - \varepsilon_{z(i-2)}} \cong \frac{\partial\varepsilon_\theta}{\partial\varepsilon_z}.$$

We also define a new parameter which will prove useful later. The volumetric modulus is defined in a manner analogous to Young's modulus but applies ·to volumetric strain instead of axial strain. It is the slope of the volumetric stress-strain curve.

$$V_i = \frac{\Delta\sigma_z}{\Delta\varepsilon_V} = \frac{\sigma_{z(i+2)} - \sigma_{z(i-2)}}{\varepsilon_{V(i+2)} - \varepsilon_{V(i-2)}} \cong \frac{\partial\sigma_z}{\partial\varepsilon_V}.$$

To determine the initial value of Young's modulus, E_0, and Poisson's ratio, v_0, the appropriate low strain portion of a plot of each versus axial strain is fit with a straight line by least squares. This fit is evaluated at zero strain to yield the initial values. The reproducibility of Young's modulus was often a fraction of a percent, with some values differing by several percent. The initial values of Poisson's ratio are all very similar around 0.21. The spread for the entire set of measurements is about $\pm 3\%$.

Yield strength, S_y, is taken as the axial compressive stress amplitude at which the observed axial strain departs from an extrapolation of the initial slope by 0.2% in strain amplitude. This is analogous to the commonly reported 0.2% tensile offset stress from uniaxial tensile tests. Strain at yield, ε_y, is the axial strain at this stress. Yield strengths repeat to approximately $\pm 1\%$.

We identify a stress associated with dilatancy. Traditionally, the quantity C' is defined in rock mechanics as the stress amplitude at the onset of dilatancy after BRACE et al. (1966). It is taken as the stress at which the volumetric stress-strain curve departs from linear. C' is functionally very difficult to determine. We borrow this nomenclature but modify the definition slightly. Our C'' strength, $S_{C''}$, is the axial compressive stress amplitude at which the rate of change of volumetric strain with respect to axial stress goes to zero,

$$\frac{\partial\varepsilon_V}{\partial\sigma_z} = 0.$$

C'' is therefore the point at which the volume of our specimens reaches a minimum. Numerically C' and C'' would be very similar for these carbides. As C'' is approached the volumetric modulus, defined previously, increases rapidly toward plus infinity. The volumetric modulus behavior is used to identify C''. The instability is easy to observe in the numerical data. We observed C'' to lie with a few percent of the ultimate stress. The strain at C'', $\varepsilon_{C''}$, is the corresponding axial strain. Values of the C'' strength typically repeat to about $\pm 0.5\%$.

Ultimate strength, S_u, is the magnitude of the maximum axial compressive stress achieved in the test. It can be a local stress amplitude maximum in the case of post-peak deformation, the stress amplitude at failure in the case of monotonic increasing stress-strain curves, or the maximum stress amplitude achieved at the termination of a test without failure at 5% axial strain amplitude. The strain at ultimate stress, ε_u, is the axial strain associated with the ultimate strength. In the case of post-peak deformation it is not the maximum amplitude strain achieved in the test. The ultimate strength is remarkably reproducible. Except for the most brittle grades, it typically repeats to about $\pm 0.3\%$.

Table 3

Elastic parameters, strengths, and associated strains. Symbols are defined in the text above

Manf.	Grade	No	Initial		Yield		Dilatancy		Ultimate		Failure	
units			E_0 GPa	v_0	S_y GPa	ε_y	$S_{C''}$ GPa	$\varepsilon_{C''}$	S_u GPa	ε_u	S_f GPa	ε_f
Federal	FC3M	75	637	0.21	5.02	−0.010	no dilatancy		6.13	−0.016	6.13	−0.016
Federal	FC10M	79	598	0.21	4.10	−0.009	5.59	−0.021	5.65	−0.027	5.55	−0.038
Federal	FC12M	81	545	0.22	3.23	−0.008	4.73	−0.025	4.76	−0.031	*	
Federal	FC12M	82	550	0.22	3.24	−0.008	4.74	−0.024	4.77	−0.030	4.61	−0.074
Federal	FC12M	83	555	0.19	3.21	−0.008	4.72	−0.024	4.75	−0.028	4.75	−0.028
Fujilloy	TN05	45	635	0.21	5.79	−0.011	7.34	−0.021	7.53	−0.030	7.51	−0.030
Fujilloy	TD05	39	690	0.20	5.68	−0.010	6.84	−0.019	6.86	−0.020	6.85	−0.022
Fujilloy	TD05	40	684	0.21	5.71	−0.010	6.83	−0.019	6.85	−0.019	6.85	−0.019
Fujilloy	TJ07	42	634	0.21	5.48	−0.011	6.49	−0.018	6.52	−0.020	6.51	−0.020
Hertel	K05	12	632	0.21	5.28	−0.010	6.54	−0.018	6.75	−0.025	6.75	−0.025
Hertel	KF1	10	611	0.21	5.10	−0.010	exploratory run — test terminated					
Hertel	KF1	7	624	0.22	5.08	−0.010	6.46	−0.019	6.60	−0.026	6.26	−0.045
Kennametal	KF306	48	638	0.21	5.07	−0.010	6.74	−0.021	6.86	−0.029	6.62	−0.041
Kennametal	KF306	49	637	0.21	5.07	−0.010	6.77	−0.022	6.87	−0.027	6.64	−0.042
Kennametal	K313	60	624	0.21	5.03	−0.010	6.65	−0.023	6.70	−0.027	*	
Kennametal	K313	61	625	0.21	5.04	−0.010	6.64	−0.023	6.70	−0.027	6.61	−0.050
Kennametal	K68	72	626	0.21	4.95	−0.010	6.30	−0.021	6.35	−0.025	6.03	−0.043
Kennametal	KF308	51	610	0.22	4.64	−0.010	6.27	−0.023	6.37	−0.030	*	
Kennametal	KF308	52	605	0.21	4.72	−0.010	6.25	−0.022	6.37	−0.031	6.20	−0.047
Kennametal	KF310	54	585	0.22	4.64	−0.010	6.36	−0.025	6.44	−0.032		
Kennametal	KF310	55	568	0.22	4.77	−0.010	6.40	−0.025	6.47	−0.032	6.30	−0.051
Kennametal	K96	69	633	0.21	4.57	−0.009	5.83	−0.020	5.89	−0.025	5.47	−0.050
Kennametal	KF312	57	540	0.22	4.05	−0.010	5.70	−0.024	5.77	−0.032	5.72	−0.040
Kennametal	KF315	63	510	0.22	3.71	−0.009	5.48	−0.025	5.61	−0.038	*	
Kennametal	KF315	64	543	0.22	3.51	−0.009	5.54	−0.026	5.66	−0.037	5.65	−0.039
Kennametal	KF315	65	535	0.22	3.58	−0.009	5.51	−0.026	5.63	−0.039	5.55	−0.053
Kennametal	K3109	66	564	0.22	2.58	−0.007	3.76	−0.021	3.97	−0.048	*	
Kennametal	K3109	67	557	0.22	2.56	−0.007	3.81	−0.024	4.06	−0.090	4.06	−0.090

Maker	Grade	No.			σ	slope	σ	slope	σ	slope	σ	slope
Sandvik	3UF	18	624	0.21	7.33	−0.014	no dilatancy		8.13	−0.017	8.13	−0.017
Sandvik	3UF	19	610	0.21	7.21	−0.014	no dilatancy		7.81	−0.016	7.81	−0.016
Sandvik	6UF	21	613	0.21	6.16	−0.012	7.93	−0.024	8.01	−0.026	8.01	−0.026
Sandvik	6UF	22	617	0.21	6.15	−0.012	no Pb pad—premature failure					
Sandvik	8UF	24	602	0.22	5.85	−0.012	no Pb pad—premature failure					
Sandvik	8UF	25	585	0.21	5.71	−0.012	7.51	−0.027	7.52	−0.027	7.51	−0.027
Sandvik	H6F	27	636	0.21	4.94	−0.010	6.38	−0.021	6.43	−0.025	6.18	−0.041
Sandvik	H10F	30	597	0.22	3.95	−0.009	5.65	−0.023	5.72	−0.030	*	
Sandvik	H10F	31	578	0.22	4.05	−0.009	5.66	−0.023	5.71	−0.031	5.48	−0.059
Sumitomo	F0	1	619	0.21	6.50	−0.013	no dilatancy		8.02	−0.021	8.02	−0.021
Sumitomo	F0	2	629	0.21	6.41	−0.012	8.28	−0.024	8.31	−0.025	8.31	−0.025
Sumitomo	F1	4	583	0.22	5.46	−0.011	7.34	−0.025	7.47	−0.029	7.47	−0.029
Toshiba	F	37	616	0.21	6.02	−0.012	7.78	−0.024	7.90	−0.027	7.90	−0.027
Toshiba	F	36	613	0.21	6.02	−0.012	7.75	−0.024	7.86	−0.026	7.86	−0.026
Toshiba	G1F	33	656	0.21	5.20	−0.010	6.19	−0.017	6.23	−0.019	5.79	−0.033
Toshiba	G1F	34	656	0.20	5.14	−0.010	6.18	−0.018	6.19	−0.019	5.71	−0.034
Maraging Steels:												
VascoMax	C-350	90	190	0.29	2.40	−0.015	2.55	−0.026	2.58	−0.051	*	
VascoMax	C-300	91	188	0.30	2.08	−0.013	2.16	−0.022	2.17	−0.040	*	
VascoMax	C-250	92	188	0.32	1.75	−0.011	1.87	−0.021	2.08	−0.050	*	

bad strain gage—test terminated

* Run terminated at 5% axial strain amplitude with specimen intact.

Failure strength, S_f, is the magnitude of the axial compressive stress at which catastrophic failure of the specimen occurs. In the case of relatively brittle grades which do not reach a local stress maximum, it coincides with the ultimate strength. In tests with post-peak deformation, the failure strength is lower than the ultimate strength. Strain at failure, ε_f, is the axial strain at which catastrophic failure occurs. It is the maximum amplitude strain achieved in the test. Failure strength shows behavior similar to ultimate strength with slightly more scatter, around $\pm 1\%$.

Table 4 applies to two nonlinear rheological models which were fit to the stress-strain data. Both are useful for materials whose stress-strain curves do not demonstrate an abrupt reduction in slope at yield.

The first of these is the power law strain hardening relation:

$$-\sigma_z = A(-\varepsilon_{zp})^h,$$

where σ_z is the axial stress, and $\varepsilon_{zp} = \varepsilon_{z\,\text{total}} - \sigma_z/E_0$ is the plastic axial strain (LUBLINER, 1990, p. 74). The quantity σ_z/E_0 is the elastic strain taken as an extrapolation of the initial slope of the stress-strain curve. A and h (the coefficient of strain hardening) are adjustable parameters. The negative signs arise as a consequence of describing compressive stresses and strains as negative. A and h are determined from a least squares straight line fit to $\log(-\sigma_z)$ versus $\log(-\varepsilon_{zp})$ over a plastic strain range of approximately -10^{-3} to -10^{-2}. The maximum amplitude strain in the fit domain is selected subjectively to achieve a satisfying fit. Over the selected strain range, the fits are very good. The range of values for the coefficient of strain hardening for these carbides is about $0.14-0.23$. At plastic strain amplitudes in excess of $\sim 1.0\%$, $\log(-\varepsilon_{zp}) \cong -2$, the log-log plot shows substantial departure from linear; the plastic strain is larger in amplitude than is predicted by this relation. This pattern was also observed by HARA and IKEDA (1972). The region at which the power-law relation breaks down is also the region at which the specimen volume begins to increase.

The second relation explored is the original Ramberg-Osgood nonlinear stress-strain equation (RAMBERG and OSGOOD, 1943):

$$\varepsilon_z = \frac{\sigma_z}{Y_0} + \frac{3}{7}\frac{\sigma_{0.7}}{Y_0}\left(\frac{\sigma_z}{\sigma_{0.7}}\right)^n,$$

where σ_z is the axial stress, ε_z is the total axial strain. Y_0 (the initial Young's modulus), $\sigma_{0.7}$ (the uniaxial stress at a secant modulus of $0.7Y_0$), and n (the hardening index) are adjustable parameters. They are determined by a nonlinear least-squares fit to the stress-strain data using the Levenberg-Marquardt algorithm, (e.g. PRESS *et al.*, 1988, p. 542). The Ramberg-Osgood equation has proven very useful for modeling continuum plasticity. By its functional form, it is monotonic increasing and therefore does not accommodate post-peak deformation. Note that while the quantity Y_0 serves the functional role of an initial Young's modulus, its value as determined by the fit is typically slightly different from E_0 as defined previously. While the Ramberg-Osgood equation was originally developed empiri-cally, it has a basis in dislocation theory. With a slight modification it describes the stress over and above a linear limit stress as proportional to the square root of the increase in dislocation density associated with plastic strain (LEMAITRE *et al.*, 1990, p. 166). In the original formulation, this linear stress limit was taken as zero. While this is physically unrealistic for most materials, the difference is (usually) numeri-cally inconsequential. For a discussion of the Ramberg-Osgood equation as applied here in its original form see, for instance, LUBLINER (1990, p. 74) or SHANLEY (1957, p. 204).

As in the power-law case, the maximum amplitude axial strain for the fit must be limited in order to achieve satisfying fits. Limits were selected which produce a maximum axial strain residual to the fit of 5×10^{-4}, about 2.5% of the maximum axial strain amplitude in the fit domain. With this criterion, the stress-strain curves are fit to a maximum total axial strain amplitude of $\sim 2.0\%$.

Figure 4 shows examples of axial stress-strain curves (filled circles) and their associated Ramberg-Osgood equation fits including the two extreme strength grades

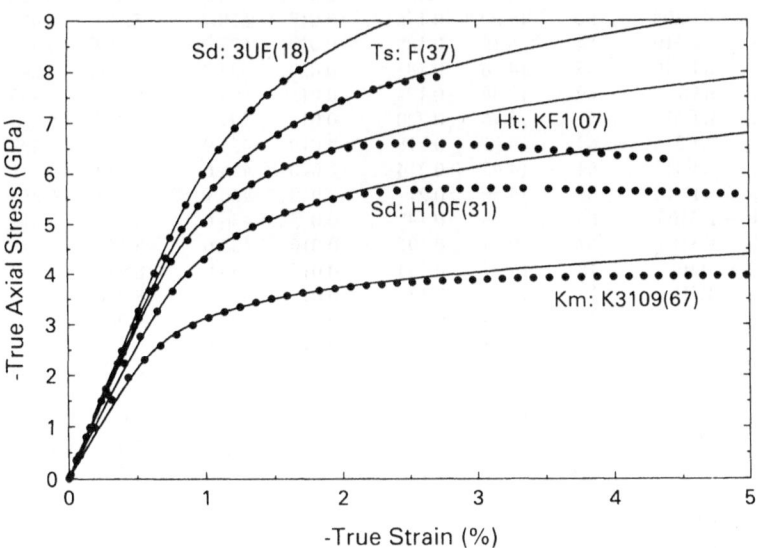

Uniaxial Compr. Stress-Strain Curves

Figure 4

Typical axial stress-strain curves (filled circles, representative points) and Ramberg-Osgood equation fits (solid lines). The curves presented represent the extremes of strength behavior plus three intermediate examples. These examples include stress-strain curves which are monotonic increasing to failure and curves which exhibit post-peak behavior. The Ramberg-Osgood equation was fit to approximately 2% axial strain amplitude. Departures beyond that level are associated with dilatancy. Manufacturer abbreviations are as in Figure 2. The numbers in parenthesis indicate test numbers.

Table 4

Rheological model fit parameters and fit axial strain ranges. Symbols are defined in the text above. Values are given with more significant figures than are physically warranted for computational consistency

Manf. units	Grade	No.	Power Law			Ramberg-Osgood Equation			
			A GPa	h	$\varepsilon_{z\,fit}$	Y_0 GPa	$\sigma_{0.7}$ GPa	n	$\varepsilon_{z\,fit}$
Federal	FC3M	75	15.26	0.177	−0.016	621.0	−5.756	5.974	−0.016
Federal	FC10M	79	14.06	0.195	−0.018	525.1	−5.011	7.496	−0.020
Federal	FC12M	81	14.03	0.232	−0.015	454.1	−4.113	7.341	−0.020
Federal	FC12M	82	14.71	0.238	−0.013	453.6	−4.149	7.509	−0.020
Federal	FC12M	83	14.63	0.238	−0.014	454.0	−4.131	7.420	−0.020
Fujilloy	TN05	45	15.09	0.153	−0.021	626.0	−6.629	6.753	−0.020
Fujilloy	TD05	39	13.93	0.143	−0.016	663.3	−6.385	8.405	−0.018
Fujilloy	TD05	40	13.65	0.140	−0.017	662.7	−6.389	8.387	−0.018
Fujilloy	TJ07	42	14.92	0.162	−0.015	608.4	−6.201	8.559	−0.018
Hertel	K05	12	14.81	0.166	−0.017	603.9	−6.081	7.592	−0.020
Hertel	KF1	10	15.54	0.180	−0.018	599.1	−5.853	6.191	−0.019
Hertel	KF1	7	14.64	0.170	−0.019	587.5	−5.940	7.539	−0.020
Kennametal	KF306	48	16.81	0.193	−0.017	602.4	−5.985	6.472	−0.020
Kennametal	KF306	49	15.92	0.182	−0.019	602.1	−6.003	6.500	−0.020
Kennametal	K313	60	15.62	0.182	−0.017	595.3	−5.869	6.765	−0.020
Kennametal	K313	61	15.46	0.180	−0.018	596.0	−5.867	6.741	−0.020
Kennametal	K68	72	14.22	0.171	−0.016	590.1	−5.701	7.960	−0.020
Kennametal	KF308	51	14.71	0.183	−0.018	568.0	−5.498	6.826	−0.020
Kennametal	KF308	52	14.20	0.176	−0.018	558.7	−5.581	7.491	−0.021
Kennametal	KF310	54	15.18	0.189	−0.019	527.2	−5.674	7.733	−0.023
Kennametal	KF310	55	14.76	0.183	−0.020	534.3	−5.659	7.241	−0.022
Kennametal	K96	69	13.46	0.174	−0.015	594.3	−5.191	7.405	−0.018
Kennametal	KF312	57	15.76	0.220	−0.017	509.2	−4.847	5.820	−0.020
Kennametal	KF315	63	14.71	0.218	−0.019	454.9	−4.700	6.402	−0.023
Kennametal	KF315	64	14.45	0.214	−0.020	456.6	−4.719	6.321	−0.023
Kennametal	KF315	65	14.66	0.217	−0.019	460.1	−4.692	6.256	−0.022
Kennametal	K3109	66	9.28	0.198	−0.015	436.6	−3.172	7.811	−0.020
Kennametal	K3109	67	9.17	0.197	−0.016	432.6	−3.167	7.778	−0.021
Sandvik	3UF	18	16.77	0.134	−0.017	636.8	−8.333	7.062	−0.017
Sandvik	3UF	19	16.91	0.138	−0.016	617.8	−8.280	7.019	−0.016
Sandvik	6UF	21	16.53	0.159	−0.021	593.7	−7.266	7.540	−0.025
Sandvik	8UF	24	15.23	0.154	−0.015	604.6	−6.657	6.402	−0.016
Sandvik	8UF	25	15.11	0.156	−0.023	555.8	−6.772	8.164	−0.026
Sandvik	H6F	27	14.38	0.170	−0.018	581.1	−5.838	8.095	−0.020
Sandvik	H10F	30	15.07	0.208	−0.016	520.2	−4.952	6.843	−0.020
Sandvik	H10F	31	15.18	0.210	−0.016	515.3	−4.972	7.029	−0.020
Sumitomo	F0	1	16.91	0.155	−0.021	643.0	−7.407	6.106	−0.020
Sumitomo	F0	2	17.02	0.157	−0.023	611.8	−7.557	7.423	−0.025
Sumitomo	F1	4	16.79	0.181	−0.020	556.6	−6.561	6.797	−0.025
Toshiba	F	37	15.30	0.150	−0.025	608.5	−6.917	7.023	−0.025
Toshiba	F	36	15.29	0.150	−0.024	607.4	−6.921	7.008	−0.025
Toshiba	G1F	33	13.72	0.156	−0.015	624.4	−5.856	8.868	−0.017
Toshiba	G1F	34	13.96	0.161	−0.015	649.0	−5.704	6.690	−0.015
Maraging Steels:									
VascoMax	C-350	90	3.05	0.039	−0.020	189.1	−2.818	29.46	−0.024
VascoMax	C-300	91	2.58	0.035	−0.016	186.7	−2.134	33.23	−0.017
VascoMax	C-250	92	2.30	0.044	−0.019	182.7	−1.812	27.66	−0.020

and three intermediate examples. The fits are universally good to $\sim 2\%$ axial strain amplitude, or to failure if it occurs before 2%. The departure beyond 2% is associated with dilatancy. Monotonic increasing behavior is modeled fairly well at larger strains. Post-peak behavior is not accommodated.

The application of uniaxial stress produces strain in both the direction of application and in transverse directions. The power-law relation and the Ramberg-Osgood equation provide descriptions of the stress-strain relation only in the direction of applied force. Some description of the transverse strains must be assumed to apply these descriptions in structural analyses.

The conventional assumption for isotropic continua is as follows: the longitudinal strain, ε_z for our cylindrical geometry, is made up of an elastic contribution, ε_{ze}, and a plastic contribution, ε_{zp},

$$\varepsilon_z = \varepsilon_{ze} + \varepsilon_{zp},$$

where

$$\varepsilon_{ze} = \frac{\sigma_z}{E_0}.$$

The transverse strain, ε_θ for our cylindrical geometry, is then given by the sum of an elastic contribution with the initial Poisson's ratio and an incompressible plastic contribution with Poisson's ratio equal to 0.5,

$$\varepsilon_\theta = \nu_0 \varepsilon_{ze} + 0.5\varepsilon_{zp}.$$

This expression includes the infinitesimal strain approximation that volume is conserved with Poisson's ratio equal to 0.5. While this turned out not to be a very satisfying approximation for the calculation of volumetric strain over the entire observed range, it does work well for the tangential strain. We can test the above expression numerically since we determine all of the relevant quantities: σ_z, ε_z, E_0, ν_0, and ε_θ. We have calculated transverse strains based on this assumption for both our observed axial stress-strain data and for the Ramberg-Osgood fits to that data.

Figure 5 shows an example of the Ramberg-Osgood equation fit to axial strain data for Kennametal KF308 to 2.1% axial strain amplitude. The fit is represented by solid curves and the data by individual plotted points for the curve labeled "Axial." Tangential strains calculated from the Ramberg-Osgood fit and the transverse strain assumption are shown in the solid curve labeled "Tan." The filled circles are actual data. The divergence beyond about 0.7% tangential strain is associated with dilatancy and is intrinsic to the Ramberg-Osgood equation as discussed previously. The transverse strain assumption holds very well to the same strain amplitude as is appropriate for the power law and Ramberg-Osgood equations themselves. For the Ramberg-Osgood equation, strain residuals for both longitudinal and transverse strains were less than 2.5% of the strain for axial strain amplitudes up to $\sim 2\%$. Similarly, the volumetric strain is well described by the

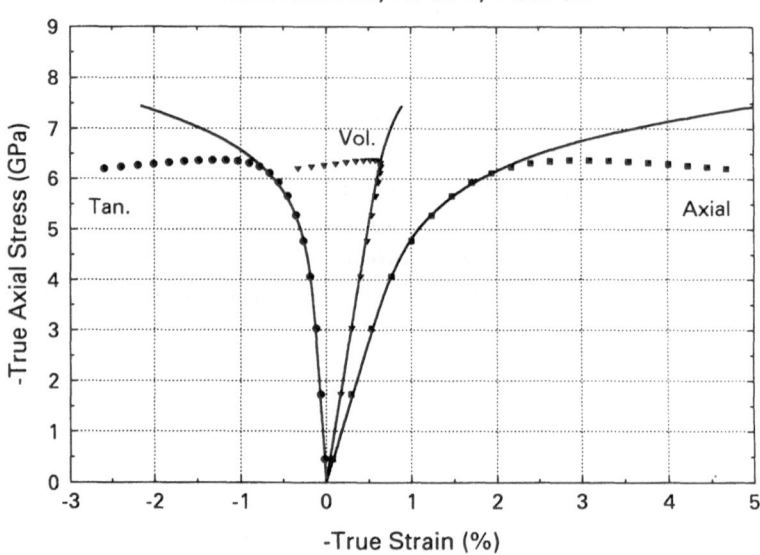

Figure 5

An example of the stress-strain curves based on the Ramberg-Osgood fit and the transverse strain assumption (solid lines) compared with the observed values (filled symbols, representative points) for a grade which exhibits substantial post-peak behavior. The quality of the fit and the divergence above 2% axial strain amplitude is evident in the axial curve. The volumetric curve is calculated from the other two for both the data and for the model. Prior to dilatancy, the model is very good.

Ramberg-Osgood fit and the transverse strain assumption over the limited strain range where the Ramberg-Osgood fit is appropriate. Until the significant microcrack volume is created, these descriptions of the rheology of cemented tungsten carbide are extremely good. These results are typical.

Various derivatives of the data in Figure 5 are shown in Figure 6. Here we show the Young's modulus, the volumetric modulus, and the Poisson's ratio as a function of axial strain. Representative data are plotted as filled symbols; the fits as curves. The Ramberg-Osgood equation typically underestimates the initial Young's modulus slightly, and then matches the data quite well to the onset of dilatancy. In the dilatant domain the observed Young's modulus becomes slightly negative (postpeak deformation) while the fit values remains positive as dictated by the functional form of the equation. The instability in the volumetric modulus, at about 2.1% axial strain amplitude, can be seen as the volumetric modulus points disappear out the top of the plot and return from the bottom. The Poisson's ratio is quite well modeled to about 1.7% axial strain amplitude where the data exhibit an inflection and the volumetric modulus starts to increase rapidly. This may be an indication of the onset of dilatancy. At strain amplitude larger than ~2% the Poisson's ratio

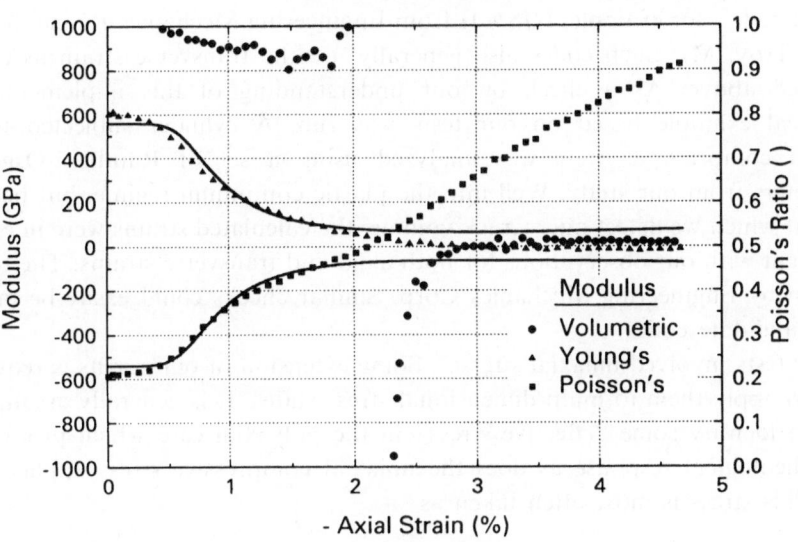

WC Stress-Strain Derivatives
Kennametal, KF308

Young's modulus, Poisson's ratio, and the volumetric modulus as a function of axial strain for a typical test. Representative observed values are plotted as filled symbols. Modeled values from the Ramberg-Osgood equation and the transverse strain assumption are shown as curves. The initial Young's modulus is typically underestimated slightly as a consequence of the functional form of the Ramberg-Osgood equation. Otherwise these derivative fits are generally quite good until the onset of dilatancy approaching ~2% axial strain amplitude. Our determination of dilatancy, C'' in the text, is made from the volumetric modulus. It trends toward plus infinity as the condition of volume conservation is approached and reappears from negative infinity as the volume begins to increase. Beyond this point, the modeled and observed values depart dramatically as the model is appropriate to continuum plasticity, not to dilatancy.

increases nearly linearly to a value approaching 1 at 5% strain amplitude. Specimen volume increases in this region.

The power-law relation and the Ramberg-Osgood equation fit the stress-strain data well over very nearly the same range. These materials have initial Young's moduli about 600 GPa and compressive yield strengths about 6 GPa. Thus the axial strain amplitude at yield is ~1.0%. As a consequence, the fit plastic axial strain amplitude of ~1.0% used for the power-law relation occurs at a total axial strain amplitude of ~2.0% used for the Ramberg-Osgood equation fits. The failure of both relations is approximately coincident with dilatancy.

Application to Structural Analysis

The Ramberg-Osgood equation is implemented in some 3-dimensional finite element codes, for example, NISA II from Engineering Mechanics Research Corporation, Troy, MI. Such codes also generally use the transverse strain assumption described above. As a check on our understanding of this implementation a numerical example based on our tests was run. A cylinder subjected to large uniaxial compressive stress was analyzed using a set of Ramberg-Osgood fit parameters from our study. Well into the plastic continuum regime, but below the strain at which we detect microcrack volume, the calculated strains were in excellent agreement with our observations for both axial and transverse strains. The test case was run by Engineering Mechanics Corp. Similar checks could easily be made on other candidate codes.

Our tests involved uniaxial stresses. Some extension of our results is required in order to apply them to multi-dimensional stress states. It is generally assumed that one can identify some "effective stress" in the polyaxial case which produces the same rheological response as does the uniaxial compressive stress applied in our tests. This stress is most often taken as

$$\sigma_{\text{eff}} = \frac{1}{\sqrt{2}} [(\sigma_1 - \sigma_2)^2 + (\sigma_2 - \sigma_3)^2 + (\sigma_3 - \sigma_1)^2]^{1/2},$$

where σ_{eff} is referred to literally as the effective stress. This formulation is developed from the concept of octahedral shear stress and appears in the Von Mises yield criterion, see for instance SEELY and SMITH (1967, pp. 63 and 90). It has proven highly reliable for reasonable ductile materials. It is employed in typical 3-dimensional finite element codes including NISA II. In the absence of extensive polyaxial testing, making such an assumption is the best we can do, but we must do it carefully. In the case of highly brittle materials, there are obvious cases where this assumption fails dramatically. For instance, if carbide specimens were to be loaded in uniaxial tension they would fail erratically at low stresses and strains. This case is completely misrepresented by our data and relatively useless structurally. We anticipate that in most cases where carbides are reasonable structural candidates, the formulation above will prove to be entirely reasonable.

From 1) the satisfying fits obtained to our data, 2) the verification of the finite element code implementation, and 3) a certain amount of experience with carbides in high-pressure applications we conclude that the Ramberg-Osgood equation, the transverse strain assumption, and the concept of effective stress may be used to perform accurate analysis of highly stressed carbide structures. The exact strain limit to which these analyses apply may be difficult to assess. To the extent that the relevant rheology is modeled realistically by our uniaxial compression tests, such calculations should be very good to compressive strain amplitudes of ~2.0%. This is the point beyond which significant microcrack volume occurs under uniaxial compression. In the polyaxial compressive stress fields typical of many high-pressure designs, however, strain amplitudes in excess of 2% might occur without significant microcrack volume. Polyaxial compression inhibits fracture. The intrinsic functional form of the Ramberg-Osgood equation and the transverse strain assumption lead to a very plausible rheological description beyond our ~2% fit range for a continuum without significant microcrack volume. By ~2% strain amplitude in our tests, and in the rheological model, the Young's modulus is practically zero and the Poisson's ratio is nearly 0.5. In normal continuum plastic behavior one would expect these two quantities to smoothly approximate these limiting values as the strain increases. The model approximates this behavior very well and may apply effectively in the continuum regime to significantly larger strain amplitudes. Figure 6 shows the observed and modeled values of Young's modulus and Poisson's ratio for a typical test. Under large polyaxial compression the model, solid curves in Figures 4 and 6, is likely to provide a more realistic description of continuum rheology than does our observed data for axial strain amplitudes in excess of ~2%.

All of the failures in our uniaxial compression tests were cataclysmic. The specimens "exploded" into a very large number of small fragments, few larger than ~1 mm, most much smaller. We suspect that this is a consequence of a well developed microcrack fabric in the specimens before failure and the large available stored energy in the press. The more brittle grades generally formed a larger number of smaller fragments. In most high-pressure applications such behavior is inhibited by polyaxial compression. Most carbide failures in high-pressure applications are not of the cataclysmic form we observed. They more typically involve plastic deformation beyond acceptable limits and/or a small number of major fractures. These fractures often leave the structure more or less intact, even if unsuitable for further use. The uniaxial compression tests which we have made are relatively nonconservative compared with most high-pressure applications. Two of the three principal stresses in our tests are not compressive. Other things being equal, one should expect better performance from carbides in polyaxial compression than was observed in our tests. This is encouraging since we generally observed an attractive combination of strength and ductility. It also implies that the uniaxial test should be a sensitive indicator of the relative behavior of the various grades examined.

A note of caution is in order. The principal aim of this study was to provide rheological information on various commercial carbide grades and to provide a means of applying this information realistically to structural analysis. In applying these data and methodologies, it must be born in mind that the actual performance of real structures may depend on many conditions beyond the state of stress and strain as calculated from our results. Creep was observed in these materials at room temperature and high stresses. Strain recovery was also observed on unloading. Initial creep rates were around $10^{-7} s^{-1}$ after loading over 5 ks (1.4 hours) to 5%

axial compressive strain amplitude, a highly dilatant condition. This creep rate was reduced many fold over an additional 2 ks at constant stress. Creep was not investigated at axial strain amplitudes below 2% in the continuum regime. The data we present are strictly applicable only to time scales comparable to our loading times. At significantly longer times, strains may be slightly greater and microfracture damage may be significantly increased. Creep effects have not been quantified in this study. We anticipate, however, that our relatively low strain rate tests provide a reasonable approximation to the quasi-static behavior of these materials. No effort was made to investigate the effects of static fatigue, low cycle fatigue, high cycle fatigue, or surface finish. No information was developed on the effects of loading rates, unloading rates, or reverse yielding. The possible effects of elevated temperatures and deleterious chemical environments were not studied. All of these phenomena, while beyond the scope of this study, can strongly affect the performance of highly stressed cemented tungsten carbide components in real applications. In short, the art of high stress design with carbides will probably remain, in large part, an art.

The art may be rationalized to some extent, however. Our tests provide a substantial basis for this process. They provide 1) a framework for understanding the relative behavior of different grades in various high-pressure apparatus, 2) an applicable catalog of the rheologies of many relevant carbide grades on which to base grade selections, and 3) a basis for the structural analysis of present and candidate high-pressure designs and the comparison of various grades in these designs.

Conclusions

A suite of uniaxial compression tests has been made on twenty-six grades of commercial tungsten carbide cermets of current interest in high-pressure apparatus operation. A test method has been developed that yields highly reproducible stress-strain curves to very high stresses and large plastic strains. All but the most brittle of the tested grades exhibit dilatancy whose onset occurs with a few percent of the ultimate stress. Two rheological modes associated with continuum plasticity have been fit to the data. They both describe our results very well to the onset of dilatancy. Implementation of one of these, the Ramberg-Osgood equation, in finite element code reproduces our uniaxial results in terms of both longitudinal and transverse strains. These results should be appropriate for analysis of highly stressed carbide structures as they are used in high-pressure apparatus. Polyaxial compression in these apparati is likely to increase the effective ductility of carbide components over that observed in uniaxial tests and to extend the applicable range of the fit models.

Appendix A

Axial Stress-strain Curves Grouped by Manufacturer

The axial stress-strain curves are presented together for the various tested grades of each manufacturer. This enables direct comparison within each group. The manufacturers are identified at the top of the plots. Individual grades are labeled on the plots themselves. A diamond at the end of a curve indicates catastrophic failure. Grades without a diamond failed catastrophically at an axial strain amplitude greater than 5%. Failure strains are listed in Table 3. Within the KF3xx suite of grades from Kennametal only the cobalt weight fraction is varied, from 6 to 15%. Other composition and grain size remain constant.

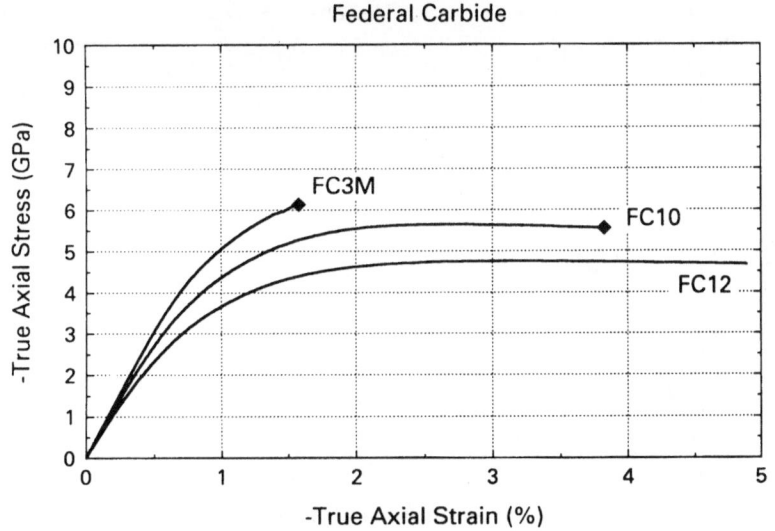

Figure A1

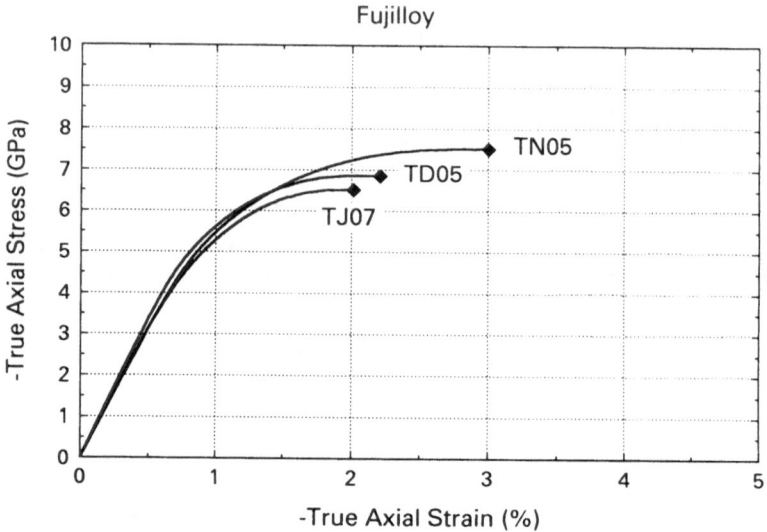

Figure A2

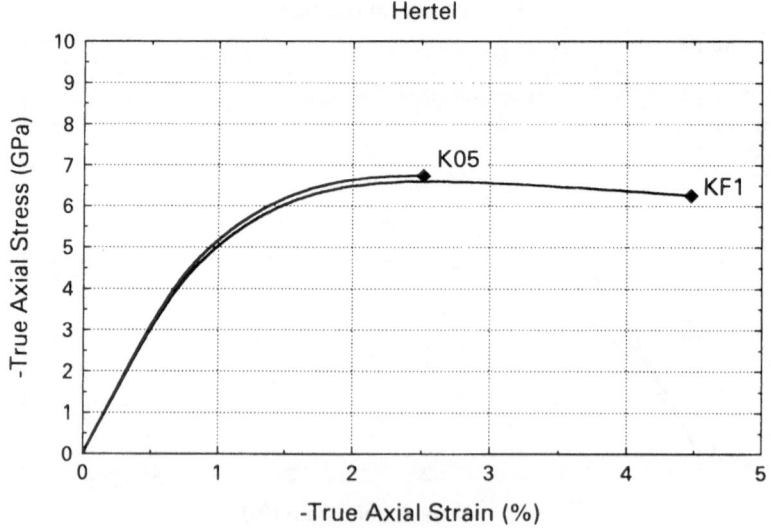

Figure A3

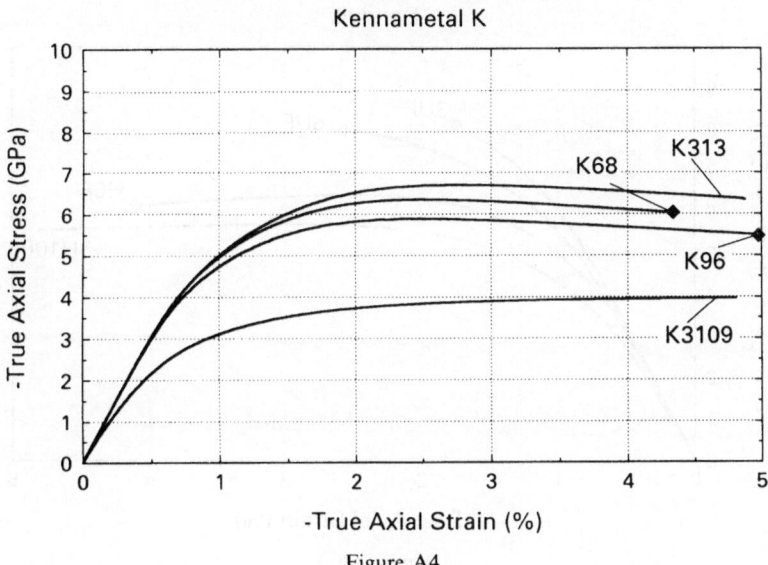

Figure A4

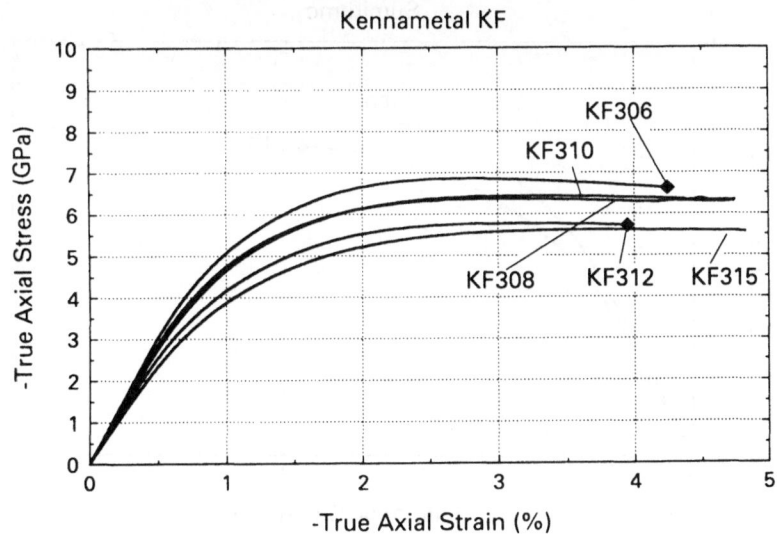

Figure A5

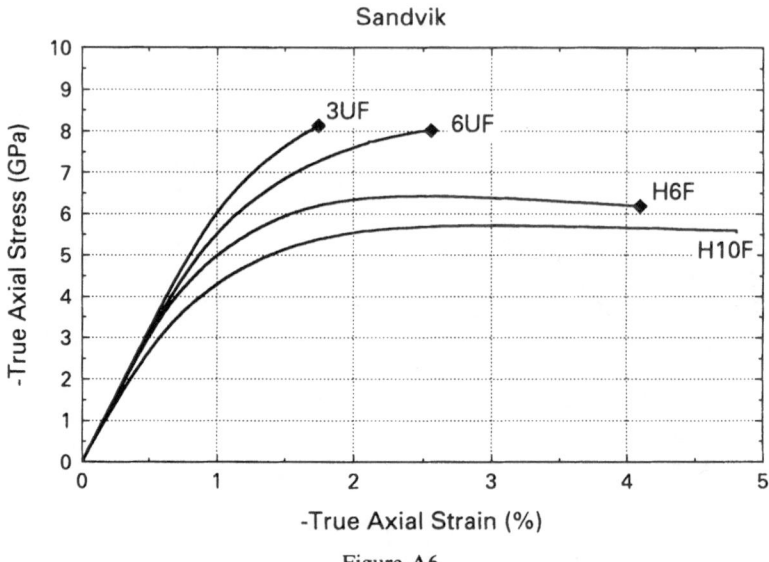

Figure A6

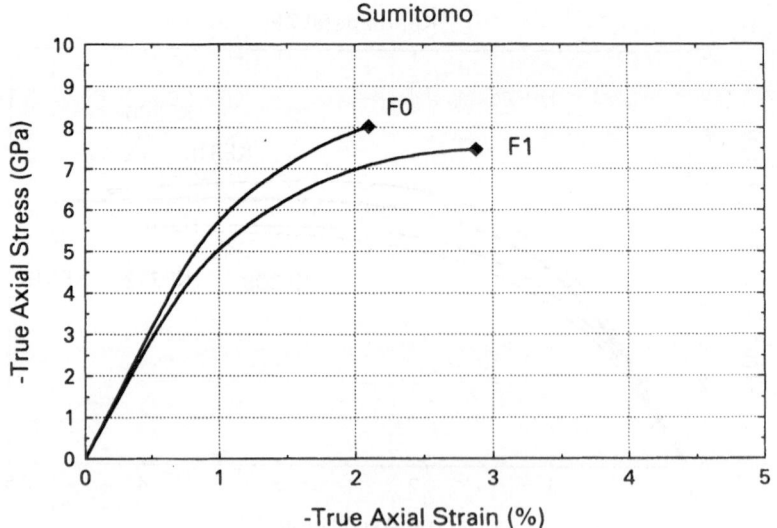

Figure A7

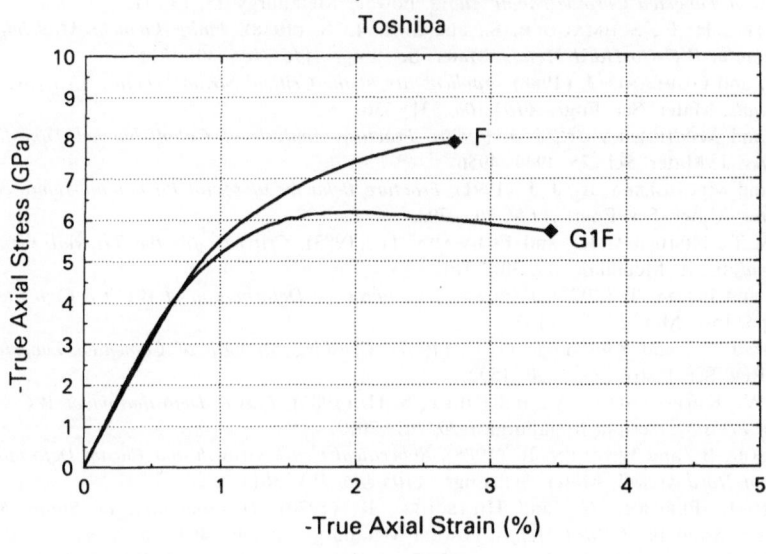

Figure A8

Acknowledgments

This work was supported by the National Science Foundation through grant number EAR–9018911 and by a consortium of institutions involved in high-pressure research. These supporting institutions were: The Geophysical Laboratory, The Lamond-Doherty Geological Observatory, The Lawrence-Livermore National Laboratory, The State University of New York at Stony Brook, The University of Alberta, and The University of Texas at Dallas. All of the carbide test specimens and anvils were donated by the manufacturers. Confirming finite element calculations were run by Jim Wang of the Engineering Mechanics Research Corporation, Troy, MI. The support of all of these organizations and associated individuals is gratefully acknowledged. The careful and thoughtful efforts of the reviewers, David Walker and Terry Tullis, are much appreciated.

REFERENCES

BRACE, W. F., PAULDING, B. W., and SCHOLZ, C. (1966), *Dilatancy in the Fracture of Crystalline Rocks*, J. Geophys. Res. *71*, 3939–3953.

DOI, H., FUJIWARA, Y., and MIYAKE, K. (1969), *Mechanism of Plastic Deformation and Dislocation Damping of Cemented Carbides*, Trans. Met. Soc. of AIME *245*, 1457–1470.

EXNER, H. E. (1979), *Physical and Chemical Nature of Cemented Carbides*. International Metals Reviews *4*, 149–173.

EXNER, H. E., and GURLAND, J. (1970), *A Review of Parameters Influencing Some Mechanical Properties of Tungsten Carbide-cobalt Alloys*, Powder Metallurgy *13*, 13–31.

FISCHMEISTER, H. F., SCHMAUDER, S., and SIGL, L. S. (1988), *Finite Element Modeling of Crack Propagation in WC-Co Hard Metals*, Mater. Sci. Engr. *A105/106*, 305–311.

GODSE, R., and GURLAND, J. (1988), *Applicability of the Critical Strain Fracture Criterion to WC-Co Hard Metals*, Mater. Sci. Engr. *A105/106*, 331–336.

HAN, D., and MECHOLSKY Jr., J. J. (1990), *Fracture Analysis of Cobalt-bonded Tungsten Carbide Composites*, J. Mater. Sci. *25*, 4949–4956.

HAN, D., and MECHOLSKY Jr., J. J. (1991), *Fracture Behavior of Metal Particulate-reinforced WC-Co Composites*, Mater. Sci. Engr. *A144*, 293–302.

HANABUSA, T., NISHIOKA, K., and FUJIWARA, H. (1983), *Criterion for the Triaxial X-ray Residual Stress Analysis*, Z. Metallkde. *74*, 307–313.

HARA, A., and IKEDA, T. (1972), *Behavior of Compressive Deformation of WC-Co Cemented Carbide*, Trans. Jpn. Inst. Met. *13*, 129–133.

HAYGARTH, J. C., and KENNEDY, G. C. (1967), *Crushing Strength of Cemented Tungsten Carbide Pistons*, Rev. Sci. Instru. *38*, 1590–1592.

JAYARAM, V., KRONENBERG, A., and KIRBY, S. H. (1986), *Plastic Deformation of WC-Co at High Confining Pressure*, Scripta Metallurgica *20*, 701–705.

JOHANNESSON, B., and WARREN, R. (1988), *Subcritical Crack Growth and Plastic Deformation in the Fracture of Hard Metals*, Mater. Sci. Engr. *A105/106*, 353–361.

JOHANSSON, I., PERSSON, G., and HILTSCHER, R. (1970), *Determination of Static and Fatigue Compressive Strength of Hard Metals*, Powder Metallurgy *13*, 449–464.

KERPER, M. J., MONG, L. E., STIEFEL, M. B., and HOLLEY, S. F. (1958), *Evaluation of Tensile, Compressive, Torsional, Transverse, and Impact Tests and Correlation of Results for Brittle Cermets*, J. Res. Nat. Bureau of Standards *61*, 149–169.

KRAWITZ, A. D., REICHEL, D. G., and HITTERMAN, R. L. (1989), *Residual Stress and Stress Distribution in a WC-Ni Composite*, Mater. Sci. Engr. *A119*, 127–134.

KRAWITZ, A. D., ROBERTS, R., and FABER, J., *Residual stress relaxation in cemented carbide composites*. In *Proc. 2nd Int. Conf. on the Science of Hard Materials* (ed. Almond, E. A., Brookes, C. A., and Warren, R.) (Adam Hilger Ltd., 1986) pp. 577–589.

LAUGIER, M. T. (1988), *Elevated Temperature Properties of WC-Co Cemented Carbides*, Mater. Sci. Engr. *A105/106*, 363–367.

LEMAITRE, J., and CHABOCHE, J., *Mechanics of Solid Materials* (Cambridge University Press, Cambridge, 1990).

LUBLINER, J., *Plasticity Theory* (Macmillan Publishing Co., New York, 1990).

NABARRO, F. R. N., and VEKINIS, G. (1988), *Pre-compression, Internal Stresses and Coercivity in WC-Co*, Mater. Sci. Engr. *A105/106*, 337–342.

PATERSON, M. S., *Experimental Rock Deformation, The Brittle Field* (M. S., Springer-Verlag, New York, 1978).

PELEPELIN, V. M. (1965), *Effect of Plastic Deformation of the Physicomechanical Properties of Tungsten Carbide-cobalt Hard Alloys*, Poroshkovaya Metallurgica *35*, 76–82.

PELEPELIN, V. M. (1967), *Variation in Density and Coefficient of Transverse Deformation of Hard Alloys*, Poroshkovaya Metallurgica *59*, 108–110.

PRESS, W. H., FLANNERY, B. P., TEUKLSKY, A. S., VETTERLING, W. T., *Numerical Recipes in C, The Art of Scientific Computing* (Cambridge University Press, Cambridge, 1988).

ROWCLIFFE, D. J. JAYARAM, V., HIBBS, M. K., and SINCLAIR, R. (1988), *Compressive Deformation and Fracture in WC Materials*, Mater. Sci. Engr. *A105/106*, 299–303.

SARIN, V. K., and JOHANNESSON, T. (1975), *On the Deformation of WC-Co Cemented Carbides*, Metal Science *9*, 472–476.

SCHMID, H. G., MARI, D., BENOIT, W., and BONJOUR, C. (1988), *The Mechanical Behavior of Cemented Carbides at High Temperatures*, Mater. Sci. Engr. *A105/106*, 343–351.

SEELY, F. B., and SMITH, J. O., *Advanced Mechanics of Materials* (John Wiley and Sons, Inc., New York, 1967).

SHANLEY, F. R., *Strength of Materials* (The Maple Press, York, PA 1957).

SPIEGLER, R., and FISCHMEISTER, H. F. (1992), *Prediction of Crack Paths in WC-Co Alloys*, Acta Metall. Mater. *40*, 1653 1661.

SURESH, S. (1988), *The Failure of Hard Materials in Cyclic Compression: Theory, Experiments and Applications*, Mater. Sci. Engr. *A105/106*, 323 329.

VANDEPUT, R. R., and MASTRANTONIS, N. (1988), *A Comparison of the Strength of WC-Co Measured by Ring and Transverse Rupture Strength Specimens*, Mater. Sci. Engr. *A105/106*, 423–428.

VASEL, C. H., KRAWITZ, A. D., DRAKE, E. F., and KENIK, E. A. (1985), *Binder Deformation in WC-(Co, Ni) Cemented Carbide Composites*, Metall. Trans. A *16A*, 2309–2317.

VEKINIS, G., and LUYCKX, S. B. (1987), *The Effects of Cyclic Precompression on the Magnetic Coercivity of WC-6wt%Co*, Mater. Sci. Engr. *96*, L21–L23.

(Received February 19, 1993, revised October 1993, accepted November 22, 1993)

PAGEOPH, Vol. 141, No. 2/3/4 (1993)

0033–4553/93/040579–21$1.50 + 0.20/0

The Large Volume Multi-anvil Press as a High *P-T* Deformation Apparatus

GILLES Y. BUSSOD,[1,2] TOMOO KATSURA[1,3] and DAVID C. RUBIE[1]

Abstract — The rheological properties of mantle materials are being investigated up to pressures of 16 GPa and temperatures of 1600°C for times up to 24 h, using a new sample assembly for the 6–8 multi-anvil apparatus. Al_2O_3 pistons, together with a liquid confining medium, are used to generate deviatoric stress in the specimen. Strain rates are estimated by monitoring the relative displacement of the guide blocks of the multi-anvil apparatus, scaled to the total axial strain of the sample. The applied stress on the sample is estimated using grain size piezometry. Strain rates and flow stresses of approximately 10^{-4} to 10^{-6} s^{-1} and 50 to 250 MPa respectively, are presently attainable.

Preliminary results on San Carlos olivine single crystals, partially dynamically recrystallized to a grain size of 10 to 300 μm, indicate that the effective viscosity of polycrystalline olivine is consistent with values obtained from olivine single crystal creep laws. Assuming a dislocation creep mechanism ($n \approx 3.5$) with (010)[001] as the dominant slip system, the data are best fit using a creep activation volume of 5 to 10×10^{-6} m^3 mol^{-1}.

Key words: High pressure, deformation, olivine, multi-anvil, mantle rheology, activation volume.

Introduction

Knowledge of the rheological behavior of upper and lower mantle mineral assemblages is essential for understanding the dynamical behavior of the deep earth and the formulation of reliable models for mantle convection. Due to significant advances in recent years, we now have a reasonable understanding of the rheology of $(Mg,Fe)_2SiO_4$ olivine, the dominant silicate mineral in the earth's upper mantle. Classical experimental deformation apparatus however, are restricted to a relatively low pressure range (0.1 MPa–3 GPa), and the dependence of the rheological behavior of olivine on depth, represented by an activation volume in the constitutive equations, has been the subject of relatively few studies (ROSS *et al.*, 1979; GREEN and BORCH, 1987, 1990; BORCH and GREEN, 1989). Many phase transi-

[1] Bayerisches Geoinstitut, Universität Bayreuth, D-95440 Bayreuth, Germany.
[2] Presently at: Los Alamos National Laboratory, EES-4, Los Alamos, NM 87545, U.S.A.
[3] Presently at: Institute for Study of the Earth's Interior, Okayama Univ., Misasa, Tottori-ken 682-01, Japan.

tions are known for mantle minerals (RINGWOOD and MAJOR, 1966; IRIFUNE and RINGWOOD, 1987; ITO and TAKAHASHI 1987, 1989; KATSURA and ITO, 1989), which must greatly complicate mantle rheology as a funtion of depth. Olivine transforms to modified spinel (ca. 400 km), and subsequently to a spinel structure (ca. 520 km) and finally dissociates to $(Mg,Fe)SiO_3$ perovskite and magnesiowüstite (ca. 660 km). Pyroxenes gradually dissolve into garnet (300–520 km), forming majorite. $CaSiO_3$ perovskite exsolves from majorite (550–660 km), and the remaining majorite dissolves into $(Mg,Fe)SiO_3$ perovskite (660–720 km). As a consequence, the rheology of mineral assemblages representing 95 vol.% of the earth's mantle is unknown.

The most widely employed apparatus for high-temperature and high-pressure rock mechanics studies and which comes closest to reproducing the actual *P-T* conditions of the earth's upper mantle is the Griggs-Blacic piston-cylinder type apparatus, originally designed by D. T. Griggs and co-workers in the early 1960s, in which deformation tests are routinely performed at 1.5 GPa confining pressure (CARTER *et al.*, 1961; CARTER, 1964; GRIGGS and BLACIC, 1964; GRIGGS, 1967). In a piston cylinder design, the maximum total pressure is limited by the tensile strength of the pressure vessel and the compressive strength of the pistons. The maximum pressure at which deformation can be performed routinely is currently 3 GPa (GREEN and BORCH, 1990; GREEN, 1992) and a further increase in confining pressure is not likely until new materials of much greater strength and toughness are developed.

The large-volume split-sphere multi-anvil apparatus has an inherent advantage over the piston cylinder design in that it can attain much higher confining pressures because parts under high stress are loaded in compression and/or are laterally supported by gaskets. However, before quantitative stress-strain relations can be determined at high pressures and temperatures, it will be necessary to develop *in situ* stress measurement capabilities. Several attempts to exploit the deviatoric component inherent to the multi-anvil apparatus have been reported previously (WANG *et al.*, 1988; GREEN *et al.*, 1990; WEIDNER *et al.*, 1992), in addition to attempts to minimize the deviatoric stress (LIEBERMANN and WANG, 1992; RUBIE *et al.*, 1993).

In this work, an initial series of experiments were undertaken to explore the use of the 6–8 type multi-anvil as a constant displacement-rate deformation apparatus. These experiments confirm that experimental plastic deformation of deep mantle phases is possible, much as was done previously for quartz in the cubic anvil apparatus (GRIGGS, 1936; GRIGGS *et al.*, 1960). For this purpose, a modified pressure assembly has been designed which permits the uniaxial compression of specimens at confining pressures up to 16 GPa and temperatures to 1600°C.

Experimental Assembly Design and P-T Calibration

For nearly 30 years, the multi-anvil apparatus has now been used successfully as a "pseudo-hydrostatic" solid-medium press, to attain pressures up to 40 GPa (ITO *et al.*, 1984; OHTANI *et al.*, 1989; LIEBERMANN and WANG, 1992). In this design, an octahedron of the solid pressure medium is inserted into a nest of tungsten carbide or sintered diamond inner cubic anvils with truncations of various edge lengths. In the present study octahedral pressure assemblies with edge lengths of 18.0 mm (18M) and 14.0 mm (14M) are used with tungsten carbide (Toshiba Grade F) anvils (truncation edge lengths of 11.0 mm and 8.0 mm, respectively) to generate confining pressures of up to 10 GPa and 16 GPa, respectively. 5 mm wide pyrophyllite strip gaskets, 2.5 mm thick (14M assembly) and 3.0 mm thick (18M assembly) respectively are used as pressure seals, together with thin cardboard backings.

Confining pressure was calibrated at room temperature using transitions of bismuth (Bi I–II: 2.55 ± 0.01 GPa, Bi III–V: 7.7 ± 0.3 GPa; LIU and BASSETT, 1986) and the ZnS semiconductor to metal transition (15.6 ± 1.0 GPa; PIERMARINI and BLOCK, 1975; YAGI and AKIMOTO, 1976). The olivine-spinel transformation in the Mg_2SiO_4-Fe_2SiO_4 system (AKAOGI et al., 1989; KATSURA and ITO, 1989), and the coesite-stishovite (YAGI and AKIMOTO, 1976) and quartz-coesite trans-formations (BOYD and ENGLAND, 1960; BOHLEN and BOETTCHER, 1982) were also

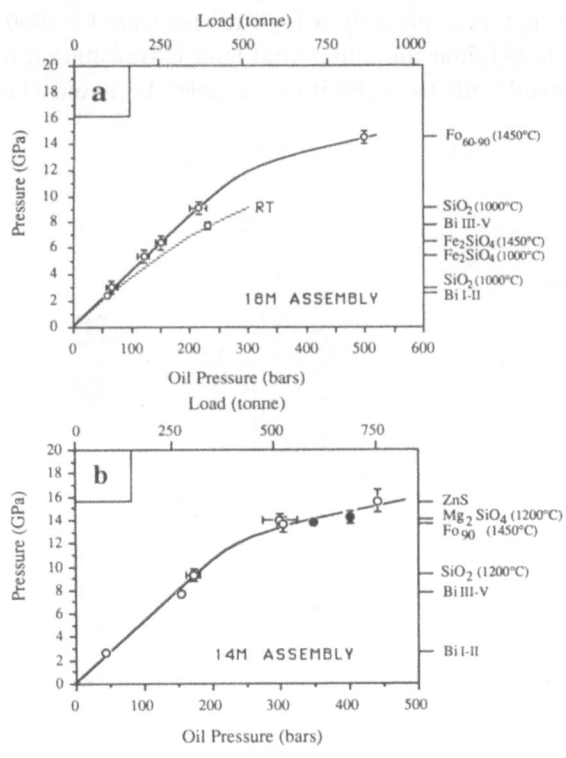

Figure 1
Pressure calibrations for 6 8 type multi-anvil apparatus for (a) 18M and (b) 14M assemblies. Solid symbols represent calibrations in the new deformation assemblies. Note that above approximately 250 bars oil pressure, the efficiency of pressure generation is substantially reduced. Bismuth data at ambient temperature not included in regression fit (18M).

used for calibrations of the 18M and 14M assemblies (Fig. 1) at 1000°C and 1200°C, respectively. The calibration for the 18M assembly was further constrained at higher pressure by using the high-pressure transitions in four separate olivine solid-solution powders in a single experimental charge (Fo_{90}, Fo_{80}, Fo_{70} and Fo_{60}), reacted for 3 h at 1450°C and 500 bars ram oil pressure (Fig. 1a). The $\alpha - \beta$ transition (Fo_{90}) at 1450°C was used as a calibration point for the 14M assembly (Fig. 1b). Within error, these results are in agreement and consistent with the thermochemically calculated and experimentally determined phase diagrams (AKAOGI *et al.*, 1989; KATSURA and ITO, 1989).

Cross sections of the sample assemblies used are represented schematically in Figure 2. A cylinder of single-crystal olivine (1.8–2.3 mm long × 1.1 mm diameter), is jacketed in rhenium foil 0.05 mm thick, and enclosed in a sodic alumino-boro-silicate capillary glass tube [70%SiO_2, 16%Na_2O, 8%Al_2O_3, 3.5%CaO, 2%B_2O_3 by weight], with a glass transition temperature of 770°C at 1 atmosphere. Because the capillary tube remains solid (glass) at ambient temperature, the sample compresses uniformly, preserves a cylindrical geometry and does not become canted during pressurization. At high pressure and at high temperature (> 1000°C), the confining medium becomes liquid, and the differential load is transmitted uniaxially via hard Al_2O_3 pistons coaxial with the cylindrical sample. To avoid chemical contamina-

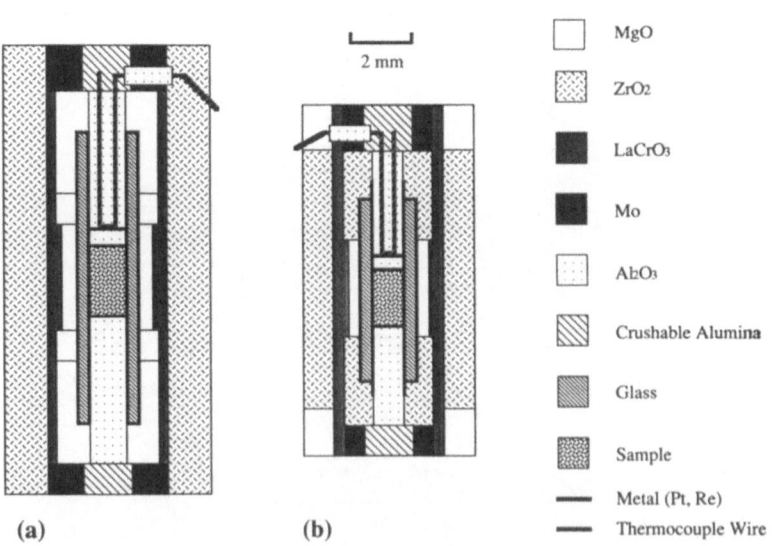

Figure 2
New 18M and 14M deformation assemblies for the 1200 ton 6–8 type multi-anvil press. These cylindrical assemblies are inserted into MgO (+ 5% Cr_2O_3) octahedra with 18 mm and 14 mm edge lengths, respectively. Sample is deformed uniaxially between two Al_2O_3 pistons and surrounded by a silicate liquid at high *P* and *T*. These assemblies are functional up to 1600°C at 14 GPA (18M) and 16 GPA (14M), respectively.

tion, the sample is separated top and bottom from the pistons by Re shims (0.025 mm thick). The capillary glass tube (confining medium) is enclosed in a platinum tube and is of sufficient length that the ends of the tube are out of the hot spot region of the furnace and thus remain solid. This design constitutes a seal which restricts the liquid confining medium to the sample area and prevents the degradation of the $LaCrO_3$ furnace by the silicate liquid. Attempts were made to use NaCl as the confining medium in lieu of the capillary glass, but all of these failed due to furnace corrosion by the salt (cf. Table 1).

Temperature was measured using a $W_{97}Re_3/W_{75}Re_{25}$ thermocouple consisting of 0.127 mm diameter wires contained axially in the top alumina piston. The thermocouple junction was made mechanically by crossing the two wires. The thermocouple is separated from the sample by a solid end piece 1.0 mm diameter and 0.3 mm thick, to avoid indentation of the sample by the TC junction. Molybdenum rings are used as electrical contacts between the furnace and WC anvils (Fig. 2).

Because of the importance of temperature in thermally activated deformation processes, efforts were made to minimize thermal gradients over the length and width of the sample during deformation. The metal jackets (Pt and Re) are used in conjunction with a stepped furnace design, along with thermally insulating stabilized zirconia sleeves and end-pieces (14M). The longitudinal and radial thermal gradients in the 14M sample chamber were estimated by reacting a 1:1 (wt.%) mixed powder of end-member diopside and enstatite ($CaO-MgO-SiO_2$ system), with a 2 wt.% barium oxide–borate flux added (wt. fraction: 0.78 $BaO-0.22$ B_2O_3) for five hours at 10 GPa and 1450°C. Temperatures recorded by the pyroxenes were determined using compositions from 30 coexisting pairs, distributed over the length and width of a cross-sectional area of the sample, using the empirically calibrated pyroxene solvus (WELLS, 1977). The results are presented in Figure 3. Within the precision and accuracy of this method ($\pm70°C$), no temperature gradient is discernible, and the apparent absolute temperature obtained ($1480 \pm 30°C$) compares well with the measured temperature (1450°C), utilizing the $W_{97}Re_3/W_{75}Re_{25}$ thermocouples. The results are consistent with small temperature variations across the sample ($\leq 20°C$ at 1400°C), measured in multi-anvil assemblies with similar furnace designs (KANZAKI, 1987; YASUDA et al., 1990). Temperature measurements are uncorrected for the possible effect of pressure on the thermocouple emf (GETTING and KENNEDY, 1970).

The chemical environment of the samples was estimated following the procedure outlined in RUBIE et al. (1993). An infrared absorption spectrum of the starting olivine single crystal indicates a low H/Si ratio ($\approx 10^{-5}$). It was not possible to determine the H/Si ratio after high-pressure deformation experiments, but it is also assumed to be low because the silicate liquid confining medium is expected to act as a hydrogen getter.

The presence of ubiquitous microscopic inclusions of enstatite in the sample implies the silica activity [a_{SiO_2}] was buffered (O'NEILL and WALL, 1987).

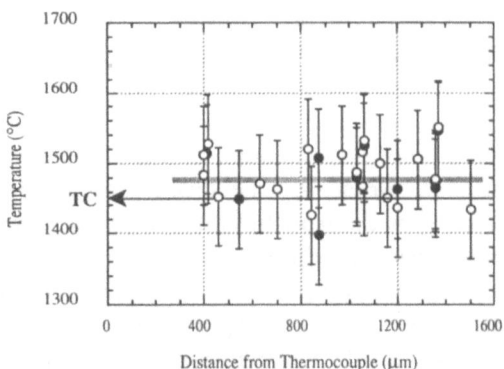

Figure 3
Estimate of temperature profile from a 5 h experiment at 10 GPa, 1450 C, for the 14M deformation assembly using pyroxene equilibria (WELLS, 1977). The abscissa represents distance measured axially along the sample assembly (parallel to σ_1). The data represent temperature estimates from pyroxene pairs located in the central area of the sample (open circles) as well as along the sample edges (solid symbols), and cover the entire width and length of the specimen. Solid line terminated by arrow indicates the value and relative location of the $W_{97}Re_3/W_{75}Re_{25}$ thermocouple. A linear regression fit to all 30 coexisting pyroxene pairs gives an average temperature of 1480 ± 30 C (shaded line).

To estimate the fO_2 of the run conditions, platinum shims were placed at the top and bottom of specimens MA 538 and MA 696 which were deformed at 1600 C and 6 GPa. The Fe contents in the coexisting platinum and olivine were used to evaluate fO_2 using the reaction:

$$Fe_2SiO_{4[Ol]} = 2Fe_{[Pt]} + SiO_2 + O_2. \tag{1}$$

The value obtained for both experiments, $\log fO_2 = -2.9$ (± 0.3) log bar units, is similar to the value of $\log fO_2 = -2.76$ (± 0.5) log bar units calculated from the data of O'NEILL (1987) for the Ni-NiO buffer at identical P-T conditions.

Experimental Procedure

The geometry of the pressure assembly is such that the long axis of the sample cylinder is coaxial with the axis of the loading ram (Fig. 4). This simple but essential configuration is compatible with the use of a 6–8 type multi-anvil apparatus as a deformation apparatus. Since the sample strain cannot be measured *in situ*, a relation must be established between the displacement of the loading ram of the press and the deformation or strain of the sample. In order to obtain this relation, over 30 experiments were attempted to pressures up to 16 GPa, in which the samples were recovered at various stages of the pressurization, heating and deformation. A single experiment at 23 GPa, using the smaller 7M assembly with 3 mm edge truncations, indicates that higher pressure experiments are also possible.

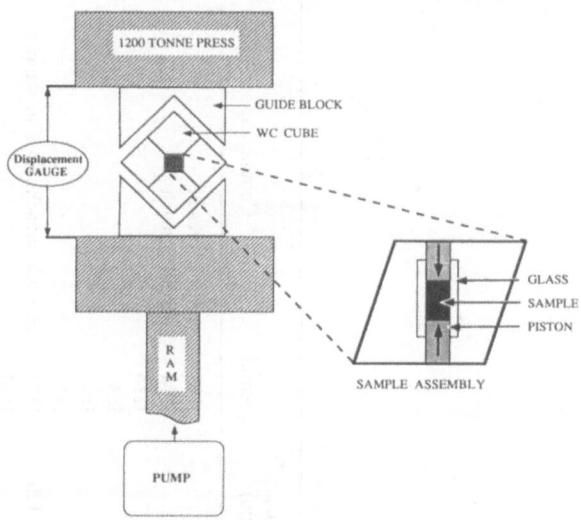

Figure 4

Schematic representation of the geometry of the ram-guide block-WC cube-sample assembly system for the 1200 ton 6 8 type Sumitomo press used in experiments.

The conditions and results of experiments restricted to the olivine stability field are summarized in Table 1. The single crystals used were originally oriented using X-ray diffraction so that the [010] olivine crystallographic axis was subparallel ($\pm 10°$) to the compression axis σ_1. This corresponds to the "strongest" orientation for olivine as the principal slip plane (010) is oriented perpendicular to σ_1.

The total axial sample strain was determined for each experiment using optical microscopy measurements of the initial (single crystal) and final sample lengths. Because the total sample strain at the end of each experiment must be accurately known, only single crystals or fully dense compacted cylindrical samples could be used in the developmental stages of this procedure. Compaction effects, associated with powders or porous synthetic aggregates, can greatly distort the sample geometry and make geometric measurements or the determination of total strain inaccurate and irreproducible.

The run procedure consists of increasing the loading ram oil pressure at room temperature until the desired operating pressure is reached (Fig. 5). Once at pressure, the temperature is raised to the desired value at a rate of 75°C/min. Prior to the high P-T deformation stage, the sample, which now consists of a strongly deformed olivine single crystal, is allowed to relax at high temperature (Fig. 6). During the subsequent deformation stage (Fig. 6, stage C), monitoring of the sample strain rate is implemented by measuring the rate of advancement of the guide block (see below). The hydraulic pumping system automatically advances the guide block to maintain a preset rate of increase in oil pressure and compensates for a slow effective pressure leak due to gasket flow.

Table 1

Multi-anvil deformation experiments

MA #	O.P.[1] (bars)	P_c[2] (GPa)	T[3] (°C)	$\varepsilon_s/\varepsilon_{oct}$[4] (%)	$\bar{D}_g$[5] (μm)	$\dot{\varepsilon}_s$[6] (sec^{-1})	σ[7] (MPa)	t_{relax}/t_{deform}[8] (min)	Assembly	Comments
465	140	6.0	1600 (20)	22.2/20.4	35 (4)	8(2) × 10^{-6}	87	63	18M	capsule breach
486	140	6.0	1600**	20.9/20.7	31 (4)	7(2) × 10^{-6}	95	54	18M	capsule breach
505	140	6.0	300 (5)	15.8/16.1				0	18M	
514	140	6.0	1600 (10)	24.0/20.9				13	18M	
523	140	6.0	1600 (20)	23.5/21.6				13	18M	
530	140	6.0	1600 (10)	29.0/21.8				55	18M	
538	140/200*	6.0$^+$	1600**	35.6/21.2	287 (120)	3(1) × 10^{-5}	>20(GBM)	13/120	18M	capsule breach; glass in sample
634	140	6.0	1450**	22.5/21.4	34 (11)	7(1) × 10^{-6}	89	60	18M	
639	140/160*	6.0$^+$	1450**	?/21.2				20/40	18M	
651	140/160*	6.0$^+$	1450 (5)	23.1/21.6	14 (7)	5(1) × 10^{-5}	180	33/20	18M	
656	140/160*	6.0$^+$	1450 (5)	27.2/21.5	12 (9)	4(1) × 10^{-5}	203	27/40	18M	
696	140/160*	6.0$^+$	1600 (5)	53.9/21.3				28/20	18M	capsule breach; sample melted
699	140/160*	6.0$^+$	1450 (5)	28.0/21.8	13 (8)	5(1) × 10^{-5}	190	36/20	18M	
736	140	6.0	1450**	23.5/22.2	37 (13)	4(1) × 10^{-6}	83	60	18M	
738	140	6.0	1450 (5)	25.3/21.2	43 (9)	2(1) × 10^{-6}	74	176	18M	

Sample	[1]	[2]	[3]	[4]	[7]	[6]	[5]	[8]	[9]	Comments
743	140	6.0	1450**	25.2/21.6	92 (36)	$3(1) \times 10^{-6}$	>45(GBM)	177	18M	capsule breach; glass in sample
700	310	12.5	50	26.1/25.3				3	14M[9]	blow out
705	310	12.5	1450**	34.3/24.1				13	14M[9]	sample lost during TS preparation
712	310	12.5	1450 (5)	28.0/25.8	106 (24)	$5(1) \times 10^{-6}$	>40(GBM)	180	14M[9]	capsule breach; NaCl in sample
768	310/350*	12.5+	1450 (5)	27.8/23.9				180/20	14M	
789	290/330*	12.0+	1450 (5)	25.2/25.3	10 (2)	$9(1) \times 10^{-5}$	235	181/20	14M	
926	350	13.5	1450 (5)	18.5/24.5				2	14M	β-phase field

(1) Oil pressure in bars. Varied in some cases* in order to conduct 'higher strain-rate' deformation experiments. The final pressures of relaxation and deformation stages are shown.

(2) Confining pressure in GPa. During deformation the actual pressure is assumed equal to or slightly higher (+) than the initial pressure.

(3) Sample mean temperature in °C. Experiments in which the thermocouple failed were power controlled**. Uncertainty in parentheses includes temperature variations.

(4) Uncorrected final longitudinal sample (ε_s) and octahedron strains (ε_{oct}).

(5) Mean recrystallized grain size in μm. Sample population is $N = 200$ except samples MA 538 ($N = 40$), MA 712 ($N = 120$) and MA 743 ($N = 50$). Errors in parentheses represent one standard deviation.

(6) Sample strain rates. Uncertainties in parentheses represent errors in sample length measurements only.

(7) Flow stress in MPa estimated from grain size piezometry. Samples exhibiting grain growth by diffusion-enhanced grain boundary migration are marked (GBM). The stress estimates based on these samples are unreliable and represent a lower limit.

(8) Times corresponding to stages B (t_{relax}) and C (t_{deform}) of Figure 6.

(9) Assemblies in which the capillary glass was replaced by compact NaCl powder.

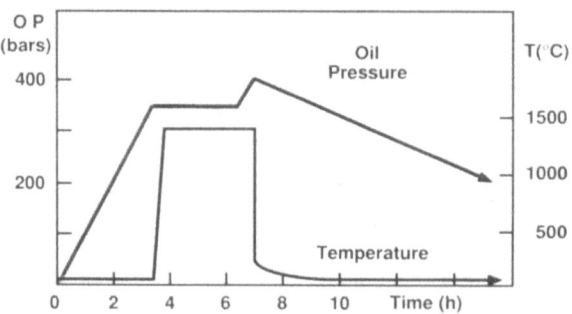

Figure 5
Experimental procedure for the multi-anvil deformation experiments. (a) Ram oil pressure (OP) versus
time (gray line), and (b) temperature versus time (solid line). See text for details.

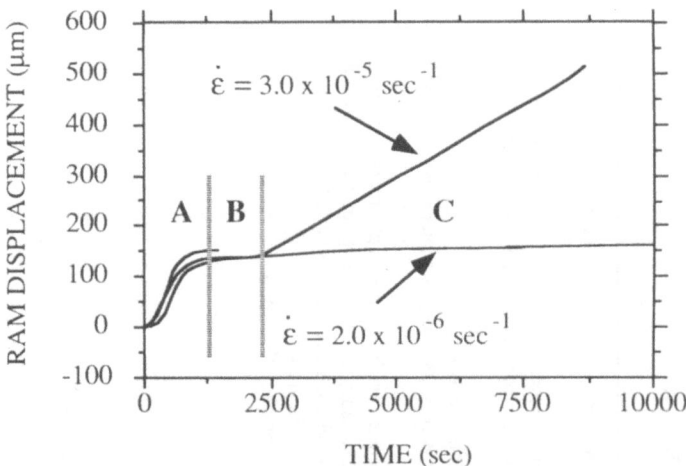

TIME (sec)

Figure 6
Examples of ram displacement versus time for several experiments. Regions A, B and C represent
various stages of the experiments. Stage A corresponds to the increase of temperature at high pressure
(e.g., MA 926). Stage B corresponds to a relaxation time (variable) of the assembly and sample at
high temperature prior to stage C represented by a prolonged relaxation at constant oil pressure (e.g.
MA 738; $\dot{\varepsilon} \approx 10^{-6} \, s^{-1}$) or a deformation experiment induced by further increasing the oil pressure (e.g.
MA 538; $\dot{\varepsilon} \approx 10^{-5} \, s^{-1}$). The accessible range of strain rates is approximately two orders of magnitude
$(10^{-4}-10^{-6} \, s^{-1})$ depending on the original sample length and the rate of increase of oil pressure.

During deformation (stage C, Fig. 6), the minimum possible strain rate
($\dot{\varepsilon} \approx 10^{-6} \, s^{-1}$), is achieved by maintaining a constant oil pressure. Higher strain
rates (e.g., $\dot{\varepsilon} \approx 10^{-6}-10^{-4} \, s^{-1}$), are achieved by increasing the oil pressure (0.5–
2.0 bar/min), so as to produce a relatively fast guide block advancement rate.
Increasing oil pressure at high temperature probably does not substantially increase
the confining pressure as the efficiency of pressure generation is reduced due to flow
of the gasket material. That this is the case is suggested by experiments using the

14M assembly, in which samples were annealed just below the $\alpha - \beta$ transition boundary (350 bars oil pressure; MA 926, Table 1) and subsequently deformed by further pumping (c.f., MA 768 and MA 789; Table 1). In these experiments no β phase was formed.

Following deformation, the sample is quenched to $< 200 °C$ in $1-2$ sec by cutting the power to the furnace at high pressure and then decompressed slowly at 0.5 GPa h^{-1} to avoid blow outs.

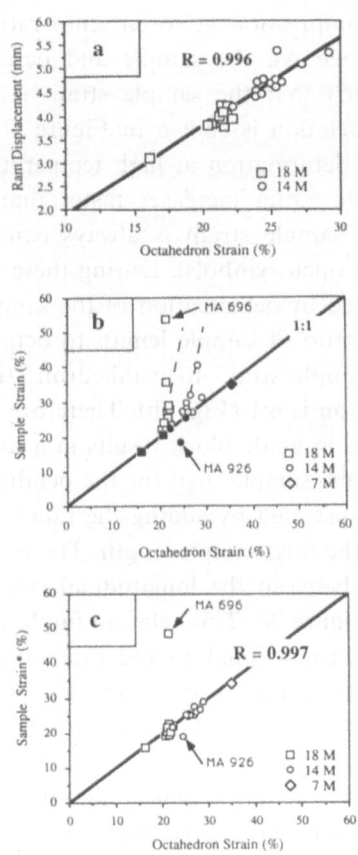

Figure 7

(a) Measured final axial strain of MgO octahedra versus maximum ram displacement at both room temperature and high temperature (Fig. 6; stages A, B and C) for 14M (open circles) and 18M (open squares) assemblies. Data are not corrected for compressibility. (b) Measured final longitudinal sample strain versus final coaxial octahedron strain for 7M (diamond), 14M (circles) and 18M (squares) assemblies. Departure from linear correlation found in samples quenched during stage A of Figure 6 (solid symbols), is a consequence of high-temperature deformation (see text). Experiments MA 696 (melted) and MA 926 (transformed to β phase) are anomalous (cf. Table 1). Dashed lines represent a sample to octahedron strain ratio of approximately 6:1. (c) Corrected longitudinal sample strain (*) versus octahedron strain (see text). Linear regression does not include samples MA 696 and MA 926.

The displacement of the guide block is monitored by an external displacement gauge as the ram is advanced during compression at experimental conditions. The ram displacement vs. time curve (Fig. 6, stages B and C) demonstrates that during the high-temperature deformation stage the ram advancement rate is constant.

The total longitudinal strains of the octahedron and the sample, in a direction coaxial with the multi-anvil ram, are determined by measuring their respective lengths before and after each experiment. The relation between the ram advancement rate and the total coaxial strain of the octahedron assemblies is approximately linear (Fig. 7a). During compression at room temperature and during the initial heating period (Fig. 6, stage A), the sample and octahedral pressure assembly deform homogeneously such that the sample strain is equal to the octahedron strain; this one-to-one correlation is shown in Figure 7b (solid symbols). During subsequent relaxation and deformation at high temperature (Fig. 6, stages B and C), the longitudinal sample strain increases faster than the coaxial octahedron strain, such that the final sample strain is always equal to or greater than the octahedron strain (Fig. 7b; open symbols). During these stages, the overall octahedron strain is accommodated by deformation of the sample and the Al_2O_3 pistons remain rigid. Because the ratio of sample length to octahedron length is approximately 1/6, the ratio of sample strain to octahedron strain that develops during high-temperature deformation is 6:1 (Fig. 7b). Therefore, under these conditions, a displacement of the advancing guide block results in a longitudinal strain approximately 6 times greater for the sample than for the octahedron. The validity of this interpretation can be demonstrated by adding the ram displacement, which occurs during stages B and C, to the final sample length. The result of this correction is to restore the linear relation between the longitudinal strain of the sample and the octahedron, as shown in Figure 7c. This relation implies that at high temperature the sample strain rate is proportional to the rate of advancement of the ram. Because the ram advancement rate is constant (Fig. 6), the implication is that sample strain rates are also constant and can be varied over a range of at least two orders of magnitude (10^{-4} to $10^{-6}\,s^{-1}$).

Stresses were estimated from the olivine grain size of the deformed samples, determined using optical and backscattered electron microscopy. With some exceptions, dynamic recrystallization occurs by a subgrain rotation mechanism (see following section). The recrystallized grain sizes for each specimen were determined using a linear intercept method from a population of 200 grains, adjacent to porphyroclasts of the residual single crystal (Table 1). The olivine piezometers of KARATO *et al.* (1980), BUSSOD and CHRISTIE (1991) and VAN DER WAL *et al.* (1993) indicate that the grain size is a strong function of stress during the dynamic recrystallization of olivine, but independent of temperature and the presence or absence of a fluid phase. These experimentally determined relations were combined

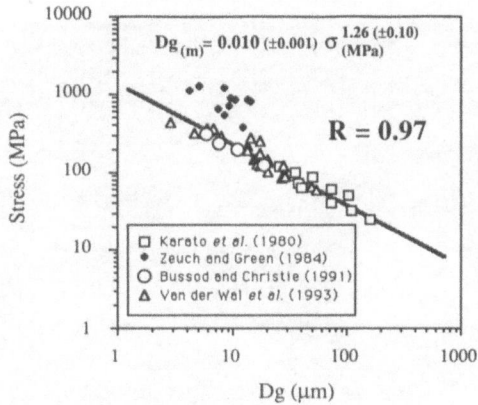

Figure 8

Grain size piezometer. Applied stress (MPa) versus dynamically recrystallized grain sizes (μm) from dry olivine single crystals (KARATO *et al.*, 1980), synthetic dunite (ZEUCH and GREEN, 1984), hypersolidus lherzolite (BUSSOD and CHRISTIE, 1991) and 'wet' and 'dry' dunites (VAN DER WAL *et al.*, 1993). The regression has a correlation coefficient of $R = 0.97$ and does not include the ZEUCH and GREEN (1984) data for which stresses were overestimated (VAN DER WAL *et al.*, 1993).

into a single piezometer involving the rotation mechanism, of the form:

$$D_{g(m)} = 0.010(\pm 0.001)\sigma_{(MPa)}^{1.26(\pm 0.10)} \tag{2}$$

with a correlation coefficient of $R = 0.97$, where D_g is the recrystallized grain size (in meters) and σ is the applied stress (in MPa). This piezometer is robust and favored over dislocation density as it is both statistically significant and applies to natural and synthetic polycrystalline aggregates and recrystallized single crystals, over a wide range of temperatures (900 to 1650°C), with and without a fluid phase present (Fig. 8). There also appears to be no pressure dependence over the range of experimental conditions (300 to 1500 MPa), and assuming the recrystallization mechanism involves grain boundary bulging and subgrain rotation, limited static recrystallization appears to have no significant effect (VAN DER WAL *et al.*, 1993).

Experimental Results

Samples were recovered at various stages of the procedure outlined above. Cylindrical single crystals of San Carlos olivine, recovered after increasing the pressure at room temperature, were plastically and cataclastically deformed to strains of 16 to 20% at confining pressures of 6 GPa to 13.5 GPa. Optical

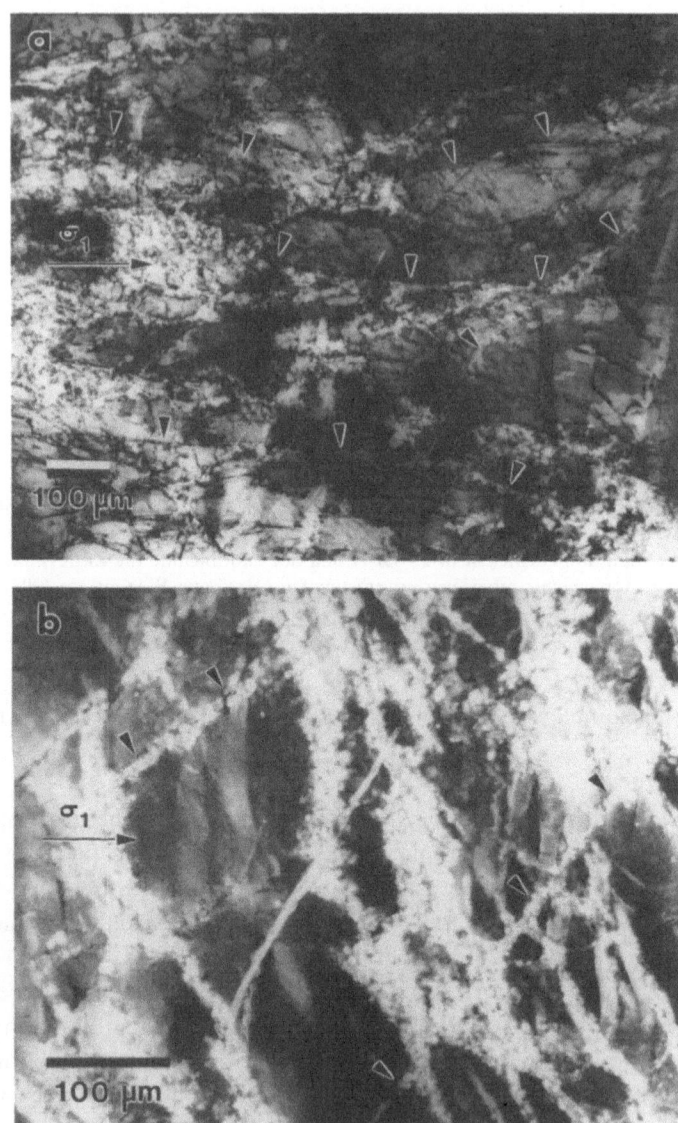

Figure 9

Photomicrographs of olivine single crystals (crossed nicols). The maximum principal compressive stress σ_1 is EW (arrow). (a) Sample cataclastically (indicated) and plastically deformed at high pressure and ambient temperature (MA 505). (b) Sample after 13 min. at 1600°C and 6 GPa (MA 523). Grain growth in cataclastic zones and dynamic recrystallization along single crystal edges are visible (arrows).

microscopy (Fig. 9a) shows that the strain was accommodated plastically by kink band formation with kink band boundaries nearly perpendicular to σ_1, together with cataclasis and crushing along narrow conjugate fractures, indicating that stresses in excess of the compressive strength value were attained (>2.5 GPa).

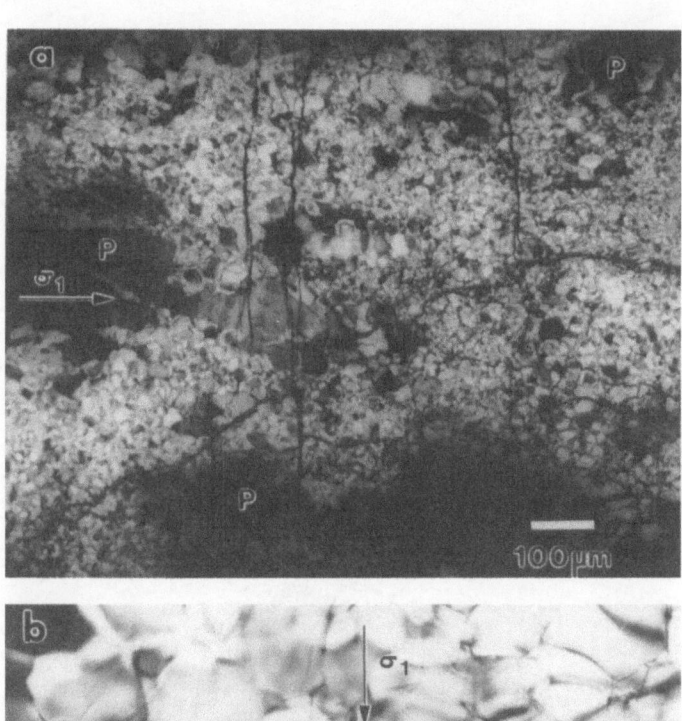

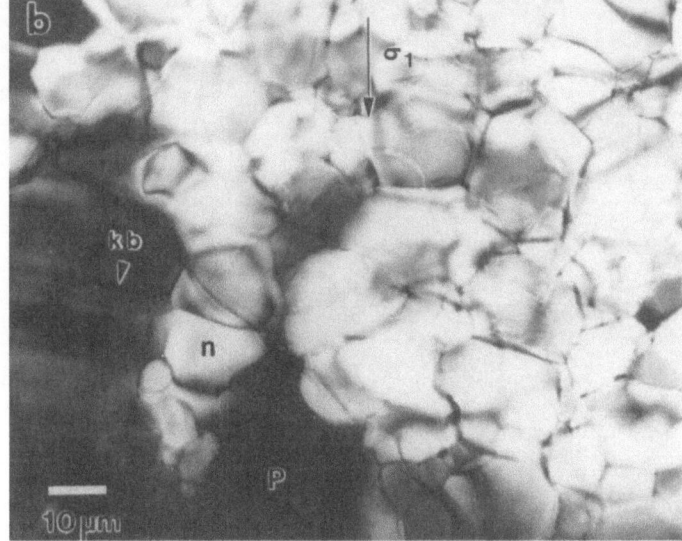

Figure 10

Photomicrographs of deformed and dynamically recrystallized sample at 1450 C and 6 GPa for 60 min (MA 634). (a) Sample is recrystallized and residual single crystal (**P**) represents less than 30% of the sample. σ_1 is EW. (b) Same sample as above showing detail of recrystallization. Porphyroclasts (**P**) exhibit kink bands (**kb**) nearly perpendicular to σ_1 (oriented NS), with grain size reduction along grain edges by subgrain rotation and the formation of neoblasts (**n**).

Following the increase in temperature at high pressure, both sample and sample assembly deform in response to the flow of the gasket material (Fig. 6, stage A). Recrystallization and grain growth occur immediately upon attainment of high temperature, and are principally localized near severely damaged regions (cataclas-

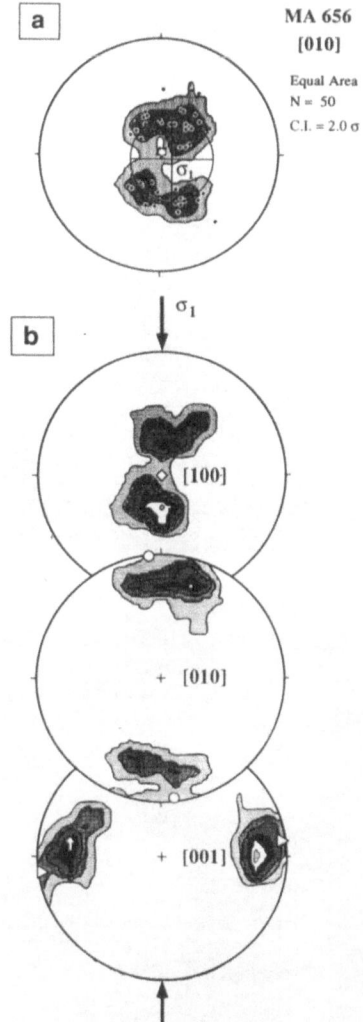

Figure 11

5-axis universal stage measurements of sample MA 656 deformed at 6 GPa, 1450 C and $4 \times 10^{-5} s^{-1}$.
Kamb contour plots of crystallographic axes are represented on lower hemisphere equal area projections.
The data base includes 50 olivine neoblasts recrystallized from the San Carlos single crystal starting material.
Shown are the original single crystal orientations of [100] (diamonds), [010] (circles) and [001] (triangles).
(a) The applied maximum principal compressive stress (σ_1) is oriented vertically (cross hairs). [010] are
aligned along small circle ≈ 30 to σ_1 (small circle). Counting and significance intervals are 2σ and 1σ,
respectively. (b) The applied stress (arrow) is oriented north-south. Counting and significance intervals
are 2σ.

tically deformed) of the single crystal and representing less than 20 areal % of the
sample in thin section. Inspection of specimens recovered after 15 minutes at
temperatures of 1450–1600 C, a time which allows the mechanical and thermal
adjustment of the sample assembly (stage A of Fig. 6), shows that the sample and

sample assembly are strained an additional 6%. During this time, grain size reduction occurs along the porphyroclast margins (Fig. 9b). With some exceptions, samples recovered after three hours at 1600°C are nearly 80% recrystallized, with an average grain size of $30 \pm 5\ \mu$m. Similarly, samples deformed at 1450°C are 60–70% recrystallized to an average grain size of $20 \pm 14\ \mu$m (Fig. 10).

Petrofabric analyses of selected samples suggest dynamic recrystallization by subgrain rotation (Fig. 11). This situation represents a dislocation deformation mechanism ($n \approx 3$), whereby the grain size is inversely proportional to the applied stress (KARATO et al., 1980; CHOPRA and PATERSON, 1984; BUSSOD and CHRISTIE, 1991). The [010] axes of the recrystallized olivine grains (neoblasts) are obtained along a small circle 30° from the maximum principal compressive stress, σ_1 (Fig. 11a). Thus (010) can act as the principal slip plane. From this configuration one may conclude that both the 'easy' and 'hard' slip systems in olivine are operative, (010)[100] and (010)[001] respectively (Fig. 11b). The dominant slip system however, cannot be determined from petrofabric analysis alone and further analysis by TEM is underway.

Discussion

Dynamic recrystallization, grain growth and primary recrystallization are all processes which can affect grain size and therefore the texture of the specimens. Experimental data on annealing times for grain growth and dislocation recovery in olivine suggest that given nominally dry conditions, several hours are required at 1400 to 1500°C and 8 GPa for a fine-grained olivine aggregate to recrystallize to a microstructure with a $20–40\ \mu$m grain size (KARATO, 1989; KARATO et al., 1993). The unusually high degree of recrystallization exhibited by the samples after one hour can be explained by the very high initial stresses transmitted by the assembly to the sample as pressure is increased at ambient temperature. Grain size reduction during stages B and C of Figure 6, high initial stresses and the absence of grain size dependence with time, all suggest that dislocation recovery is the dominant mechanism responsible for the grain size of the experiments. The initial high stresses result in a very high strain energy, stored in dislocations in the remaining single crystal or $\geq 80\%$ of the sample. When temperature is increased, this energy acts as a driving force for dynamic recrystallization. Exceptions involve samples in which the inner Re jacket failed, allowing the confining liquid to penetrate. These specimens show evidence of diffusion-induced grain boundary migration and exhibit substantially larger grain sizes which were not used for piezometry (MA 538, MA 743 and MA 712; Table 1).

The most reliable stress-strain data from experiments at 1450°C are compared with low-pressure olivine constitutive relations (BAI et al., 1991). Data for natural and synthetic polycrystalline aggregates of olivine generally compare well with the

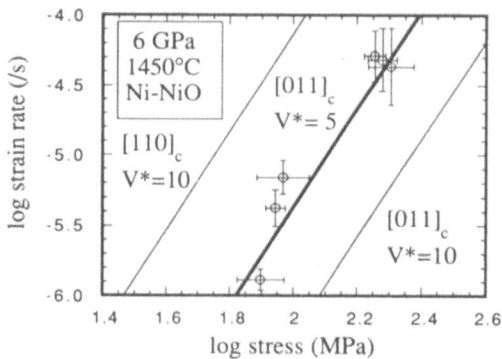

Figure 12

Stress strain-rate constitutive relation for polycrystalline olivine at 6 GPa and 1450 C. Shown are the stress-strain-rate relations from Bai *et al.* (1991) for the 'hard' $[011]_c$ and 'easy' $[110]_c$ single crystal olivine slip systems, buffered at Ni-NiO, assuming an activation volume of $V^* = 10$ cm³ mol⁻¹ (thin lines) and $V^* = 5$ cm³ mol⁻¹ (heavy line). Errors bars include error in sample strain determination and stress estimates (Table 1).

single crystal data at low pressures. Assuming an activation volume of $V^* = 10.0 \times 10^{-6}$ m³ mol⁻¹, the data at 6 GPa and 1450 C are bracketed by both the 'easy' $[110]_c$ and 'hard' $[011]_c$ orientations, favoring the (010)[100] and (010)[001] slip systems. Typically however, the deformation of 'nominally dry' polycrystalline olivine aggregates is controlled by the 'hard' slip system. If this assumption is correct the high-pressure data at 1450 C are best fit assuming an activation volume of $V^* = 5.0 \times 10^{-6}$ m³ mol⁻¹ (Fig. 12).

This value is lower than the value commonly cited for olivine and determined at lower pressures, however it is similar to the value of $V^* \approx 6.0 \times 10^{-6}$ m³ mol⁻¹ determined for edge dislocation recovery at high pressure (KARATO *et al.*, 1993). Although preliminary, the data for olivine at 12 GPa confining pressure is also consistent with a low activation volume. This observation suggests that the activation volume in olivine may decrease with increasing pressure or is lower than previously reported on the basis of low-pressure data.

Conclusions

From these results we conclude that (1) the multi-anvil press may be used as a 'constant strain-rate' solid medium deformation apparatus to pressures and temperatures of at least 16 GPa, and 1600 C, and (2) the determination of zeroth-order values for viscosity is possible at high pressures. Quantitative stress–strain-rate constitutive relations will only be possible with the introduction of new technological developments permitting the *in situ* measurement of stress. Our preliminary results for olivine (Fo₉₀) at 6 and 12.5 GPa are consistent with values obtained from olivine single crystal creep laws, assuming the hard olivine slip system dominates. This result implies an activation volume for creep of 5 to 10×10^{-6} m^3 mol^{-1}, a value substantially lower than estimates obtained from lower pressure data (ROSS et al., 1979; GREEN and BORCH, 1987, 1990; BORCH and GREEN, 1989). We are currently extending this data base and investigating the rheological behavior of β-Mg$_{1.8}$Fe$_{0.2}$SiO$_4$ at 13.5 to 16 GPa and 1450°C.

Acknowledgements

The authors acknowledge the Bayerisches Geoinstitut and its staff for the support provided, particularly during the costly developmental stages of this study. We thank David Kohlstedt, Tracy Tingle and an anonymous reviewer for their constructive criticisms. This contribution would not have been possible without the exceptional technical expertise of Georg Herrmannsdörfer, Detlef Krauße, Herbert Küfner and Hubert Schulze. This work is dedicated to Edward Schreiber and Andrew Gratz.

REFERENCES

AKAOGI, M., ITO, E., and NAVROTSKY, A. (1989), *Olivine-modified Spinel-spinel Transitions in the System* Mg_2SiO_4-Fe_2SiO_4: *Calorimetric Measurements, Thermochemical Calculation, and Geophysical Application*, J. Geophys. Res. *94*, 15671–15685.

BAI, Q., MACKWELL, S. J., and KOHLSTEDT, D. L. (1991), *High-temperature Creep of San Carlos Olivine: I. Steady-state Flow Laws*, J. Geophys. Res. *96*, 2441–2463.

BOHLEN, S. R., and BOETTCHER, A. L. (1982), *The quartz $\Leftrightarrow$ coesite Transformation: A Precise Determination and the Effects of Other Components*, J. Geophys. Res. *87*, 7073–7078.

BORCH, R. S., and GREEN, H. W. II (1989), *Deformation of Peridotite at High Pressure in a New Molten Salt Cell: Comparison of Traditional and Homologous Temperature Treatments*, Phys. Earth Planet. Interiors *55*, 269–276.

BOYD, F. R., and ENGLAND, J. L. (1960), *The Quartz-coesite Transition*, J. Geophys. Res. *65*, 749–756.

BUSSOD, G. Y., and CHRISTIE, J. M. (1991), *Textural Development and Melt Topology in Spinel Lherzolite Experimentally Deformed at Hypersolidus Conditions*, J. Petrol. Special Volume: Orogenic Lherzolites and Mantle Processes, 17–39.

CARTER, N. L. (1964), *Experimental Deformation and Recrystallization of Quartz*, J. Geology *72*, 687–733.

CARTER, N. L., CHRISTIE, J. M., and GRIGGS, D. T. (1961), *Experimentally Produced Deformation Lamellas and Other Structures in Quartz*, J. Geophys. Res. *66*, 2518–2519.

CHOPRA, P. N., and PATERSON, M. S. (1984), *The Role of Water in the Deformation of Dunite*, J. Geophys. Res. *89*, 7861–7876.

GETTING, I. C., and KENNEDY, G. C. (1970), *Effect of Pressure on the Emf of Chromel-alumel and Platinum-platinum 10% Rhodium Thermocouples*, J. Applied Phys. *41*, 4552–4562.

GREEN, H. W. II (1992), *Accurate Stress Measurement at Pressures to 5 GPa*, Trans. Am. Geophys. Union *73*, 556.

GREEN, H. W. II, and BORCH, R. S. (1987), *The Pressure Dependence of Creep*, Acta Metall. *35*, 1301–1305.

GREEN, H. W. II, and BORCH, R. S., *High pressure and temperature deformation experiments in a liquid confining medium*. In *The Heard Volume* (eds. Duba, A. G. et al. 1990) pp. 195–200.

GREEN, H. W. II, YOUNG, T. E., WALKER, D. and SCHOLZ, C. H. (1990), *Anticrack-associated Faulting at Very High Pressure in Natural Olivine*, Nature, *348*, 720–722.

GRIGGS, D. T. (1936), *Deformation of Rocks under High Confining Pressures. I. Experiments at Room Temperature*, J. Geology *44*, 541 577.

GRIGGS, D. T. (1967), *Hydrolytic Weakening of Quartz and Other Silicates*, Royal Astronomical Soc. Lond., Geophys. J. *14*, 19–31.

GRIGGS, D. T., and BLACIC, J. D. (1964), *The Strength of Quartz in the Ductile Regime*, Am. Geophys. Trans. *45*, 102 103.

GRIGGS, D. T., TURNER, F. J., and HEARD, H. C., *Deformation of rocks at 500 C to 800 C*. In *Rock Deformation* (eds. Griggs, D. T., and Handin, J.), (Geol. Soc. Amer. Mem. 79, 1960), pp. 39 104.

IRIFUNE, T., and RINGWOOD, A. E., *Phase transformations in primitive MORB and pyrolite compositions to 25 GPa and some geophysical implications*. In *High-pressure Research in Mineral Physics* (eds. Manghnani, M. H., and Syono, Y.), Geophys. Monogr. Ser. 39 (AGU, Washington, DC 1987) pp. 231 242.

ITO, E., and TAKAHASHI, E., *Ultrahigh-pressure phase transformations and the constitution of the deep mantle*. In *High-pressure Research in Mineral Physics* (eds. Manghnani, M. H., and Syono, Y.), Geophys. Monogr. Ser. 39 (AGU, Washington, DC 1987) pp. 221 230.

ITO, E., and TAKAHASHI, E. (1989), *Postspinel Transformations in the System Mg_2SiO_4-Fe_2SiO_4 and Geophysical Implications*, J. Geophys. Res. *94*, 10637 10646.

ITO, E., TAKAHASHI, E., and MATSUI, Y. (1984), *The Mineralogy and Chemistry of the Lower Mantle: An Implication of the Ultrahigh-pressure Phase Relations in the System MgO-FeO-SiO_2*, Earth and Planet. Sci. Lett. *67*, 230–248.

KANZAKI, M. (1987), *Physical Properties of Silicate Melts at High Pressures*, Ph.D. thesis, Geophysical Institute, University of Tokyo, Tokyo, Japan.

KARATO, S. (1989), *Grain Growth Kinetics in Olivine Aggregates*, Tectonophys. *168*, 255 273.

KARATO, S., RUBIE, D. C., and YAN, H. (1993), *Dislocation Recovery in Olivine under Deep Upper Mantle Conditions: Implications for Creep and Diffusion*, J. Geophys. Res. *98*, 9761 9768.

KARATO, S., TORIUMI, M., and FUJII, T. (1980), *Dynamic Recrystallization of Olivine Single Crystals during High-temperature Creep*, Geophys. Res. Lett. *7*, 649 652.

KATSURA, T., and ITO, E. (1989), *The system Mg_2SiO_4-Fe_2SiO_4 at High Pressures and Temperatures: Precise Determination of Stabilities of Olivine, Modified Spinel, and Spinel*, J. Geophys. Res. *94*, 15663–15670.

LIEBERMANN, R. C., and WANG, Y., *Characterization of sample environment in a uniaxial split-sphere apparatus*. In *High-pressure Research: Application to the Earth and Planetary Sciences* (eds. Syono Y., and Manghnani, M. H.), Geophys. Monogr. Ser. 67 (AGU, Washington, DC 1992) pp. 19 31.

LIU, L., and BASSETT, W. A., *Elements, Oxides, Silicates* (Oxford University Press, 1986).

OHTANI, E., KAGAWA, N., SHIMOMURA, O., TOGAYA, M., SUITO, K., ONODERA, A., SAWAMOTO, H., YONEDA, M., TANAKA, S., UTSAMI, W., ITO, E., MATSUMURO, A., and KIKEGAWA, T. (1989), *High-pressure Generation by a Multiple Anvil System with Sintered Diamond Anvils*, Rev. Sci. Instrum. *60*, 922–925.

O'NEILL, H. St. C. (1987), *The Free Energies of Formation of NiO, CoO, Ni_2SiO_4 and Co_2SiO_4*, Am. Mineral. *72*, 280 291.

O'NEILL, H. St. C., and WALL, V. J. (1987), *The Olivine-orthopyroxene-spinel Oxygen Geobarometer, the Nickel Precipitation Curve, and the Oxygen Fugacity of the Earth's Upper Mantle*, J. Petrol. *28*, 1169–1191.

PIERMARINI, G. J., and BLOCK, S. (1975), *Ultrahigh Pressure Diamond Anvil Cell and Several Semi-conductor Phase Transitions Pressures in Relation to the Fixed Point Pressure Scale*, Rev. Sci. Instrum. *46*, 973–979.

RINGWOOD A. E., and MAJOR, A. (1966), *Synthesis of Mg_2SiO_4-Fe_2SiO_4 Spinel Solid Solutions*, Earth Planet. Sci. Lett. *1*, 241–245.

ROSS, J. V., AVÉ LALLEMANT, H. G., and CARTER, N. L. (1979), *Activation Volume for Creep in the Upper Mantle*, Science *203*, 261–263.

RUBIE, D. C., KARATO, S., YAN, H., and O'NEILL, H. St. C. (1993), *Low Differential Stress and Controlled Chemical Environment in Multianvil High-pressure Experiments*, Phys. Chem. Minerals, *20*, 315–322.

VAN DER WAL, D., CHOPRA, P., DRURY, M., and FITZGERALD, J. (1993), *Relationships between Dynamically Recrystallized Grain Size and Deformation Conditions in Experimentally Deformed Olivine Rocks*, Geophys. Res. Lett. *20*, 1479–1482.

WANG, Y., LIEBERMANN, R. C., and BOLAND, J. N. (1988), *Olivine as an in situ Piezometer in High Pressure Apparatus*, Phys. Chem. Minerals *15*, 493–497.

WEIDNER, D. J., WANG, Y., and VAUGHAN, M. T. (1992), *X-ray Strain Measurement and Strength of Materials at High Pressure and Temperature*, Trans. Am. Geophys. Union *73*, 556.

WELLS, P. R. A. (1977), *Pyroxene Thermometry in Simple and Complex Systems*, Contrib. Mineral. Petrol. *62*, 129–139.

YAGI, T., and AKIMOTO, S. (1976), *Direct Determination of Coesite-stishovite Transition by in situ X-ray Measurements*, Tectonophys. *35*, 259–270.

YASUDA, A., FUJII, T., and KURITA, K., *Melting relations of an anhydrous abyssal basalt at high pressure*. In *Dynamic Processes of Material Transport in the Earth's Interior* (ed. Marumo, F.) (Kluwer Academic, Dordrecht 1990) pp. 327–337.

ZEUCH, D. H., and GREEN, H. W. II (1984), *Experimental Deformation of a Synthetic Dunite at High Temperature and Pressure. I. Mechanical Behavior, Optical Microstructure and Deformation Mechanisms*, Tectonophysics *110*, 233–262.

(Received April 23, 1993, revised October 7, 1993, accepted October 19, 1993)

PAGEOPH, Vol. 141, No. 2/3/4 (1993)

0033-4553/93/040601-11$1.50 + 0.20/0

The Use of Sintered Diamond Anvils in the MA8 Type High-pressure Apparatus

Tadashi Kondo,[1] Hiroshi Sawamoto,[1] Akira Yoneda,[1] Manabu Kato,[1] Akihito Matsumuro,[2] Takehiko Yagi[3] and Takumi Kikegawa[4]

Abstract —A new multi-anvil type high-presure apparatus has been developed using sintered diamond anvils to generate pressures over 30 GPa and temperatures up to about 2000°C. A maximum sample volume of about 1 mm³ is available in this system. The pressure was confirmed by dissociation of forsterite into Mg-perovskite and periclase. The basic techniques and problems in utilizing sintered diamond in the MA8 type high-pressure apparatus are discussed with an emphasis on the future prospect of incorporating simultaneous X-ray diffraction observation.

Key words: 6–8 double stage, synchrotron, lower mantle condition.

1. Introduction

In recent years, relatively large blocks of sintered diamond have become available commercially. Consequently, high-pressure apparatus using sintered diamond as an anvil material have been developed intensively in many laboratories in Japan. Utsumi *et al.* (1992) have summarized the recent advances in certain types of apparatus with sintered diamond anvils. Shimomura et al. (1992) described a DIA type apparatus (MAX90) which extended the pressure and temperature range to 15 GPa and 1500°C. Funamori and Yagi (1993) succeeded in generating pressures reaching 36 GPa and temperatures up to 1900 K in a Drickamer anvil cell. These new high-pressure apparati were all combined with synchrotron radiation for simultaneous X-ray observation. The use of sintered diamond has produced promising results, and it will be an important anvil material for the next generation of high-pressure apparatus.

[1] Department of Earth and Planetary Sciences, Nagoya University, Chikusa, Nagoya 464, Japan.
[2] Department of Mechanical Engineering, Nagoya University, Chikusa, Nagoya 464, Japan.
[3] Institute for Solid State Physics, University of Tokyo, Roppongi, Minato, Tokyo 106, Japan.
[4] Photon Factory, National Laboratory for High Energy Physics, Oho, Tukuba, Ibaraki 305, Japan.

The MA8 system (the 6–8 multi-anvil system originally proposed by KAWAI and ENDO, 1970) has many advantages, such as relatively uniform pressure distribution, small temperature gradient, and, especially, a large volume for the sample cell. Therefore, the MA8 system has been used to study phase relations, for single crystal synthesis and for other studies of mineral behavior in multi-component systems. However, experiments under lower mantle conditions have been difficult because the strength of the anvil material (usually tungsten carbide) limits the maximum attainable pressure. Since OHTANI *et al.* (1989) successfully generated pressures to 41 GPa at room temperature, using the MA8 system with the anvils (0.5 mm truncation) made of sintered diamond (5.0 mm cubes), much larger sintered diamond blocks have been developed. These can be used for high pressure apparatus with internal heating capability. IRIFUNE *et al.* (1992) developed a hybrid MA8 system consisting of four tungsten carbide cubes and four sintered diamond cubes (10.0 mm with truncated edge length of 1.0–1.5 mm, nonconductive, Advanced Diamond Composite) to generate 1500°C by utilizing tungsten carbide anvils as electric power leads at pressures over 30 GPa. Although this is a practical means of generating higher pressure and high temperature with *in situ* X-ray, it is more desirable to redesign the power leads and to replace all the tungsten carbide anvils with sintered diamond anvils, thereby utilizing fully the intrinsic strength of diamond.

We have worked on the MA8-type high-pressure apparatus in order to achieve pressures higher than those reached in the DIA-type apparatus for sample volumes larger than those in the Drickamer anvil cell. In this paper we report the performance of the MA8 system with eight sintered diamond anvils with truncation edge lengths of 2.0 mm and details of the experimental technique. The basic design of this system was described in KONDO *et al.* (1993).

2. Physical and Mechanical Properties of Sintered Diamond

The anvil material (SYNDIE15151, De Beers Diamond Co. Ltd., 10.0 mm in height and 15.0 mm in diameter; sintered core size of cylinder) was originally produced for wire-drawing die material. This is a composite material made of diamond powder (grain size ranging from 50 μm to 100 μm) with cobalt as a binder. It is sintered in the stability field of diamond. The electrical resistivity of each anvil is about 2 Ωm. Consequently, the anvil can be used as an electrical lead for various measurements. However, it is not appropriate for power leads in heating experiments because the electrical resistance of the heating elements is comparable to that of the anvils. The density (3.819 g/cm^3) is larger than that of pure diamond (3.515 g/cm^3) because of the presence of the cobalt binder. The P-wave velocity is 16.4 km/s at 50 MHz and the S-wave velocity is 11.0 km/s at 20 MHz according to acoustic velocity measurements by means of the pulse-echo-overlap method at ambient conditions. The calculated bulk modulus is 410 GPa, which is almost the same as that of single crystal diamond (420 GPa). The compressive strength estimated by KONDO et al. (1993) is more than 12 GPa, twice that of the strongest tungsten carbide grades. In addition, unlike single-crystal diamond, sintered diamond has no characteristic cleavage plane. Thus, the properties of sintered diamond appear to be ideal for an anvil material.

The supplied cylinders of sintered diamond were shaped into cubes by means of an electric discharge process. The surfaces were polished to produce cubes of 10.00 ± 0.01 mm edge length. One corner of each cube was truncated to an edge size of 2.0 mm. Because of the brittleness of the material, the accuracy of anvil size is important in attaining sufficiently good alignment to generate stable high pressures, as experienced in the early development of the single-crystal diamond anvil cell (DAC).

3. Experimental Setup and Results

3.1. High-pressure Generation

The MA8 double-stage system used in this study consists of eight sintered diamond anvils as the second stage (Figure 1a) and six tungsten carbide anvils with 18 mm truncation as the first stage (Figure 1b). The assembly is driven by a 1200-ton cubic press installed at the Department of Mechanical Engineering, Nagoya University. The octahedral pressure medium (6.0 mm in edge length) is made of semisintered magnesia with about 30% porosity. We also tried semisintered alumina (porosity 35%) as a pressure medium for more efficient pressure generation. Pyrophyllite gaskets were used with dimensions of 1.9 mm (thickness) × 2.0 mm (width). Usually we prepare a set of six short size gaskets and six long size gaskets to fill the gap between the gasket ends. However, in this work twelve gaskets with the same dimension and six edge supporters (1.9 mm × 1.9 mm × 1.9 mm) were used to obtain good symmetry. Although paper spacers were set on anvil corners at the beginning, eventually no spacer was used because the gaskets have sufficient size to fix the small anvils. The MA8 system of the second stage was covered with mica sheets for electrical insulation from the first stage anvils.

The best way to precisely calibrate the generated pressures is by in situ X-ray observation of lattice constants of standard materials such as sodium chloride and gold, for which the equations of state are well-known. The transparency to X-rays of the sintered diamond used in this study is relatively good (more than 30%) despite the presence of metal in the material. It is common practice to test the suitability of gaskets and pressure cells using in situ X-ray observation. However, it could not be made in this work because of unavailability of a synchrotron.

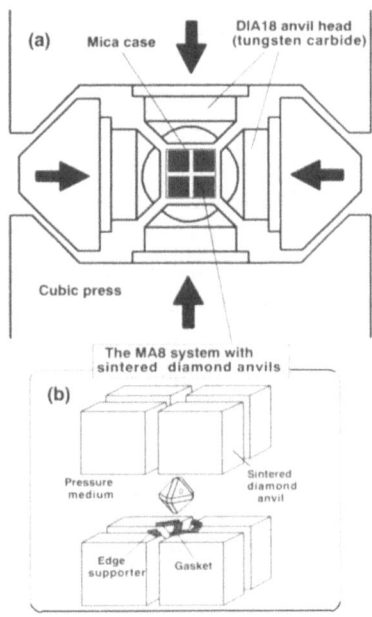

Figure 1

(a) Configuration of a DIA-type first stage (tungsten carbide with 18 mm truncation) and the second-stage sintered diamond anvil assemblage. (b) Details of the sintered diamond anvil assemblage.

Therefore, pressure calibration was carried out by measurement of electrical resistance changes in ZnS (15.6 ± 0.6 GPa) and GPa (23 ± 1 GPa). While the exact transition pressures are still debated, these values are used for consistency with other laboratories. We can easily recognize the semiconductor-metal transition point by a sudden decrease in electrical resistance. Generated pressures above the GPa point are usually estimated by extrapolation of the pressure calibration curve obtained from the GPa point, mainly because there are no established pressure standard points above this pressure.

We have tried Fe-V alloys as calibrants above 25 GPa. ENDO *et al.* (1987) determined the $\alpha - \varepsilon$ transition pressures of four Fe-V alloys corresponding to the minimum points of the electrical resistance by X-ray diffraction. They also cross-checked the transition pressures by the ruby fluorescence method and by X-ray diffraction in the DAC. Whereas the transition pressures in Fe-V alloys are still somewhat uncertain, they are the best resistance standards for estimating the generated pressure at present. In order to obtain materials with α ε transition pressures at 25, 27.5 and 30 GPa, alloys of three different compositions: Fe-11.6wt%V, Fe-14.74%V and Fe-16.46%V are prepared. They are rolled to foils of 10 μm in thickness for actual use. The four wire method was used to raise the sensitivity of the resistivity measurement. The pressure calibration curve acquired in

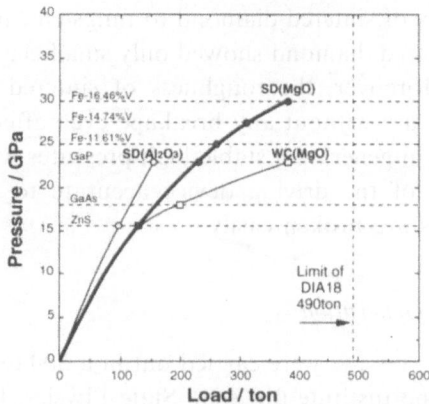

Figure 2

Comparison of sintered diamond and tungsten carbide as material of the second stage anvils (truncated edge length of 2.0 mm) by pressure calibration curves. The generated pressure with sintered diamond anvils and magnesia pressure medium is shown by hatched circles, SD(MgO). Open circles, SD(Al₂O₃), show the generated pressure with sintered diamond system with alumina pressure medium. Open squares, WC(MgO), show the result of comparative experiments using tungsten carbide anvils with magnesia pressure medium under the same condition. Vertical broken line indicates the maximum load attainable for the present MA8 system, limited by the anvil size of sintered diamond and the strength of tungsten carbide used as the first stage anvils.

the present study is shown by a thick solid line in Figure 2. Note that a smooth calibration curve is found, passing through the five fixed points including the Fe-V alloys. The loads required for the semiconductor-metal transition in ZnS and GPa were 130 ton and 230 ton, respectively. Reproducibility of the load to generate the pressure of the GPa point is within 5 ton. The minimum points of electric resistance in three Fe-V alloys were detected at loads of 260 ton, 310 ton and 380 ton, respectively.

The efficiency of pressure generation is compared in three different cases; the two sintered diamond systems with magnesia (hatched circles) and alumina (open circles) pressure media, and the tungsten carbide system (open squares) with the same gasket and magnesia pressure medium. Among the materials tested as pressure media, alumina has the highest efficiency for pressure generation. However, the available length of pyrophyllite gasket (maximum 10 mm) was not sufficient to confine the pressure when the alumina pressure medium was used, and sometimes blowouts occurred even in the course of increasing load. A radical modification of the gasket design would probably be needed in order to utilize a harder pressure medium than magnesia in the present system.

The vertical broken line shows the maximum load available to this system, which is limited by the strength of the first-stage tungsten carbide anvils. We note that a pressure of 30 GPa can be attained with a load of only 380 ton, and recognize the superiority of sintered diamond to tungsten carbide above 15 GPa. In this pressure range, sintered diamond showed only small elastic deformation and no plastic deformation. Moreover, the toughness of sintered diamond was demonstrated by the repeated use without any breakage even after blowouts. One of the most important factors in generating stable high pressures with sintered diamond is to keep the alignment of the driving device accurate to 10 μm. Otherwise, the sintered diamond anvils are broken easily.

3.2. High Temperature Generation

Heating tests under pressure were carried out in a 500 ton cubic press (CAPRICORN-1) installed at the Institute for Solid State Physics, University of Tokyo. A furnace assembly newly designed for this study is shown in Figure 3. Sheet heaters (1.5 mm × 1.5 mm × 0.2 mm) were made from a sintered mixture of tungsten carbide and diamond powder (50%/50% by volume). Temperatures were measured by two 0.1 mm W3%Re-W25%Re thermocouples, one at the sample center and the other at the bottom of a sintered diamond anvil. Various materials were employed for a sample container. In most cases a boron nitride (BN) box was used, with a sample volume of 1.0 (in diameter) × 0.5 (in height) mm^3. Platinum or graphite containers were also used; in this case, thin sheets made of magnesia (0.1–0.2 mm in thickness) were put around the sample container for electrical insulation. Most parts were made individually by the compact computer-controlled three-dimensional milling machine CAMM-3 (Roland Co. Ltd.) to achieve high accuracy in the sizes of various parts for the cell.

Several problems were encounterred in designing the present furnace assembly for the sintered diamond anvil system. The most serious one was the electrical leads

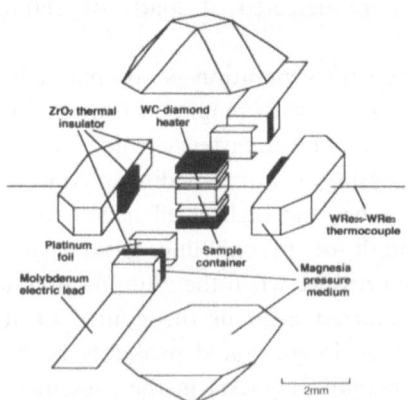

Figure 3
A pressure cell with the furance assembly newly designed in the present study. Shaded parts are zirconia as thermal insulator.

used to supply power to the furnace. Platinum and molybdenum foils (50 μm in thickness and 2.0 mm in width) were tested as electric leads. Platinum foil was chemically stable at high temperatures and easy to deform. However, its ductile nature was found unsuitable for electric leads. Since the leads were set between anvil and gaskets, the extrusion of gaskets was promoted by the ductility of platinum foil, especially at high pressures. Therefore, the efficiency of high-temperature generation at higher pressure was lower than at lower pressure because of variation in the electrical resistance in the platinum foils due to deformation. Although molybdenum has a high melting point (2600°C), it tends to oxidize at high temperatures. It is also difficult to make good contact with the heater because of the brittle nature of molybdenum. As a compromise, molybdenum leads are used in combination with platinum foil contacts (see Figure 3).

Other serious problems were the high thermal conductivity of the diamond and the lack of information about the strength of sintered diamond at high temperatures. Sheets (0.2 mm in thickness) of zirconia (essentially nonporous) were placed around the sample container and heater to achieve high thermal insulation from the sintered diamond anvils. We constrained the cell design for the heating experiments to keep the temperature at the bottom of sintered diamond anvils below 150°C, in order to protect the tungsten carbide anvils.

Figure 4a shows the result of the highest temperature run at 27.5 GPa (true pressure would be different due to heating). At a temperature of 1960°C the electric resistance of the furnace circuit increased, probably because the local temperature was above the melting point of platinum used for electric contact between heating element and power leads. Figure 4b shows the result of a heating experiment at the highest pressure of 31 GPa. A temperature of 1500°C was maintained stably for one

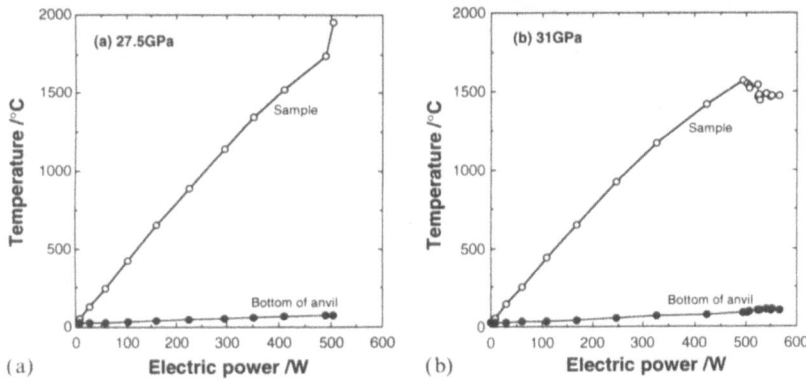

Figure 4

Examples of the relations between generated temperature and suplied electric power. Open circles show temperatures in the sample, solid circles temperatures at the bottom of sintered diamond anvil. (a) The highest temperature achieved is 1960 C at 27.5 GPa. (b) The result of heating test at the highest pressure of 31 GPa. Last twelve symbols show the fluctuation of temperature around 1500°C for one hour.

hour. The temperature limit is not set by the properties of sintered diamond, and the maximum temperature attainable could be extended beyond 2000°C by improving the electric leads with a suitable material and design.

3.3. Pressure Calibration at High Temperatures

KATO *et al.* (1992) reported the result of *in situ* X-ray observation of Mg_2SiO_4 at high pressures in MAX80 (SHIMOMURA *et al.*, 1985) combined with a double stage MA8 system, in which one tungsten carbide anvil was replaced by sintered diamond for use as an X-ray window. They also reported the variation of generated pressure at high temperatures. The pressure in the cell at constant load may in some cases drop during heating because of extrusion of the pressure medium, whereas it may in other cases increase by thermal pressure.

We checked the pressure condition at high temperatures by dissociation of forsterite or natural olivine (San Carlos; $Fo_{88}Fa_{12}$). ITO and TAKAHASHI (1989) determined the phase boundary between Mg_2SiO_4 spinel and an assemblage of Mg-perovskite and periclase up to 25 GPa. The Clausius-Clapeyron curve of this phase boundary has a negative slope and therefore the transition temperature is expected to be lower at higher pressures. Some samples recovered after heating were analyzed by micro-focus X-ray diffraction. The identified phases are plotted in Figure 5 against the pressure values estimated on the basis of pressure calibration at room temperature. The perovskite phase was found even at the low temperature of 850°C and a pressure of 26 GPa. Only 300 tons were required to synthesize a perovskite sample of 1 mm³ in the present system, compared to the large force of over 1000 tons needed in most multi-anvil systems. However γ spinel was found at

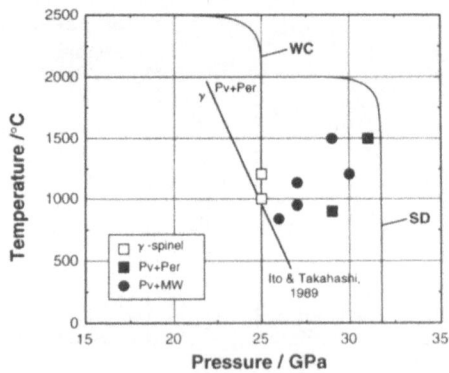

Figure 5
The available region of pressures and temperatures by the present system (SD) compared with that of tungsten carbide system (WC). The identified phase of recovered sample is also plotted. Solid line shows the phase boundary between spinel and assemblage of Mg-perovskite and periclase determined by ITO and TAKAHASHI (1989). Starting materials were pure forsterite (squares) and natural olivine, $Fo_{88}Fa_{12}$ (circles). Solid symbols show postspinel phase. Open squares show the γ-spinel phase.

1200°C and 25 GPa, indicating that pressure reduction of 1–2 GPa has occurred in the cell during heating, if Ito and Takahashi's phase boundary is correct. There are few established phase boundaries usable for pressure calibration at high temperatures in the present pressure range, whereas the pressure calibration at room temperature cannot be relied on at high temperatures. We emphasize the need for phase boundaries to be established for pressure calibration use above 25 GPa at high temperatues by *in situ* X-ray observation. The MA8 system with sintered diamond anvils has the potential to cover a wide range of pressures and also to produce temperatures more stable and uniform than in the DAC, as shown by the hatched area in Figure 5 for the estimated range of pressure and temperature attained in the present system. Therefore, the incorporation of the X-ray observation method in this system would be the best one for determining precise phase relations above 25 GPa.

The possibility of *in situ* X-ray observation with the present furnace assembly was examined by synchrotron radiation at KEK. A starting material of forsterite packed closely in a BN container was quenched from 31 GPa and 1500°C. A pair of zirconia sheets were removed from the recovered sample cell and were replaced by amorphous boron cemented by epoxy resin to provide a window for the X-ray beam. An incident X-ray beam passed successfully through the magnesia pressure medium and between a couple of sheet heaters in this test cell as shown by the result of X-ray diffraction with the energy dispersive method (Figure 6). One possibility

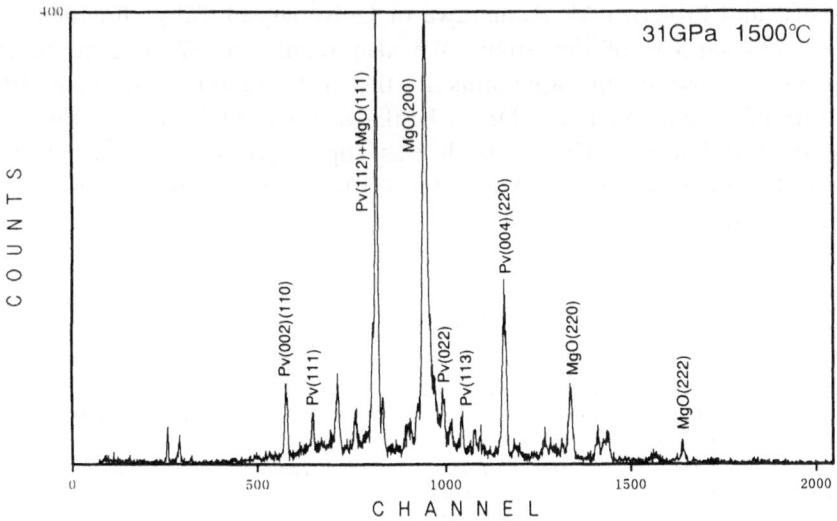

Figure 6

X-ray diffraction profile of a recovered sample. Starting material was synthesized forsterite powder. Sample was quenched from 31 GPa and 1500 C. X-ray diffraction analysis was done by the energy dispersive method using synchrotron radiation source at National Laboratory for High Energy Physics. We can recognize perovskite phase dissociated from forsterite in the pattern.

for improving the present furnace design is the use of a material with low atomic number for thermal insulation, needed for efficient X-ray observation *in situ*.

4. Conclusion

A high-pressure apparatus has been developed to generate pressures over 30 GPa and temperatures to about 2000°C with a maximum sample volume of 1 mm^3, for which *in situ* X-ray diffraction can be made. Sintered diamond is the most promising material for high-pressure generation in a large volume. The breakage of anvils is rare if the anvil-press alignment is accurately adjusted, even when blowout of the gasket takes place. The MA8 system with sintered diamond anvils would surely provide high-quality data on materials under lower mantle conditions. The technical advancement achieved up to the present demands available synchrotron radiation and considerably larger sintered diamond blocks to attain more stable generation and reliable measurement of higher pressures in a large sample volume.

Acknowledgments

We are very grateful to Professors H. Aoki, N. Fujii and Y. Fukao of Nagoya University, and Professor M. Kumazawa of University of Tokyo for their encouragement and support of this study. We also thank Dr. W. Utsumi for useful discussion and use of the apparatus at the Institute for Solid State Physics, University of Tokyo. We thank Dr. S. Urakawa for technical discussion and help at the Photon Factory. This research was supported by the Grant-in-Aid for Scientific Research on Priority Areas (03232101), Ministry of Education, Science and Culture, Japan.

References

ENDO, S., TOYAMA, N., ISHIBASHI, A., CHINO, T., and FUJITA, F. E., *Determination of α ‑ ε Transition in Fe-V Alloy.* In *High-pressure Research in Mineral Physics* (eds. Syono, Y., and Manghnani, M. H.) (Terrapub 1987) pp. 29–33.

FUNAMORI, N., and YAGI, T. (1993), *High Pressure and High Temperature in situ X-ray Observation of MgSiO₃ Perovskite under Lower Mantle Conditions,* Geophys. Res. Lett. *20,* 387–390.

IRIFUNE, T., UTSUMI, W., and YAGI, T. (1992), *Use of a New Diamond Composite for Multi-anvil High-pressure Apparatus,* Proc. Japan Acad. *68B,* 161–166.

ITO, E., and TAKAHASHI, E. (1989), *Postspinel Transition in the System Mg₂SiO₄/Fe₂SiO₃ and Some Geophysical Implications,* J. Geophys. Res. *94,* 10637–10646.

KATO, T., OHTANI, E., KAMAYA, N., SHIMOMURA, O., and KIKEGAWA, T., *Double-stage multi-anvil system with a sintered diamond anvil for X-ray diffraction experiment at high pressures and high temperatures.* In *High-pressure Research: Application to Earth and Planetary Sciences* (eds. Syono, Y., and Manghnani, M. H.) (Terrapub. 1992) pp. 33–36.

KAWAI, N., and ENDO, S. (1970), *The Generation of Ultrahigh Hydrostatic Pressures by a Split Sphere Apparatus*, Rev. Sci. Instrum. *41*, 1178–1181.

KONDO, T., SAWAMOTO, H., YONEDA, A., KATO, M., MATSUMURO, A., and YAGI, T. (1993), *Ultrahigh Pressure and High-temperature Generation by MA8 System with Sintered Diamond Anvils*, High Temp. High Pressures *25*, 105–112.

OHTANI, E., SHIMOMURA, O., TOGAYA, M., SUITO, K., ONODERA, A., SAWAMOTO, H., YONEDA, A., TANAKA, S., UTSUMI, W., ITO, E., MATSUMURO, A., and KIKEGAWA, T. (1989), *High Pressure Generation by a Multiple Anvil System with Sintered Diamond Anvils*, Rev. Sci. Instrum. *60*, 922–925.

SHIMOMURA, O., YAMAOKA, S., YAGI, T., WAKATSUKI, T., TSUJI, K., FUKUNAGA, O., KAWAMURA, H., AOKI, K., and AKIMOTO, S., *Multi-anvil type high-pressure apparatus for synchrotron radiation*. In *Solid State Physics under Pressure, in Recent Advance with Anvil Devices* (ed. Minomura, S.) (KTK/Reidel, Tokyo/Dordrecht 1985) pp. 351–356.

SHIMOMURA, O., UTSUMI, W., TANIGUCHI, T., KIKEGAWA, T., and NAGASHIMA, T., *A new high pressure and high-temperature apparatus with sintered diamond anvils for synchrotron radiation use*. In *High-pressure Research: Application to Earth and Planetary Sciences* (eds. Syono, Y., and Manghnani, M. H.) (Terrapub. 1992) pp. 3–11.

UTSUMI, W., YAGI, T., LEINENWEBER, K., SHIMOMURA, O., and TANIGUCHI, T., *High-pressure and high-temperature generation using sintered diamond anvils*. In *High-pressure Research: Application to Earth and Planetary Sciences* (eds. Syono, Y., and Manghnani, M. H.) (Terrapub. 1992) pp. 37–42.

(Received March 16, 1993, revised October 27, 1993, accepted November 6, 1993)

Advances in High-pressure Calorimetry, Diffusion, Sealing and Calibration

PAGEOPH, Vol. 141, No. 2/3/4 (1993)

0033–4553/93/040615–15$1.50 + 0.20/0

Differential Scanning Calorimetry in a Piston-cylinder Apparatus: Design and Calibration

ROBERT P. RAPP[1,2] and ALEXANDRA NAVROTSKY[1]

Abstract —We have designed and calibrated a piston-cylinder cell assembly suitable for conducting *in situ* measurements of enthalpies of phase transitions at elevated pressures by heat-flux differential scanning calorimetry (DSC). The high-pressure DSC detector consists of a Pt-Pt13%Rh thermopile wrapped around a frame of fired pyrophyllite. Four thermocouple junctions, arranged radially around the sample capsule, are connected in series, with four reference thermocouple junctions located 3–4 mm above the sample and embedded in thermally inert ceramic. A W-W25%Re control thermocouple is situated directly above the top of the sample; the whole detector assembly is enclosed in a 1.5 mm thick cylindrical ceramic sleeve located at the center of a 8–10 mm long "hot-zone" in the tapered graphite furnace. Using this detector design and cell assembly, we have observed the thermal signal associated with the fusion of Au at 0.5 and 1.2 GPa, and have calculated a calibration factor (K) for this detector based on the gold melting curve of MIRWALD and KENNEDY (1979). Detector sensitivity decreases by a factor of four over this pressure-temperature interval. The reproducibility of the enthalpy of fusion of gold at 0.5 GPa suggests that detector geometry is reproducible from one experiment to the next, and thus confirms the viability of this particular detector design for quantitative DSC measurements. Subsequent experiments will assess the dependence of (K) on temperature and pressure by measuring the enthalpies of fusion of additional metals (e.g., Ag, Cu, Al, Ge) and salts (e.g., NaCl, CsCl).

Key words: Experimental techniques, calorimetry, high-pressure, mineral physics, phase transitions.

Introduction

Pressure is an intensive variable of particular significance to solid earth sciences because it strongly affects the stability of many important phases within the crust and the mantle. The concept of metamorphic facies and the techniques of geobarometry have as their basis the pressure dependence of transformation reactions between mineral assemblages in systems of constant composition. Dehydration-melting reactions within the lower crust (e.g., CLEMENS and VIELZEUF, 1987; RAPP and WATSON, 1993), as well as solid-solid phase transitions in the mantle (e.g., the olivine-spinel transformation; RUBIE, 1984) are strongly pressure-dependent.

---

[1] Department of Geological and Geophysical Sciences, Princeton University, Princeton, New Jersey 08544 and the NSF Science and Technology Center for High-Pressure Research, U.S.A.

[2] Now at: Mineral Physics Institute, Department of Earth and Space Sciences, University at Stony Brook, Stony Brook, NY 11791, U.S.A.

BASIC PRINCIPLES OF DTA/DSC

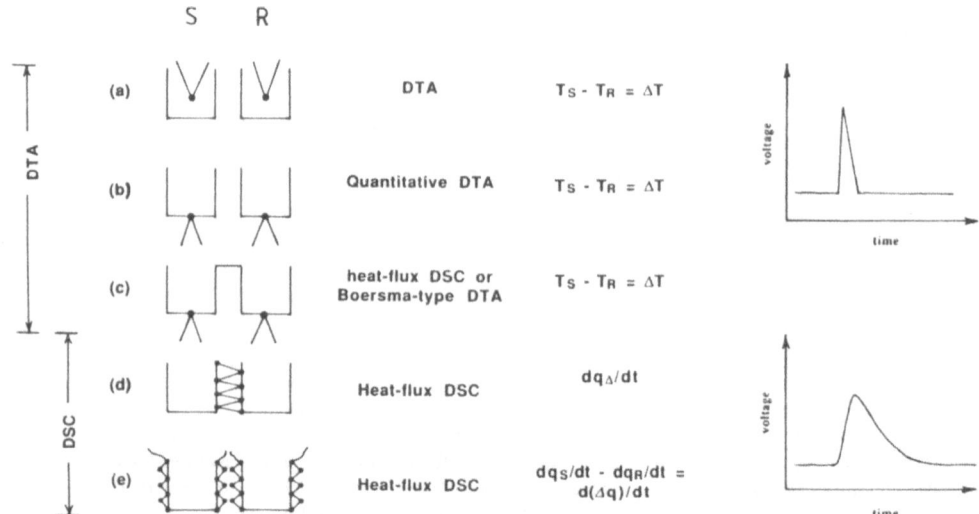

Figure 1
Schematic representation of detector design and nature of instrument signal for DTA and heat flux DSC detectors (after MACKENZIE, 1980).

Differential thermal analysis (DTA) has been used for many years to locate exactly melting curves of simple compounds (e.g., WILLIAMS and KENNEDY, 1969) and elements (e.g., LUEDEMANN and KENNEDY, 1968; MIRWALD and KENNEDY, 1979), and phase transitions in minerals (e.g., high-low quartz inversion; COHEN and KLEMENT, 1967), at conditions of elevated pressure attainable in the piston-cylinder apparatus (i.e., up to ≈ 6 GPa), following the general design of COHEN et al. (1966). In DTA the temperature difference between the sample and reference material is measured (Figs 1a,b,c). DTA in the piston-cylinder measures the difference in temperature between two thermocouples (one next to or in the sample, and the other, the reference junction, placed in contact with an inert material) as a phase transition is traversed by increasing temperature approximately isobarically. The primary objective in piston-cylinder DTA experiments has been the determination of transformation boundaries in pressure-temperature space. In these DTA experiments, the goal has been to determine the peak position (temperature of transition) accurately; thus a very sharp DTA signal has been desirable.

To convert the DTA signal to a quantifiable parameter directly related to the heat effect associated with a phase transition, that is, to do differential scanning calorimetry (DSC), additional criteria must be met (see MACKENZIE, 1980, for a review). The area under the calorimetric signal must be proportional to the total heat effect, and this proportionality constant (calibration factor) must be constant from one experiment to the next. This is possible if the geometry of the detector is constant and reproducible, and the detector has high sensitivity. In heat flux DSC (Figs. 1d,e; PETIT et al., 1961), the sample and reference are thermally connected through a conduction path that receives heat from the surroundings, and a series of thermocouples (thermopile) record an e.m.f. proportional to the temperature difference between sample and reference (as opposed to sample and reference material temperature measured at a single point, as in DTA). This temperature difference, directly proportional to the heat flux, is recorded as a curve of e.m.f. versus time, whose integral (area) is proportional to the enthalpy.

The use of a thermopile, a set of thermocouples in series that measure the temperature difference between sample and reference in a number of locations (e.g., PETIT et al., 1961), has the advantage of greater sensitivity and of measuring heat flow, not at one point, but at a number of locations symmetrically positioned around the sample. The latter we believe to be a major potential advantage in the high-pressure environment, since it may tend to cancel out uncertainties resulting from small geometric changes, which can be exaggerated if only one thermocouple is used. However, the signal from a thermopile is intrinsically less sharp in the time domain than that for a single thermocouple (see Fig. 1), hence the time constant for heat flow becomes important. Thus issues of "baseline stability" become more important (see below). We believe that, in the piston-cylinder environment, DTA is useful to locate the temperature of a transition, and DSC to measure its enthalpy.

We have designed a heat flux DSC detector for use in a standard piston-cylinder apparatus. It consists of a thermopile of four Pt-Pt13%Rh thermocouples connected in series, wrapped around a cylindrical ceramic frame containing a sample and reference chamber. We have calibrated this detector against the fusion enthalpy of gold, in an attempt to develop the piston-cylinder heat flux DSC technique to the stage where quantitative *in situ* measurements of the energetics of geologically important phase transitions can be made at elevated temperature *and* pressure. Such capabilities should be pertinent to the study of the energetics of any kinetically rapid phase transition occurring in the pressure-temperature range of the piston-cylinder, roughly 0.5–2.5 GPa and 500–1400°C. In this paper, we describe the design and testing of this instrument.

Experimental Procedure

General Approach

The first and most important consideration in the design of a piston-cylinder assembly for DSC experiments at elevated pressures is that the measurement of the enthalpy of a given phase transition be accurate and reproducible. In turn, this requires a thermal and pressure environment that is itself reproducible. The quality of the measurements is therefore critically dependent upon initial characterization of the deformation behavior of the detector, and the temperature and pressure distribution within the cell. Therefore, initial experiments assessed the extent to which the overall integrity of the detector was maintained over the course of an experiment. These experiments also investigated the time required for a stable baseline signal after the initial setup of an experiment, the time required for the baseline signal to stabilize after an experiment, and the response of the detector and its signal to adjustments in temperature and pressure. We define a stable baseline as a detector signal that varies less than about ± 0.002 mV (2μV) over the course of 30 minutes at constant temperature. An unstable baseline during a temperature scan through a phase transition will increase the signal-to-noise ratio, increase the possibility of sudden and significant shifts of the baseline during a scan, and thereby decrease the precision of the measurement.

The detector described in the next section proved capable of providing stable baseline signals on a digital voltmeter at elevated pressure and temperature, maintained its overall geometry, and sustained temperature cycles of $100-200°C$ repeatedly without failure. Pressure changes associated with these temperature cycles are less than 2% of the total pressure, and have little apparent effect on the DSC signal; this has been confirmed by monitoring the baseline as temperature is increased 100 to 200°C, over a range in which no phase transition occurs. Apparently, any effect on detector e.m.f. due to this slight increase in pressure is less than normal fluctuations in the instrument signal; a stable baseline signal is generally maintained during temperature scans of $100-200°C$, despite the slight change in the pressure of the assembly. We chose to calibrate this detector using the melting curves of metals (Au, Ag, Cu) as a reference because the thermophysical constants necessary to calculate the enthalpy of fusion as a function of pressure and temperature (thermal expansivity, heat capacity, molar volume, compressibility for both liquid and solid) are well-known.

Pressure Cell Assembly for Piston-cylinder DSC

The detector frame and thermopile (Figs. 2, 2a)

The DSC detector (Figs 2 and 2a) body consists of a hollow cylindrical frame of fired pyrophyllite with a series of zigzag grooves cut on the outer surface (Fig. 2a), which accommodate four Pt-Pt13%Rh thermocouples connected in series. The overall length of the detector is 9 mm, with a 4 mm deep, 3 mm diameter sample chamber on the bottom and a 3 mm deep, 1.0 mm diameter access hole on the top for the control thermocouple. The walls of the detector are approximately 1.0 mm thick. At present, each detector is hand-made from easily machined pyrophyllite, which is fired to a hard ceramic for 24 hours at 1000°C. This material shrinks about 2% upon firing. The detector thermopile consists of a crown of four thermocouple junctions oriented around the sample chamber at the bottom of the frame, and four junctions imbedded in inert Al_2O_3 cement and positioned at the top. A separate W-W26%Re control thermocouple is positioned at the center of the detector frame. The dimensions of the detector represent a balance between keeping the location of

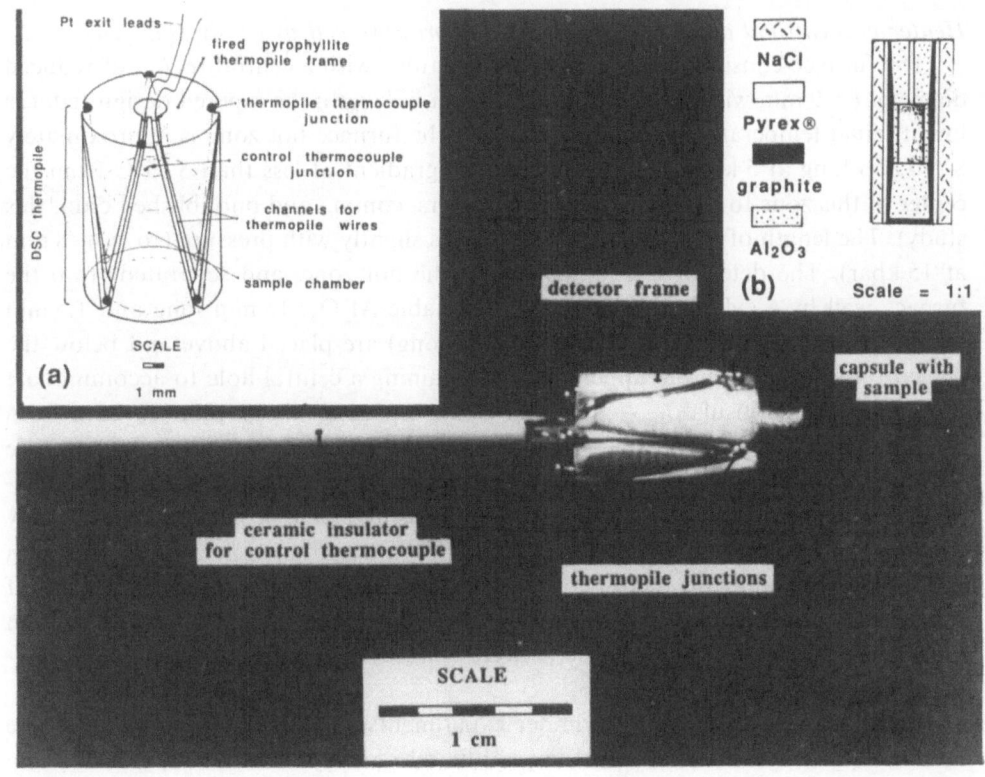

Figure 2

Assembly for heat flux DSC experiments in a piston-cylinder apparatus. The photograph shows the sample capsule in position within the detector frame, the thermopile wrapped around the detector frame, and the ceramic insulator for both control and thermopile thermocouple leads. Insets: (a) Sketch of piston-cylinder DSC detector frame and thermopile configuration, (b) cross-sectional profile of 3/4" pressure cell assembly, drawn to scale.

the reference thermopile junctions, encased in inert ceramic and Al_2O_3 cement, close enough to the sample chamber such that a minimal temperature difference exists between these two regions, yet far enough away so that any heat effects associated with a phase transition are not registered at the reference junctions. The thermopile junctions around the sample chamber are separated from the sample capsule by a thin layer (<0.5 mm) of Al_2O_3 cement. The control thermocouple monitors temperature in the center of the furnace 'hot spot', and is separated from the sample capsule by a 1.5 mm thick ceramic wall. Thermocouple access to the interior of the pressure cell is accommodated by a four hole, 1.5 mm diameter ceramic tube, slotted in the side to permit access of the two platinum exit leads of the thermopile to the cell exterior. The control thermocouple is connected to a Eurotherm ^{R} model 818 temperature controller and SCR, and the thermopile leads are connected to a Keithley ^{R} 181 nanovoltmeter. The analog voltage output from the thermopile is converted to a digital signal by a National Instruments IEEE-488 GPIB board interfaced with an IBM-compatible PC.

Heater element and other components of the pressure cell assembly (Fig. 2b)

The furnace consists of a tapered graphite tube, with a central region of reduced diameter (≈ 9 mm wide) approximately 12 mm in length; the tapered design reduces longitudinal temperature gradients such that the furnace hot zone is approximately 8–10 mm long at 5 kbar, with a temperature gradient of less than 3–5°C from the center of this zone to either end (B. Mysen, pers. comm., and unpublished data, this study). The length of this "hot zone" decreases slightly with pressure (to ≈ 6–8 mm at 15 kbar). The detector is situated within the hot zone, and separated from the furnace wall by a cylindrical sleeve of machinable Al_2O_3, 15 mm long and 1.5 mm thick. Spacers of machinable Al_2O_3 (15 mm long) are placed above and below the detector assembly, with the upper spacer containing a central hole to accommodate the thermocouple insulator. A tube of Pyrex® separates the graphite furnace from the NaCl outer sleeve, and the whole assembly is wrapped in 10 mil lead foil. We have conducted experiments with outer sleeves of both talc and NaCl, but anticipate that most future experiments will be made using NaCl, because of its extremely low friction coefficient, high purity, and lack of a phase transition (e.g., dehydration of talc may contaminate thermocouples) within the pressure-temperature range of the apparatus. In general, we have used arc-welded platinum capsules to contain samples, although several experiments were conducted using pressure-welded tantalum capsules.

A major concern in piston-cylinder experiments is the effect of friction on the hydrostatic transmission of pressure; although ideally inner components of a piston-cylinder assembly should be made of materials with low coefficients of friciton, we have used "hard" material (i.e., fired pyrophyllite ceramic) for the detector because we wanted to minimize the amount of deformation of the thermopile frame. One might incorporate low-friction spacer materials (such as NaCl or boron nitride) in combination with the ceramic detector frame, but this introduces the possibility of contamination of the thermocouples (both W-W26%Re and Pt-Pt13%Rh) with salt or boron. Our primary concern at this stage in the development of the piston-cylinder DSC technique is to maintain the geometry of the detector frame.

Setup of a piston-cylinder DSC experiment

The pressure cell assembly described above was inserted into a 3/4" pressure vessel, that itself was then placed in a 150 ton end-loaded piston-cylinder apparatus capable of attaining pressures of 2.5 GPa and temperatures of 1400°C. Both control and thermopile thermocouples were connected to their respective monitoring instruments. The cell assembly was pressurized at room temperature to 0.02–0.05 GPa, and then heated to approximately 200°C. The assembly was maintained at this temperature for 1 hour to insure that the detector was providing a signal, before the pressure was increased to approximately 0.2 GPa and temperature increased at a rate of 50°C/minute. Once the temperature had exceeded 700°C, the pressure was increased to a level slightly above that of the experiment, and the temperature continued to rise until it was approximately 100°C below the temperature of the phase transition (e.g., melting point of gold at elevated pressure). Slight relaxation of the assembly brought the pressure down to the final experimental pressure (i.e., the "hot piston out" procedure was followed) over the course of several hours.

A common problem encountered in early experiments during setup was shearing of the thermocouple wires and shorting of the thermopile against the sample capsule, resulting in a null signal from the detector. The procedure outlined above has reduced the experimental failure rate from >50% to about <25% of all runs. The time required for a stable baseline signal (i.e., <2 μV variation in the signal over 30 minutes) to be established after the initial setup of an experiment was determined from preliminary experiments to be from 6 to 8 hours.

Overall Performance of the DSC Detector in Preliminary Experiments

Initial experiments aimed at assessing the overall variability of the detector design examined the melting of gold at 0.5 GPa; the procedure involved increasing the temperature (at rates varying between 1°C/min to 20°C/min) isobarically in the vicinity of the expected phase transition (for the fusion of gold, ≈ 1092°C at 0.5 GPa; MIRWALD and KENNEDY, 1979). We found that at an optimal scan rate of 10°C/minute, the detector recorded a sharp change in the slope of a voltage versus time curve, with an inflection at the gold melting point; this phenomenon was observable on 3–5 separate scans through the transition before the detector failed. At slower scan rates, the profile of the transition peak was broad and diffuse, and at faster scan rates the peak was presumably "washed out" by the rapid rate of heating. Approximately 0.5–1.0 hours were required for a stable baseline to be re-established after an individual scan. Although the discontinuity associated with fusion was sharp and occurred at the same temperature in each scan at the optimal scan rate, the return to the baseline after the transition was not always complete (i.e., anomalous baseline shifts were sometimes observed). For a given experiment, approximately 1/3 to 1/2 of all scans through a transition displayed such a baseline shift. This may be related to differences in heat capacity and thermal conductivity of gold in the solid and the liquid state, but it may also indicate slight changes in geometry, either in the thermopile or the control thermocouple.

The cell assembly was recovered after each of these experiments and cut into longitudinal sections for *post-mortem* examination; a representative sample is shown in Figure 3. The overall cylindrical geometry of the detector assembly, and the relative positions of the thermopile junctions, the sample capsule, and the control thermocouple have been maintained over the course of the experiment. Having demonstrated that a reasonably steady baseline signal can be maintained at elevated pressure and temperature, that a reproducible reaction peak can be detected as a phase transition is traversed, and that the detector geometry remains intact through multiple scans, the transition from qualitative DTA to quantitative

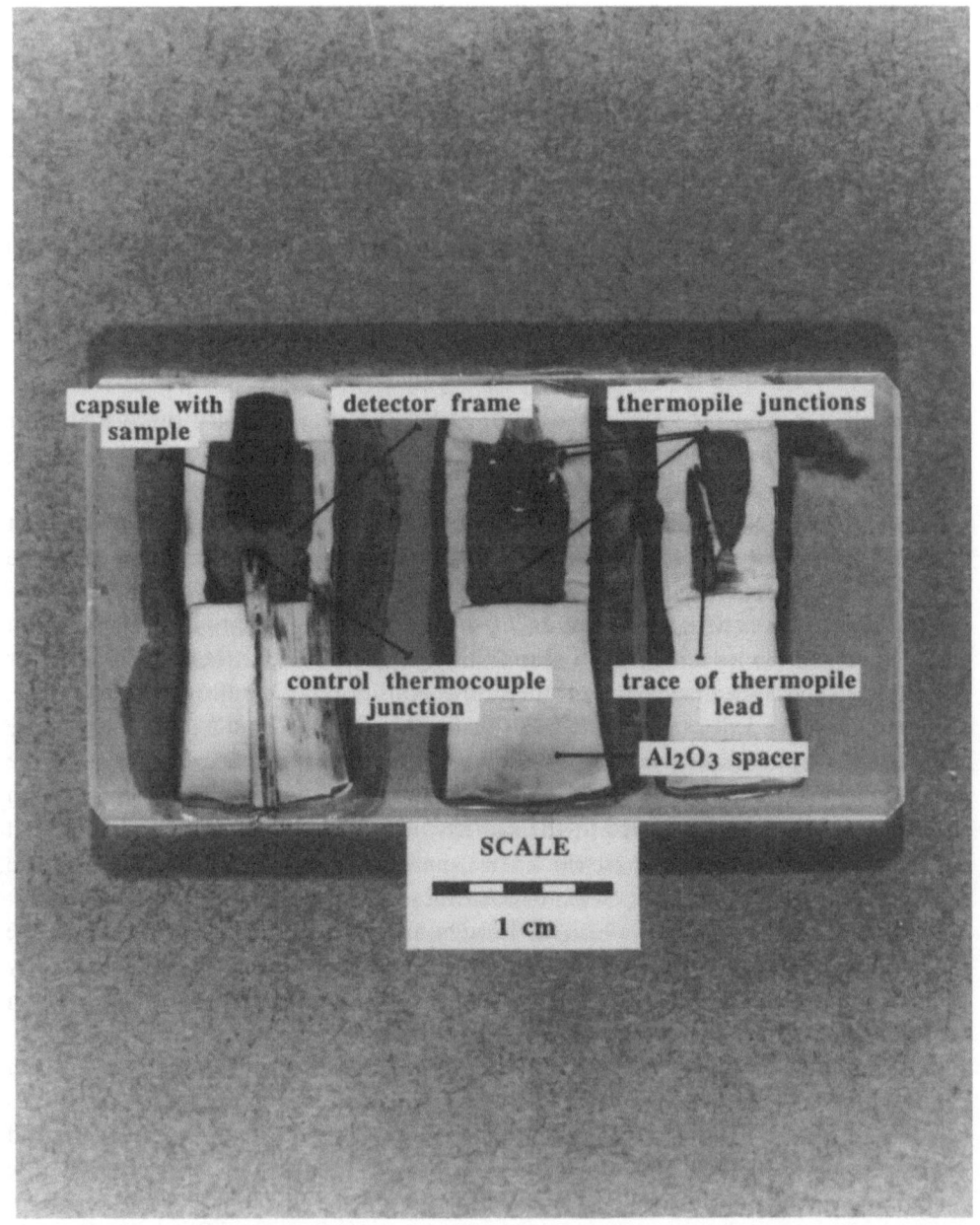

Figure 3
Polished longitudinal sections through the upper 2/3 of the piston-cylinder DSC cell assembly, excluding graphite furnace, Pyrex^R sleeve and NaCl sleeve; section on right-hand side is closest to outer surface, section on left-hand side is right down the middle of the assembly.

DSC measurements next requires calibration of the detector against the calculated or known fusion enthalpy vs. pressure curve of a suitable standard. This calibration factor must be determined empirically over a range of pressure and temperature conditions.

The Fusion of Gold and Detector Calibration

DSC Calibration Using the Enthalpy of Fusion of Gold

We conducted a number of piston-cylinder DSC experiments which measured the response of our detector to the fusion of gold at 0.5 and 1.2 GPa (see Table 1). These data represent both replicate experiments at the same pressure (e.g., runs Au5, Au8, and Au3) as well as repeated scans through the melting point in a single experimental run (e.g., runs Au3E1-Au3E3). The experiment at 0.5 GPa, using tantalum capsules to contain the sample (Au5E1-Au5E4), instead of platinum, emphasizes the importance of using exactly the same assembly components from one experiment to the next; counts per unit mass are approximately 25% higher in

Table 1

Summary of piston-cylinder DSC detector calibration experiments, gold fusion

Expt. #	Pressure (GPa)	Mass (grams)	Counts (mV-sec)		Counts/unit mass (mV-sec/gm)
Au3E1[a]	0.5	0.4472	4.046		9.047
Au3E2[a]	0.5	0.4472	3.067		6.858
Au3E3[a]	0.5	0.4472	3.457		7.730
			mean	=	7.88
			(±1 s.e.)		±0.63
Au5E1[b]	0.5	0.1462	1.616		11.053
Au5E2[b]	0.5	0.1462	1.795		12.278
Au5E3[b]	0.5	0.1462	1.347		9.213
Au5E4[b]	0.5	0.1462	2.095		14.330
			mean	=	11.72
			(±1 s.e.)		±1.07
Au8E1[c]	0.5	0.3886	2.565		6.600
Au9E1[c]	1.2	0.5151	10.957		21.272
Au9E2[c]	1.2	0.5151	7.936		15.407
Au9E3[c]	1.2	0.5151	11.971		23.240
			mean	=	19.97
			(±1 s.e.)		2.35

[a] Talc pressure medium, platinum sample capsule.
[b] NaCl pressure medium, tantalum sample capsule.
[c] NaCl pressure medium, platinum sample capsule.

these experiments than those in which the sample was contained in platinum. Platinum is our capsule material of choice because it can be easily welded shut, preventing migration of molten sample (or volatile constituents) and thermocouple contamination. Although the two materials presumably have different friction coefficients, comparable results were obtained using outer sleeves of both talc and NaCl (i.e., compare experiment Au8 with Au3). Scans through the melting point were made by increasing temperature at a rate of 10°C/min, beginning at approximately 50°C below the melting point and continuing approximately 50°C beyond. Mismatch of the thermal properties of the sample and fired pyrophyllite ceramic reference material result in an initial displacement from zero (BROWN, 1988), and ramps in temperature are generally accompanied by a sloping baseline. A representative scan through the gold melting point is shown in Figure 4 for experiment Au8E1 at 0.5 GPa.

The features of interest in the DSC curve of Figure 4 are the deviations of the signal from the detector baseline which correspond to a phase transition; a linear approximation to the signal on either side of the reaction peak defines the baseline used in calculating the integrated area under the peak. Although a reasonably linear trend can be proposed for the profile in Figure 4, other scans displayed unstable (i.e., nonlinear) baselines which were displaced at some point during the tempera-

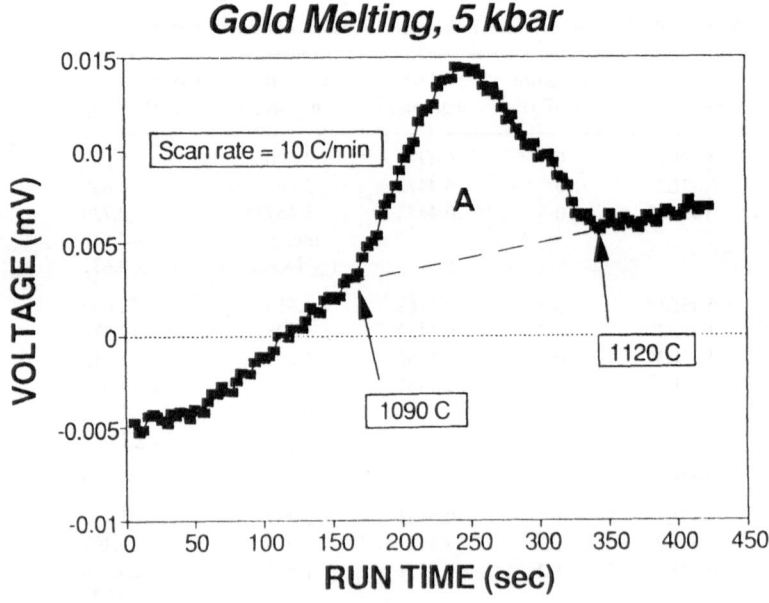

Figure 4

Sample DSC scan through the gold melting transition at 0.5 GPa. The area under the curve and above the heavy dashed line (labelled A) from the point labelled 1090°C to the point labelled 1120°C is equal to A in equation (1), and is proportional to ΔH_f.

ture scan through the transition. These can be corrected for with extrapolated approximations to the original baseline from the initial onset of the transition, in order to estimate the area under the reaction endotherm or exotherm (BROWN, 1988; GUTTMAN and FLYNN, 1973), and calculate an enthalpy of transition. We are currently exploring possible sources of this variability.

Once a suitable baseline has been defined, the area under the endotherm (for melting), A, can be assumed to be proportional to the enthalpy change, ΔH, for the transition, according to the expression:

$$\Delta H = A \times K/m \tag{1}$$

where m is the sample mass and K is the calibration factor. For the data in Table 1, m is obtained by weighing the sample prior to its emplacement in the detector, and A is determined by cubic interpolation of the data points in the transition peak. The parameter K must be determined by comparison of a known or calculated fusion enthalpy with the measured peak area.

Pressure and Temperature Dependence of the Enthalpy of Fusion

The progression from DTA to quantitative DSC requires calibration over a range of conditions, followed by verification of the accuracy of the calibration factor by means of secondary standards. In order to accomplish this, it is necessary to know the enthalpy of fusion as a function of pressure and temperature for a number of suitable standards. We have calculated $\Delta H_{\text{melting}}$ for a number of metals and a sample silicate ($CaMgSi_2O_6$, diopside) along their melting curves using the following relationship:

$$\Delta H_{f,T,P} = \Delta H^0_{f,T} + \int_1^P \Delta V \, dP \tag{2}$$

where

$$\Delta H^0_{f,T} = \Delta H^0_{f,T^0} + \int_{T^0}^T \Delta C_p \, dT. \tag{3}$$

The volume term is approximated using the following relationships:

$$V^P_T = V^0_T(1 - \beta P) \tag{4}$$

where:

$$V^0_T = V^0_{298}[1 + \alpha(T - 298)] \tag{5}$$

where V^0_{298} is the molar volume at the standard state of 1 bar and 298 K, β is the isothermal compressibility, α is thermal expansivity, and T is temperature in Kelvin. Equation (4) then becomes

$$V^P_T = V^0_{298}(1 - \beta P)[1 + \alpha(T - 298)]. \tag{6}$$

Table 2

Thermophysical parameters used to calculate pressure dependence of ΔH_{fusion}

Property	Au	Cu	Ag	CaMgSi$_2$O$_6$
1. ΔH_{fusion}				
(1 bar, T_m, kJ mol^{-1})	12.7	13.0	11.3	77.4
2. Melting point, 1 bar (°C)	1064	1085	962	1392
3. Thermal expansivity				
(α, 10^{-6} K^{-1})				
a. liquid	70.0	100.0	95.0	60.0
b. solid	14.2	16.5	19.2	18.8
4. Isothermal compressibility				
(β, 10^{-12} Pa^{-1})				
a. liquid	8.8	10.4	14.0	42.9
b. solid	5.9	7.3	9.7	8.9
5. Molar volume (cm^3·mol^{-1})				
a. solid (1 bar, 298 K)	10.2	7.1	10.4	66.1
b. liquid (1 bar, T_m)	11.3	7.9	11.6	83.0
6. Isobaric heat capacity				
(C_p, J mol^{-1} K^{-1})				
a. liquid	29.3	31.4	30.5	357.3
b. solid	25.4	24.4	25.4	268.3

If we let $t = T - 298$, ΔV_T^P in the volume integral term from equation (2), along the melting curve at a given pressure, then becomes

$$(\Delta V)_T^P = \Delta V_{298}^0 + (V_l\alpha_l - V_s\alpha_s)t - (V_l\beta_l - V_s\beta_s)P - (V_l\alpha_l\beta_l - V_s\alpha_s\beta_s)Pt \quad (7)$$

where the subscripts (s) and (l) denote solid and liquid, respectively.

In order to calculate the enthalpy of fusion $\Delta H_{f,T,P}$ from equation (2) at any point along the fusion curve, the following thermophysical parameters must be known for both liquid and solid phases: thermal expansivity, isothermal compressibility, isobaric heat capacity, and the molar volume. The reference state enthalpy of fusion must also be known. These parameters are listed in Table 2 for the metals gold, copper, silver, and also for diopside for comparison.

The pressure-temperature dependence of the fusion enthalpies for these substances has been calculated using equations (2)–(7) and is represented in Figure 5 up to 2.0 GPa; $\Delta H_{f,T,P}$ has been normalized to the 1 bar fusion enthalpy. The metals show similar behavior along the melting curve; fusion enthalpy increases slightly and approximately linearly with increasing pressure, whereas diopside shows a much stronger increase with pressure.

Calibration Factors for the Detector at 0.5 and 1.2 GPa

Using the combined data from Table 1 and Figure 5, we have calculated the calibration factor (K) for the piston-cylinder DSC for an assembly using NaCl as

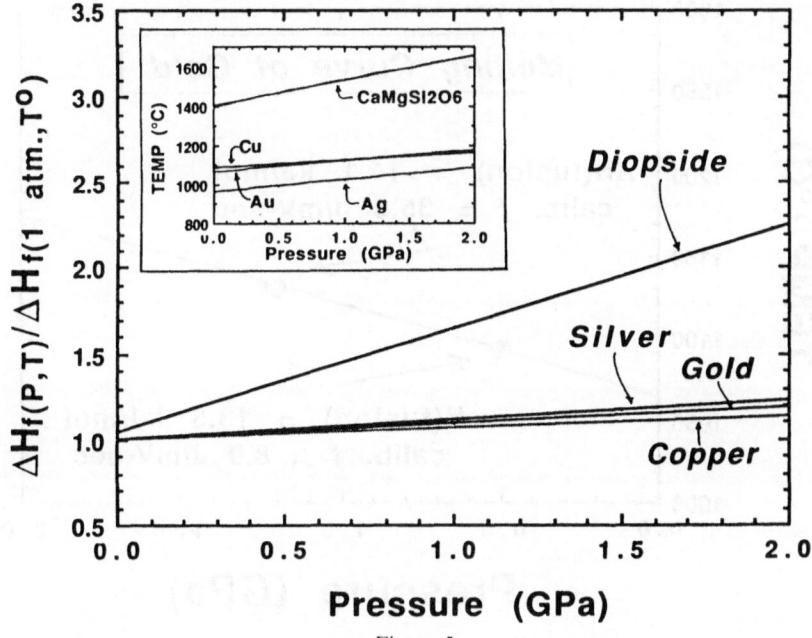

Figure 5

Relationship between $\Delta H_{f(P,T)}$, normalized to $\Delta H_{f(0)}$ (enthalpy of fusion at 1 bar), and pressure along the fusion curve for Au, Ag, Cu, and diopside, calculated using equations (2)–(5). Inset shows the melting curves as a function of pressure and temperature.

the pressure-transmitting medium, and platinum capsules as sample containers. The calibration factor at 0.5 GPa is determined by averaging the data obtained in experiments Au3 and Au8 from Table 1, and disregarding the data from the experiment made using tantalum as the capsule material (i.e., Au5). Incidentally, the duration of the experiments is short enough such that diffusion of platinum into the gold sample is insignificant, and our results are therefore unaffected by impurities of platinum in gold (e.g., the diffusivity of Pt in Au at 900°C is $\approx 6.7 \times 10^{-11}$ cm^2 sec^{-1}; BARRER, 1941). The melting curve of gold between 0 and 2.0 GPa from MIRWALD and KENNEDY (1979) is shown in Figure 6, along with our calculated value for (K) at 0.5 and 1090°C and 1.2 GPa and 1132°C. Keep in mind that these calibration factors are derived from both replicate experiments at the same pressure and repetitive scans through the transition with the same cell. It is clear that detector sensitivity is reduced by about a factor of four at the higher pressure *and* temperature. It is not yet clear whether this is primarily an effect of pressure, temperature, or both. The likelihood of a higher thermal conductivity for the ceramic thermopile frame at higher temperatures and of greater heat transfer by radiation suggests that the decreased detector sensitivity at higher pressures could be the result of the higher temperature of gold melting. A strong temperature dependence for K has been observed in DSC experiments at 1 atmosphere and

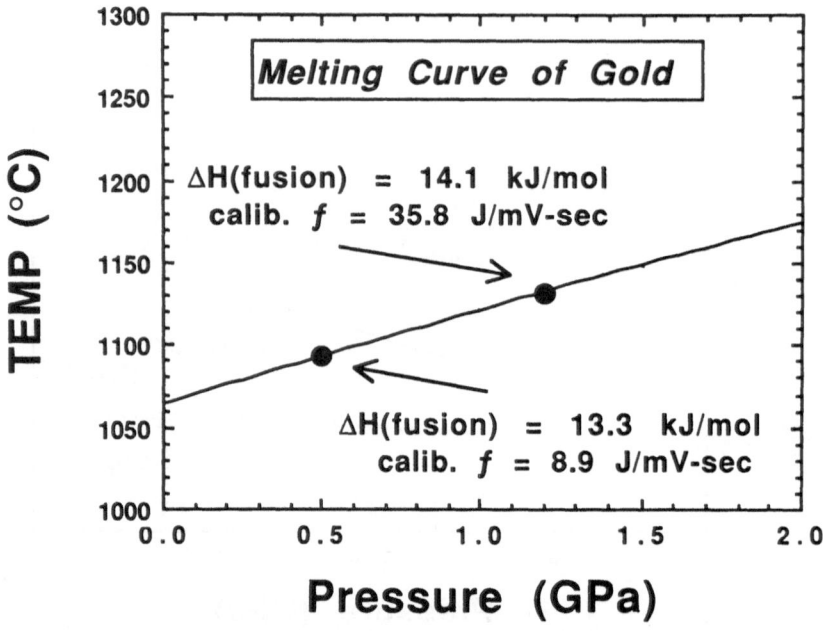

Figure 6

Calibration factors (K) for the piston-cylinder DSC detector based on the enthalpy of fusion of gold at 0.5 and 1.2 GPa. Fusion curve for gold is from MIRWALD and KENNEDY (1979).

temperatures up to 820°C (e.g., DUBRAWSKI, 1991). The future course of study will examine the relative effects of pressure and temperature on detector sensitivity, by calibrating against the heats of fusion of other calibration materials (i.e., other metals and/or salts). Since the accuracy of any measurement of a transition enthalpy is only as good as the calibration, we reiterate the importance of detector calibrations over the intended range of pressure and temperature conditions under which the piston-cylinder heat flux DSC detector will be used. The ability of this detector to maintain its geometry, and its reproducibility, suggest that this design may be suitable for a more comprehensive series of calibrations. After these calibrations are completed, possible applications of the piston-cylinder DSC technique include the study of amphibole dehydration reactions and order-disorder phenomena in dolomites.

Acknowledgements

The authors wish to thank L. Topor and C. L. Bennett for scientific and technical contributions. Constructive official reviews by J. Ganguly and D. Jenkins are gratefully acknowledged. Critical comments by K. Bose are also appreciated. This research was supported by the NSF Science and Technology Center for High Pressure Research (CHiPR).

REFERENCES

BARRER, R. M., *Diffusion in and Through Solids* (Cambridge University Press, 1941) p. 275.

BROWN, M. E., *Introduciton to Thermal Analysis: Techniques and Applications* (Chapman and Hall, London, 1988).

COHEN, L. H., KLEMENT, Jr., W. H., and KENNEDY, G. C. (1966), *Investigation of Phase Transformations at Elevated Temperatures and Pressures by Differential Thermal Analysis in Piston-cylinder Apparatus*, J. Phys. Chem. Solids 27, 179.

COHEN, L. H., and KLEMENT, Jr. W. H. (1967), *High-low Quartz Inversion: Determination to 35 Kilobars*, J. Geophys. Res. 72, 4245–4251.

CLEMENS, J. D., and VIELZEUF, D. (1987), *Constraints on Melting and Magma Production in the Crust*, Earth Planet Sci. Lett. 86, 287–306.

DUBRAWSKI, J. V. (1991), *Differential scanning calorimetry and its application to mineralogy and the geosciences*. In *Thermal Analysis in the Geosciences* (eds. Smykatz-Kloss, W., and Warne, S. St. J.) (Springer-Verlag, Berlin 1991) pp. 16–59.

GUTTMAN, C. M., and FLYNN, J. H. (1973), *On the Drawing of the Baseline for Differential Scanning Calorimetric Calculation of Heats of Transition*, Anal. Chem. 45, 408–410.

LUEDEMANN, H. D., and KENNEDY, G. C. (1968), *Melting Curves of Lithium, Sodium, Potassium, and Rubidium to 80 Kilobars*, J. Geophys. Res. 73, 2795–2805.

MACKENZIE, R. C. (1980), *Differential Thermal Analysis and Differential Scanning Calorimetry: Similarities and Differences*, Anal. Proc. June 1980, 217.

MIRWALD, P. W., and KENNEDY, G. C. (1979), *The Melting Curve of Gold, Silver, and Copper to 60-kbar Pressure: A Reinvestigation*, J. Geophys. Res. 84, 6750–6756.

PETIT, J. L., SICARD, L., and EYRAUD, L. (1961), *Dispositif simple d'analyse enthalpique differentielle*, C. R. Acad. Sci. 252, 1741.

RAPP, R. P., and WATSON, E. B. (1993), *Water-deficient Partial Melting of Metabasalt and Continental Growth*, J. Petrology (in press).

RUBIE, D. C. (1984), *The Olivine → Spinel Transformation and the Rheology of Subducting Lithosphere*, Nature 308, 505–508.

WILLIAMS, D. W., and KENNEDY, G. C. (1969), *Melting Curve of Diopside to 50 Kilobars*, J. Geophys. Res. 74, 4359–4366.

(Received May 14, 1993, revised October 14, 1993, accepted October 19, 1993)

PAGEOPH, Vol. 141, No. 2/3/4 (1993)

0033–4553/93/040631–12$1.50 + 0.20/0

A Thin Film Approach for Producing Mineral Diffusion Couples

CRAIG S. SCHWANDT,[1] RANDALL T. CYGAN[1] and HENRY R. WESTRICH[1]

Abstract — Few diffusion coefficient values have been measured for silicate minerals at pertinent geologic conditions because of experimental restrictions. Until recently, analysis of diffusion couples was conducted principally with electron microprobes which have rather poor spatial resolution (micrometer scale). Ion microprobe analyses, however, eliminate many of the previous experimental restrictions; in depth profile mode they have excellent spatial resolution (tens of angstroms) and diffusion couples can be analyzed normal to the interface. Diffusion couples analyzed by ion microprobe must be well-defined and uniform; previous methods using solution precipitates to form the diffusion couples were heterogeneous and had limited success. A new approach, the thermal evaporation of ^{25}MgO under high vacuum onto a crystalline substrate (oxide, silicate), produces a 1000 Å thick ^{25}MgO$_x$ ($x < 1$) thin film. This method yields an excellent diffusion couple for low-temperature diffusion experiments. Diffusion anneal experiments using this approach for garnet provide a Mg self-diffusion coefficient of $D = 0.60 \pm 0.09 \times 10^{-21}$ m^2/s at 1000°C ($\log fO_2 = -11.3$, $P = 1$ atm, $X_{Almandine} = 0.24$).

Key words: Thin film, diffusion coefficients, ion microprobe.

Introduction

Understanding and modeling geochemical, nuclear waste, and materials science systems require accurate diffusion coefficient data for silicate minerals. Diffusion coefficients and the concentration gradient of the species determine the migration rates of a component through a material. Commonly, a silicate glass or mineral of one composition is placed in contact with a sample having a different composition, forming a diffusion couple that will generate concentration gradients. The couple is annealed at temperature for a given time and then penetration profiles in response to diffusion are measured. Diffusion follows the Arrhenius-type behavior typical for thermally-activated kinetic processes. Therefore, diffusion data should be determined for a range of geologically-relevant temperatures to evaluate the temperature dependence of the diffusion process. Minerals that have especially small diffusion coefficients often exhibit compositional zoning that can yield thermal history

---

[1] Geochemistry Department, Sandia National Laboratories, Albuquerque, New Mexico 87185-0750, U.S.A.

information if appropriate diffusion models are used and the temperature dependence of diffusion is known. Natural garnets commonly exhibit compositional zoning that can typically be modeled as a diffusion process, however, only limited diffusion data are available and only for restricted conditions (FREER, 1979, 1981; ELPHICK *et al.*, 1985; CYGAN and LASAGA, 1985; CHAKRABORTY and GANGULY, 1992). The preservation of compositional zoning in garnets, but not in the minerals that coexist with garnet, suggests garnet possesses some of the smallest silicate mineral diffusion coefficients. Accordingly, we developed a method for measuring diffusion coefficients in garnet, as they should be the most difficult to determine at temperatures representative of metamorphic conditions in the crust of the earth. A technique that works for garnet should be suitable for other minerals with larger diffusion coefficients.

Previous investigators (FREER, 1979; ELPHICK *et al.*, 1985; CHAKRABORTY and GANGULY, 1992) conducted cation diffusion experiments on garnet at temperatures of 1000°C and higher to experimentally compensate for the slow diffusion rates and inherently-small penetration distances. Also, the spatial resolution of electron microprobes requires diffusion penetration distances of tens of micrometers, thereby limiting diffusion experiments to high temperatures for moderate laboratory anneal times. The existing high-temperature data are limited in extent, and have inconsistent temperature trends; activation energies range from 100 to 285 kJ/mole (FREER, 1979; ELPHICK *et al.*, 1985; CHAKRABORTY and GANGULY, 1992) with very limited information regarding compositional and oxygen fugacity dependencies (c.f. BUENING and BUSECK, 1973; RYERSON *et al.*, 1989). Therefore, extrapolation of the high-temperature data to lower temperatures is questionable.

The introduction of the ion microprobe provided a means of resolving spatial information to the tens-of-angstroms level in depth profiling mode (WILSON *et al.*, 1989). Shorter experimental penetration distances permit diffusion experiments to be conducted at relatively low temperatures and shorter anneal durations. CYGAN and LASAGA (1985) took advantage of this approach and conducted self-diffusion experiments in the temperature range of 750°C to 900°C. Self-diffusion is the migration of an isotope due to an isotopic gradient, rather than a chemical gradient. Self-diffusion coefficients are approximately the same order of magnitude as chemical diffusion coefficients (CYGAN and LASAGA, 1985; CHAKRABORTY and GANGULY, 1992). CYGAN and LASAGA (1985) investigated magnesium self-diffusion using diffusion couples formed by dissolving ^{25}MgO in oxalic acid and precipitating $^{25}MgC_2O_4$ on the surfaces of single-crystal garnets. The ^{25}Mg isotope was enriched to 97.8 percent of total magnesium in the oxide. The natural abundance of ^{25}Mg is 10 percent of total magnesium. This approach, although successful, produced a ^{25}MgO source that was very heterogeneous on the garnet surface (Fig. 1a). The oxalate decomposes to the oxide during the initial heating period of the diffusion anneal yielding a nonuniform distribution of the source material and subsequent difficulty in obtaining reproducible depth profile analyses.

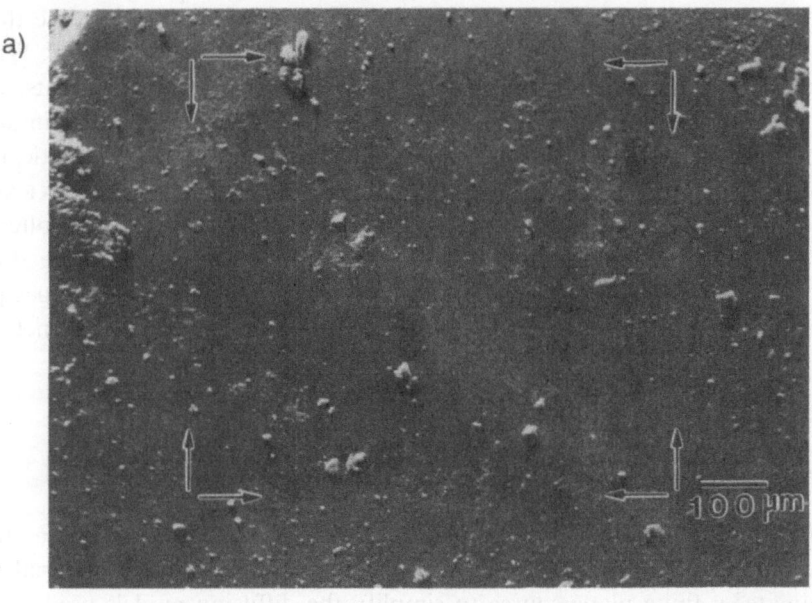

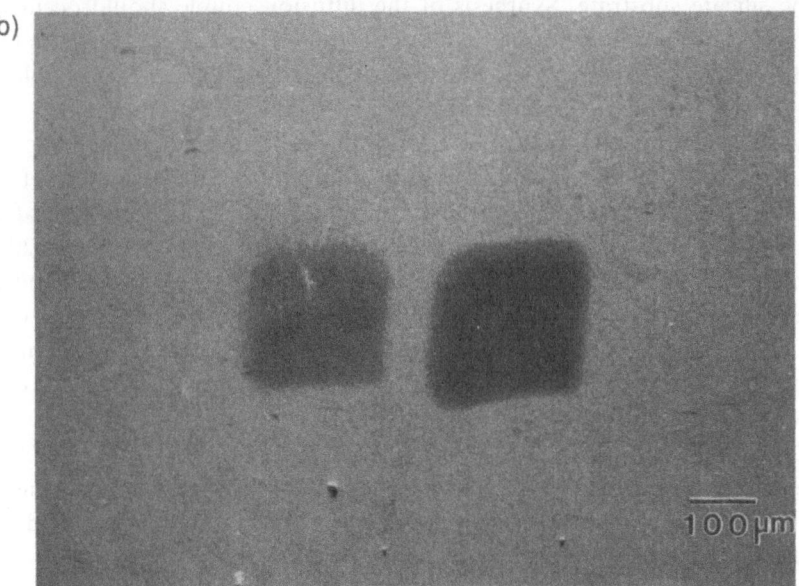

Figure 1
a) SEM photomicrographs of garnet surface with ^{25}MgO layer formed by precipitation of ^{25}Mg oxalate and diffusion anneal. Large amounts of MgO are observed in the 750 μm × 750 μm ion microprobe crater. b) SEM photomicrographs of ^{25}MgO thin film on garnet. The uniform ion microprobe craters are 150 μm × 150 μm. Scale bars are 100 μm.

The inconsistencies of the previous high-temperature data underscore the need for determining accurate diffusion data in the range of 700°C to 1000°C. Whereas, the ion microprobe provides the analytical means for these experiments, an improved technique for creating diffusion couples that provides and maintains the uniformity of the diffusion interface is needed, thereby improving analytical precision. Because precipitation techniques (CYGAN and LASAGA, 1985; CHAKRABORTY *et al.*, 1992) have serious limitations, we initially explored the use of molten salts, precipitated metal chlorides, and evaporative thin films as alternative diffusion couple preparation methods. The molten salt and metal chloride techniques proved to be inappropriate, as the presence of these compounds corrodes the garnet surface during the diffusion anneal.

Evaporative Thin Film Techniques

Self-diffusion experiments impose several restrictions on the choice of a diffusion couple. Foremost, it is important that the distribution of source material closely approximates a finite plane source to simplify the diffusion models used to obtain the diffusion coefficients. Second, to avoid disruption or destruction of the diffusion interface, the source material must not chemically react, outside of ion exchange, with the silicate substrate. Synthesis of the diffusion couple should occur at a low enough temperature to avoid premature diffusion or reaction. Finally, the technique must be efficient to minimize expense when utilizing enriched stable isotope compounds that can cost as much as $10 to $20 per milligram.

Although various physical vapor deposition techniques are available to produce thin films, not all of the techniques are appropriate for preparing garnet diffusion couples. JAOUL *et al.* (1981) and HOULIER *et al.* (1990) utilized radio-frequency sputtering techniques to produce an amorphous thin film (100 Å to 200 Å) for examining silicon diffusion in olivine. However, thin film dimensions and the cost of producing an enriched film severely limit the use of this technique for examining diffusion in garnet. Two other vapor deposition techniques, electron beam evaporation and thermal evaporation (SURDASHAN, 1989; OHRING, 1992; GEORGE, 1992), were considered for creation of an oxide thin film on a garnet substrate to form a diffusion couple. Both methods use a vacuum evaporation apparatus which includes a cryogenic pump, a quadrupole residual gas analyzer, and a film-thickness monitor. The vacuum chamber is pumped down to 1.1×10^{-4} MPa and typically increases to 1.3×10^{-4} MPa during thin film deposition.

The vacuum evaporator can be set to operate in either electron beam or thermal-resistive evaporation mode (Fig. 2). Samples are mounted in an inverted position above either evaporation source. The distance between the source and samples is approximately 40 cm. This configuration provides a uniformly thick thin film over an area several square centimeters (OHRING, 1992; GEORGE, 1992). An

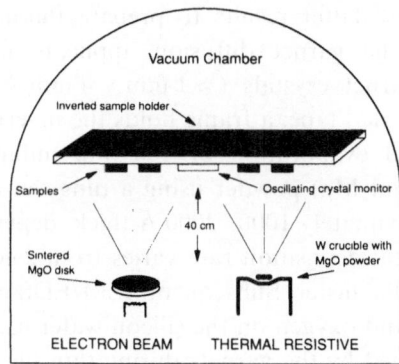

Figure 2
Schematic of the vacuum evaporation apparatus.

oscillating quartz crystal positioned in the plane of the sample holder monitors film thickness during either mode of operation. Silicon wafers act as control samples for convenient analysis of the thin films with standard energy dispersive spectrometry (EDS) and scanning electron microscope (SEM) imaging techniques. Silicon wafers from both evaporation methods were examined to determine which was more appropriate.

Electron beam bombardment of a pressed MgO pellet, using a 10 kV gun potential and about 200 mA of current, produces uniform thin films of high quality MgO about 2000 Å thick in approximately 30 seconds. However, this technique requires a source pellet with a mass of at least 10 grams, and is prohibitively expensive if using enriched stable isotope materials. We also explored the possible use of sintered pellets comprised of an oxide mixture, with a pyrope composition ($3MgO \cdot 2Al_2O_3 \cdot 3SiO_2$), for evaporation to produce thin films with the appropriate pyrope garnet compositions. However, evaporation of the oxide mixture was nonstoichiometric and produced a silica-rich thin film.

The thermal-resistive evaporation technique requires less material and is therefore more economical. A residual gas analyzer monitored the composition of the vapor throughout the evaporation process. Thermal evaporation of a mixture of oxides with pyrope composition ($3MgO \cdot 2Al_2O_3 \cdot 3SiO_2$) in tungsten crucibles also failed to produce thin films with the appropriate composition. Results of the vapor analysis and subsequent EDS analysis of the thin film/silicon wafer assemblages confirm the absence of Mg evaporation from the oxide mixture.

Given that garnet-like thin films were not obtained, we investigated the thermal evaporation of pure MgO. Although the melting point of pure MgO is 2852°C, under high vacuum the MgO sublimes at a much lower temperature. Sintered MgO pellets appear to evaporate nonuniformly as a result of uneven heating of the pellets. In contrast, MgO in powder form evaporates uniformly.

We made use of these latter results to prepare thermally-evaporated $^{25}MgO_x$ $(x < 1)$ thin films for the garnet diffusion application. Polished, gem-quality, single-crystal, pyrope garnet crystals (≈ 1 mm $\times 4$ mm $\times 4$ mm) are mounted to glass slides with double-sided tape; a frame holds the inverted glass slides above the thermal source. Thermal evaporation of 100–140 milligrams of ^{25}MgO (^{25}Mg enriched to 94.5% of total Mg) powder using a tungsten crucible yields thin films of $^{25}MgO_x$ $(x < 1)$ approximately 1000–2000 Å thick, depending on the current and length of time. The typical deposition rate varies from 2 to 15 Å/s and results in a thin film with a submetallic luster. Subsequent SEM/EDS examination confirms the presence of magnesium and oxygen on the silicon wafer used as a control substrate. The temperatures sustained by the garnets during thin film production are too low ($< 40°C$) to produce any premature self-diffusion of ^{25}Mg into the garnet. Survival of the double-sided tape, and previous depositions that utilized thermocouples with the vacuum chamber, support the estimate of low temperatures at the sample holder.

^{25}MgO Thin Films as Diffusion Source Material

Secondary ion mass spectrometry (SIMS) depth-profile analysis with an ion microprobe confirms that the ^{25}Mg content of the thin film is equivalent to the enrichment of the source powder (94.5%). Therefore, the thermal evaporation process does not result in observable isotopic fractionation with the formation of the oxide thin film. Ion microprobe sputtering and surface contact profilometry indicate the film thickness increases to approximately 3000 Å during annealing. The change in thickness is consistent with the approximate difference in molar volumes of Mg metal and MgO. Thermodynamic calculations suggest the Mg-MgO transition occurs at very reducing conditions, much more reducing than those of the self-diffusion experiments ($\log fO_2 = -10$ to -18 depending on the anneal temperature). Therefore, the magnesium metal present in the thin film is likely to be metastable and will tend to oxidize upon annealing. The luster of the thin film changes from submetallic to nonmetallic during experimental anneal. These observations are consistent with oxidation of the thin film during experimental anneal and suggest that the ^{25}MgO thin film is nonstoichiometric before the diffusion anneal. OHRING (1992) and GEORGE (1992) also believe that thermal evaporation is ineffective in producing stoichiometric thin films. The thin film is probably in the form of $^{25}MgO_x$ $(x < 1)$ before the diffusion anneal.

SEM observation of the thin film before the diffusion anneal indicates the film is smooth and uniform (Fig. 3a). During diffusion anneal the $^{25}MgO_x$ $(x < 1)$ film recrystallizes into "islands" (Figs. 3a–f) that appear to grow slowly with time as a function of temperature. Longer diffusion anneal times and higher temperatures produce fewer but larger islands. Mechanisms of island generation and coalescence

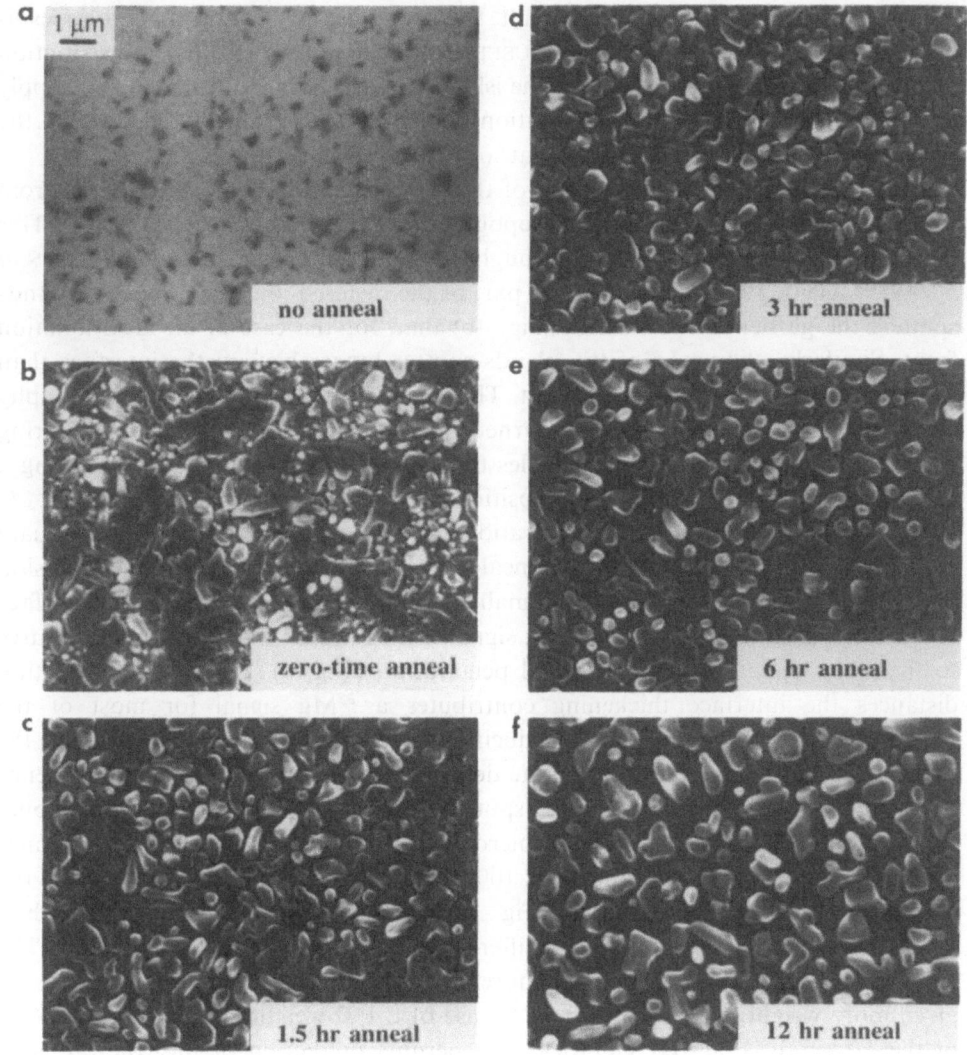

Figure 3

Secondary electron photomicrographs of ^{25}MgO thin film garnet diffusion anneals at 1000 C, log fO$_2$ = −11.3, for different anneal durations. The 1 μm scale bar in a) is representative of that for photomicrographs a–f).

are not completely understood, but island formation is a common phenomenon of sintered thin films on substrate materials (HOULIER *et al.*, 1990; OHRING, 1992). Island formation is a combination of: 1) nucleation and growth of crystallites; 2) the result of Ostwald ripening or the minimization of surface-free energy of the thin film material; and 3) coalescence as the result of collisions between islands as they execute random motion. It is difficult to distinguish which of the processes is dominant (OHRING, 1992). TANNHAUSER (1956) demonstrated that a distribution of source material similar to that of the island material provides an adequate supply of source ^{25}Mg throughout the duration of the diffusion anneals, and yields the same effective diffusion coefficient that an ideal thin film would.

The ablation or sputtering process of the ion microprobe removes material from the sample according to the original topography of the sample surface (Fig. 4). This topographic phenomenon persists throughout the depth profile analysis (WILSON *et al.*, 1989; KING, 1992). Therefore, the part of the rastered ion beam between islands sputters the garnet surface earlier (Fig. 4b) than for the case of an ideal ablation front. Similarly, the peaks of the islands sputter longer beyond the interface than they would for a planar ablation front. This type of hummocky sputter topography effectively thickens the thin film/garnet interface. Additionally, the sputtering process does not produce depth profiles that are step-like even when penetrating a perfect planar interface into a compositionally different substrate (WILSON *et al.*, 1989). Because the diffusional penetration distance is proportional to the square root of time $(d \approx (4Dt)^{1/2})$ short anneal times will produce very small diffusion distances. If the distances are too small (< 500 Å), then the thickened interface contributes a greater amount of ^{25}Mg signal from residual thin film islands relative to the ^{25}Mg signal due to diffusional penetration (Fig. 4). For small penetration distances the interface thickening contributes a ^{25}Mg signal for most of the penetration distance, ultimately producing an artificial time dependence for the diffusion coefficients (Fig. 5). The time dependence is the result of this interference from the thickened interface and the sputtering process. However, with diffusional penetration distances of at least 0.1 micrometer obtained with longer anneal times, the diffusion coefficients obtained for 1000°C anneals asymptotically approach the "true" diffusion coefficient value (Fig. 5). The diffusion coefficients (and their associated standard deviations) for different anneal times follow a $1/t$ curve. The longer anneal times which provide more accurate diffusion coefficient values, are given more weight in the $1/t$ curve fit by use of a $1/D$ weighting factor. The choice of the $1/t$ curve is entirely empirical and permits prediction of the true diffusion

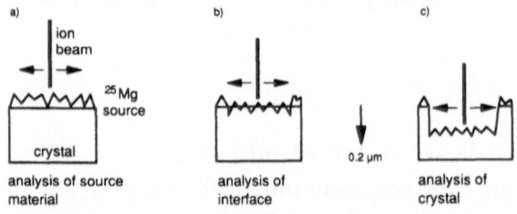

Figure 4

Schematic diagram of sputtering process for annealed thin film on crystalline substrate diffusion couples. Note the retention of the original topography throughout the sputtering process.

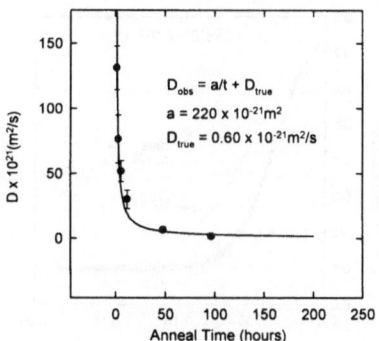

$D_{obs} = a/t + D_{true}$

$a = 220 \times 10^{-21} m^2$

$D_{true} = 0.60 \times 10^{-21} m^2/s$

Figure 5

Log D versus anneal time at 1000°C for a ^{25}MgO thin film on pyrope. Error bars, shown when larger than symbol size, equal one standard deviation associated with replicate analyses for each anneal time. The weighted curve fit provides an estimate for the limiting value of the self-diffusion coefficient, D.

coefficient given infinite anneal time. Longer anneal times provide deeper diffusional penetration distances, which can be analyzed more accurately, as the interferences contribute less to the measured profile and therefore provide better evaluations of the diffusion coefficients.

Pyrope garnet ($X_{Almandine} = 0.24$) samples for our diffusion experiments are held at the anneal temperature and oxygen fugacity for 24 hours to remove surface defects that may have formed during the polishing process and to equilibrate the bulk point defect structure to the conditions of the diffusion anneal (Ryerson et al., 1989). After the preanneal, a ^{25}MgO thin film is applied to the garnets, and then the diffusion couples are annealed at 1000°C and atmospheric pressure; a $CO-CO_2$ gas mixture provides a $\log fO_2 = -11.3$ (QFM buffer). Diffusion anneal times ranged from 1.5 hours to 4 days. Replicate analyses were obtained for each anneal time. A best fit value was determined for the Mg self-diffusion coefficient for each replicate analysis using a thin film diffusion model:

$$C_x = C_b + (C_s - C_b)\exp(-x^2/4Dt)$$

where C_x is ^{25}Mg concentration at a distance x, C_b is the bulk concentration of the garnet, and C_s is the surface concentration (CRANK, 1975). The uncertainty in the diffusion coefficient for the best fit of each replicate is about two percent relative. However, because there is more variation among the replicates, the means and standard deviations for each of the anneal times are used to represent the uncertainty in the diffusion coefficient measurement (see Fig. 5). The standard deviations for each anneal time are roughly 15 percent relative, but absolute values for the standard deviations decrease with increasing anneal time. Therefore, the $1/t$ approach can be used to estimate the diffusion coefficient at infinite time and the relative uncertainty should not exceed 15 percent.

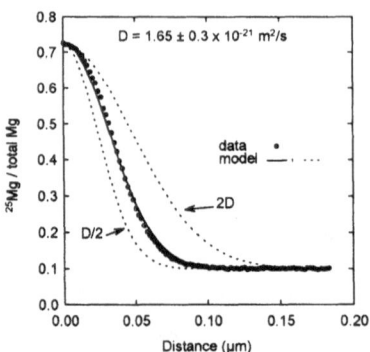

Figure 6
Best fit of the thin film diffusion model for the 96.5 hour 1000°C anneal ($\log fO_2 = -11.3$).

The depth profile for the 4-day diffusion experiment is shown in Figure 6 with the best fit of the model providing a Mg self-diffusion coefficient of $1.65 \pm 0.30 \times 10^{-21}$ m²/s. The fit in Figure 6 is quite good ($r^2 = 0.97$), supporting the use of the simple diffusion model to simulate the observed diffusion profile. The residuals, portions of data with a poor match to the diffusion model, are probably related to the remnant interface thickening not yet corrected by the asymptotic fit presented in Figure 5. The analytical precision determined from replicate analyses is much less than the envelope formed by the $D/2$ and $2D$ curves also presented in Figure 6. The $1/t$ curve fit for the diffusion coefficients, obtained for the various anneal times (Fig. 5), yields a limiting value of $0.60 \pm 0.09 \times 10^{-21}$ m²/s for the Mg self-diffusion coefficient at 1000°C. The values for the Mg self-diffusion coefficient, both measured and empirically fit, are smaller by 3 orders of magnitude than the extrapolated value from the data of CYGAN and LASAGA (1985), though garnet composition and oxygen fugacity conditions are different. The present data are about 1.5 to 2 orders of magnitude smaller than the extrapolated values from the data of CHAKRABORTY and GANGULY (1992), again for different garnet compositions. Comparison of the present data with these earlier studies suggests that the present experimental approach is viable.

Conclusions

Precipitation of tracer compounds from solution produces a nonuniform distribution of the source material that complicates subsequent analysis of diffusion profiles. The use of molten salts or the precipitation of metal chlorides as diffusion source material generates a disrupted diffusion couple interface that complicates interpretation of the diffusion profiles. Thermal vacuum evaporation is the most economical option for the preparation of enriched stable isotope thin films. Electron beam vacuum evaporation is a better choice when source materials are more readily available.

The evaporative thin film technique, used in conjunction with ion microprobe analysis, allows for the experimental determination of diffusion coefficients at much lower temperatures than previously performed. The improved uniformity of the diffusion interface permits precise measurement of much shorter diffusional penetration depths than previous studies. Evaporative thin film methods eliminate numerous problems experienced in making viable diffusion couples by providing a uniform source with no destruction of the couple interface. This type of diffusion couple is especially amenable to depth-profile mass analysis with an ion microprobe.

Acknowledgements

The Geosciences Program of the U.S. Department of Energy, Office of Basic Energy Sciences, supported this research under contract DE–AC04–94AL85000. We thank George Edgerly and Ted Neil for their assistance with the vacuum evaporator and ion microprobe, respectively. The critical comments of two anonymous reviewers significantly improved the final paper.

REFERENCES

BUENING, D. K., and BUSECK, P. R. (1973), *Fe-Mg Lattice Diffusion in Olivine*, J. Geophys. Res. *78*, 6852–6862.

CHAKRABORTY, S., and GANGULY, J. (1992), *Cation Diffusion in Aluminosilicate Garnets: Experimental Determination in Spessartine-almandine Diffusion Couples, Evaluation of Effective Binary Diffusion Coefficients, and Applications*, Contributions to Mineralogy and Petrology *111*, 74–86.

CHAKRABORTY, S., RUBIE, D. C., and ELPHICK, S. C. (1992), *Mg Tracer Diffusion in Aluminosilicate Garnets at 800°C, 1 Atm. and 1300°C, 8.5 GPa*, Trans. Am. Geophys. Union, EOS *73*, 43, 567.

CRANK, J., *The Mathematics of Diffusion* (Oxford University Press, Oxford 1975).

CYGAN, R. T., and LASAGA, A. C. (1985), *Self-diffusion of Magnesium in Garnet at 750° to 900°C*, Am. J. Science *285*, 328–350.

ELPHICK, S. C., GANGULY, J., and LOOMIS, T. P. (1985), *Experimental Determination of Cation Diffusivities in Aluminosilicate Garnets: Experimental Methods and Interdiffusion Data*, Contributions to Mineralogy and Petrology *90*, 36–44.

FREER, R. (1979), *An Experimental Measurement of Cation Diffusion in Almandine Garnet*, Nature *280*, 220–222.

FREER, R. (1981), *Diffusion in Silicate Minerals and Glasses: A Data Digest and Guide to the Literature*, Contributions to Mineralogy and Petrology *76*, 440–454.

GEORGE, J., *Preparation of Thin Films* (Marcel Dekker, New York 1992).

HOULIER, B., CHERAGHMAKANI, M., and JAOUL, O. (1990), *Silicon Diffusion in San Carlos Olivine*, Phys. Earth Planet. Int. *62*, 329–340.

JAOUL, O., POUMELLEC, M., FROIDEVAUX, C., and HAVETTE, A., *Silicon diffusion in forsterite: A new constraint for understanding mantle deformation*, In *Anelasticity in the Earth* (eds. Stacey, F. D., and Paterson, M. S.) (American Geophysical Union, Geodynamics Series, Volume 4, 1981) pp. 95–100.

KING, B. V., *Sputter depth profiling*, In *Surface Analysis Methods in Materials Science* (eds. O'Connor, D. J., Sexton, B. A., and Smart, R. St. C.) (Springer Verlag, Berlin 1992) pp. 97 116.

OHRING, M., *The Materials Science of Thin Films* (Academic Press, San Diego 1992).

RYERSON, F. J., DURHAM, W. B., CHERNIAK, D. J., and LANFORD, W. A. (1989), *Oxygen Diffusion in Olivine: Effect of Oxygen Fugacity and Implications for Creep*, J. Geophys. Res. *94*, 4105 4118.

SURDASHAN, T. S., *Surface Modification Technologies: An Engineers Guide* (Marcel Dekker, New York 1989).

TANNHAUSER, D. S. (1956), *Concerning a Systematic Error in Measuring Diffusion Constants*, J. Appl. Phys. *27*, 662.

WILSON, R. G., STEVIE, F. A., and MAGEE, C. W., *Secondary Ion Mass Spectrometry: A Practical Handbook for Depth Profiling and Bulk Impurity Analysis* (J. Wiley and Sons, New York 1989).

(Received March 3, 1993, revised July 22, 1993, accepted October 17, 1993)

PAGEOPH, Vol. 141, No. 2/3/4 (1993)

0033-4553/93/040643-10$1.50 + 0.20/0
© 1993 Birkhäuser Verlag, Basel

Determination of Phase Transition Pressures of ZnTe under Quasihydrostatic Conditions

Keiji Kusaba,[1] Laurence Galoisy,[1] Yanbin Wang,[1]
Michael T. Vaughan[1] and Donald J. Weidner

Abstract—Pressure behavior of ZnTe at room temperature was studied using an X-ray energy dispersive method on a DIA type cubic anvil apparatus (SAM-85) at NSLS-X17B1. By using powdered polyethylene, the sample and NaCl for a pressure scale were held under quasihydrostatic conditions, which were confirmed by X-ray diffraction method. Two high-pressure phase transitions were confirmed using X-ray powder diffraction simultaneously with electrical resistance measurements. The phase transition pressures under quasihydrostatic conditions were determined to be 9.6 GPa, at which the resistance increased, and 12.0 GPa, which was the midpoint of a large resistance decrease. Errors in the pressure determinations were estimated to be less than 0.2 GPa. These pressure values may depend on grain size and anisotropic stress effects on the calibrant. From X-ray observation of ZnTe, the bulk modulus of the zinc blende structure was calculated to be $K_0 = 51(3)$ GPa and $K'_0 = 3.6(0.8)$, and the first transition at 9.6 GPa was found to have about 9% volume change. It was consistent with an anomaly in the pressure generating curves.

Key words: High pressure, pressure calibration, ZnTe, synchrotron X-ray source, DIA type cubic anvil apparatus.

Introduction

Many II–VI semiconducting compounds with the zinc blende structure have been expected to transform to metallic phases with the rock salt structure under high pressure. For some of them, the phase transitions were observed using change in the electrical resistance (SAMARA and DRICKAMER, 1962) and others by *in situ* X-ray observations (MARIANO and WANEKOIS, 1963; OWEN *et al.*, 1963; SMITH and MARTIN, 1965). However ZnTe, which also has the zinc blende structure, exhibits different behavior under high pressure. On the basis of resistance changes, ZnTe is known to undergo at least two high-pressure phase transitions below 15 GPa (SAMARA and DRICKAMER, 1962; OHTANI *et al.*, 1979, 1980a; ENDO *et al.*, 1982). These two-phase transitions were recognized by *in situ* X-ray observations,

---

[1] Center for High Pressure Research, Department of Earth and Space Sciences, State University of New York at Stony Brook, Stony Brook, N.Y. 11794, U.S.A.

which showed that neither high-pressure phase had the B1 type structure. These observations were insufficient to determine the structures (OHTANI *et al.*, 1980a), because only poor X-ray data had been obtained using a laboratory X-ray source.

The semiconductor-metal transitions in some II–IV compounds, such as ZnS and ZnSe, have been used for pressure calibration in high-pressure studies. The phase transition pressures of ZnS and ZnSe were confirmed by an optical method using a diamond anvil apparatus (PIERMARINI and BLOCK, 1975), however those of ZnTe have never been confirmed using optical methods because the lower high-pressure phase is opaque (OHNO *et al.*, 1983). ZnTe may be a good calibrant material with two transition pressures previously reported to be at 8.5 and 13 GPa (OHTANI *et al.*, 1980a). An investigation of the phase transition pressures using an *in situ* method is reported here. Since the development of MAX80 at the Photon Factory at KEK, Tsukuba, Japan (KEK-PF) (SHIMOMURA *et al.*, 1985), a combination of a synchrotron X-ray source and a large volume high-pressure device has been one of the most useful methods for high-pressure research.

The aims of the present study are to obtain high quality X-ray powder diffraction patterns of the two high-pressure phases, confirming the presence of the phase transitions with simultaneous electrical resistance measurements, and to determine the phase transition pressures using *in situ* X-ray diffraction measurements on a NaCl pressure standard (DECKER, 1971).

Experiment

High-pressure experiments were carried out by using a DIA type cubic anvil apparatus (SAM-85) (WEIDNER *et al.*, 1992), using sintered diamond anvils with 4 mm truncations capable of generating pressures reaching about 15 GPa (SHIMO-MURA *et al.*, 1992). A 6 mm boron-epoxy (4:1 by weight) cube was used for a pressure medium. A 2 mm diameter vertical hole was drilled through the pressure medium to hold the sample assembly. Preformed gaskets (1.4 mm × 1.4 mm size) made of pyrophylite were used in order to prevent blowouts.

X-ray diffraction patterns were taken by an energy dispersive method at $2\theta = 5.30°$ and $7.25°$ on the superconducting wiggler port (X17) of the National Synchrotron Light Source at Brookhaven National Laboratory, Upton, NY, U.S.A. The useful energy range was approximately 35–120 keV. The incident X-ray beam size was limited using slits to 0.3 mm in the horizontal direction and 0.1 mm in the vertical direction, corresponding to the accuracy of the X-ray observation system. Diffracted X-rays from a small diffraction volume were collected by germanium solid state detector. The length of the diffraction volume along the X-ray direction for a given 2θ was controlled by two vertically defining receiving slits of 0.1 mm width. This length was calculated to be about 2 mm for $2\theta = 5.3°$ and about 1 mm for $2\theta = 7.25°$.

Table 1

Summary of experimental conditions

Run	2θ value of solid state detector	Thickness of sample disks used for X-ray observations	Measurement method for resistance
A	5.299	0.8 mm	four wires
B	7.250	0.8 mm	four wires
C	5.281 and 7.483 *	0.8 mm	none
D	5.284	0.4 mm	two wires

* 2θ value of 7.483 was selected only to take an X-ray pattern of HPPI at 11.5 GPa.

Chemical reagents of ZnTe (99.99%) and NaCl (for a pressure marker) were each mixed with powdered polyethylene with a weight ratio of 1:1 and 2:1, respectively. Polyethylene is added to the samples because: (1) The material is soft enough to hold the sample and NaCl under quasihydrostatic conditions in the high-pressure experiments. No pressure-induced phase transitions have been observed below 20 GPa (MARSH, 1980). (2) Like the boron-epoxy pressure medium, it is made of only light elements (carbon and hydrogen), which means that absorption effects are small and can be ignored. It can be added to the sample to increase the thickness for X-ray observation without appreciable absorption. (3) It has only two remarkable X-ray diffraction lines, and their d values are greater than 3.5 Å at room conditions (YAGI et al., 1979).

Both mixtures were formed into disks of 2 mm diameter. One run used a thickness of 0.4 mm for both disks and the other three used 0.8 mm. The disk sizes were larger than the calculated diffraction volume, so the volume was filled by the sample or NaCl, insuring the accuracy of X-ray data.

In order to measure electrical resistance changes, a pure powdered ZnTe disk of 0.1 mm thickness was inserted between the polyethylene mixed disks. The electrical resistance was measured using both a two wire and a four wire method. This disk was expected to be held under quasihydrostatic conditions. The grain size in this disk was estimated to be several micrometers. Experimental conditions for all runs are summarized in Table 1.

Results and Discussion

Table 2 shows the d values of NaCl at room conditions and at the maximum pressure condition in this present study. The *111* diffraction line is omitted in the table because it overlaps a diffraction line of polyethylene under high pressure. An overall accuracy of the present study, using the X-ray system on SAM-85 at NSLS-X17B1, could be estimated from the results at room pressure. It showed that

Table 2

Observed and calculated d values of NaCl at room conditions and the maximum pressure in the Run D

h	k	l	Room Conditions			The Maximum Pressure		
			$d_{obs}/Å$	$d_{calc}/Å$	$\Delta d/d_{obs}$	$d_{obs}/Å$	$d_{calc}/Å$	$\Delta d/d_{obs}$
2	0	0	2.8194	2.8201	−0.00028	2.5570	2.5532	0.00149
2	2	0	1.9947	1.9941	0.00030	1.8041	1.8054	−0.00072
3	1	1	1.7014	1.7006	0.00047	1.5400	1.5396	0.00026
2	2	2	1.6286	1.6282	0.00046	1.4727	1.4741	−0.00095
4	0	0	1.4093	1.4100	−0.00050	1.2785	1.2766	0.00149
4	2	0	1.2609	1.2612	−0.00024	1.1414	1.1418	−0.00035
4	2	2	1.1512	1.1513	−0.00009	1.0417	1.0423	−0.00058

From each 7 diffraction lines, the unit cell parameters of NaCl and pressure were calculated by minimizing the sum of $(\Delta d/d_{obs})$ to be 5.6401(19) Å at 0.000(24) GPa and 5.1063(46) Å at 14.629(230) GPa, respectively.

an average observation error of d values was less than 0.03%. The error under the maximum high pressure was calculated to be 0.12%, which was four times as great as that at room conditions. It could be explained by nonhydrostatic effects for NaCl (WEIDNER *et al.*, 1992), which caused only 1.6% (0.23 GPa) of a pressure determination error at 14.63 GPa. No remarkable line broadening was observed even under the maximum pressure. These two facts showed that NaCl was held under quasihydrostatic conditions at even the maximum pressure and that nonhydrostatic effects for NaCl could be ignored for the present study.

Figure 1 shows compression behavior of ZnTe vs. pressure, which was based on the volume of NaCl. The data during both the compression and decompression

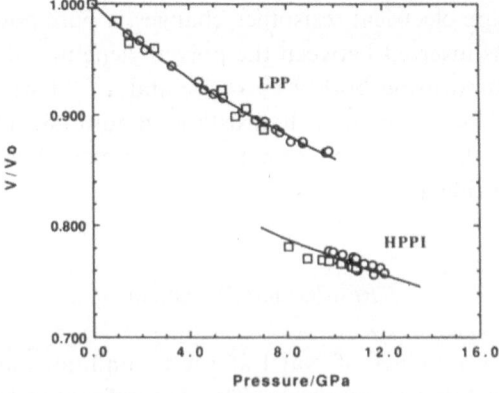

Figure 1

Compression behavior of ZnTe up to 12 GPa. Circles show data collected during compression and squares show data collected during decompression. LPP; Low Pressure Phase with the zinc blende structure. HPPI; High Pressure Phase I with a hexagonal cell.

processes were consistent with each other. This shows that the nonhydrostatic effects for ZnTe are negligibly small in this study, the same as that for NaCl. From both compression and decompression data, the bulk modulus of the zinc blende type ZnTe was calculated to be $K_0 = 51(3)$ GPa and $K'_0 = 3.6(0.8)$ by a least squares method using the Birch-Murnaghan equation.

Cell pressures vs. loads of all runs are shown in Figure 2. The figure shows two anomalies at around 9.6 and 12 GPa, which could be explained by volume changes at phase transitions in ZnTe. This is likely because all the other materials within the cell assembly have no phase transitions with volume changes in this pressure range and because the amplitudes were greater in those runs with greater volumes of ZnTe.

Typical electric resistance change vs. pressure behavior is shown in Figure 3, in which three anomalies (around 6, 9.6 and 12.0 GPa) were observed in the compression process; only one anomaly was observed in the decompression process. The behavior of electrical resistance changes was similar to that in the previous *in situ* studies (OHTANI *et al.*, 1979, 1980a). In particular, a very sharp anomaly around 9.6 GPa was observed, as shown in Figure 3. It shows that a ZnTe calibrant disk was held under quasihydrostatic conditions.

In the compression process, the anomaly at the lowest pressure had already been explained by a change of the band gap in the zinc blende type ZnTe (ONODERA *et al.*, 1985). Present X-ray observation data reconfirmed that no structural changes occur in ZnTe near the anomaly, because no new diffraction lines appeared and the compression behavior was continuous (Fig. 1).

The pressures at which the two higher pressure anomalies occurred were consistent with Figure 2. The two electrical resistance changes were confirmed to be

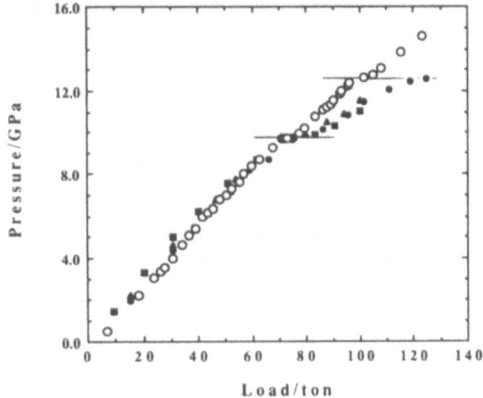

Figure 2

Pressure generating curves for all experimental runs. In run D shown by open circles, the amount of ZnTe in the pressure medium was almost half as much as in the other runs shown by solid symbols (A; triangles, B; squares and C; circles).

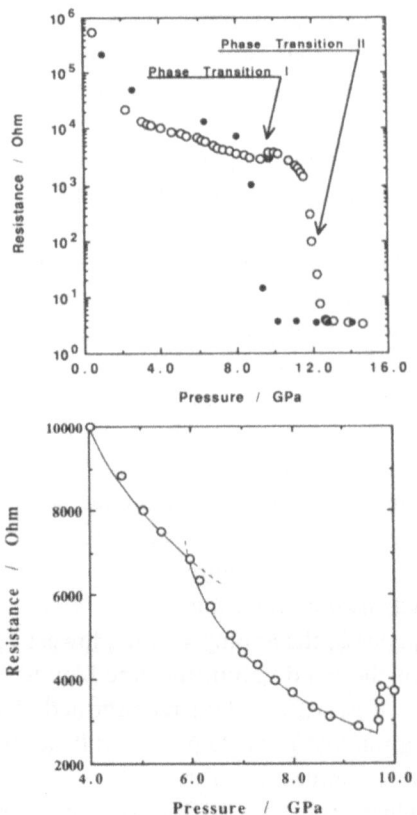

Figure 3
Electric resistance change vs. pressure for ZnTe in Run D. Open and solid circles show data collected during compression and decompression, respectively. Top; for whole pressure range. Bottom; expanded figure around the anomalies at 6 and 9.6 GPa.

due to structural transformations in ZnTe by X-ray diffraction patterns of the low pressure phase (LPP) with the zinc blende structure and high-pressure phases I and II (HPPI and HPPII), as shown in Figure 4. From the X-ray diffraction patterns of ZnTe, the first transition was observed to occur at 9.6 GPa and to have a narrow mix region of LPP and HPPI. At 10.3 GPa, a very small amount of LPP was observed with HPPI. The HPPII appeared at 12.1 GPa. Even at the maximum pressure (14.6 GPa) of the present study, ZnTe was found to be a mixture of HPPI and II.

In the decompression process, the HPPII completely reverted to the HPPI between 9.3 and 8.8 GPa and finally to LPP between 8 and 7 GPa; however only one anomaly was observed in the electrical measurements, as shown in Figure 3.

In this study, the lower phase transition pressure was determined to be 9.6 GPa, at which the electrical resistance started increasing (Fig. 3). The higher phase

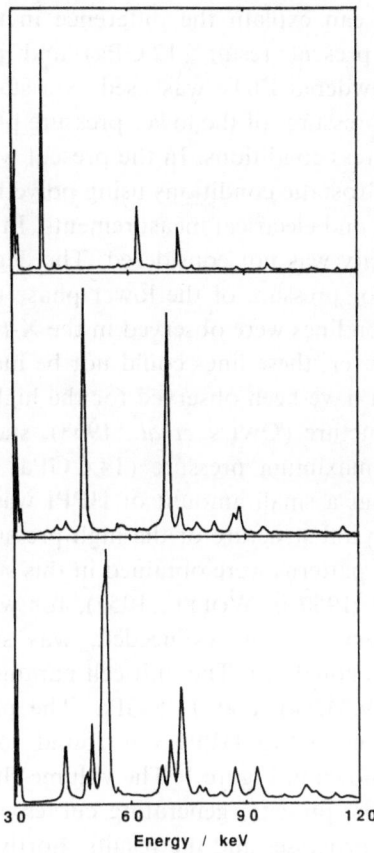

Figure 4

X-ray diffraction patterns of ZnTe at room temperature, taken at 5.284° for 2θ values in Run D. The doublet lines around 31 keV are the Te Kβ X-ray fluorescence lines. Top; Low Pressure Phase (LPP) with the zinc blende structure, before compression at room conditions. Middle; High-Pressure Phase I (HPPI) at 11.2 GPa. Bottom; High-Pressure Phase II (HPPII) and a small amount of HPPI at 14.1 GPa.

transition pressure was 12.0 GPa, which corresponds to the midpoint of a large resistance decrease. These pressure values differ from those in the previous work (8.5 and 13.0 GPa: OHTANI et al., 1980a); however the definitions of the phase-transition pressures were the same as the previous study.

According to OHTANI et al. (1980b), in which the pressure was determined by forces on a high-pressure device, the pressure for the higher pressure transition depended upon the grain size of the starting material: when a single crystal was used for the starting material, the phase transition pressure was maximum (13 GPa). This was consistent with their previous work. The pressure was observed to shift to a lower pressure by using a smaller grain size as a starting material. The pressure of the lower pressure phase transition was independent of such a grain size effect.

The grain size effect can explain the difference in the pressure of the phase transition between the present result (12 GPa) and previous study (13 GPa), because in this study, powdered ZnTe was used as a starting material.

The difference in the pressures of the lower pressure phase transition may be due to the difference in the stress conditions. In the present work, the sample and NaCl were held under quasihydrostatic conditions using polyethylene. This condition was confirmed by both X-ray and electrical measurements. In the previous *in situ* study, the degree of hydrostaticity was not considered. The degree of hydrostaticity may play an important role for pressure of the lower phase transition.

Twenty-four diffraction lines were observed in the X-ray patterns of almost pure HPPI at 11.5 GPa. However, these lines could not be indexed on any of the cubic or tetragonal cells, which have been observed for the high-pressure phases of CdTe with the zinc blende structure (OWEN *et al.*, 1963), such as the NaCl and β-Sn structures. Even at the maximum pressure (14.6 GPa) of the present study, the X-ray pattern showed that a small amount of HPPI was still mixed with HPPII.

We considered the crystal structure of the high-pressure phase I, of which pure X-ray powder diffraction patterns were obtained in this study. A trial application of Ito's method (ITO, 1949, 1950 in WOLFF, 1957), for which a large number of *d* values with high accuracy are always needed, was successful in indexing the diffraction lines on a hexagonal cell. The unit cell parameters were calculated to be $a = 4.045(1)$ Å and $c = 9.342(4)$ Å at 11.5 GPa. The phase transition from LPP with the zinc blende structure to HPPI was found to have about 9% volume decrease at 9.6 GPa, as shown in Figure 1. The volume change is consistent with the anomaly at 9.6 GPa in the pressure generating curves, as shown in Figure 2. We shall report the indexing processes and the results shortly (KUSABA and WEIDNER, 1993).

This success is another indication of the accuracy of *d* value data we can obtain using SAM-85. No successful application of Ito's method to any high pressure *in situ* X-ray study has been reported. For the next step, we now propose a crystal structural model for the high-pressure phase I, by considering intensities of diffraction lines. The results will be reported with the determination of unit cell parameters.

Conclusion

Two phase transition pressures in ZnTe were determined to be 9.6(0.2) and 12.0(0.2) GPa by simultaneous *in situ* X-ray and electrical resistance measurements. This shows some of the advantages of a combination of a synchrotron X-ray source and a DIA type high-pressure device: 1) We can compress a much larger volume than the X-ray diffracting volume. 2) This volume can be held under quasihydrostatic conditions. These results were confirmed by simultaneous X-ray and electrical resistance data. This method enables us to obtain high accuracy X-ray data easily and in combinations with other physical measurements.

We must point out that the present results of 9.6 and 12.0 GPa were obtained in our own experimental conditions; using powdered ZnTe for an electrical resistance measurement and holding the sample under quasihydrostatic conditions. These two factors play important roles in the pressure of the phase transition. So pressure calibration, using present values, must be done with a consideration of the two experimental factors.

Acknowledgments

The authors thank Prof. N. Hamaya, Ochanomizu Univ., who assisted in the experiments at NSLS–X17B1. They express gratitude to Profs. O. Shimomura, KEK-PF and A. Onodera, Osaka Univ. for their comments on the data analysis. They also thank Prof. R. C. Liebermann, SUNY at Stony Brook, for his encouragement.

REFERENCES

DECKER, D. L. (1971), *High-pressure Equation of State for NaCl, KCl and CsCl*, J. Appl Phys. *42*, 3239–3244.

ENDO, S., YONEDA, A., ICHIKAWA, M., TANAKA, S., and KAWABE, S. (1982), *High Pressure Study of Transition in ZnTe by Manganin Coil Method*, J. Phys. Soc. Jan. *51*, 138–140.

KUSABA, K., and WEIDNER, D. J., *Structural consideration of high pressure phase I in ZnTe*. In *AIRAPT/APS Conference* (abstract) (ed. Samara, G.) (Colorado Springs 1993).

MARIANO, A. N., and WANEKOIS, E. P. (1963), *High Pressure Phases of Some Compounds of Groups II–VI*, Science *142*, 672–673.

MARSH, S., *LASL Hugoniot Data* (Univ. California Press, California 1980).

OHNO, Y., ENDO, S., KOBAYASHI, M., and NARITA, S. (1983), *Pressure Dependence of the Absorption Edge in ZnTe*, Phys. Lett. *95A*, 407–408.

OHTANI, A., ONODERA, A., and KAWAI, N. (1979), *Pressure Apparatus of Split-octahedron Type for X-ray Diffraction Studies*, Rev. Sci. Instrum. *50*, 308–315.

OHTANI, A., MOTOBAYASHI, M., and ONODERA, A. (1980a), *Polymorphism of ZnTe at Elevated Pressure*, Phys. Lett. *75A*, 435–437.

OHTANI, A., MOTOBAYASHI, M., ONODERA, A., SHIMOMURA, O., and FUKUNAGA, O., *Behaviors of ZnTe under high pressure*. In Programme and Abstracts of Papers for *The 21st High Pressure Conference of Japan* (Tokyo 1980b), *in Japanese*.

ONODERA, A., OHTANI, A., SEIKE, T., MOTOBAYASHI, M., SHIMOMURA, O., and KAWAMURA, H., *Physical properties of some chalcogenides under pressure; multi-anvil type high-pressure apparatus for synchrotron radiation*. In *Solid State Physics Under Pressure in Recent Advance with Anvil Devices* (ed. Minomura, S.) (KTK/Reidel, Tokyo/Dordrecht 1985) pp. 141–144.

OWEN, N. B., SMITH, P. L., MARTIN, J. E., and WRIGHT, A. J. (1963), *X-ray Diffraction at Ultra-high Pressure*, J. Phys. Chem. Solids *24*, 1519–1524.

PIERMARINI, G. J., and BLOCK, S. (1975), *Ultrahigh Pressure Diamond-anvil Cell and Several Semiconductor Phase Transition Pressures in Relation to the Fixed Point Pressure Scale*, Rev. Sci. Instrum. *46*, 973–979.

SAMARA, G., and DRICKAMER, H. G. (1962), *Pressure Induced Phase Transitions in Some II-IV Compounds*, J. Phys. Chem. Solids *23*, 457 461.

SHIMOMURA, O., YAMAOKA, S., YAGI, T., WAKATSUKI, M., TSUJI, K., FUKUNAGA, O., KAWAMURA, H., AOKI, K., and AKIMOTO, S., *Multi-anvil type high pressure apparatus for synchrotron radiation*. In *Solid State Physics Under Pressure in Recent Advance with Anvil Devices* (ed. Minomura, S.) (KTK/Reidel, Tokyo/Dordrecht, 1985) pp. 351 356.

SHIMOMURA, O., UTSUMI, W., TANIGUCHI, T., KIKEGAWA, T., and NAGASHIMA, T., *A new high-pressure and high-temperature apparatus with sintered diamond anvils for synchrotron radiation use*. In *High-pressure Research: Application to Earth and Planetary Sciences* (eds. Syono, Y., and Manghnani, M. H.) (Terra/AGU, Tokyo/Washington, D.C. 1992) pp. 3 11.

SMITH, P. L., and MARTIN, J. E. (1965), *The High-pressure Structures of Zinc Sulphide and Zinc Selenide*, Phys. Lett. *19*, 541 543.

WEIDNER, D. J., VAUGHAN, M. T., KO, J., WANG, Y., LIU, X., YEGANEH-HAERI, A., PACALO, R. E. G., and ZHAO, Y., *Characterization of stress, pressure, and temperature in SAM85, a DIA type pressure apparatus*. In *High-pressure Research: Application to Earth and Planetary Sciences* (eds. Syono, Y., and Manghnani, M. H.) (Terra/AGU, Tokyo/Washington, D.C. 1992) pp. 13 17.

WOLFF, P. M. (1957), *On the Detemination of Unit-cell from Powder Diffraction Patterns*, Acta Cryst. *10*, 590 595.

YAGI, T., JAMIESON, J., and MOORE, P. B. (1979), *Polymorphism in MnF_2 (Rutile type) at High Pressures*, J. Geophys. Res. *84*, 1113 1115.

(Received March 23, 1993, revised June 3, 1993, accepted October 17, 1993)

PAGEOPH, Vol. 141, No. 2/3/4 (1993) 0033 4553/93/040653 05$1.50 + 0.20/0

An Improved Sealing System for Triaxial Sample Columns

TED KOCZYNSKI[1] and ERICH SCHOLZ[2]

Abstract—A new method for making gas-tight seals for moderate temperature duty on triaxial deformation apparatus sample columns is described. This includes the modification of the piston and closure plug to enable rapid and inexpensive changes to the loading column.

Key words: Rock deformation technique, high pressure seals.

Introduction

Over the years, many techniques have been devised for effecting gas-tight sealing of metal jackets to sample columns for use in high temperature/pressure triaxial deformation apparatuses. The jackets are intended to isolate the sample from the invasion of confining medium. Jackets are constructed from materials such as annealed copper, lead or other soft metals in foil or tubing form.

One commonly applied sealing method uses a "jam" ring or a parallel ring with an interference fit to squeeze a tubular jacket over a tapered piston end (HANDIN, 1953; PATERSON, 1970). These devices are often cumbersome to use, requiring tedious and time-consuming assembly. Another inherent problem with the jam-ring technique is cost, since complex machining of the piston, plug, and jam rings is required.

Some experiments, such as those designed to study the diagenesis of sedimentary rocks, may require the use of brine as a pore fluid. This fluid may corrode the internal surface of the copper jacketing, making it necessary to electroplate the inside of the jacket with a corrosion resistant material such as nickel. When attempting to seal an electroplated jacket with the traditional tapered ends and jam-rings, the plating commonly fractures, providing a path for argon gas to leak into the pore fluid system.

[1] Rock Mechanics, Lamont Doherty Earth Observatory, Palisades, N.Y. 10964, U.S.A.
[2] RockWorks, P.O. Box 468, Piermont, N.Y. 10968, U.S.A.

This paper will describe the modification and successful testing of factory stock tube fittings to seal a 25 mm diameter sample for use with an argon gas confining pressure in excess of 150 MPa and sample temperatures to 350°C.

Triaxial Apparatus

The new sealing scheme was used in a triaxial deformation apparatus at the Lamont-Doherty Earth Observatory. This apparatus (developed by T. Engelder and C. Scholz) is similar to the apparatus described by SCHOLZ and KOCZYNSKI (1979). The main pressure vessel is designed for internal heating at sample temperatures up to 400°C with gas confining medium. The vessel has a 6.35 cm inside diameter and is oriented with the column axis vertical. It is rated for 736 kN axial force and 1.0 GPa maximum confining pressure. Samples are top loaded into the pressure vessel.

Sample deformation is achieved via a hydraulic cylinder loading a piston that completes the lower end of the sample column. Both the confining and pore pressure systems utilize piston intensifiers that are also hydraulically actuated. The pore pressure system consists of two independent 300 MPa intensifiers with pressure vessels that may contain distilled water or brine. By using a dual pore pressure approach, with pore pressure applied at each end of the sample, both the ambient pressure and the differential across the sample can be maintained with precision. All systems are under closed loop servo-control.

Improved Column Seal

For our initial design, we used a 25 mm diameter sample with an overall sample column length of 200 mm between the face of the piston and the top closure plug. One end of the sample column used for these tests is shown in Figure 1. There are two primary improvements. The first is the addition of a set of four small tapped holes in both the piston and top closure. We used the ANSI standard thread form, 4–40 UNC–2B. These are used to secure the sample column via end caps such as the modified fitting in Figure 1. This modular approach allows rapid changeover of columns as experimental parameters, or sample diameters change. All ends have face seal O-ring grooves to effect sealing between confining and pore pressures. Construction is identical on each end of the column.

The second improvement is the use of a commercially available hydraulic fitting as a column end. This allows for rapid assembly and provides a remarkably reliable seal at low cost. Our experiments have proven the seal capable of leakfree operation at 150 MPa confining and 50 MPa pore pressures with a sample temperature of 350°C. We have not measured the temperature at the ferrule or O-ring seal but

COLUMN SEAL SYSTEM

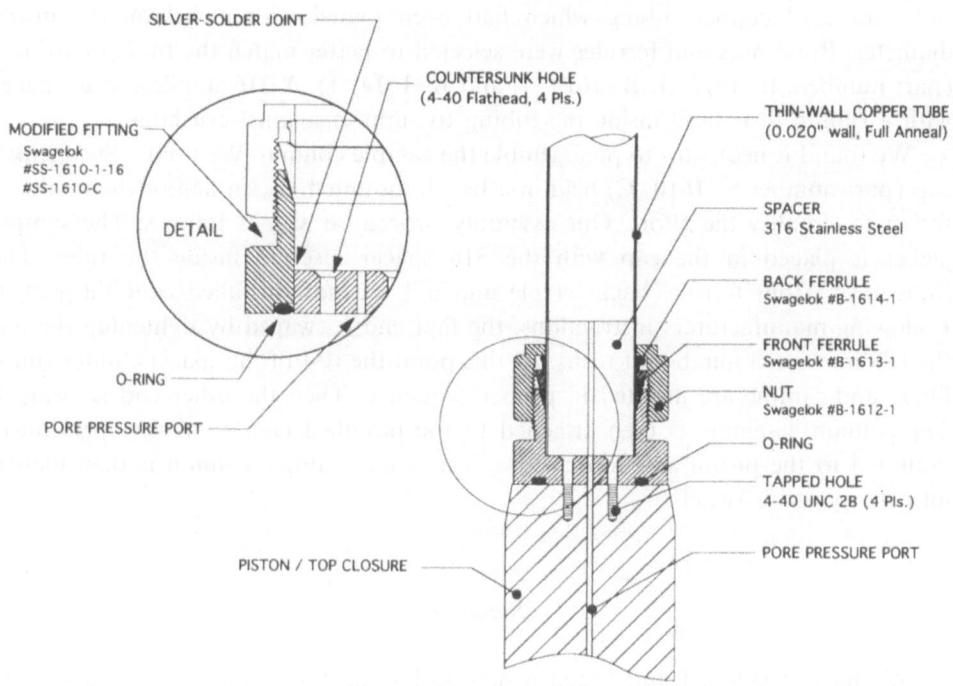

Figure 1

Illustration of improved column seal based on Swagelok® hydraulic fittings with a detail showing the necessary modifications.

estimate that it is near or below 50 C. However, with proper selection of the O-ring elastomer, we expect the system could be used for much higher temperatures. The test durations were approximately 48 hrs, during which the volume of the pore system was monitored for invasion of confining medium.

Application

The new sample column ends consist of modified Swagelok® tube fittings (part number SS 1610 1 16). The pipe thread is removed and the inside diameter and O-ring groove prepared on a lathe. A 316-stainless steel plug is machined with countersunk holes to match the pattern in the piston/top closure and for the pore pressure port. Then the plug is silver soldered flush into the bottom of the fitting (see detail in figure).

The remainder of the seal uses standard parts requiring no modification. The jacket is made from 1 inch (25.4 mm) outside diameter, 0.02 inch (0.5 mm) wall fully annealed copper tubing which had been plated with nickel on the inside diameter. Brass nuts and ferrules were selected to better match the tubing hardness (part numbers B−1612−1, B−1613−1, and B−1614−1). A 316 stainless steel spacer with a sliding fit is used inside the tubing to support against crushing.

We found it necessary to preassemble the sample column. We used a Swagelok " cap (part number S−1610−C) held in a bench mounted vise in lieu of the modified fitting to simplify the effort. Our assembly procedure was as follows: The sample jacket is placed in the cap with the 316 spacer inserted inside the tube. The Swagelok® front ferrule, back ferrule and nut are now installed over the jacket. Following manufacturers instructions, the first end is swaged by tightening the nut the recommended number of turns. At this point the rest of the spacers, miter rings, filters and sample are inserted in proper sequence. Then the other end is swaged. The column assembly is then attached to the modified ends that were previously mounted to the piston and top closure. This entire sample column is then loaded into the pressure vessel.

Summary

We have developed and tested a new technique for sealing sample jackets to loading columns. It uses factory stock parts that are available in several sizes and materials. With suitable modification of these parts together with a modular approach, one can achieve high-pressure gas seals quickly and efficiently. Ironically, the fittings are being used backwards; normally the excess pressure is on the inside of the fitting.

Acknowledgements

This work was supported by the United States Geological Survey grant U.S.G.S. 14−34−92−62161. We would like to thank John Sindt of the L.D.E.O. instrument laboratory for assistance in fabrication, S. L. Karner for help in testing, and both him and C. H. Scholz for critical reviews. We would also like to thank the Swagelok Co. of Solon, Ohio; however, we would like the reader to bear in mind that the Swagelok Co. does not necessarily approve or condone use of their product in this application.

Lamont-Doherty Earth Observatory Contribution No. 5140.

REFERENCES

HANDIN, J. W. (1953), *An Application of High Pressure in Geophysics: Experimental Rock Deformation*, Am. Soc. Mech. Eng. Trans. *75*, 315–325.

PATERSON, M. S. (1970), *A High-pressure, High-temperature Apparatus for Rock Deformation*, Int. J. Rock Mech. Min. Sci. *7*, 517–526.

SCHOLZ, C. H., and KOCZYNSKI, T. (1979), *Dilatancy Anisotropy and the Response of Rock to Large Cyclic Loads*, J. Geophys. Res. *84*, 5525–5534.

(Received August 12, 1993, revised October 27, 1993, accepted November 15, 1993)